SOLID MECHANICS AND ITS APPLICATIONS

Volume 36

Series Editor: G.M.L. GLADWELL

Solid Mechanics Division, Faculty of Engineering
University of Waterloo
Waterloo, Ontario, Canada N2L 3G1

Aims and Scope of the Series

The fundamental questions arising in mechanics are: *Why?*, *How?*, and *How much?* The aim of this series is to provide lucid accounts written by authoritative researchers giving vision and insight in answering these questions on the subject of mechanics as it relates to solids.

The scope of the series covers the entire spectrum of solid mechanics. Thus it includes the foundation of mechanics; variational formulations; computational mechanics; statics, kinematics and dynamics of rigid and elastic bodies; vibrations of solids and structures; dynamical systems and chaos; the theories of elasticity, plasticity and viscoelasticity; composite materials; rods, beams, shells and membranes; structural control and stability; soils, rocks and geomechanics; fracture; tribology; experimental mechanics; biomechanics and machine design.

The median level of presentation is the first year graduate student. Some texts are monographs defining the current state of the field; others are accessible to final year undergraduates; but essentially the emphasis is on readability and clarity.

For a list of related mechanics titles, see final pages.

Inelastic Behaviour of Structures under Variable Loads

Edited by

ZENON MRÓZ

Institute of Fundamental Technological Research,
Polish Academy of Sciences, Warsaw,
Poland

DIETER WEICHERT

Université des Sciences et Technologies de Lille,
Villeneuve d'Ascq,
France

and

STANISLAW DOROSZ

Institute of Fundamental Technological Research,
Polish Academy of Sciences, Warsaw,
Poland

KLUWER ACADEMIC PUBLISHERS

DORDRECHT / BOSTON / LONDON

1995

A C.I.P. Catalogue record for this book is available from the Library of Congress

ISBN 0-7923-3397-7

Published by Kluwer Academic Publishers,
P.O. Box 17, 3300 AA Dordrecht, The Netherlands.

Kluwer Academic Publishers incorporates
the publishing programmes of
D. Reidel, Martinus Nijhoff, Dr W. Junk and MTP Press.

Sold and distributed in the U.S.A. and Canada
by Kluwer Academic Publishers,
101 Philip Drive, Norwell, MA 02061, U.S.A.

In all other countries, sold and distributed
by Kluwer Academic Publishers Group,
P.O. Box 322, 3300 AH Dordrecht, The Netherlands.

Printed on acid-free paper

Printed in the Netherlands

CONTENTS

PREFACE

This collection of papers is a state of the art presentation of theories and methods related to the problem of the behaviour of mechanical structures under variable loads beyond their elastic limit. In particular, the problems of shakedown, ratchetting, transient and asymptotic cyclic states are addressed.
The volume is composed of four chapters devoted to material modelling for cyclic loading conditions; general theory of accommodated states of structures; effects of changes of the geometry on the inelastic structural response; and numerical techniques with applications to particular engineering problems. It was aimed to provide a unified approach in order to understand both inelastic material and structural response under variable loading conditions. The attempt to extend the classical shakedown theory of Melan and Koiter to geometrically non-linear problems is presented in several papers. The industrial application of cyclic plasticity to the analysis and the design of pressure bellows, compensators, turbine disks, or flange connections under thermal and pressure cycles illustrates the great potential of the numerical techniques developed for this purpose using mostly min-max approaches. The treatment of railway problems and the analysis and optimisation of pavements are further examples of important areas of applications. Emphasis was laid on approaches that take into account the fact that loading histories are often not precisely known. Therefore, the center of interest lies in other than step by step calculation methods.
The volume evolved from the EUROMECH COLLOQUIUM 298 "Inelastic Behaviour of Structures under Variable Loads", held near Warsaw, September 14-18, 1992. However, the content does not coincide with the conference program which contained 52 presentations covering more broadly inelastic response of structures. Some volume contributors did not participate in this colloquium which was devoted, just as this book, to the memory of Professor Jan Andrzej König, prominent Polish scientist in the area of inelastic analysis of structures, and author of fundamental contributions to the theory of shakedown theory.
The editors wish to heartily thank both the participants of the Euromech Colloquium 298 and the authors of papers for their valuable contributions to this important area of structural mechanics. They also thank all their colleagues at the Laboratoire de Mécanique de Lille and the Institute of Fundamental Technological Research for their general assistance in the preparation of the final version of the volume.

Zenon MRÓZ, Dieter WEICHERT, Stanislaw DOROSZ

IN MEMORIAM JAN ANDRZEJ KÖNIG

born 16 May 1937 in Warsaw
died 8 December 1990 in Warsaw

Professor of Mechanics
Institute of Fundamental Technological Research
Polish Academy of Sciences

We are greatly saddened by the loss of our colleague and friend, Jan Andrzej König.

Andrzej graduated in civil engineering in 1960 and in mathematics in 1963. He started his research career at the Institute of Fundamental Technological Research in 1960 - all his later achievements, degrees and nominations were related to his work at the Institute.

Andrzej is well known worldwide for his contributions in the field of shakedown analysis of structures. In fact, he pioneered a number of developments in this theory - only five years after its publication, his monograph book is now justly considered a classic treatise on the subject. Among Andrzej's specific contributions we remember his original and influential papers on methods of assessment of plastic deformation at shakedown, shakedown-based optimization, novel numerical techniques for shakedown analysis, methods for evaluation of residual stresses in inelastic structures subject to variable repeated loads, formulations of shakedown theory for problems with nonlinear hardening laws, inadaptation assessment accounting for nonlinear geometrical effects, experimental methods for shakedown evaluation, thermal effects in shakedown analysis and many others. During his 30 years at the Institute, Andrzej was also a constant source of novel ideas for many other important branches of contemporary nonlinear mechanics. This, together with his friendship, are today sorely missed among mechanicians in and outside Poland; his untimely death is a great loss to the entire applied mechanics community.

Andrzej's professional accomplishments, impressive as they were, shed little light on him as a person. Andrzej was a colleague and a close friend of mine for more than 20 years. We cooperated closely on very many different occasions - doing research and publishing papers together, being involved in the administration of a huge, five year long govermental research project on mechanics (of which Andrzej was the chairman and I his deputy), editing a book dedicated to the late Professor Antoni Sawczuk, organizing conferences, lecturing at symposia and courses, and finally, a fact which is not widely known and clearly outsteps mechanics, being together on the all-Poland election committee for a candidate running for presidency in the recent election campaign in Poland. We also spent literally hundreds of hours on talking to each other in private. All these contacts made me think of Andrzej as somebody of truly exceptional personality - his knowledge of mechanics was deep and thorough, his judgements always clear and unbiased, his loyalty and friendship natural and full of genuine generosity.

It was impossible not to like him - it came as something natural that recently, shortly before his death, during the election of a council to represent the Institute's staff, Andrzej got by far the greatest number of votes and was then consequently elected the chairman of the body - in fact it would have been simply incomprehensible to entrust somebody other than him with the task of representing all of us at the Institute.

His influence on us was direct as well as indirect - his gentleness and low-key approach to every problem, however difficult, meant that it was often only much later that we were able to fully appreciate his arguments or suggestions.

Andrzej never strived for credit or recognition for his accomplishments. He will always be remembered as an example of modesty, fully devoted to research, never awaiting special attention and recognition. I hope that at least some of the spirit and style of Andrzej has rubbed off on us - it is a rare, loyal and selfless spirit that enriched all those close to him when he was alive, and enriches us still as we remember him today.

Michał Kleiber

LIST OF MAIN PUBLICATIONS BY J.A. KŐNIG

1965 Przybliżone określanie odkształceń płyt i powłok plastycznych w oparciu o niestowarzyszone prawo płynięcia, Rozprawy Inżynierskie **13**, 55-65

1965 Zagadnienia teorii dostosowywania się konstrukcji sprężysto-plastycznych (Problems of shakedown of Elastic-Plastic Structures), Ph.D. thesis, IPPT, Warszawa.

1966 Theory of shakedown of elastic-plastic structures, Arch. Mech. Stos. **18**, 227-238.

1967 Shakedown of strainhardening structures, First Canadian Cong. Appl. Mech., Quebec.

1968 Shakedown theory of plates, Arch. Mech. Stos. **21**, 623-636.

1969 A shakedown theorem for temperature dependent elastic moduli, Bull. Acad. Polon. Sci. Sér.Sci. Tech. **17, 1**61-165.

1970 Podstawowe twierdzenia z zakresu teorii dostosowywania się konstrukcji sprężysto-plastycznych do obciążeń zmiennych w czasie (Basic theorems in the theory of shakedown of elastic-plastic structures to variable loads), Mech. Teoret. Stos. **8**, 149-158.

1971 A method of shakedown analysis of frames and arches, Internat. J. Solids and Structures **7,** 327-344.

1971 Optymalne projektowanie płyt na obciążenia cykliczne (Optimum design of plates for cyclic loadings), Arch. Bud. Masz. **18,** 575-589.

1971 O przystosowaniu się konstrukcji z materiału przejawiającego wzmocnienie (On shakedown of structures in a material exhibiting strainhardening), Report IFTR PAS No 18.

1971 Adaptacja konstrukcji, gdy stałe sprężyste zależą od temperatury (Adaptation of structures in the case of temperature-dependent elastic moduli), Report IFTR PAS No 39.

1971 On shakedown design of structures with temperature depending elastic moduli, First SMiRT Conf., Berlin 1971, paper L5/7.

1971 Ocena maksymalnych ugieć ram sprężysto-plastycznych (Estimate of maximum deflections of elastic-plastic frames), Report IFTR PAS No 39, (with BRZEZIŃSKI, R.).

1972 Evaluation of shakedown deflections of framed structures, by linear programming, Proc. Symp. Plastic Analysis of Structures, Jassy (with BRZEZIŃSKI, R.).

1972 On necessity of shakedown analysis of framed structures, Regional IABSE Conf., Warsaw, (with BRZEZIŃSKI, R.) .

1972 Shakedown deflections, a finite element approach, Teoret, i Priloz. Mech. **3**, No 2 65-69.

1972 O projektowaniu konstrukcji z uwzględnieniem przystosowania w przypadku, gdy stale sprężyste zależą od temperatury (On structural design in the case of temperature-dependent elastic moduli), Rozpr. Inż. **20**, 423-434.

1972 On shakedown criterion, Proc. Int. Symp. Foundations of Plasticity, Warsaw, vol. 2, 56-60.

1972 Analiza plastyczna konstrukcji (Plastic Analysis of Structures), Ossolineum, Wrocław, (with SAWCZUK, A. and JANAS, M.).

1973 Deflection bounding at shakedown under thermal and mechanical loadings, Second SMiRT Conf, Berlin, paper **L**7/3.

1974 Obliczanie ram i belek na przystosowanie (Shakedown analysis of frames and beams), Zesz. Problemowe Mostostal No 5, Warszawa (with SAWCZUK, A. and PAPROCKA-GARLICKA, W.).

1974 Projektowanie konstrukcji sprężysto-palstycznych przy obciążeniach zmiennych (Design of elastic-plastic structures for variable loads), Dr. Sc. thesis, Report IFTR PAS No 24.

1975 Analysis of incremental collapse of framed structures accounting for axial forces, Proc. Symp. Theor. Probl. Steel Structs, Bratislava 116-122.

1975 On optimum shakedown design, IUTAM Symp. Optimization in Structural Design, Warsaw 1972, In: Optimization in Structural Design, editors A. Sawczuk and Z. Mróz, Springer, Berlin-Heildelberg-New York , 405-414.

1976 Adaptation of rigid-workhardening discrete structure subjected to load and temperature cycles and second-order geometric effects. Comput. Methods Appl. Mech. Eng. **8**, 37-50 (with MAIER, G.).

1976 A finite element formulation for shakedown problems using a yield criterion of the mean, Comp. Meth. Appl. Mech. Eng. **8**, 179-192 (with Nguyen Dang Hung)

1977 Quelques problems recent dans la théorie del'adaptation, Symp. Franco-Polonais Problemes de Rhéologie et de Mécanique des Sols, (ed. W.K. Nowacki) PWN, Warszawa.

1977 A new method of shakedown analysis, Symposium franco-polonais: Problemes non-lineaires de mecanique, Cracovie, 343-347.

1978 On a new method of shakedown analysis, Bull. Acad. Polon. Sci. Sér. Sci Tech. **26**, 165-171 (with KLEIBER, M.).

1979 Bauschinger adaptation in the case of continuum, In: Some problems in elastoplasticity, Ossolineum, Wrocław-Warszawa-Kraków-Gdańsk 21-27.

1979 On upper bounds to shakedown loads, Z. Angew. Math. Mech. **59,** 349-354.

1979 On the incremental collapse criterion accounting for temperature dependence of yield-point stress, Arch. Mech. Stos. **31**, 317-325.

1980 On stability of the incremental collapse process, Arch. Inż. Ląd. **26**, 219-229.

1981 Deflections of elastic-plastic hyperstatic beams under cyclic loading, Arch. Mech., **33**, 611-624 (with DOROSZ, S., SAWCZUK, A., BIEGUS, S., KOWAL, Z., AND SEIDEL, W.).

1981 Shakedown analysis of elastoplastic structures, a review of recent developments, Nucl. Eng. Design **66**, 81-95, (with MAIER, G.).

1982 Shakedown criteria in the case of loading and temperature variations, J.Mech. Teor. Appl. **21**, special issue, 92-108.

1982 On exactness of the kinetical approach in the structural shakedown and limit analysis, Ing.-Archiv. **52** , 318-326.

1982 On some recent developments in the shakedown theory, Advances in Mech. **5**, 237-258.

1983 On shakedown of strainhardening structures, ZAMM **63**, T177-T178.

1984 Experimental investigation on shakedown of portal frames, Eng. Trans. **32**, 349-359, (with ALI, M.H., and MAHRENHOLTZ, O.).

1984 Incremental shakedown analysis in the case of thermal effects, Int.I.Num.Meth. Engng, **20**, 1567-1573, (with KLEIBER, M.).

1984 Experimental shakedown analysis of space truss nodes, Eng. Trans., **32**, 545-561, (with KARCZEWSKI, J.A.).

1984 Shakedown of tubes, a theoretical analysis and experimental investigations, In: Metal Forming and Impact Mechanics, W. Johnson Commemorative Volume (S.R.Reid Ed.) Pergamon Press, Oxford, New York, Toronto, Sydney, Frankfurt, 155-172, (with MAHRENHOLTZ, O., and LEERS, K.).

1985 Experimental investigations of shakedown of tubes, In: Plasticity Today, (Sawczuk, A. and Bianch, G., Eds.) Elsevier Appl. Sci. Bubl., London- New York, 259-275, (with LEERS, K., KILLE, W., and MAHRENHOLTZ, O.).

1985 Analysis of stability of incremental collapse of skeletal structures, J. Struct. Mech. **13**, 301-321, (with SIEMASZKO, A.).

1985 An iterative method of evaluation of elastic-plastic deflections of framed structures. Ing. Arch. **55** 201-212, (with DOROSZ, S.).

1987 Shakedown of elastic-plastic structures, Elsevier & PWN-Polish Scientific Publishers, Warsaw, 1987.

1988 Strainhardening effects in shakedown process, Ing. Archiv. **58** 58-66, (with SIEMASZKO, A.).

1988 Method of substitutive plastic rotation applied to elastic-plastic structures with axial forces, Ingenieur Archiv **58**, 229-239, (with DOROSZ, S.).

1990 Shakedown analysis in the case of imposed displacement, Mechanika Teoretyczna i Stosowana, **1-2**(28), 101-108. (with PYCKO, S.).

1990 Micromechanical modelling and computational of shakedown with nonlinear kinematic hardening including examples for 2-d problems. In Proc. CSME Mechanical Engineering Forum, Toronto, (with STEIN, E., ZHANG, G., and MAHNKEM, R.).

1990 Geometric effects in shakedown of optimum structures, In: Kleiber M., König J.A., editors, Inelastic solids and structures. Antoni Sawczuk Memorial Volume, Pineridge Press, 503-515, (with SIEMASZKO, A.).

1991 Shakedown optimization accounting for nonlinear geometrical effects, ZAMM, **71**, 294-296, (with SIEMASZKO, A.).

1991 Steady plastic cycles on reference configuration in the presence of second-order geometric effects. European Journal of Mechanics, **10** (6), 503-574, (with PYCKO, S.).

1992 Elastic plastic structures subjected to variable repeated imposed displacement and mechanical loads, Int. J. Plast., **8**, 603-618. (with PYCKO, S.).

1992 Shakedown with nonlinear strain-hardening including structural computation using finite element method, Int. J. Plasticity 8, 1-31 (with E. Stein and G. Zhang).

MATERIAL MODELLING FOR CYCLIC PLASTICITY AND DAMAGE

THE MODIFIED MULTISURFACE HARDENING MODEL AND ITS APPLICATION TO CYCLIC LOADING PROBLEMS

Zenon MRÓZ and Piotr RODZIK
Polish Academy of Sciences
Institute of Fundamental Technological Research
UL. Świętokrzyska 21, 00-049 Warsaw, Poland

Abstract: A complex interaction of cyclic loading histories in elasto-plastic material response can be described by using the multisurface hardening model with proper evolution rules for consecutive surfaces corresponding to specified hardening moduli. After model formulation, the parameters are identified for stainless steel 316, and applied to predict cyclic response for proportional and non-proportional loading. The stress and strain evolution near the circular hole in a cyclically loaded disk is then studied by using the multisurface model.

INTRODUCTION

A quantitative description of inelastic material response under a variety of loading conditions poses a challenging task in formulating constitutive models. There are numerous formulations, mostly based on the assumption of anisotropic hardening, specifying evolution of state variables with proper account for memory rules. Recent experiments for uniaxial and multiaxial cyclic loading provide a lot of new data on material hardening and evolution of material state in the course of loading. This information should be accounted for in formulating hardening rules in order to make the model applicable to complex loading histories. In particular, such effects as strain amplitude dependent hardening, ratcheting for non-symmetric stress cycles, non-proportional hardening, etc., should be incorporated into the model.

In this paper, we shall discuss an improved version of the multisurface hardening model originally proposed by MRÓZ [1967,1969] for metals, and then extended to soils, MRÓZ et al. [1981]. A set of nested surfaces is assumed to represent the field of hardening moduli, and evolution of surfaces occurs in the course of plastic straining. The memory of specific loading events is then naturally incorporated into the model

Z. Mróz et al. (eds.), Inelastic Behaviour of Structures under Variable Loads, 3–18.

through positions of centers of consecutive surfaces. The subsequent developments used either two or an infinite number of surfaces, thus providing continuous variation of hardening moduli, cf. DAFALIAS and POPOV [1976], MRÓZ [1981], TRĄMPCZYŃSKI and MRÓZ [1992]. Extensive reviews of constitutive models of this kind can be found in articles by MRÓZ [1986] and OHNO [1990].

In the next section, we shall briefly discuss the model equations and apply the model to simulate stainless steel response for uniaxial and multiaxial cyclic loading. The evolution of cyclic plasticity in a disk weakened by a circular hole is studied with the help of the presented model.

MEMORY RULES AND EVOLUTION OF MATERIAL STATE

The hardening state of material is usually specified by a set of state variables evolving with the macroscopic plastic strain. Such evolution is usually characterized by a property of *convergence* (or asymptotic stability) and is related to asymptotic behaviour of deformation processes induced by the same loading programmes, but starting from different initial states. This property is also fundamental for structures of elastic-perfectly plastic or hardening materials, for which the elastic shakedown theorem is proved. In fact, the difference of initial states may be associated with different prior deformation histories in the elasto-plastic regime involving different hardening and different residual stress distributions.

The existence of steady cycles independent of initial states was proved by FREDERICK and AMSTRONG [1966] for elastic-perfectly plastic and elastic-viscoplastic materials or composites made of such materials. This property was next demonstrated by MRÓZ [1972,1973] for elastic, kinematically hardening plastic or viscoplastic materials, and also by AINSWORTH [1977], who constructed bounding solutions. By using generalized standard models proposed by HALPHEN and NGUYEN [1975], POLIZZOTTO et al. [1990] discussed the convergence of transient states to steady cycles and properties of cyclic states. In fact, it can be proved that the distance norm between two processes measured in terms of elastic energy of stress and strain differences $\Delta\boldsymbol{\sigma} = \boldsymbol{\sigma}_2 - \boldsymbol{\sigma}_1$, $\Delta\boldsymbol{\epsilon} = \boldsymbol{\epsilon}_2 - \boldsymbol{\epsilon}_1$, decreases during the subsequent elastic deformation process , that is

$$E(t) = \int \frac{1}{2}\Delta\boldsymbol{\sigma}\cdot\boldsymbol{C}\Delta\boldsymbol{\sigma}dV_e = \int \frac{1}{2}\Delta\boldsymbol{\sigma}\cdot\Delta\boldsymbol{\epsilon}^e dV_e > 0,$$

$$\dot{E}(t) \leq 0 \tag{1}$$

where $\boldsymbol{C}$ denotes the elastic material compliance and the integral is calculated over the representative element domain. The states $\boldsymbol{\sigma}_1$, $\boldsymbol{\epsilon}_1$, and $\boldsymbol{\sigma}_2$, $\boldsymbol{\epsilon}_2$ correspond to the same boundary tractions and displacements, and the deformation processes start from different initial states. In particular, the form of steady hysteresis loops does not depend on the initial state, nor on the initial fixed prestrain kept unaltered during the subsequent cyclic tests.

This general convergence (or asymptotic stability) property may facilitate the formulaion of constitutive models, since we may ignore the prior history in considering asymptotic states depending only on actual cyclic loading. Some metals, in fact, exhibit such property, though this is not generally valid. For instance, when an annealed specimen is subjected to alternating tension of prescribed stress or strain amplitude, it usually hardens, and the hysteresis loops become thinner until the steady cyclic state is reached. On the other hand, metals initially cold worked may soften during cyclic deformation and the stress amplitude may decrease in a test with prescribed strain amplitude. For some metals, the stress amplitude in the steady state does not depend on the initial state, but only on the applied strain amplitude. For instance, COFFIN [1962] observed this property on 2S-aluminum, comparing cyclic response of annealed and prestrained specimens.

However, for some structural, high strength, and stainless steels such convergence property does not occur exactly, and the memory of larger plastic prestrain remains incorporated, thereby affecting the subsequent response. The experimental study of back stress evolution, summarized by TRĄMPCZYŃSKI and MRÓZ [1992], indicates that the isotropic hardening portion depends only on plastic strain or stress amplitude. On the other hand, the back stress evolution is affected by prior maximal back stress value, and the hardening state attained by the material cannot be removed by subsequent deformation history of lower amplitude. Such effects as amplitude dependent hardening and maximal prestress effect should therefore be incorporated into the model. The effect of non-proportional loading is associated with additional hardening which can be removed by subsequent proportional loading.

The multisurface model discussed in this paper allows for incorporation of such effects as maximal prestress, amplitude-dependent and non-proportional hardening. In fact, the maximal surface reached during the loading history provides a natural memory measure of such prestress. Similarly, the loading amplitude is specified by the diameter of loading surface within which the loading cycle operates. In the next section, all three effects will be accounted for, thus providing a description applicable to both transient and steady cyclic states. The asymptotic stability property discussed in this section is not assumed in formulation of the model.

MULTISURFACE HARDENING RULE FORMULATION

The multisurface formulation is based on the concept of nested surfaces corresponding to different hardening moduli. The active loading surface is assumed to translate toward the consecutive external surface, so that when the stress point reaches this surface, both surfaces are tangential and the contacted surface becomes the active loading surface, cf. MRÓZ [1967]. This model is very convenient in applications since all hardening moduli can be identified from simple loading-unloading tests. However, some effects such as strain ratcheting during proportional and non-proportional loading cannot be properly described by the original formulation. In fact, it is based on the assumption that all external surfaces with respect to the active loading surface do not translate until they became engaged by the loading surface. This assumption can now be modified. The isotropic hardening rules of nested surfaces should also be introduced. Let us discuss in more detail these modifications in more detail.

Consider the l-th active loading surface specified by the equation:

$$F^{(l)}(\boldsymbol{S}, \boldsymbol{\alpha}^{(l)}, R^{(l)}) = \|\boldsymbol{S} - \boldsymbol{\alpha}^{(l)}\| - R^{(l)} = 0 \tag{2}$$

and the associated flow rule

$$d\boldsymbol{e}^p = \frac{1}{H^{(l)}}\boldsymbol{n}(d\boldsymbol{S} \cdot \boldsymbol{n}) \tag{3}$$

where $\boldsymbol{S}$ and $\boldsymbol{e}^p$ denote the stress and strain deviators, $\boldsymbol{\alpha}^{(l)}$ is the back stress associated with the l-th surface, $\boldsymbol{n}$ denotes the normalized stress gradient of the loading surface, $\|\boldsymbol{n}\| = (\boldsymbol{n} \cdot \boldsymbol{n})^{\frac{1}{2}} = 1$, and $H^{(l)}$ is the hardening modulus. The subsequent nested surface is described by a similar equation

$$F^{(l+1)}(\boldsymbol{S}, \boldsymbol{\alpha}^{(l+1)}, R^{(l+1)}) = \|\boldsymbol{S} - \boldsymbol{\alpha}^{(l+1)}\| - R^{(l+1)} = 0 \tag{4}$$

For any stress point $\boldsymbol{S}^{(l)}$ on the surface $F^{(l)} = 0$ there is an associated point $\boldsymbol{S}^{(l+1)}$ on the surface $F^{(l+1)} = 0$ such that

$$\frac{\boldsymbol{S}^{(l+1)} - \boldsymbol{\alpha}^{(l+1)}}{R^{(l+1)}} = \frac{\boldsymbol{S}^{(l)} - \boldsymbol{\alpha}^{(l)}}{R^{(l)}} \tag{5}$$

and

$$\boldsymbol{S}^{(l+1)} = \boldsymbol{\alpha}^{(l+1)} + \frac{R^{(l+1)}}{R^{(l)}}(\boldsymbol{S}^{(l)} - \boldsymbol{\alpha}^{(l)}) \ , \tag{6}$$

$$\boldsymbol{S}^{(l)} = \boldsymbol{\alpha}^{(l)} + R^{(l)}\boldsymbol{m} \ , \quad \boldsymbol{S}^{(l+1)} = \boldsymbol{\alpha}^{(l+1)} + R^{(l+1)}\boldsymbol{m} \tag{7}$$

where $\boldsymbol{m}$ is the unit vector along the vector $\boldsymbol{S}^{(l)} - \boldsymbol{\alpha}^{(l)}$. Assume that the relative motion of the point $\boldsymbol{S}^{(l)}$ with respect to $\boldsymbol{S}^{(l+1)}$ occurs along the line connecting these points. We have

$$d\boldsymbol{S}^{(l)} = d\boldsymbol{\alpha}^{(l)} + dR^{(l)}\boldsymbol{m} \ , \quad d\boldsymbol{S}^{(l+1)} = d\boldsymbol{\alpha}^{(l+1)} + dR^{(l+1)}\boldsymbol{m} \ , \tag{8}$$

$$d\boldsymbol{S}^{(l)} - d\boldsymbol{S}^{(l+1)} = d\mu(\boldsymbol{S}^{(l+1)} - \boldsymbol{S}^{(l)}) \tag{9}$$

$$d\boldsymbol{\alpha}^{(l)} = d\boldsymbol{\alpha}^{(l+1)} + (dR^{(l+1)} - dR^{(l)})\boldsymbol{m} + d\mu(\boldsymbol{S}^{(l+1)} - \boldsymbol{S}^{(l)}) \tag{10}$$

Thus, when $d\boldsymbol{\alpha}^{(l+1)} = 0$, $dR^{(l+1)} = dR^{(l)}$, the center of the loading surface $F^{(l)} = 0$ translates along the vector connecting $\boldsymbol{S}^{(l)}$ and $\boldsymbol{S}^{(l+1)}$. Here $d\mu$ is a scalar proportionality factor to be determined from the consistency condition $dF^{(l)} = 0$, thus

$$d\mu = \frac{d\boldsymbol{S} \cdot \boldsymbol{m} - d\boldsymbol{\alpha}^{(l+1)} \cdot \boldsymbol{m} - dR^{(l+1)}}{(\boldsymbol{S}^{(l+1)} - \boldsymbol{S}^{(l)}) \cdot \boldsymbol{m}} \tag{11}$$

Equation (10) provides the translation rule for the active loading surface $F^{(l)} = 0$. In the former formulation, the external surfaces to $F^{(l)} = 0$ were assumed to remain fixed before they become active. Now, however, we assume that the evolution of material state is represented by proper translation rules of external surfaces. We state the following rule: *the translation of the external surface* $F^{(k)} = 0,\ l < k < n$, *is governed by hardening and recovery processes, the first inducing translation toward the associated point on the consecutive surface* $F^{(k+1)} = 0$, *the other requiring translation to the center of the surface* $F^{(k+1)} = 0$. The center of the boundary surface $F^{(n)} = 0$ is assumed to be fixed, thus $d\boldsymbol{\alpha}^{(n)} = 0$. This translation rule is affected by cyclic loading conditions and can be expressed as follows

$$\begin{aligned} &d\boldsymbol{\alpha}^{(k)} = d\boldsymbol{\alpha}^{(k+1)} + \chi_h \boldsymbol{\eta}_h^{(k)} + \chi_r \boldsymbol{\eta}_r^{(k)} , \quad k = l+1,\ l+2, ..., n-1, \\ &d\boldsymbol{\alpha}^{(n)} = 0 \end{aligned} \tag{12}$$

where χ_h, χ_r are material functions specifying the rate of evolution and $\boldsymbol{\eta}_h^{(k)}$, $\boldsymbol{\eta}_r^{(k)}$ are translation vectors specifying the orientation of evolution due to cyclic hardening. The effect of non-proportionality on the hardening rate is introduced in the isotropic hardening. Assume the following translation vectors $\boldsymbol{\eta}_h^{(k)}$ and $\boldsymbol{\eta}_r^{(k)}$, namely

$$\boldsymbol{\eta}_h^{(k)} = \boldsymbol{S}^{(k+1)} - \boldsymbol{S}^{(k)} \ , \boldsymbol{\eta}_r^{(k)} = \boldsymbol{\alpha}^{(k+1)} - \boldsymbol{\alpha}^{(k)} \tag{13}$$

where $\boldsymbol{S}^{(k+1)}$ and $\boldsymbol{S}^{(k)}$ are the points on $F^{(k+1)} = 0$ and $F^{(k)} = 0$ associated with the same normal as that at the stress point (Fig. 1), so we have

$$\boldsymbol{S}^{(k+1)} = R^{(k+1)} \boldsymbol{m} + \boldsymbol{\alpha}^{(k+1)} \ , \quad \boldsymbol{S}^{(k)} = R^{(k)} \boldsymbol{m} + \boldsymbol{\alpha}^{(k)}, \tag{14}$$

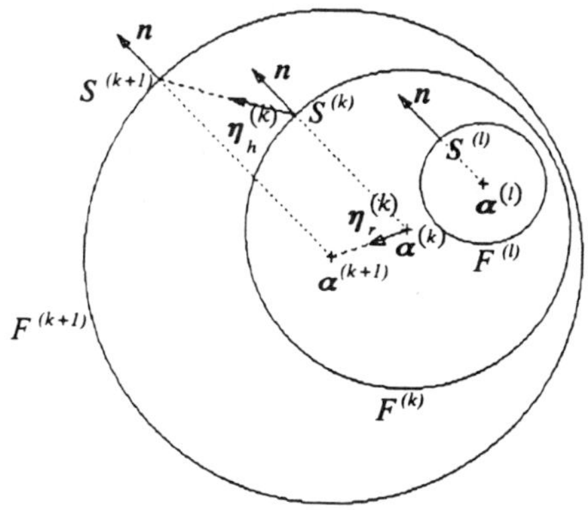

Fig. 1. Graphical illustration of the additional translation rule.

In order to specify the evolution rule (12), let us now propose the forms of the functions χ_h and χ_r .

First, let us introduce the measures of amplitude ξ_a and maximal prestress ξ_m

$$\xi_a = \frac{R^{(l_u)}}{R^{(1)}} , \quad l_u = \left\{ l : F^{(l)} = 0 \ and \frac{\partial F^{(l)}}{\partial \boldsymbol{S}} \cdot d\boldsymbol{S} < 0 \right\}$$

$$\xi_m = \frac{R^{(l_m)}}{R^{(1)}} , \quad l_m = \max \left\{ l : F^{(l)}(\tau) = 0, \ \tau = [0, t] \right\} \tag{15}$$

Thus, the amplitude is specified by the radius of active loading surface at the instant of unloading, and the maximal prestress is measured by the largest radius of the loading surface reached during the past loading history. Obviously, at the initial state we have

$$l_u = l_m = 0 \rightarrow R^{(l_u)} = R^{(l_m)} = 0 , \quad \xi_a = \xi_m = 0 , \quad \chi_h = \chi_r = 0 \tag{16}$$

The evolution rules of the cyclic hardening function χ_h and the recovery function χ_r are assumed in the incremental form

$$d\chi_h = A(d\boldsymbol{e}^p \cdot d\boldsymbol{e}^p)^{\frac{1}{2}}(\chi_h^{as} - \chi_h) \tag{17}$$

$$d\chi_r = C(d\boldsymbol{e}^p \cdot d\boldsymbol{e}^p)^{\frac{1}{2}}(\chi_r^{as} - \chi_r) \tag{18}$$

where the asymptotic values of the cyclic hardening function and recovery function reached for a constant amplitude cycle are assumed as

$$\chi_h^{as} = B(\xi_a)^L \quad and \quad \chi_r^{as} = D(\xi_m)^M \tag{19}$$

and A, B, C, D, L and M denote the material parameters.
The model requires specification of 6 parameters: A, B, C, D, L and M. They can be identified from uniaxial tests of constant strain or stress amplitude.
Assume the isotropic hardening to develop only for the non-proportional loading history and to be expressed in a simple form

$$R^{(k)} = (1 + \beta^{(k)}\psi)R_0^{(k)} , \quad k = 1, \ 2, ..., \ n \tag{20}$$

where $\beta^{(k)} = \frac{R_0^{(k)} - R_0^{(1)}}{R_0^{(n)} - R_0^{(1)}}$ are the scaling factors varying within the range [0,1] and depending on surface sizes. For the elastic surface there is $\beta^{(1)} = 0$ which gives $R^{(1)} = R_0^{(1)}$; for the boundary surface there is $\beta^{(n)} = 1$ which gives $R^{(n)} = (1 + \psi)R_o^{(n)}$. The material function ψ depends on loading non-proportionality and loading amplitude, growing from its initial value $\psi = 0$ to an asymptotic value ψ^{as}.
The evolution rule for function ψ is assumed in the incremental form

$$d\psi = Pd\theta(\psi^{as} - \psi) \tag{21}$$

where P is a material constant, θ is the non-proportionality factor and ψ^{as} denotes the asymptotic value of ψ. It is assumed that ψ^{as} depends on the radius of maximal loading surface reached during the process of non-proportional loading, or the actual amplitude surface during the non-proportional unloading, that is

$$\psi^{as} = Y\left(\frac{R_0^{(m)}}{R_0^{(1)}}\right)^N, \quad m = \begin{cases} l_m, & \text{if } l_u = 0 \\ l_u, & \text{if } l_u > 0 \end{cases} \tag{22}$$

where Y, N are the material parameters.
The non-proportionality factor is defined as follows

$$d\theta = \sin^2(\boldsymbol{\alpha}^{(l)}, d\boldsymbol{\alpha}^{(l)}) = 1 - \left(\frac{d\boldsymbol{\alpha}^{(l)} \cdot \boldsymbol{\alpha}^{(l)}}{\|d\boldsymbol{\alpha}^{(l)}\| \, \|\boldsymbol{\alpha}^{(l)}\|}\right)^2 \tag{23}$$

where $\boldsymbol{\alpha}^{(l)}$ is the back stress of the active surface $F^{(l)} = 0$ and $d\boldsymbol{\alpha}^{(l)}$ denotes the increment of the back stress $\boldsymbol{\alpha}^{(l)}$.
The isotropic hardening is thus specified by 3 constants P, Y and N which can be identified from non-proportional (for instance, circular) loading programs of varying amplitudes.

MODELLING OF UNIAXIAL AND MULTIAXIAL CYCLIC RESPONSE

Material constants occuring in the model were identified using the experimental results reported by TANAKA et al. [1985] and MURAKAMI et al. [1989] for stainless steel SS316 loaded at room temperature. For this material parameter, identification requires the following data: proportional loading curve; two uniaxial cyclic loading tests of constant strain or stress amplitudes; one uniaxial cyclic test with step-wise decreasing amplitude and two non-proportional (e.q. circular) tests. The identification has to be performed by computer generation of model predictions (the initial set of model surfaces is assumed arbitrarily on the basis of proportional loading curve). For the multisurface modelling, we assumed 7 surfaces with the following radii and internal plastic hardening moduli:

$R^1 = 128$ MPa $\quad H^0 = 203$ GPa
$R^2 = 195$ MPa $\quad H^1 = 402$ GPa
$R^3 = 230$ MPa $\quad H^2 = 101$ GPa
$R^4 = 260$ MPa $\quad H^3 = 30$ GPa
$R^5 = 285$ MPa $\quad H^4 = 19$ GPa
$R^6 = 320$ MPa $\quad H^5 = 9$ GPa
$R^7 = 400$ MPa $\quad H^6 = 5$ GPa

and for this set of surfaces the identified cyclic hardening parameters are as follows
A=70, B=5.5 10^{-5}, L=6, C=30, D=2 10^{-4}, M=4, P=4 10^{-4}, Y=3.3 10^{-3}, N=6.

For this set of parameters the model was used for calculation of material response for some uniaxial and multiaxial loading tests. Figures 2 to 8 present comparison of predicted and experimental results for some simple uniaxial and multiaxial tests.
Figure 2 presents experimental and predicted cyclic curves for the stainless steel SS316 for constant plastic strain amplitudes. The modified multisurface model predicts the experimental curves fairly well.

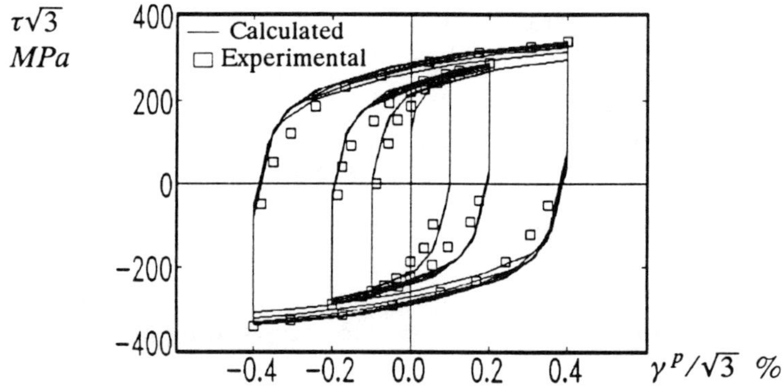

Fig. 2. Simulation of the constant plastic strain amplitude tests, $\Delta\varepsilon^p$ = 0.1, 0.2 and 0.4% .

Figure 3 presents transient stress amplitudes corresponding to increasing specified cyclic strain amplitudes. As before, descriptions provide good approximations to experimental amplitude curves.
Figure 4 presents the predicted amplitude curves for decreasing strain amplitudes and the experimental amplitude curve.
Figure 5 provides predicted stress paths for the specified circular strain path. The effect of non-proportional hardening is clearly seen.
Figure 6 presents the comparison of the experimental and the predicted amplitude curves for step-wise increasing amplitude for the specified circular strain path.
Figure 7 presents the cyclic ratcheting for uniaxial non-symmetrical cycling in the range [0,330MPa]. A decreasing strain accumulation is observed for all cycles.
Figure 8 presents the axial ratcheting curves for an alternating shear stress of specified amplitude superposed on a constant axial stress. Whereas the previous multisurface model predicted linear growth of axial strain, the present modification provides the effect of decreasing ratcheting in the course of plastic straining.

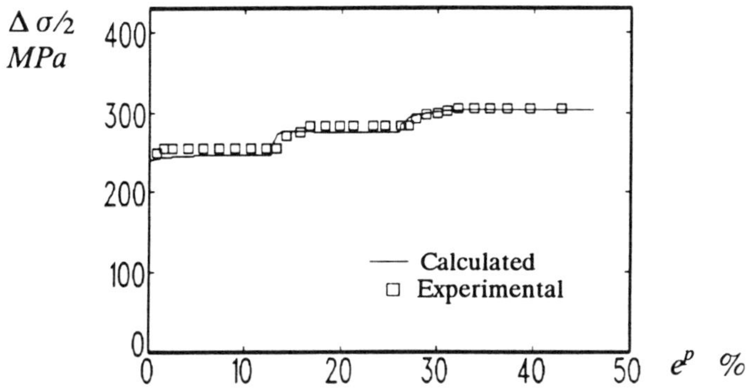

Fig. 3. Simulation of the step-wise increasing strain amplitude test $\Delta\varepsilon = 0.2 - 0.3 - 0.4\%$.

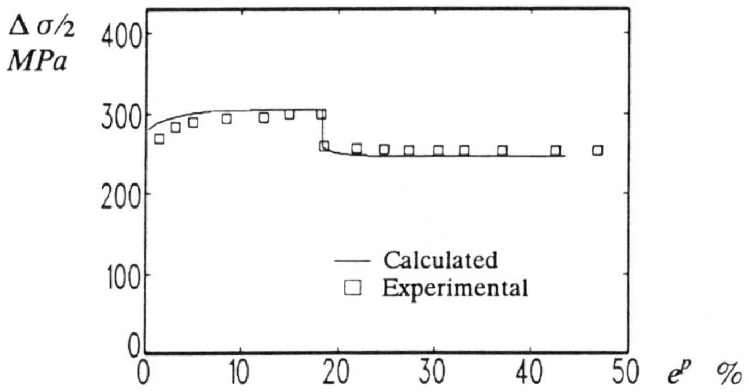

Fig. 4. Simulation of the step-wise decreasing strain amplitude test $\Delta\varepsilon = 0.4 - 0.2\%$.

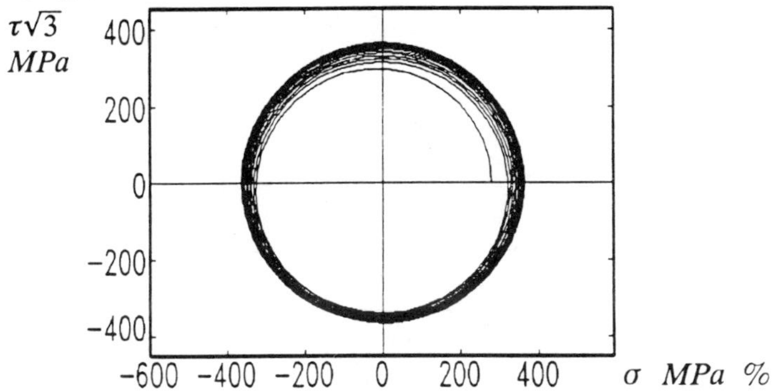

Fig. 5. Effect of loading non-proportionally: stress response prediction for the circular strain path.

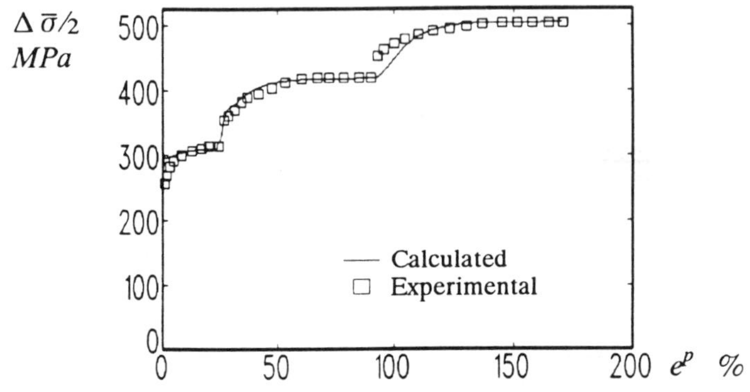

Fig. 6. Simulation of the step-wise increasing strain amplitude circular test $\Delta\bar{\varepsilon} = 0.2 - 0.3 - 0.4\%$.

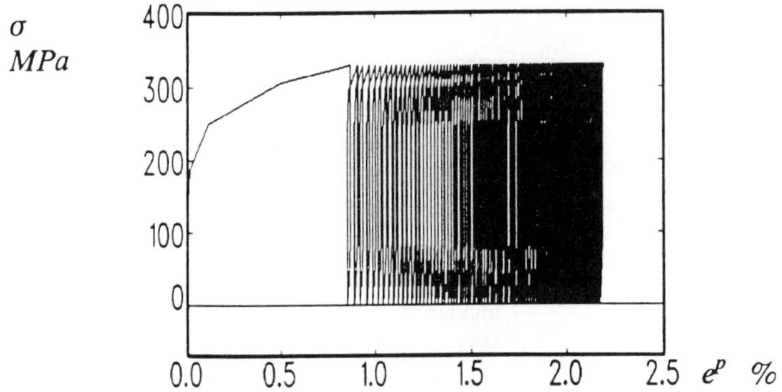

Fig. 7. Uniaxial ratcheting simulation for stress range [0,330MPa].

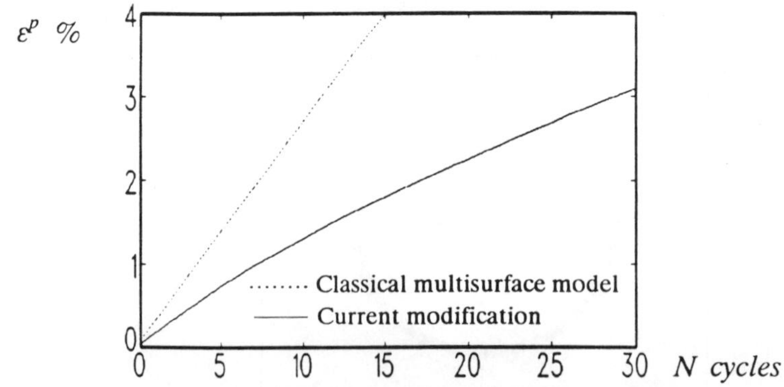

Fig. 8. Plastic strain accumulation for non-proportional ratcheting test obtained for original and current multisurface formulations.

It is seen that for the multisurface model the original version could not describe such material behavior. On the other hand, the present modified formulation provides predictions consistent with experimental data.

THE STRESS CONCENTRATION NEAR THE CIRCULAR HOLE IN A CYCLICALLY LOADED PLATE

The present model was included into the FEM programme reported by HINTON and OWEN [1980] for plane stress isoparametric elements. The local incremental stress-strain relation was assumed in the form

$$\dot{\boldsymbol{\sigma}} = \boldsymbol{C}^{ep} \dot{\boldsymbol{\epsilon}} \tag{24}$$

where $\boldsymbol{C}^{ep}$ is the tangent elasto-plastic stiffness matrix. The matrix has the form

$$\boldsymbol{C}^{ep} = K \mathbf{1} \times \mathbf{1} + 2G(\boldsymbol{I} - \frac{1}{3} \mathbf{1} \times \mathbf{1}) - \frac{2G}{1 + \frac{H^l}{2G}} \boldsymbol{n} \times \boldsymbol{n} \tag{25}$$

where $\boldsymbol{I}$ is a fourth-rank tensor, $I_{ijkl} = \frac{1}{2}(\delta_{ik}\delta_{jl} + \delta_{il}\delta_{jk})$; $\boldsymbol{n}$ is the external unit vector normal to the active loading surface $F^{(l)} = 0$; the tensor product $(\boldsymbol{n} \times \boldsymbol{n})_{ijkl} = n_{ij} n_{kl}$; $H^{(l)}$ is a plastic hardening modulus connected with the active loading surface; G and K are the elastic shear and bulk moduli.

The local elasto-plastic stiffness matrix specified by (25) is valid only for loading processes when the active loading surface continues to be active (the consecutive loading surface is not reached). Generally, the global stiffness matrix (assembled for a structure) is valid only for a loading process (loading step) when the active surface remains active at each integration point. If for any integration point the active surface changes during a loading step, the step length has to be decreased.

As a benchmark analysis, the problem of determining the plane stress-strain history for a cyclically loaded plate weakend by a circular hole was selected. Figure 9 presents a scheme of this plate. Calculations were performed for the plate dimensions: a=40mm, b=80mm, d=20mm and thickness g=1mm. It was assumed that the plate was made of stainless steel SS316. Typically, only a quarter of the plate was used in the analysis. The FEM mesh for this part is shown in Figure 10.

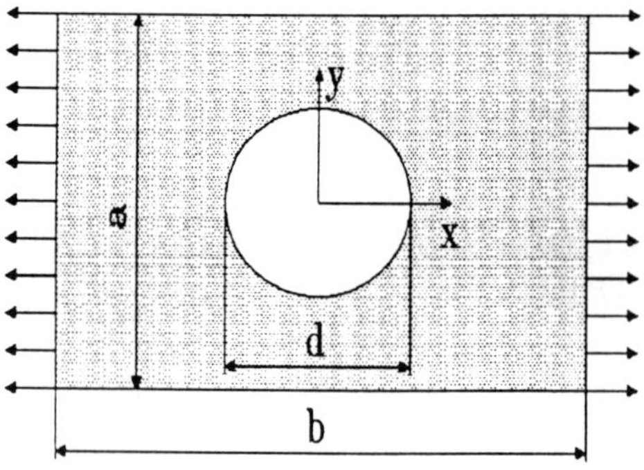

Fig. 9. Scheme of the studied disk.

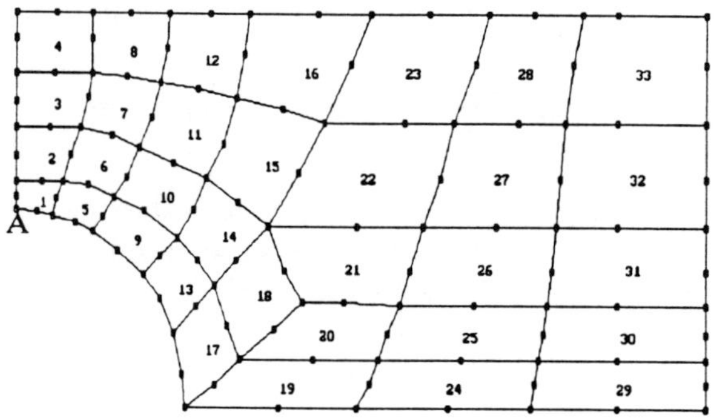

Fig. 10. FEM mesh used in calculations.

Three cyclic tension-compresion tests were carried out: a symmetrical one in the range [-2420N, 2420N] (loading was distributed as shown in Figure 9), and two non-symmetrical programmes in the ranges [-1210N, 2420N] and [0, 2420N]. Figures 11, 12 and 13 present stress-strain histories obtained for these loading programmes at the most stressed point (denoted by A in Figure 10). For the symmetrical loading, the stress concentration grows from cycle to cycle (symmetrical hardening is observed). For non-symmetrical programmes the relaxation of the hysteresis curve is observed, and the material response tends to become symmetric. Evolution of stress intensity near the notch root for the first loading and some selected cycles are shown in Figures 14 a-d. Stress maps are calculated using the algorithm proposed by ZIENKIEWICZ [17]. It is seen that only regions of greatest values of stress intensity which lie very

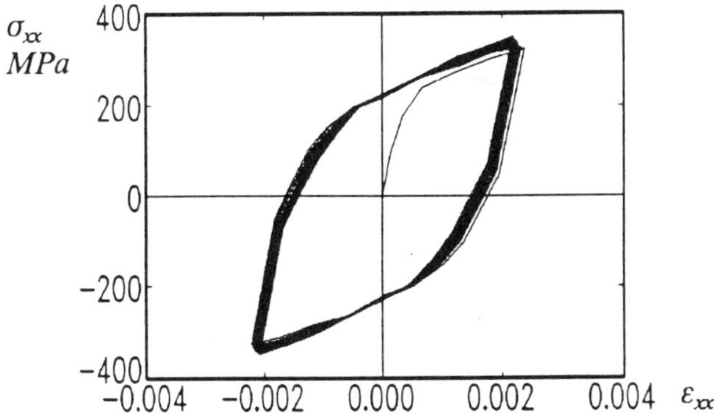

Fig. 11. Stress-strain history at notch root for symmetrical loading in the range [-2420N, 2420N].

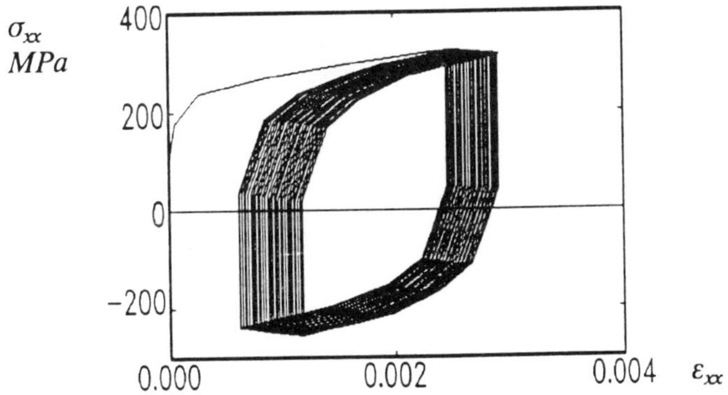

Fig. 12. Stress-strain history at notch root for non-symmetrical loading in the range [-1210N, 2420N].

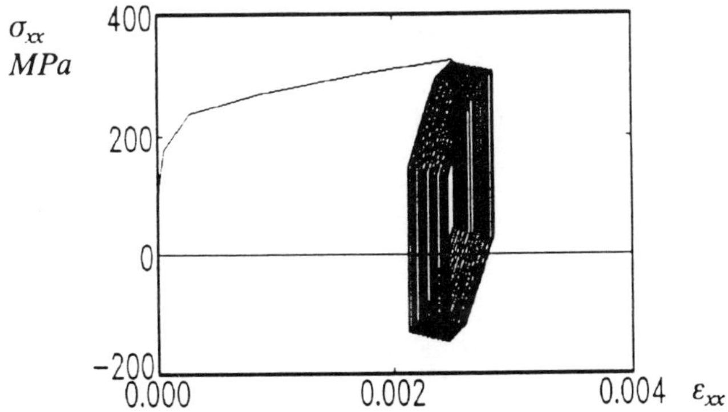

Fig. 13. Stress-strain history at notch root for non-symmetrical loading in the range [0N, 2420N].

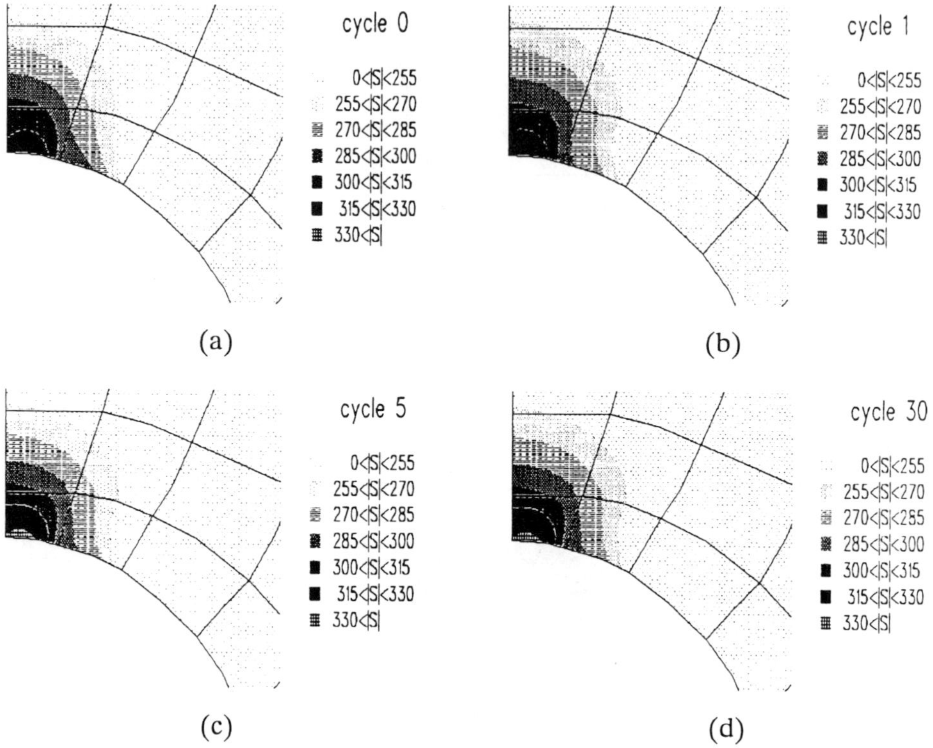

Fig. 14. Stress distribution near the notch root for the end of selected cycles: 0,1,5 and 30.

near to the notch root change visibly. The regions of smallest values are nearly constant during all the cycles.

After several cycles, a stable regime is reached in the structure, and the stress-strain response becomes cyclic.

CONCLUDING REMARKS

A simple modification for a well known constitutive model with multisurface description of back stress was presented. The possibility of quantitative modelling of cyclic material response was thus extended to such phenomena as amplitude dependent cyclic hardening or softening, and sensitivity to amplitude variation for both uniaxial and multiaxial loading programmes.

The material functions and parameters were identified for stainless steel SS316 and the predicted results were compared with experimental ones. A reasonably good correlation of theoretical results and experimental data was observed.

ACKNOWLEDGMENT: The present research was conducted with support from The Science Research Council (KBN), Grant No 31014-91-01.

REFERENCES

1962 COFFIN L.F. Jr., Low-cycle fatigue: a review, Appl.Mat. Res.,**1** , 129-141.

1966 FREDERICK C.O. and AMSTRONG P.J., Convergent internal stresses and steady cyclic states of stress, Journ. Strain Anal., **1**, 154-160.

1967 MRÓZ Z., On the description of anisotropic workhardening, J.Mech.Phys. Solids, **15**, 163-175.

1969 MRÓZ Z., An attempt o describe the behaviour of metals under cyclic loads using a more general work-hardening model, Acta Mechanica, **7**, 199.

1972 MRÓZ Z., On the theory of steady plastic cycles in structures, Proc. 1-st SMIRT, **6**, part L/6.

1973 MRÓZ Z., Mathematical models of inelastic material behaviour, Univ. Waterloo Press.

1975 HALPHEN B. and NGUYEN Q.S., Sur le materiaux standards généralisés, Journ. de Mécanique, **14**, 39-63.

1976 DAFALIAS Y. and POPOV E.P. , Plastic internal variables formalism of cyclic plasticity, J. Appl. Mech, **98**,645.

1977 AINSWORTH R.A., Bounding solutions for creeping structures subjected to load variations above the shake-down limit, Int.J. Solids Struct., **13**, 1-980.

1977 ZIENKIEWICZ O.C., The Finite Element Method, McGraw-Hill, Berkshire, England.

1980 OWEN D.R.J. and HINTON E., Finite Element in Plasticity, Pineridge Press Ltd, Swansea, U.K.

1981 MRÓZ Z., On generalized kinematic hardening rule with memory of maximal prestress, J.Mec. Appl. , **5**, 241-259.

1981 MRÓZ Z., NORRIS V.A. and ZIENKIEWICZ O.C., Anisotropic critical state model for soils subject to cyclic loading, Géotechnique, **31**, 451-469.

1985 TANAKA E., MURAKAMI S. and OOKA M., Effects of plastic strain amplitudes on non-propotional cyclic plasticity, Acta Mechanika, **57**, 1-182.

1986 MRÓZ Z., Phenomenological constitutive models for metals, in "Modelling of small deformations of metals", Ed. J.Zarka and J.Gittus, Elsevier Sci. Publ.

1989 MURAKAMI S., KAWAI M. and OHMI Y., Effects of amplitude-history and temperature-history on multiaxial cyclic behavior of type 316 stainless steel, J.Eng.Mat. Techn., **111**, 278-285.

1990 OHNO N., Recent topics in constitutive modelling of cyclic plasticity and viscoplasticity, ASME Appl. Mech. Revs., **43**, 283.

1990 POLIZZOTTO C., BORINO G. and FUSCHI P., On the steady state response of elastic, perfectly plastic solids to cyclic loads, in "Inelastic Solids and Structures", ed. Kleiber M. and König J.A., Pineridge Press Swonsea, 473-488.

1992 TRĄMPCZYŃSKI W. and MRÓZ Z., Anisotropic hardening model and its application to cyclic loading, Int. J. of Plasticity, **8**, 925-946.

STRUCTURES CONSISTING OF TWO-PHASE MATERIALS UNDER THERMOMECHANICAL LOADS

Thomas SIEGMUND[1], Ewald WERNER[2], Franz Dieter FISCHER[1]
[1]Christian Doppler Laboratory for Micromechanics of Materials and Institute of Mechanics,
[2]Institute of Metal Physics,
Montanuniversität, A-8700 Leoben, Austria

ABSTRACT

The cyclic thermomechanical behavior of two-phase materials is investigated. Specimens of austenitic-ferritic duplex steels with a uniaxially anisotropic microstructure deform irreversibly even under pure thermal cycling conditions. In this work the combination of cyclic thermal and monotonic external loads is studied. The analysis reveals the necessity of incorporating the thermomechanical strain increment into the material law. The results are applied to predict the thermomechanical behavior of structural parts such as expansion joints.

INTRODUCTION

In duplex stainless steels, the constituent phases austenite and ferrite have different coefficients of thermal expansion. On varying the temperature, a remarkable stress state develops which changes in sign from phase to phase. Even under pure thermal cycling, the yield strength of the phases can be exceeded. Consequently, specimens of a forged duplex steel deform irreversibly when subjected to pure thermal cycling (SIEGMUND et al. [1993]). The investigation of the thermal properties of duplex steels is of interest because they combine relatively low coefficients of thermal expansion with high mechanical strength. Applications of these steels have been proposed in heat exchangers and expansion joints (CHARLES [1991], JOLY et al. [1992]) as well as in parts of heat treatment equipment (FISCHER et al. [1990]). In all these cases, cyclic thermomechanical loading occurs. For an application of duplex steels under these conditions a correct description of the properties of the duplex steel is required which has to take into account the interaction between the constituents austenite and ferrite. Classical models proposed for the thermal behavior of two-phase materials (HASHIN [1983], SCHAPERY [1968]) are not sufficient, since these models aim to determine the overall properties of the composite. Under cyclic loading, the knowledge of the evolution of stresses and strains in the constituents is necessary; this requires an incremental constitutive material law for the duplex steel (KARAYAKA & SEHITOGLU [1993]). Furthermore, previously presented models for the thermal proper-

Z. Mróz et al. (eds.), Inelastic Behaviour of Structures under Variable Loads, 19–31.

ties do not take into account the plastic properties of the constituent phases. The present analysis tries to compensate for these shortcommings.
The uniaxial anisotropic microstructure of the forged duplex stainless steel under investigation allows the thermomechanics to be analyzed for loadings by relatively simple models assuming equal strains in longitidinal directions of both phases. Analytic expressions for an incremental constitutive law for thermomechanical loadings are derived within the context of the linear superposition of strain increments in the theory of small strains. The present model is applied to analyse the results obtained from the dilatometer experiments. This analysis is shown to be capable of revealing the mechanisms responsible for the experimentally observed thermal ratchet effect. Furthermore, an investigation of a structural part made of a duplex steel under thermomechanical cyclic loading is presented with two different types of material descriptions. An analysis that takes into account of the inelastic deformations due to internal stresses is compared to calculations neglecting these deformations.

MATERIALS AND EXPERIMENTAL OBSERVATIONS

The ferritic-austenitic duplex steel whose thermomechanical properties are investigated is a high-chromium stainless steel. During fabrication the cast steel block is hot forged to a rod which is then solution annealed and quenched in water. Forging yields a uniaxially anisotropic microstructure formed of elongated domains of the two phases both aligned parallel to the axis of the forged rod (Fig (1)). The volume fractions of ferrite (α) and austenite (γ) are $V_V^\alpha = 0.53$ and $V_V^\gamma = 0.47$, respectively. In the temperature interval 20°C – 900°C these volume fractions, and the anisotropy of the microstructure, remain nearly unchanged (SIEGMUND et al. [1993], FISCHER et al. [1990]). A single-phase ferritic and a single-phase austenitic steel of compositions similar to that of ferrite and austenite in the duplex steel are used as reference materials.
Continuous cyclic one-dimensional (1D) dilatometer tests are performed in the interval 20°C – 900°C – 20°C with the single-phase materials and the duplex steel using cylindrical specimens. The heating and cooling rates applied are $\dot{T} = 0.03, 0.06, 0.2, 1$ and 20°C/s. For the duplex steel, the specimens are taken with their axes parallel and perpendicular to the axis of the forged rod (longitudinal and transverse specimens, respectively).
Results of the dilatometer tests are shown in Fig (2) for $\dot{T} = 20$°C/s. Here the strain in the specimen's axial direction as obtained from four thermal cycles is plotted for the longitudinal (a) and transverse (b) specimens. An irreversible elongation of the longitudinal

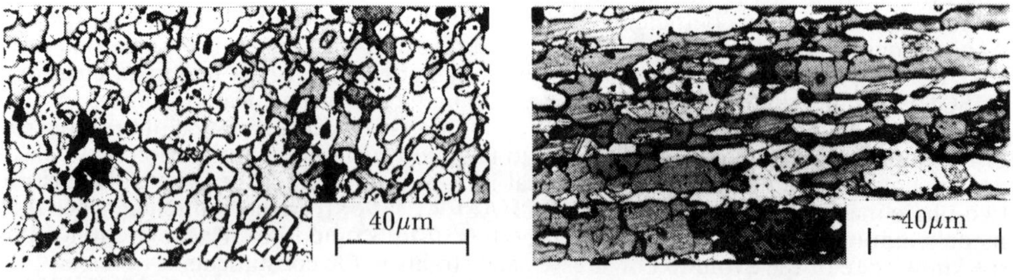

(a) (b)

Figure 1: Metallographic sections of the duplex steel: (a) perpendicular and (b) parallel to the axis of the forged rod. Ferrite is the dark phase.

specimens can be observed. Transverse specimens, on the other hand, shorten irreversibly. With the exception of the first cycle, the length change introduced by each cycle is nearly constant.

For the duplex steel Fig (3a) shows an almost linear relation between the irreversible deformation and the number N of thermal cycles applied. For the single-phase steels, however, nearly no irreversible deformation is measured. The amount of the irreversible strain is influenced significantly by $\dot{T}$ in slow thermal cycles. This is evident from Fig (3b) where the average irreversible deformation per cycle is depicted as a function of the temperature rate $\dot{T}$. The fact that $\dot{T}$ influences the irreversible axial strain from longitudinal and transverse duplex steel specimens similarly, indicates that two different mechanisms may act in this material. An isotropic contribution to the deformation arises from a shrinkage in volume of the duplex steel due to the precipitation of the intermetallic σ-phase in the high temperature part of the thermal cycles (HULL [1979]). This precipitation depends on time and hence is present only in slow thermal cycles. The deformation of the duplex steel due to the stresses generated by the difference in the thermal expansion behavior of its constituent phases depends, on the other hand, on the anisotropy of the microstructure and, therefore, is observed at all temperature rates applied (SIEGMUND et al. [1993]). It is this contribution to the irreversible strain to be analysed in detail. Thermal cycling experiments (20°C – 900°C – 20°C, $\dot{T} = 1°\mathrm{C/s}$) are also performed in 3D-dilatometer tests. In such tests it is possible to measure the deformations in the longitudinal and transverse directions simultaneously on the same specimen. The experiments reveal irreversible deformations similar to those measured in 1D-tests. The mean values of the irreversible strains per thermal cycle are $14.36 \cdot 10^{-5}$ and $-8.77 \cdot 10^{-5}$ in the longitudinal and the transverse directions, respectively. Thus, the ratio of the two deformations is approximately -0.5. This means that the observed irreversible deformations due to pure thermal cycling preserve the volume of the specimen. Plastic deformation can be made responsible for this experimental finding.

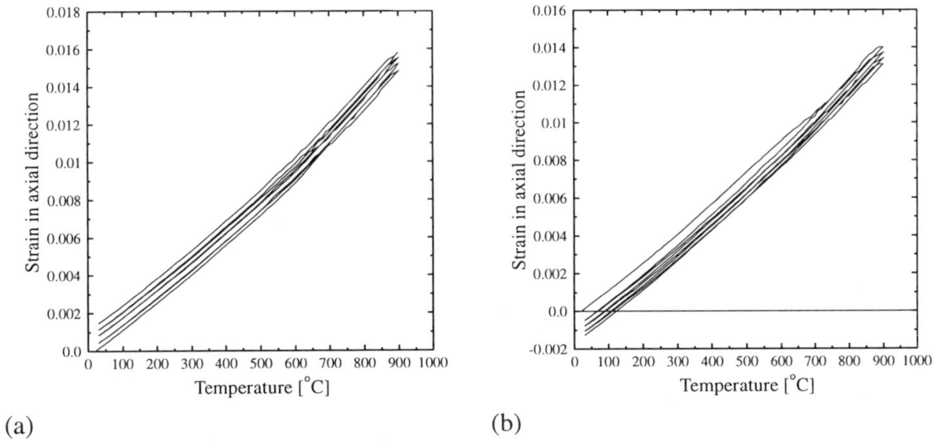

Figure 2: Results from dilatometer tests on the duplex steel. The strain in the specimen's axial direction is plotted as a function of the temperature for (a) a longitudinal specimen and (b) a transverse specimen.

THERMOMECHANICAL PROPERTIES OF THE DUPLEX STEEL

Macroscopic constitutive relations for the description of the behavior of metallic materials under thermomechanical loading can be treated by continuum-mechanics by assigning each microscopic deformation mechanism a corresponding increment of macroscopic strain. The total increment of macroscopic strain $\dot{\underline{\underline{\epsilon}}}$ is the sum of the elastic $\dot{\underline{\underline{\epsilon}}}_{e}$, thermal $\dot{\underline{\underline{\epsilon}}}_{th}$ and inelastic $\dot{\underline{\underline{\epsilon}}}_{in}$ contributions:

$$\dot{\underline{\underline{\epsilon}}} = \dot{\underline{\underline{\epsilon}}}_{e} + \dot{\underline{\underline{\epsilon}}}_{th} + \dot{\underline{\underline{\epsilon}}}_{in} \tag{1}$$

For two-phase materials under thermomechanical loading two contributions to the inelastic part of the total strain rate are considered in the following (LEBLOND et al. [1989]):

$$\dot{\underline{\underline{\epsilon}}}_{in} = \dot{\underline{\underline{\epsilon}}}_{p} + \dot{\underline{\underline{\epsilon}}}_{tm} \tag{2}$$

$\dot{\underline{\underline{\epsilon}}}_{p}$ is the strain rate of the conventional plasticity at constant temperature and contributes to $\dot{\underline{\underline{\epsilon}}}_{in}$ if the global yield strength of the material is exceeded.

In two-phase materials consisting of phases with different thermal expansion behavior, internal stresses are generated by a change in temperature. If these internal stresses are the same in magnitude as the yield strength in any of the constituent phases, plastic deformation sets in. This occurs even if no external load stress is applied.

The determination of the inelastic deformations represented here by the irreversible thermomechanical strain increment $\dot{\underline{\underline{\epsilon}}}_{tm}$ is the subject of this report. If $\dot{\underline{\underline{\epsilon}}}_{tm}$ is incorporated into the constitutive law for the duplex steel, the thermal cycling experiments can be analysed.

The properties of the duplex steel are modeled using an approximate analysis similar to that presented in (SIEGMUND et al. [1993], PEDERSEN [1990], JANSSON & LECKIE [1992], GARMONG [1974], BÖHM [1991]).

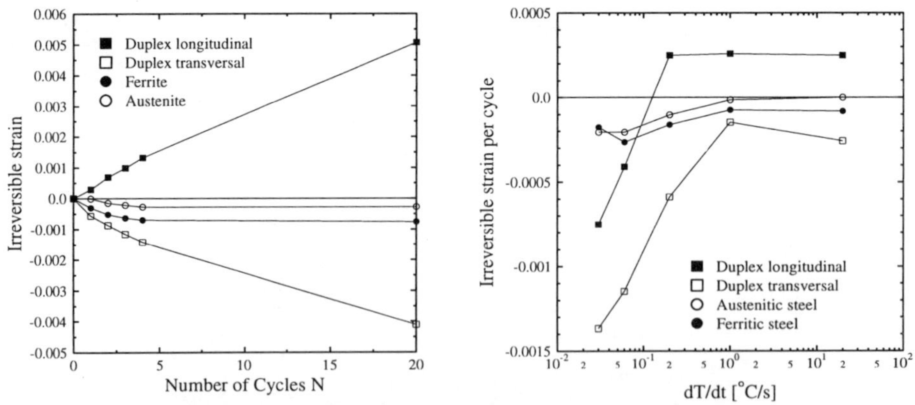

(a) (b)

Figure 3: Irreversible strain in the axial specimen direction in three different types of steels as a function of (a) the number of thermal cycles (20°C-900°C-20°C) with $\dot{T} = 20°\text{Cs}^{-1}$ and (b) the temperature rate $\dot{T}$ applied on thermal cycling.

Relatively simple but analytic expressions for $\dot{\underline{\underline{\epsilon}}}_{\rm tm}$ can be derived on the basis of classical relations (SCHAPERY [1968]) for the duplex steel's thermal properties. This method has been proved to be powerful in capturing the dominant features of the thermomechanical properties of two-phase materials. The current analysis is, however, different from the ones cited above since it recognises that in duplex steels both constituents can deform plastically. In this model, the duplex steel is idealized as a material possessing a uniaxially anisotropic microstructure with the x_3-direction as the axis of anistropy and x_1 and x_2 as the transverse directions. For a material of this type, the relation between the longitudinal strain increments is:

$$\dot{\underline{\underline{\epsilon}}}_3 = \dot{\underline{\underline{\epsilon}}}_3^\alpha = \dot{\underline{\underline{\epsilon}}}_3^\gamma \tag{3}$$

and the longitudinal Young's modulus, $E_3^{\rm d}$, and the longitudinal Poisson's ratio, $\nu_3^{\rm d}$, are evaluated from the elastic properties of the phases by:

$$E_3^{\rm d} = V_{\rm V}^\alpha E^\alpha + V_{\rm V}^\gamma E^\gamma \ , \quad \nu_3^{\rm d} = V_{\rm V}^\alpha \nu^\alpha + V_{\rm V}^\gamma \nu^\gamma \ . \tag{4}$$

E^α, E^γ and ν^α, ν^γ denote the the Young's moduli and Poisson's ratios of ferrite and austenite, respectively.
In the elastic case the instantaneous coefficients of thermal expansion for the longitudinal and transverse directions of the composite, $\hat{\alpha}_3^{\rm d}$ and $\hat{\alpha}_1^{\rm d} = \hat{\alpha}_2^{\rm d}$, are (SCHAPERY [1968]):

$$\hat{\alpha}_3^{\rm d} = \alpha^\alpha V_{\rm V}^\alpha \frac{E^\alpha}{E_3^{\rm d}} + \alpha^\gamma V_{\rm V}^\gamma \frac{E^\gamma}{E_3^{\rm d}} \ , \tag{5}$$

$$\hat{\alpha}_1^{\rm d} = \hat{\alpha}_2^{\rm d} = \alpha^\alpha V_{\rm V}^\alpha (1+\nu^\alpha) + \alpha^\gamma V_{\rm V}^\gamma (1+\nu^\gamma) - \hat{\alpha}_3^{\rm d} \nu_3^{\rm d} \tag{6}$$

Owing to the different thermal expansion behavior of the two phases thermal stresses, $\underline{\underline{\sigma}}_{\rm th}^\alpha$ and $\underline{\underline{\sigma}}_{\rm th}^\gamma$, are generated if the temperature is changed. In the uniaxially anisotropic material, the longitudinal stresses dominate, and hence the longitudinal thermal stresses have to be analysed. The application of the mechanical equilibrium condition for pure thermal loading $V_{\rm V}^\alpha \sigma_3^\alpha + V_{\rm V}^\gamma \sigma_3^\gamma = 0$ in the x_3-direction yields for the longitudinal thermal stresses

$$\begin{aligned} \sigma_{\rm th,3}^\alpha &= \int_{\rm T_0}^{\rm T} \frac{V_{\rm V}^\gamma E^\alpha E^\gamma (\alpha^\gamma - \alpha^\alpha)}{E_3^{\rm d}} \, dT \ , \\ \sigma_{\rm th,3}^\gamma &= \int_{\rm T_0}^{\rm T} \frac{V_{\rm V}^\alpha E^\alpha E^\gamma (\alpha^\alpha - \alpha^\gamma)}{E_3^{\rm d}} \, dT \ . \end{aligned} \tag{7}$$

Cooling down from an initially stressfree state at T_0 generates a compression stress state in the phase of the lower thermal expansion coefficient (ferrite) and a tension stress state in the phase of the higher thermal expansion coefficient (austenite).
If these internal stresses are the same in magnitude as the yield strength, $\sigma_{\rm y}^\alpha$, $\sigma_{\rm y}^\gamma$, in any of the constituent phases plastic deformation sets in even without any external load applied. If, in addition to a change in temperature, a longitudinal stress, σ_3, is exerted on the duplex microstructure such that $|\sigma_3| < \min(\sigma_{\rm y}^\alpha, \sigma_{\rm y}^\gamma)$, then its influence has to be considered, too. The elastic analysis for the coefficients of thermal expansion is only valid until a critical temperature is reached. Plastic deformation on thermomechanical loading sets in if the

absolute value of the sum of thermal stresses and load stresses reaches the yield strength of one of the constituents:

$$
\begin{aligned}
|\sigma_3^\alpha| &= |\int_{T_0}^{T_{crit}^\alpha} \frac{V_V^\gamma E^\alpha E^\gamma(\alpha^\gamma - \alpha^\alpha)}{E_3^d} dT + \sigma_3| = \sigma_y^\alpha \ , \\
|\sigma_3^\gamma| &= |\int_{T_0}^{T_{crit}^\gamma} \frac{V_V^\alpha E^\alpha E^\gamma(\alpha^\alpha - \alpha^\gamma)}{E_3^d} dT + \sigma_3| = \sigma_y^\gamma \ .
\end{aligned}
\tag{8}
$$

After T_{crit}^α or T_{crit}^γ are passed during cool-down, the thermal expansion behavior of the composite is dominated by the accommodation of the thermal mismatch by plastic straining. This changes the thermal behavior of the duplex steel. Assuming elastic-ideal-plastic phases, one can derive approximate solutions for the instantaneous coefficients of thermal expansion $\tilde{\alpha}_3^d$ and $\tilde{\alpha}_1^d = \tilde{\alpha}_2^d$ in the elasto-plastic stage by substituting the Young's modulus of the plastically deformed phase by its plastic tangent modulus $E_{pl} = 0$ and setting the Poisson's ratio of this phase to 0.5 (PEDERSEN [1990], BÖHM [1991]). Inserting these assumptions into eqn (5) gives for the instantaneous coefficients of thermal expansion in the elasto-plastic stage:

$$
\begin{aligned}
\tilde{\alpha}_3^d = \alpha^\gamma \ , \tilde{\alpha}_1^d = \tilde{\alpha}_2^d = \alpha^\gamma(V_V^\gamma - \frac{1}{2}V_V^\alpha) + \frac{3}{2}\alpha^\alpha V_V^\alpha \ &: \ |\sigma_3^\alpha| = \sigma_y^\alpha \ , \\
\tilde{\alpha}_3^d = \alpha^\alpha \ , \tilde{\alpha}_1^d = \tilde{\alpha}_2^d = \alpha^\alpha(V_V^\alpha - \frac{1}{2}V_V^\gamma) + \frac{3}{2}\alpha^\gamma V_V^\gamma \ &: \ |\sigma_3^\gamma| = \sigma_y^\gamma \ .
\end{aligned}
\tag{9}
$$

A subsequent heating process does not start from a stress free state any more. The stress state at the end of the cool down half of the thermal cycle ($T = T_1$) has to be considered for the heating half of the cycle. The stress state at the end of the cool-down process ($\sigma_{end,3}^\alpha$ and $\sigma_{end,3}^\gamma$) is calculated from the mechanical equilibrium condition $\sigma_3 = V_V^\alpha \sigma_3^\alpha + V_V^\gamma \sigma_3^\gamma$:

$$
\begin{aligned}
\sigma_{end,3}^\alpha &= \begin{cases} -\sigma_y^\alpha(T = T_1) & : \ |\sigma_3^\alpha| = \sigma_y^\alpha \\ \frac{\sigma_3 - V_V^\gamma \sigma_y^\gamma (T=T_1)}{V_V^\alpha} & : \ |\sigma_3^\gamma| = \sigma_y^\gamma \ , \end{cases} \\
\sigma_{end,3}^\gamma &= \begin{cases} \frac{\sigma_3 - V_V^\alpha \sigma_y^\alpha (T=T_1)}{V_V^\gamma} & : \ |\sigma_3^\alpha| = \sigma_y^\alpha \\ \sigma_y^\gamma(T = T_1) & : \ |\sigma_3^\gamma| = \sigma_y^\gamma \ . \end{cases}
\end{aligned}
\tag{10}
$$

The thermal stresses generated during the heating-up process are tensile in ferrite and compressive in austenite. Starting at T_1, the constituents are initially unloaded from $\sigma_{end,3}^\alpha$ and $\sigma_{end,3}^\gamma$. The elastic analysis for the instantaneous coefficients of thermal expansion (eqn (5)) is valid once again. Plastic deformation now sets in if one of the relations

$$
|\sigma_3^\alpha| = |\sigma_{th,3}^\alpha + \sigma_{end,3}^\alpha| = |\sigma_y^\alpha| \ , \ |\sigma_3^\gamma| = |\sigma_{th,3}^\gamma + \sigma_{end,3}^\gamma| = |\sigma_y^\gamma| \tag{11}
$$

is fulfilled with the thermal stresses evaluated from eqn (6) at the reference temperature T_1. Just as during the initial cool-down, the instantaneous coefficients of thermal expansion change from the ones given by eqn (5) to those from eqn (8), if plastic deformations of a constituent phase set in.

The resulting macroscopic inelastic deformation is represented by the irreversible thermo-mechanical strain increment $\dot{\underline{\underline{\epsilon}}}_{\mathrm{tm}}$ which is the weighted average of the mean plastic strain increments in ferrite and austenite:

$$\dot{\underline{\underline{\epsilon}}}_{\mathrm{tm}} = V_{\mathrm{V}}^{\alpha}\dot{\underline{\underline{\epsilon}}}_{\mathrm{p}}^{\alpha} + V_{\mathrm{V}}^{\gamma}\dot{\underline{\underline{\epsilon}}}_{\mathrm{p}}^{\gamma} \,. \tag{12}$$

The magnitudes of $\dot{\underline{\underline{\epsilon}}}_{\mathrm{p}}^{\alpha}$ and $\dot{\underline{\underline{\epsilon}}}_{\mathrm{p}}^{\gamma}$ are proportional to the mismatch of thermal expansion that is accommodated by plastic deformation ($\tilde{\alpha}^{\mathrm{d}} - \hat{\alpha}^{\mathrm{d}}$). On this basis $\dot{\underline{\underline{\epsilon}}}_{\mathrm{tm}}$ is:

$$\begin{aligned} \dot{\epsilon}_{\mathrm{tm},3} &= \begin{cases} 0 & : \; |\sigma_3^{\alpha}| < \sigma_{\mathrm{y}}^{\alpha} \wedge |\sigma_3^{\gamma}| < \sigma_{\mathrm{y}}^{\gamma} \\ (\tilde{\alpha}_3^{\mathrm{d}} - \hat{\alpha}_3^{\mathrm{d}})\, dT & : \; |\sigma_3^{\alpha}| = \sigma_{\mathrm{y}}^{\alpha} \vee |\sigma_3^{\gamma}| = \sigma_{\mathrm{y}}^{\gamma} \end{cases} \\ \dot{\epsilon}_{\mathrm{tm},1} = \dot{\epsilon}_{\mathrm{tm},2} &= \begin{cases} 0 & : \; |\sigma_3^{\alpha}| < \sigma_{\mathrm{y}}^{\alpha} \wedge |\sigma_3^{\gamma}| < \sigma_{\mathrm{y}}^{\gamma} \\ (\tilde{\alpha}_1^{\mathrm{d}} - \hat{\alpha}_1^{\mathrm{d}})\, dT & : \; |\sigma_3^{\alpha}| = \sigma_{\mathrm{y}}^{\alpha} \vee |\sigma_3^{\gamma}| = \sigma_{\mathrm{y}}^{\gamma} \end{cases} \end{aligned} \tag{13}$$

The conservation of volume under plastic deformation is assured so that

$$\dot{\epsilon}_{\mathrm{tm},3} = -\frac{1}{2}\dot{\epsilon}_{\mathrm{tm},1} = -\frac{1}{2}\dot{\epsilon}_{\mathrm{tm},2} \;. \tag{14}$$

For further thermomechanical cycling, the stress state at T_0 has to be considered. Hence, the initiation of inelastic deformation requires a larger change in temperature than is necessary during cool-down in the first cycle. The analysis of the first thermal cycle already provides an insight into the physical mechanisms of thermomechanical loadings of the duplex steel and, therefore, subsequent thermal cycles can be investigated on the basis of the relations given above.
The present considerations lead to an analytic description of the behavior of the duplex steel for cyclic thermal loading combined with mechanical loading in the x_3-direction. The respective relations are incorporated into a material-subroutine (DAVES [1993]) for the finite element code ABAQUS (HKS [1993]). On this basis, a simulation of the dilatometer tests is now possible, and the analysis of uniaxially loaded structural parts can be performed. A different type of approach to the same problem has been followed by the same authors applying micromechanical modeling together with the finite element method. The results obtained by this technique coincide with those from the current approach (WERNER et al. [1994]).

MATERIAL MODEL

The temperature interval investigated in the dilatometer experiments is large. Model calculations, therefore, require a correct determination of the temperature dependent material data. The fact that in two-phase materials the difference in the respective properties of the phases dominates the response of the material allows the phases of the duplex steel to be assumed as isotropic continua (SIEGMUND et al. [1993]). The properties of ferrite and austenite in the duplex steel are taken to be those of ferritic and austenitic single-phase steels of approximately the same composition as the phases possess in the duplex steel (FISCHER et al. [1990]). The temperature dependent material data of the two phases used in the calculations are given in Fig (4). The Young's moduli of austenite and ferrite are

similar ($E^\alpha \approx E^\gamma \approx$ 200GPa at 20°C). The values for σ_y^α and σ_y^γ are corrected to the appropriate grain sizes of austenite and ferrite in the duplex steel by means of the Hall–Petch relation (GLADMAN & PICKERING [1983]). Correlating the yield strength and the grain size introduces a length scale into the otherwise size-independent continuum mechanical calculation. The temperature dependence of the yield strength is distinctly different for the two types of steels and hence also for austenite and ferrite in the duplex steel. Similar data has been reported recently in (MAEHARA [1992]). The assumption of non-hardening materials is made deliberately since hardening can be annihilated due to softening by dynamic recovery during the high temperature part of the thermal cycle. Furthermore, even for isothermal cyclic deformation, it is not clear to what extent conventional material models for cyclic deformation of monophase alloys are applicable to multiphase materials. Bauschinger effects can be predicted for two-phase materials even when the constituents are isotropically hardening (BÖHM [1991], LLORCA et al. [1992]).
Figure 4a shows the longitudinal plastic properties of the duplex steel. The duplex steel's yield strength is set equal to the limit flow stress predicted from its elastic-ideal-plastic constituents ($\sigma_{y,3}^d = V_V^\alpha \sigma_y^\alpha + V_V^\gamma \sigma_y^\gamma$ (PONTE CASTAÑEDA & DEBOTTON [1992])). In Fig (4b) the longitudinal instantaneous coefficient of thermal expansion of the duplex steel is depicted as obtained from eqn (5). Conventional material models for the duplex steel which neglect inelastic deformations caused by internal stresses use this set of material data ($\sigma_{y,3}^d$, $\hat{\alpha}_3^d$, E_3^d) to characterize the properties of the duplex steel (HULL et al. [1987]).

SIMULATION OF DILATOMETER EXPERIMENTS

The material law of the previous section will now serve as a basis for the analysis of the dilatometer experiments. First quenching from a stress free state at 900°C is simulated to model the fabrication process induced residual stresses.

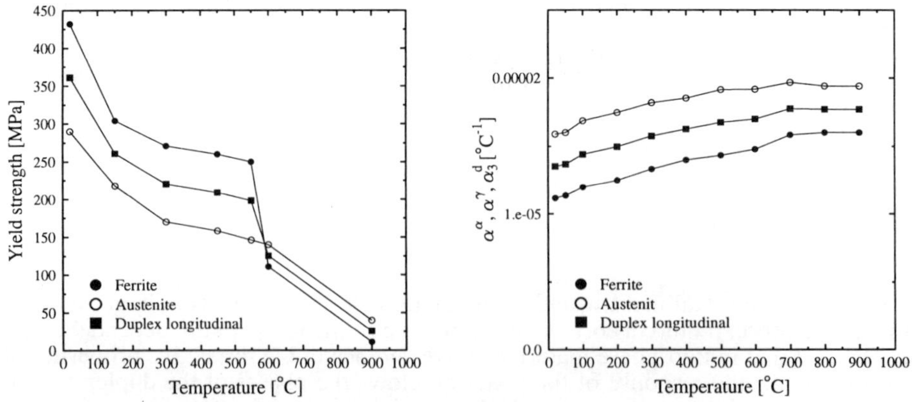

(a) (b)

Figure 4: Material properties of austenite, ferrite and the duplex steel as a function of temperature: (a) Yield strength and (b) instantaneous coefficient of thermal expansion.

Then thermal cycling 20°C-900°C-20°C is performed. The results of these calculations are presented in Fig (5). The simulated dilatometer tests exhibit the same features as are observed in the experiments, namely irreversible elongation in the longitudinal direction (compare Figs (2a) and (5a)) and irreversible contraction in the transverse direction (compare Figs (2b) and (5b)). A pronounced transient behavior is never observed and a steady state with constant irreversible strain per thermal cycle is reached by the first thermal cycle. The irreversible deformations predicted by the material model overestimate the experimentally observed values by a factor of approximately 2.5. This difference is to be attributed to the use of an ideally anisotropic microstructural model, and to uncertainties in the determination of the material data. The evolution of the irreversible thermomechanical strain together with the corresponding plastic strains in the constituents are depicted in Fig (6). The first thermal cycle starts with a nonzero residual stress state due to quenching of the duplex steel after the solution treatment. At the beginning of the nth thermal cycle, residual stresses are present due to the cool down in the preceeding $(n-1)^{\text{th}}$ thermal cycle. These residual stresses act as compression in ferrite and as tension in austenite, as also predicted by eqn (6). Upon heating-up, the phases are unloaded from their residual stress state first, and no plastic deformation occurs. When the temperature is increased further, ferrite is loaded in tension and austenite in compression. At high temperatures the mismatch in thermal expansion of the phases is accommodated by plastic elongation of ferrite, which is the phase of lower yield strength at high temperatures (Fig (4a)). Austenite, on the other hand, is not deformed plastically during heating-up, and does not contribute to the irreversible thermomechanical strain. Upon cool-down, unloading is completed after a small decrement in temperature, since the residual stress level is low at 900°C. Then austenite is subjected to tensile and ferrite to compressive stress. In the low temperature range, austenite is the phase of lower yield strength (Fig (4a)) and the thermal mismatch now is accommodated by plastic elongation of austenite. There is no plastic deformation of the ferrite during cool-down, and hence also no compensation for the elongation from the heating part of the thermal cycle. Both phases remain elongated at the end of the thermal cycle.

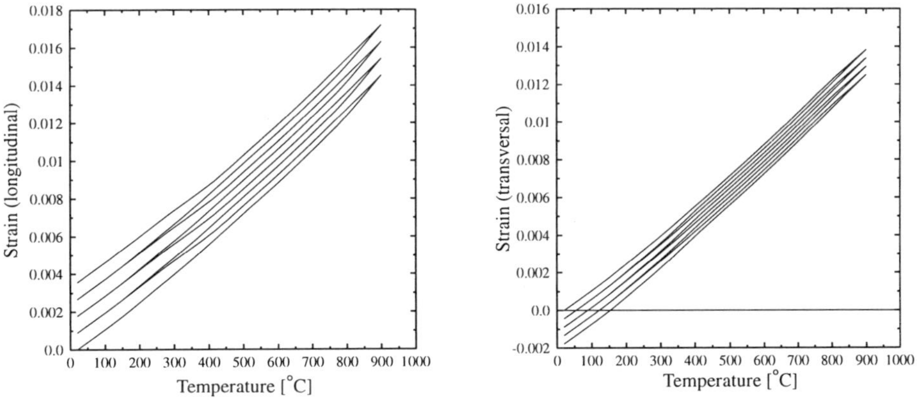

(a) (b)

Figure 5: Results from the simulations of the dilatometer tests. The strain in (a) the longitudinal direction (x_3) and (b) the transverse directions ($x_1 = x_2$) of the model is plotted as a function of the temperature.

The resulting irreversible thermomechanical strain is an elongation in the longitudinal direction. The deformation in the transverse direction is given readily from Fig (6) by applying eqn (13).

A STRUCTURAL PART UNDER THERMAL LOADING

The structural part analysed in the following is a two-bar assembly composed of a bar of the duplex steel and a bar of an austenitic single-phase steel in a parallel arrangement. The bars are of identical cross sectional area and are attached to two rigid plates which are kept parallel throughout the thermal cycle. This simple structure resembles the most important features of expansion joints used in heat exchangers. The austenitic steel is assumed to possess the same material properties as the austenitic phase of the duplex steel (Fig (4)). The properties of the duplex steel whose axis of anisotropy coincidences with the longitudinal axis of the duplex steel bar is modeled in two different ways by either considering or neglecting the irreversible thermomechanical strain. Hence, the duplex steel properties assigned are $\sigma_{\rm y}^{\alpha}, \sigma_{\rm y}^{\gamma}, \alpha^{\alpha}, \alpha^{\gamma}, E^{\alpha}, E^{\gamma}$ and $\sigma_{\rm y,3}^{\rm d}$, $\hat{\alpha}_3^{\rm d}$, $E_3^{\rm d}$, respectively. The second approach is simple, since no micromechanical analysis is needed and the material properties can be determined directly on a specimen of the duplex steel. The main features of the two models are sketched in Fig (7). Thermal cyclic loading of the expansion joint is simulated (900°C -20°C-900°C) leading to a cyclic thermomechanical loading of the duplex steel. Compressive loads are exerted on the duplex steel due to the thermal stresses generated between the austenitic single-phase steel and the duplex steel during cool-down from the initially stress free state at 900°C ($\alpha^{\gamma} < \hat{\alpha}_3^{\rm d}$). The evolution of the stresses in the austenitic single-phase steel is depicted in Fig (8a). The stresses in the duplex steel are the same in magnitude but are compressive stresses. The results obtained from a stress analysis of the expansion joint are significantly different for the two material models applied. The model without $\dot{\underline{\underline{\epsilon}}}_{\rm tm}$ predicts pure elastic response of the expansion joint during thermal cycling.

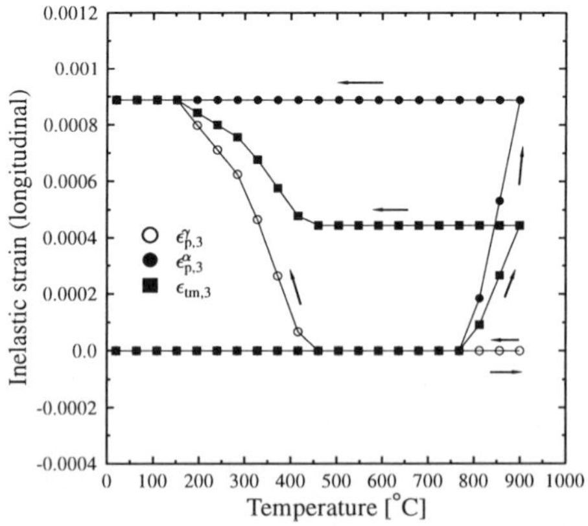

Figure 6: Evolution of the plastic strain in longitudinal direction in austenite and ferrite and the resulting irreversible thermomechanical strain in a thermal cycle 20°C-900°C-20°C.

If, however, the irreversible thermomechanical strain is taken into account, the results are quite different, and a pronounced hysteresis is predicted. A steady state is reached in the second thermal cycle. Deviations from the slope of the elastic response occur between the points A-B (first cycle) and A'-B (subsequent cycles) as well as between C-D and F-G. This behavior can be understood if the evolution of the irreversible thermomechanical strains in the duplex steel in the longitudinal direction the bar is considered. It deviates from zero at the corresponding points in the stress-temperature plot. Under the compression exerted on the duplex steel, it is ferrite that is plastified. The condition in eqn (7) is fulfilled at A for the first time, and plastic deformation of the ferrite sets in. At B, however, eqn (7) is not fulfilled any more since the yield strength of the constituents of the duplex steel increases during cool-down. The same is true between points C and D. During heating-up, inelastic deformation sets in only at point F because the residual stresses have to be overcome first. In the high temperature part (F-G), the stresses in the expansion joint decrease with increasing temperature since they are limited by the yield strength of the ferritic phase. In the following thermal cycle, inelastic deformation sets in at A' instead at A due to the residual stresses present at 900°C. The remaining part of the thermal cycle is then identical to the first thermal cycle.

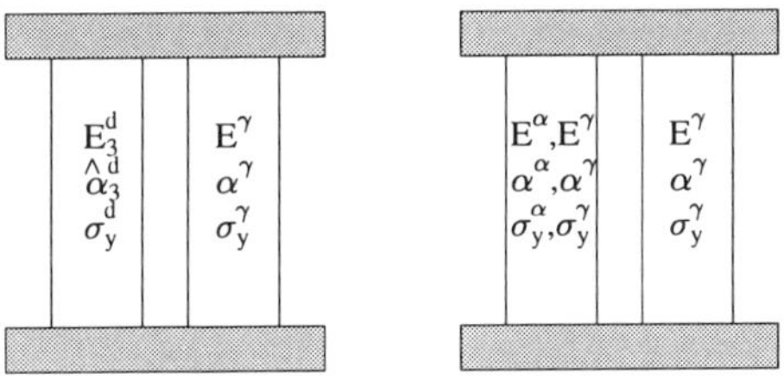

Figure 7: An expansion joint consisting of an austenitic single-phase steel bar and a duplex steel bar. Two different types of material models are indicated.

SUMMARY

The present investigation deals with the analysis of the cyclic thermomechanical behavior of a forged duplex steel with a uniaxially anisotropic microstructure. Dilatometer experiments give evidenve of a ratchet effect under the condition of pure thermal cycling. The irreversible deformations are strongly dependent on the orientation of the specimens relative to the axis of anisotropy, and on the rate of the temperature change applied. Longitudinal specimens elongate whereas transverse specimens shorten. A material model is presented which accounts for the inelastic deformations due to internal stresses by incorporating an additional inelastic strain increment within the theory of small strains. Analytic expressions for this strain increment are derived on the basis of classical results for the coefficients of thermal expansion for two-phase materials. The irreversible deformations measured in the experiments are analysed using this constitutive law and good agreement between simulation and experiment is achieved. Irreversible deformations arise because plastic deformation is necessary to accommodate the difference in thermal expansion of the phases. The deformed phase in the high temperature range is ferrite during heating-up.

In the low temperature range this phase is austenite during cool-down. This is due to the different temperature dependence of the yield strength of the two phases. An application of the constitutive law is presented, and the stresses and strains in an expansion joint subjected to thermal loading are estimated. This application of the material law points to the importance of including the inelastic deformations arising from internal stresses in the analysis of structures consisting of two-phase materials under thermal cyclic loading.

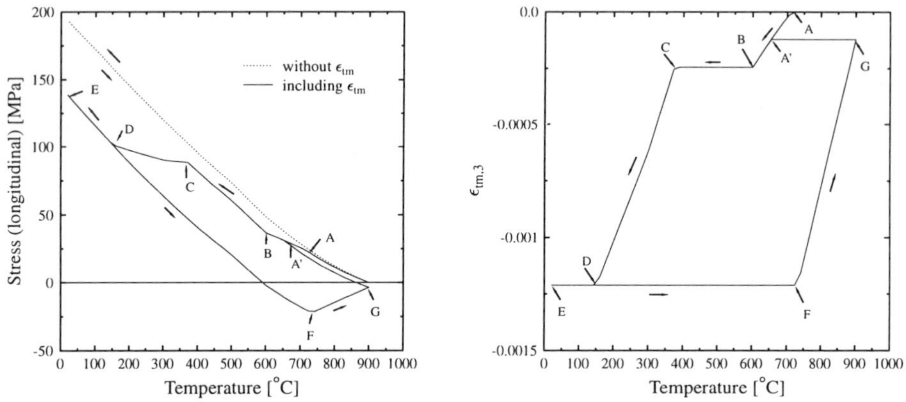

Figure 8: Analysis of the expansion joint under cyclic thermal loading: (a) Stresses in the bar made of the austenitic single-phase steel and (b) evolution of irreversible thermomechanical strain in longitudinal direction of the duplex steel bar.

REFERENCES

1993 SIEGMUND, T., WERNER, E. and FISCHER, F.D., "The Irreversible Deformation of a Stainless Duplex Steel under Thermal Cycling", Mater. Sci. Eng., A169, 125.

1991 CHARLES, J., "Proceedings of the International Conference on Duplex Stainless Steels", Beaune, France, Oct 28- 30, 1991, Les éditions de physique, Les Ulis, France, pp. 1.

1992 JOLY, P., MEYZAUD, Y. and PINEAU, A., "Micromechanisms of Fracture of an Aged Duplex Stainless Steel Containing Brittle and Ductile Phases: Developement of a Local Criterion of Fracture", in: GIOVANOLA. J.H. and ROSAKIS, A.J. (eds.), Proc. ASME Summer Mechanics and Materials Conf.: Advances in Local Fracture/Damage Models for the Analysis of Engineering Problems, Tempe, AR, April 28-May 1, 1992, AMD-Vol 137, ASME, New York, U.S.A., pp. 151–180.

1990 FISCHER, F.D., RAMMERSTORFER, F.G. and BAUER, F.J., "Fatigue and Fracture of High-Alloyed Steel Specimens Subjected to Thermal Cycling", Metall. Trans., A21, 935.

1983 HASHIN, Z., "Analysis of Composite Materials–A Survey", ASME J. Appl. Mech., 50, 481.

1968 SCHAPERY, R.A., "Thermal Expansion Coefficients of Composite Materials Based on Energy Principles", J. Compos. Mater., 2, 380.

1993 KARAYAKA, M. and SEHITOGLU, H., "Thermomechanical Deformation Modeling of Al2xxx-T4/SiC_p Composites", Acta Metall. Mater., 41, 175.

1979 HULL, F.C, "Dimensional Instabilities of CF8 Stainless Steel Castings at Elevated Temperatures", in: Symposium on Cast Metals for Structural and Pressure Containment Applcations (MPC-11), The Materials Property Council, New York, U.S.A., pp. 205–224.

1989 LEBLOND, J.B., DEVAUX, J. and DEVAUX, J.C., "Mathematical Modeling of Transformation Plasticity in Steels, I: Case of Ideal-Plastic Phases", Int. J. Plasticity, 5, 551.

1990 PEDERSEN, O.B., "Thermoelasticity and Plasticity of Composites-II. A Model System", Acta Metall. Mater., 38, 1201.

1992 JANSSON,S. and LECKIE, F.A., "Mechanical Behavior of a Continuous Fiber-Reinforced Aluminium Matrix Composite Subjected to Transverse and Thermal Loading", J. Mech. Phys. Solids, 3, 593.

1974 GARMONG, G., "Elastic-Plastic Analysis of Deformation Induced by Thermal Stress in Eutectic Composites, I: Theory, II: Thermal Expansion", Metall. Trans. 5, 2183, 2191.

1991 BÖHM, H.J., Computer Based Micromechanical Investigations of the Thermomechanical Behaviour of Metal Matrix Composites, Fortschr.-Ber. VDI-Reihe 18, Nr.101, VDI-Verlag, Düsseldorf, F.R.G., pp. 1.

1993 DAVES, W., "Mikro- und Makromechanische Simulation des Deformationsverhaltes von Stählen unter Berücksichtigung vom Umwandlungs- und Ausscheidungsvorgängen", Dissertation, Montanuniverstät, Leoben, Austria.

1993 HIBBIT, KARLSSON and SORENSEN, "ABAQUS - User's Manual", Version 5.1, Hibbit, Karlsson, Sorensen Inc., Pawtuket, U.S.A..

1994 WERNER, E., SIEGMUND, T. and FISCHER, F.D., "A Computer Study on the Thermomechanical Deformation Behavior of a Duplex Steel", Comput. Mater. Sci., submitted.

1983 GLADMAN, T. and PICKERING, F.B., "The Effect of Grain Size on the Mechanical Properties of Ferrous Materials", in: BAKER, T.N., (ed.), Yield, Flow and Fracture of Polycrystals, Applied Science Publishers, London, U.K., pp. 141–198.

1992 MAEHARA, Y., "Effect of Microstructure on Hot Deformation in Duplex Stainless Steels", Scripta metall. mater. 26, 1701.

1992 LLORCA, J., SURESH, S. and NEEDLEMAN, A., "An Experimental and Numerical Study of Cyclic Deformation in Metal-Matrix Composites", Metall. Trans., A23, 919.

1992 PONTE CASTAÑEDA, P. and DEBOTTON, G., "Homogenized Yield Strength of Two-Phase Composites", Proc. Royal Soc. Lond., A438, 419.

1987 HULL, F.C., HWANG, S.K., WELLS, J.M. and JAFFEE, R.I., "Effect of Composition on Thermal Expansion of Alloys Used in Power Generators", J. Mater. Eng., 9, 81.

Shakedown for systems of kinematic hardening materials *

E. STEIN and Y. J. HUANG

Hannover, Federal Republic of Germany

Abstract

The paper deals with the shakedown of systems with kinematic hardening materials. Analytical, numerical and experimental methods to solve shakedown problems will be introduced and discussed. Some examples are treated with these methods. Comparison shows that the results of these methods are in good agreement.

1 Introduction

The response of an elastic–plastic system subjected to cyclic loadings can be very complicated. FREDERICK & ARMSTRONG [1966] proved that a steady state of stresses in the system can always be reached as the loading proceeds, and this steady state is independent of the initial pattern of stresses. Experiments with different materials done by COFFIN et al. [1959] confirm this theory. There are three possible steady states: 1. elastic shakedown, in this case no further plastic deformation occurs, the system responds as if it were elastic; 2. alternating plasticity, the increment of plastic deformation within a loading cycle vanishes, but the system can fail due to low cycle fatigue; 3. ratchetting, the plastic deformation grows as the loading proceeds, and the system becomes unusable because of too large deformation.

Elastic shakedown (or simply, shakedown) of a system is regarded as a safe state. It is of importance to know if a system under given cyclic loadings shakes down or not. BLEICH [1932] was the first to formulate a shakedown theorem for simple static unstable structures with elastic–perfectly plastic materials. This theorem was then generalized by MELAN [1938a, 1938b] to continua with elastic–perfectly plastic and linear kinematic hardening materials. KOITER [1956] introduced a kinematic shakedown theorem for a elastic–perfectly plastic material, a dual of MELAN's static theorem. Since then extensions of the shakedown theorem dealing with thermoloadings,

*This paper is devoted to the memory of the eminent Professor J. König who contributed in a substantial way to shakedown theory and especially to our own research during his stays at our institute in Hannover

Z. Mróz et al. (eds.), Inelastic Behaviour of Structures under Variable Loads, 33–50.

dynamic loadings and geometrically nonlinear effects etc. have been carried out by different authors (CERADINI [1969]; CORRADI & MAIER [1973]; KÖNIG [1969]; MAIER [1972]; PRAGER [1956b]; ROZENBLUM [1957]; WEICHERT [1986]). A shakedown theorem for systems with nonlinear kinematic hardening materials was first formulated by NEAL [1950]. He used a MASING [1924] overlay model to describe the nonlinear kinematic behaviour of materials. NEAL's formulation is only valid for 1-d stress state problems. Recently STEIN et al. [1992a, 1992b, 1993a, 1993b, 1994] used a 3-d overlay model to describe the nonlinear kinematical behaviour of materials and formulated a corresponding shakedown theorem for this material. Based on their concept, one can determine the failure mechanism of a system under cyclic loading. For instance, a system consisting of linear kinematic hardening material fails only locally in the form of alternating plasticity.

An important task of shakedown investigations is to find a maximum (minimum) factor by which a given load domain may be enlarged, but the system still shakes down. In the sense of computational mechanics, such a shakedown problem can be treated as an optimization problem by using FEM. Generally, the dimension of the optimization problem is very large. MAHNKEN [1992] and ZHANG [1992] have used special numerical methods, namely the dual method and the reduced base technique, to solve these problems. For linear kinematic hardening, however, the dimension of the optimization problem can be reduced very effectively because the system fails only locally.

In our paper we first review the 3-d overlay material model and its static shakedown theorem. Some new relations for the overlay model will be introduced. Then the numerical algorithm for solving shakedown problems is formulated. For systems with linear kinematic hardening materials, an analytical method to determine the shakedown load limits is developed. Applying this analytical method, we solve several typical shakedown problems. At the same time, we solve these problems numerically by using Finite-Element discretization. Finally experiments with CT-specimens have been carried out. Analytical, numerical and experimental results will be compared and discussed.

2 Overlay-model for kinematic hardening materials

There exist many mathematical models for describing the kinematical hardening behaviour of materials, for example the PRAGER [1956a] linear kinematic hardening model, its modification by ZIEGLER [1959], MRÓZ [1967] multisurface model, DAFALIAS [1975] two surface model and so on. We introduce here a so-called overlay model. The idea of describing the nonlinear kinematic hardening behaviour of a material by using a overlay model is not new (e.g. BESSELING [1953], IWAN [1967], ZIENKIEWICZ [1972]), but its mathematical formulation here, its application to the shakedown problem and the consequently numerical algorithm are recent. Some new findings such as eqns. (12) and (17) have not been published before.

A macroscopic material point $\mathbf{x} \in \Omega \subset \mathbb{R}^3$ is assumed to be composed of a spectrum of microscopic elements (or microelements). Each microelement is numbered with a scalar variable $\xi \in [0,1]$. Stresses and strains are separately defined for the macroscopic material point (macrostress and macrostrain) and the microelements (microstresses and microstrains). They are denoted by $\boldsymbol{\sigma}(\mathbf{x})$, $\boldsymbol{\varepsilon}(\mathbf{x})$ and $\boldsymbol{\psi}(\mathbf{x},\xi)$, $\boldsymbol{\eta}(\mathbf{x},\xi)$, respectively.

The macrostress $\boldsymbol{\sigma}$ has to fulfill the equilibrium condition

$$\operatorname{div} \boldsymbol{\sigma}(\mathbf{x}) = \mathbf{b}(\mathbf{x}) \quad \mathbf{x} \in \Omega \tag{1}$$

In the framework of the geometrically linear continuum mechanics, the kinematic relation

$$\boldsymbol{\varepsilon}(\mathbf{x}) = \frac{1}{2}\left(\nabla \otimes \mathbf{u}(\mathbf{x}) + (\nabla \otimes \mathbf{u}(\mathbf{x}))^T\right) \quad \mathbf{x} \in \Omega \tag{2}$$

holds between the displacement $\mathbf{u}$ and the strain $\boldsymbol{\varepsilon}$.

By assuming that the stress $\boldsymbol{\sigma}$ of a macroscopic material point (macrostress) is the weighted sum of the stresses $\boldsymbol{\psi}$ of all microelements (microstresses), and the macroscopic material point and the corresponding microelements deform in a same way, we get the following static and kinematic relations:

$$\boldsymbol{\sigma}(\mathbf{x}) = \int_0^1 \boldsymbol{\psi}(\mathbf{x},\xi)d\xi\,, \tag{3}$$

$$\boldsymbol{\eta}(\mathbf{x},\xi) = \boldsymbol{\varepsilon}(\mathbf{x}) \quad \forall\ \xi \in [0,1]\,. \tag{4}$$

Futhermore, we suppose that all microelements are linear elastic, perfectly plastic and have the same temperature, the same elastic moduli, but different yield stresses $k(\xi)$. For convenience, $k(\xi)$ can be considered as a monotonic increasing function of ξ. We assume that the microstrain $\boldsymbol{\eta}$ may be decomposed into an elastic- and a plastic part. Thus the following relations can be written down for the microelements:

$$\boldsymbol{\eta}(\xi) = \boldsymbol{\eta}^E(\xi) + \boldsymbol{\eta}^P(\xi) \tag{5}$$

$$\boldsymbol{\eta}^E(\xi) = \mathbf{E}^{-1}\,\boldsymbol{\psi}(\xi) \tag{6}$$

$$\Phi(\boldsymbol{\psi}) \le k^2(\xi) \tag{7}$$

$$\dot{\boldsymbol{\eta}}^P(\xi) = \dot{\lambda}(\xi)\cdot\frac{\partial\Phi}{\partial\boldsymbol{\psi}}\,, \qquad \dot{\lambda}(\xi) \ge 0 \tag{8}$$

$$\dot{\lambda}(\xi)\,[\Phi(\boldsymbol{\psi}) - k^2(\xi)] = 0\,, \tag{9}$$

where $\mathbf{E}$ stands for the symmetrical elasticity tensor, and $\Phi(\cdot)$ the yield function. For simplicity of notation, the argument $\mathbf{x}$ of the fields will be partly omitted.

Assume that by a macrostress the microelement number ξ begins to yield, the function $k(\mathbf{x},\xi)$ is then uniquely determined by the macroscopic σ, ε–function in the one-dimensional case

$$\sigma(\xi) = \int_0^\xi k(\bar{\xi})\,d\bar{\xi} + (1 - \xi)\,k(\xi)\,. \tag{10}$$

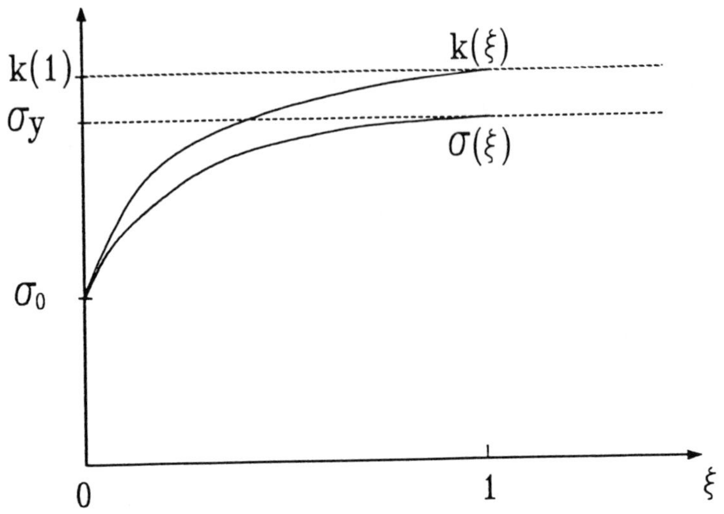

Fig. 1: Kinematic hardening of a macroscopic material point and yield stresses of microelements

It is easy to show that, like $k(\xi)$, σ is also a monotonic increasing function of the variable ξ (s. Fig 1). For $\xi = 0$, we have $\sigma = \sigma(\xi = 0) = k(\xi = 0) = k_0$. The maximum value of σ (denoted by σ_Y or K) is arrived by setting $\xi = 1$ in eqn (10)

$$\sigma_Y = \sigma(\xi = 1) = K = \int_0^1 k(\xi)\, d\xi\,. \tag{11}$$

Thus k_0 and K can be identified with the initial yield stress σ_o and the ultimate stress of the macroelement σ_Y, respectively. For the 3-dimensional case, there is a relation between $\boldsymbol{\sigma}$ and $k(\xi)$, analogous to eqn (10), namely

$$\sqrt{\Phi(\boldsymbol{\sigma})} = \int_{\bar{\xi}=0}^{\xi} k(\bar{\xi})\, d\bar{\xi} + (1 - \xi)\, k(\xi)\,. \tag{12}$$

Now we define the difference between the microstress $\boldsymbol{\psi}(\xi)$ and the macrostress $\boldsymbol{\sigma}$ as the *residual microstress* and denote it by $\boldsymbol{\pi}(\xi)$

$$\boldsymbol{\pi}(\xi) = \boldsymbol{\psi}(\xi) - \boldsymbol{\sigma}\,. \tag{13}$$

For the residual microstress defined in this way, we have

$$\boldsymbol{\pi}(\xi) = \mathbf{0}\,,\ \xi \in [0,1] \quad \text{for} \quad \Phi(\boldsymbol{\sigma}) \leq k_o^2 \tag{14}$$

and

$$\boldsymbol{\pi}(\xi) \neq \mathbf{0}\,,\ \xi \in [0,1] \quad \text{for} \quad \Phi(\boldsymbol{\sigma}) > k_o^2 \tag{15}$$

It is necessary to notice that eqn (15) holds for all $\xi \in [0,1]$, even if $\Phi(\boldsymbol{\sigma})$ may be smaller than $k(\xi)$.

Integration of (13) yields

$$\int_0^1 \boldsymbol{\pi}(\mathbf{x},\xi)d\xi = \mathbf{0} \quad \forall \mathbf{x} \in \Omega\,, \tag{16}$$

where eqn (3) has been used. Eqn (13) implies that the resultant of $\boldsymbol{\pi}(\xi)$ does not contribute to the macrostress $\boldsymbol{\sigma}$. All microstress fields which fulfill the condition (16) can be regarded as residual microstress fields.

For a kinematic hardening material, it is possible to represent the residual microstress $\boldsymbol{\pi}(\mathbf{x},\xi)$ by the backstress $\boldsymbol{\alpha}(\mathbf{x})$ which describes the translation of the initial yield surface

$$\boldsymbol{\pi}(\mathbf{x},\xi) = -\frac{\sqrt{\Phi(\boldsymbol{\sigma})} - k(\xi)}{\sqrt{\Phi(\boldsymbol{\sigma})} - k(0)}\boldsymbol{\alpha}(\mathbf{x})\,. \tag{17}$$

It is easy to show that the microstress in (17) satisfies the condition (16).

For the overlay-model described above, the following static shakedown theorem has been formulated and proved (e.g. STEIN [1992a], ZHANG [1992]):

Theorem 1 *If there exist a time-independent residual macrostress field $\bar{\boldsymbol{\rho}}(\mathbf{x})$ and a time-independent residual microstress field $\bar{\boldsymbol{\pi}}(\mathbf{x},\xi)$ satisfying*

$$\Phi[\, m\,\bar{\boldsymbol{\pi}}(\mathbf{x},0)\,] \leq [K(\mathbf{x}) - k_0(\mathbf{x})]^2 \qquad \forall \mathbf{x} \in \Omega\,, \tag{18}$$

such that, for all possible loads within the load domain, the condition

$$\Phi\left[m[\boldsymbol{\sigma}^E(\mathbf{x},t) + \bar{\boldsymbol{\rho}}(\mathbf{x}) + \bar{\boldsymbol{\pi}}(\mathbf{x},0)]\right] \leq k_0^2(\mathbf{x}) \tag{19}$$

is fulfilled $\forall\, \mathbf{x} \in \Omega$ and $\forall\, t > 0$, where $m > 1$ is a safety factor against inadaptation, then the total plastic energy dissipated within an arbitrary load path contained within the load domain is bounded, and the system consisting of the proposed material will shake down.

The static shakedown theorem 1 is formulated by using the residual microstress $\boldsymbol{\pi}$. Consider the relation between the backstress $\boldsymbol{\alpha}$ and the residual microstress $\boldsymbol{\pi}$ in eqn (17), we can formulate a static shakedown theorem directly in terms of the backstress $\boldsymbol{\alpha}$, i.e.:

Theorem 2 *If there exist a time-independent residual macrostress field $\bar{\boldsymbol{\rho}}(\mathbf{x})$ and a time-independent field $\bar{\boldsymbol{\alpha}}(\mathbf{x})$ satisfying*

$$\Phi[\, m\,\bar{\boldsymbol{\alpha}}(\mathbf{x})\,] \leq [K(\mathbf{x}) - k_0(\mathbf{x})]^2 \qquad \forall \mathbf{x} \in \Omega \tag{20}$$

such that for all possible loads within the load domain the condition

$$\Phi\left[m[\boldsymbol{\sigma}^E(\mathbf{x},t) + \bar{\boldsymbol{\rho}}(\mathbf{x}) - \bar{\boldsymbol{\alpha}}(\mathbf{x})]\right] \leq k_0^2(\mathbf{x}) \tag{21}$$

is fulfilled $\forall\, \mathbf{x} \in \Omega$ and $\forall\, t > 0$, then the system will shake down.

For the formulation of the static shakedown theorems 1 and 2 only the values of k_0 and K have been used. That means that the shakedown limits for systems of the considered material do not depend on the particular shape of the function $k(\xi)$, and therefore do not depend on the particular σ–ε curve, but solely on the magnitudes of k_0 and K.

For $K(\mathbf{x}) = k_0(\mathbf{x})$ (an elastic- perfectly plastic material) we have $\bar{\boldsymbol{\alpha}} = \mathbf{0}$ due to (20) and theorem 1 reduces to the MELAN [1938a, 1938b] theorem for an elastic- perfectly plastic material. For $K(\mathbf{x}) \rightarrow +\infty$ (materials with unlimited kinematic hardening) the constraint (20), imposed on the backstresses $\bar{\boldsymbol{\alpha}}(\mathbf{x})$, can never become active and therefore may be dropped. In this case we get the static shakedown theorem by MELAN for a linear, unlimited kinematic hardening material.

3 Shakedown as an optimization problem

One of the purposes of the shakedown investigation for a system is to find a maximum (or minimum) factor, by which the load domain is allowed to enlarge while the system will still shake down. Mathematically it is an optimization problem. In this section we will formulate the optimization problem in a discretized form by using the finite element method. Special attention will be paid to materials with linear kinematic hardening.

3.1 Discretization of shakedown problems

As mentioned in the last section, the shakedown limit of a system consisting of the presented material model depends only on the values k_0 and K. For this reason the given function $k(\xi)$ may be replaced by a step-function of $\xi \in [0, 1]$. The step-function has to be chosen such that its minimum is equal to k_0 and its area equal to K. That is

$$K = \sum_{l=1}^{m} \Delta\xi_l \, \bar{k}_l \,, \qquad m \geq 2 \tag{22}$$

where m is the number of intervals of the step-function, and $\bar{k}_l$ is the value of the step-function on the l-th interval with the length $\Delta\xi_l$.

For a plane stress problem, the microelements may be incorporated in a natural way: the thickness t of the plate is divided into several ($m \geq 2$) layers with the thicknesses $t_l, l = 1, 2, ..., m$. Each layer behaves as an elastic, perfectly plastic material and has a corresponding yield stress $\bar{k}_l$ (one value of the step-function). The thicknesses of the layers have to be chosen such that

$$\frac{t_l}{t} = \Delta\xi_l \qquad \forall \; l \in [1, 2, \ldots, m] \,. \tag{23}$$

Thus, an unique relation between the layers of the plate and the intervals of the step-function is established. The microelements over the l-th interval are replaced by the l-th layer of the plate. The parallel connection of the microelements is realized by discretizing all layers in the same way. That is, elements which lie on top of each other have the same nodes; this guarantees that the elements of the different layers have the same kinematics.

If we divide the structure into Ne finite elements with NG Gaussian points and m layers, we get the following optimization problem

$$\beta \longrightarrow \max \tag{24}$$

$$\sum_{i=1}^{NG} \mathbf{C}_i \cdot \sum_{s=1}^{m} \frac{t_s}{t} \boldsymbol{\rho}_{i,s} = \sum_{i=1}^{NG} \mathbf{C}_i \cdot \boldsymbol{\rho}_i = \mathbf{C} \cdot \boldsymbol{\rho} = \mathbf{0} \tag{25}$$

$$\Phi(\beta \boldsymbol{\sigma}_i^E(j) + \boldsymbol{\rho}_{i,1}) \leq \bar{k}_1^2 = k_0^2 \qquad \forall\ (i,j) \in \mathcal{I} \times \mathcal{J} \tag{26}$$

$$\Phi(\boldsymbol{\pi}_{i,1}) \leq (K - \bar{k}_1)^2 = (K - k_0)^2 \qquad \forall\ i \in \mathcal{I}, \tag{27}$$

for load domains with the form of a convex polyhedron with M vertices. Here β is the maximum enlarging of the load domain allowing for shakedown. $\mathcal{I} = [1, 2, \ldots, NG]$ and $\mathcal{J} = [1, 2, \ldots, M]$ are sets containing all Gaussian points and load vertices, respectively. $\mathbf{C}$ is a system dependent matrix, and $\boldsymbol{\sigma}_i^E(j)$ is the elastic stress vector at the i-th Gaussian point corresponding to the j-th load vertex $\mathbf{P}(j)$. The eqns (25), (26) and (27) represent the discretized static equilibrium conditions for the residual macrostresses and the shakedown conditions controlled at the Gaussian points, respectively. $\boldsymbol{\rho}_{i,s}$ and $\boldsymbol{\pi}_{i,s}$ are the entire residual stress vector and the residual microstress vector in the s-th layer of the i-th Gaussian point. Between $\boldsymbol{\rho}_{i,s}$, $\boldsymbol{\pi}_{i,s}$ and the residual macrostress vector $\boldsymbol{\rho}_i$ the following relation holds

$$\boldsymbol{\rho}_{i,s} = \boldsymbol{\rho}_i + \boldsymbol{\pi}_{i,s}\,. \tag{28}$$

For $m = 1$ and $K = \bar{k}_1$ problem (24)–(27) reduces to the discretized shakedown problem for a perfectly plastic material. In this case we have $\boldsymbol{\pi}_{i,1} = \mathbf{0}$, $\boldsymbol{\rho}_i = \boldsymbol{\rho}_{i,1}$, and the constraint (27) can be dropped. We come to the other extreme case by assuming $K \to \infty$, which corresponds to an unlimited kinematic hardening material. If $K \to \infty$ the constraint (27) can never become active, $\boldsymbol{\pi}_{i,1}$ is not constrained, and the constraint (27) may be dropped.

In general the discretized shakedown problem (24)–(27) is a large sized nonlinear optimization problem. The direct use of standard optimization algorithms, such as the Sequential Quadratic Programming (SQP) method (e.g. SCHITTKOWSKI [1981]), is not effective. ZHANG [1992] has used a reduced basis technique to solve the problem (24)–(27).

3.2 Shakedown for linear kinematic hardening materials

According to subsection 3.1 the following nonlinear optimization problem has to be solved for an unlimited kinematic hardening material

$$\beta \rightarrow \max \tag{29}$$

$$\sum_{i=1}^{NG} \mathbf{C}_i \cdot \boldsymbol{\rho}_i = \mathbf{C} \cdot \boldsymbol{\rho} = \mathbf{0} \tag{30}$$

$$\Phi(\beta \boldsymbol{\sigma}_i^E(j) + \boldsymbol{\pi}_{i,1} + \boldsymbol{\rho}_i) \leq k_0^2 \qquad \forall\ (i,j) \in \mathcal{I} \times \mathcal{J}\,, \tag{31}$$

where the relation (28) between $\boldsymbol{\pi}_{i,s}$, $\boldsymbol{\rho}_{i,s}$ and $\boldsymbol{\rho}_i$ has been used. The following consideration is useful for developing solution procedures for problem (29)–(31): while the residual macrostress vectors $\boldsymbol{\rho}_i$ have to fulfill the equilibrium condition (30), the $\boldsymbol{\pi}_{i,1}$ are unconstrained (cf. subsection 3.1).

Let us introduce a new vector at each Gaussian point

$$\mathbf{y}_i := \boldsymbol{\rho}_i - \boldsymbol{\pi}_{i,1}\,. \tag{32}$$

Since $\boldsymbol{\pi}_{i,1}$ are unconstrained quantities, $\mathbf{y}_i$ are unconstrained as well. With the new vector $\mathbf{y}_i$ problem (29)-(31) can be formulated in an equivalent form, namely,

$$\beta \rightarrow \max \tag{33}$$

$$\Phi\,(\,\beta \boldsymbol{\sigma}_i^E(j) + \mathbf{y}_i\,) \leq k_0^2 \qquad \forall\ (i,j) \in \mathcal{I} \times \mathcal{J}\,. \tag{34}$$

Problem (33)–(34) has a simple structure. It was shown by STEIN et al. [1993a] that the maximum value of the load factor β_s is determined by

$$\beta_s = \min_{i \in \mathcal{I}} \bar{\beta}_i = \bar{\beta}\,, \tag{35}$$

with $\bar{\beta}_i\,(i \in \mathcal{I})$ being the solution of subproblem for the i-th Gaussian point

$$\beta_i \rightarrow \max \tag{36}$$

$$\Phi\,(\,\beta_i \boldsymbol{\sigma}_i^E(j) + \mathbf{y}_i\,) \leq k_0^2 \qquad \forall\ j \in \mathcal{J}\,. \tag{37}$$

Note that the problem (36)–(37) is obtained from the problem (33)–(34) by dropping all inequality constraints which do not contain the subscript i.

The dimension of problem (36)–(37) is very low. The number of unknowns is $NSK + 1$, with NSK being the number of stress components at a Gaussian point. For plane stress problems the number of unknowns is between 3 and 6. The number of the constraints is identical to the number of vertices of the load domain.

The vectors $\boldsymbol{\sigma}_i^E(j)$ represent vertices of the elastic domain $\mathcal{S}_i^E$ in the stress space $\mathcal{S}$, which contains all elastic stress vectors $\boldsymbol{\sigma}_i^E(t)$ at the i-th Gaussian point. By identifying the vector $(-\mathbf{y}_i)$ with the shift of the initial yield surface in $\mathcal{S}$, we can give problem (36)–(37) the following geometrical interpretation: find the maximum affine enlargement of $\mathcal{S}_i^E$ and the corresponding shift $(-\mathbf{y}_i)$ of the yield surface on condition that the enlarged elastic domain $\beta_i \mathcal{S}_i^E$ is still contained in the shifted yield surface.

In view of the above interpretation, eqn (35) implies that for a system consisting of an unlimited kinematic hardening material, the shakedown behaviour is determined by some point $\mathbf{x}_{i_p}$ in the system, where the maximum enlargement β_{i_p} of the elastic domain $\mathcal{S}_{i_p}^E$ is the smallest in comparison to all other points. Thus, $\beta_s = \beta_{i_p}$ holds. Since the shape and the size of the yield surface remains unchanged during yielding, it is evident that the shakedown load of systems consisting of unlimited kinematic hardening material cannot be infinite as to be expected for monotonic loading. The only exception is that all elastic domains $\mathcal{S}_i^E$ ($\forall\ i\ \in\ \mathcal{I}$) are straight lines in the space of principal stresses, which coincide with the principal diagonal of the space of principal stresses. This means that all elastic stresses $\boldsymbol{\sigma}^E(\mathbf{x}, t)$ are hydrostatic. However, this case is not important in practice.

4 Analytical determination of shakedown limits

The analysis in the last section tells us that for linear kinematic hardening, the shakedown behaviour of a system is dominated by some point in the system, where the maximun enlargement of the elastic domain is the smallest in comparison to other points. If the number of constraints, which is equal to the number of vertices of the load domain, is not greater than four, it is possible to solve the optimization problem (36)-(37) analytically.

Let us consider a system subjected to the biaxial loadings p_1 and p_2 which may vary independently between zero and certain maximum magnitudes $\bar{p}_1$ and $\bar{p}_2$. The maximum enlarging of the loading domain

$$0 \leq p_1 \leq \bar{p}_1 \quad ,$$
$$0 \leq p_2 \leq \bar{p}_2$$

has to be found, so that the system will still shake down.

Assume that the maximum effective stress would appear at some point A in the system, then we need to solve the optimization problem

$$-\beta \rightarrow \min. \tag{38}$$

$$\Phi[\beta\boldsymbol{\sigma}^E(j) + \mathbf{y}] - \sigma_0^2 \leq 0, \quad j = 1,\, 2,\, 3,\, 4 \tag{39}$$

only for point A. $\boldsymbol{\sigma}^E$ denotes the elastic stress at A. For simplicity, the index A in (39) is neglected.

Usually the stress at a point has six components. The problem (38)-(39) has therefore 7 unknowns β, y_{11}, y_{12}, y_{13}, y_{22}, y_{23} and y_{33}. If we work with principal stresses, $\boldsymbol{\sigma}^E$ has only 3 components, the number of unknowns of problem (38)-(39) remains 4. A plane stress problem has only 3 unknowns. This reduction makes the problem much simpler to solve.

With principal stresses, the elastic stress at point A can be expressed by

$$\sigma^E(1) = \begin{bmatrix} 0 \\ 0 \\ 0 \end{bmatrix}, \sigma^E(2) = \begin{bmatrix} S_{12} \\ S_{22} \\ S_{32} \end{bmatrix},$$

$$\sigma^E(3) = \begin{bmatrix} S_{13} \\ S_{23} \\ S_{33} \end{bmatrix} = \begin{bmatrix} S_{12}+S_{14} \\ S_{22}+S_{24} \\ S_{32}+S_{34} \end{bmatrix}, \sigma^E(4) = \begin{bmatrix} S_{14} \\ S_{24} \\ S_{34} \end{bmatrix},$$

where S_{12} denotes σ_1 (the first principal stress) at load vertex 2, S_{13} stands for σ_1 at load vertex 3 and so on.

The solutions of problem (38)-(39) must satisfy the KUHN–TUCKER conditions, i.e.

$$\frac{\partial L}{\partial \beta} = 0, \quad \frac{\partial L}{\partial y_1} = 0, \quad \frac{\partial L}{\partial y_2} = 0, \quad \frac{\partial L}{\partial y_3} = 0, \tag{40}$$
$$\lambda_j \geq 0, \quad \lambda_j g_j = 0, \quad g_j \leq 0,$$

where L is LAGRANGE's function, which is defined as

$$L(\beta, \boldsymbol{\lambda}, \mathbf{y}) = -\beta + \sum_{j=1}^{4} \lambda_j g_j, \tag{41}$$

$\boldsymbol{\lambda}$ the LAGRANGE's multiplier and

$$g_j = \Phi[\beta \boldsymbol{\sigma}^E(j) + \mathbf{y}] - \sigma_0^2. \tag{42}$$

The conditions in (40) form a nonlinear equation system with 8 unknowns: β, y_1, y_2, y_3, λ_1, λ_2, λ_3 and λ_4. By doing some modifications of these equations and making use of the symbolic computation program MACSYMA, we have solved this nonlinear equation system analytically (s. STEIN [1994]). The shakedown load factor reads

$$\beta = \begin{cases} \dfrac{2\sigma_0}{\sqrt{C_1}} & \text{for} \quad C_1 > C_2 \\ \dfrac{2\sigma_0}{\sqrt{C_2}} & \text{for} \quad C_1 < C_2 \end{cases} \tag{43}$$

with C_1 and C_2 being positive quantities depending on the elastic stresses

$$\begin{aligned} C_1 = {} & S_{12}^2 - 2S_{12}S_{14} + S_{14}^2 - S_{12}S_{22} + S_{14}S_{22} + S_{22}^2 + S_{12}S_{24} \\ & -S_{14}S_{24} - 2S_{22}S_{24} + S_{24}^2 - S_{12}S_{32} + S_{14}S_{32} - S_{22}S_{32} + S_{24}S_{32} \\ & +S_{32}^2 + S_{12}S_{34} - S_{14}S_{34} + S_{22}S_{34} - S_{24}S_{34} - 2S_{32}S_{34} + S_{34}^2, \end{aligned} \tag{44}$$

$$\begin{aligned} C_2 = {} & S_{12}^2 + 2S_{12}S_{14} + S_{14}^2 - S_{12}S_{22} - S_{14}S_{22} + S_{22}^2 - S_{12}S_{24} \\ & +S_{14}S_{24} + 2S_{22}S_{24} + S_{24}^2 - S_{12}S_{32} - S_{14}S_{32} - S_{22}S_{32} - S_{24}S_{32} \\ & +S_{32}^2 - S_{12}S_{34} - S_{14}S_{34} - S_{22}S_{34} - S_{24}S_{34} + 2S_{32}S_{34} + S_{34}^2. \end{aligned} \tag{45}$$

C_1 and C_2 cannot be zero unless all stress components are hydrostatic, but this case is of no importance. Solutions for y_1, y_2 and y_3 are

$$y_1 = -\frac{1}{2}\beta(S_{12} + S_{14}), \tag{46}$$

$$y_2 = -\frac{1}{2}\beta(S_{22} + S_{24}), \tag{47}$$

$$y_3 = -\frac{1}{2}\beta(S_{32} + S_{34}). \tag{48}$$

If the load domain has only two vertices, the solution for the shakedown load factor is much simpler; it reads

$$\beta = \frac{2\sigma_0}{\sqrt{S_{12}^2 + S_{22}^2 + S_{32}^2 - S_{12}S_{22} - S_{22}S_{32} - S_{12}S_{32}}}. \tag{49}$$

The denominator of the right-hand side of eqn (49) is the effective stress caused by the second load vertex. If it is equal to the initial yield stress σ_o, then the shakedown load factor is two. This means that in this situation the shakedown limit of a system is twice as large as its elastic limit.

It should be pointed out that the above statement is usually not true for a load domain with more than two vertices.

5 Examples

In this section we apply the analytical and numerical methods introduced previously to some examples, and compare the results.

Example 1: A square plate with a central hole

The system is subjected to biaxial loadings p_1 and p_2 as shown in Fig. 2 a; p_1 and p_2 may vary cyclically between zero and certain maximum magnitudes $\bar{p}_1$, $\bar{p}_2$. For $\bar{p}_1 > \bar{p}_2$ the point A dominates the shakedown behaviour of the system. According to (43), the shakedown load factor is

$$\beta = \frac{2\sigma_0}{\sigma_{A_1} - \sigma_{A_2}},$$

where σ_{A_1} is the elastic stress at A caused by p_1; σ_{A_2} the same stress, but caused by p_2.

Normally σ_{A_1} and σ_{A_2} can be calculated by using a numerical method, for example by using FEM or BEM. If the plate is infinite, so that $D/L \to 0$, one has

$$\sigma_{A_1} = 3p_1 \quad \text{and} \quad \sigma_{A_2} = -p_2 .$$

In this case the shakedown diagram is shown in Fig.2 b.

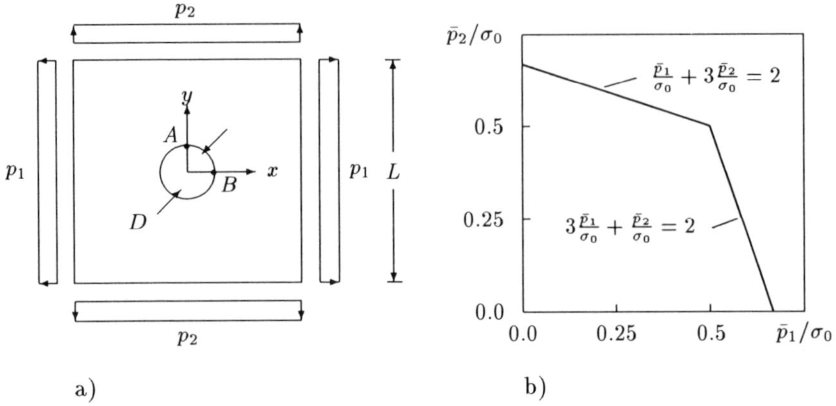

Fig. 2: a) Square plate with uniform loads p_1 and p_2 b) Shakedown limits for $D/L \to 0$

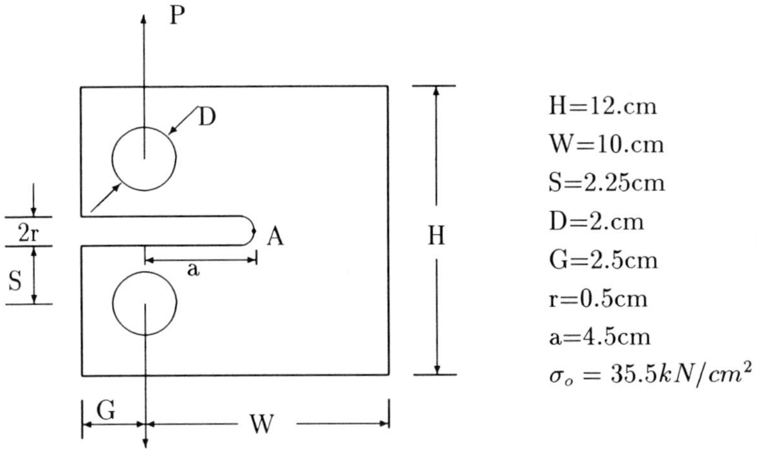

Fig. 3: A CT specimen subjected to an uniaxial loading P

In one of our previous papers (s. STEIN [1993a]) we calculated the shakedown limit for a finite plate with $D/L = 0.01$; that result agrees well with the analytical one achieved here.

Example 2: A compact tension specimen with a notch

The system shown in Fig. 3 is subjected to an uniaxial loading P allowed to vary between 0 and 1. kN. Obviously, the failure of this system is of local character. The point A at the notch root dominates the shakedown behaviour of the system.

According to the eqn (43), the shakedown limit factor reads

$$\beta = \frac{2\sigma_0}{\sigma_A}$$

with σ_o being equal to $35.5kN/cm^2$. σ_A is the elastic stress at point A which can be calculated numerically; here, however, we employ an analytical form suggested by PARIS & SIH [1965]

$$\sigma_A = \frac{2K_I}{\sqrt{\pi r}}, \tag{50}$$

where r is the notch root radius, K_I the so-called stress intensity factor calculated from fracture mechanics as if the notch were a crack. For a compact tension specimen, the stess intensity factor can be expressed as

$$K_I = \frac{P}{BW^{1/2}}[\ 29.6(a/W)^{1/2} - 185.5(a/W)^{3/2} + 655.7(a/W)^{5/2} \\ -1017(a/W)^{7/2} + 638.9(a/W)^{9/2}]\ , \tag{51}$$

where B stands for the thickness of the specimen and the meaning of other symbols is illustrated in Fig. 3. Substituting all quantities into (51) and then again into (50), we obtain $\sigma_A = 4.205kN/cm^2$. Therefore the shakedown limit load is $P = 16.885kN$.

The same problem has been computed by using adaptive FEM. The specimen is devided into 894 elements. Because of the strong stress concentration at the notch root the FE-mesh is generated adaptively, using asymptotic *a posteriori* indicators and quantitative errors. The Finite-Element computation gives a shakedown limit load $P = 16.806kN$; the difference between the numerical and analytical results is less than 1%.

The same Finite-Element program gives the elastic limit of the system $P = 8.44kN$; this value is just half of the shakedown limit of the system.

6 Shakedown experiments with CT-specimen

A CT-specimen with a notch is illustrated in Fig. 3, which in its original form (a crack instead of a notch) has been used very frequently for fracture experiments. The material is a structure steel ST-52 with an initial yield stress of $\sigma_o = 35.5kN/cm^2$. A more detailed description of the mechanical behaviour of this material can be found in the works of SCHEER [1990] and SCHEIBE [1990].

Experiments were carried out on three CT-specimens. Eleven strain gauges were pasted on each specimen. One of them was located at the notch root (point A). Theoretically the strain at point A should be measured. Technically, however, it is difficult to do that. On the one hand, the deformation at point A is not infinitesimally small, and on the other hand the strain gauge at this point is curved; it responds in an unstable fashion as loading proceeds. The measurement from this point is therefore not very reliable. A point $2mm$ far from point A, say point B, was chosen for the measurement.

For the first specimen four different cyclic loadings ($0 \rightleftharpoons 8.75kN$, $0 \rightleftharpoons 17.0kN$, $0 \rightleftharpoons 17.75kN$ and $0 \rightleftharpoons 18.5kN$) with 10, 20, 30 and 30 cylces, respectively, were applied. The loading-deformation diagrams of point B for first two loadings are shown in Fig. 4. It can been seen that, with the loading $P = 8.75kN$, the system shakes down, and with loadings of $P = 17.0kN$, $P = 17.75kN$ and $P = 18.5kN$ the system does not shake down**. This means that the shakedown limit is between $P = 8.75kN$ and $P = 17.0kN$.

In the second experiment, two loadings ($P = 15.0kN$ and $P = 18.0kN$) with 2000 cycles, were applied. The loading-deformation diagrams of point B are shown in Fig. 5. For a loading of $P = 15.0kN$ the system shakes down, and for a loading $P = 18.0kN$ the system does not shake down.

The third experiment concerned two different loadings ($P = 18.0kN$ and $P = 20.0kN$) with 2000 cycles **. For both of the loadings the system does not shake down.

Comparing the results from three experiments, we can conclude that the shakedown limit measured from point B should be between $P = 15.0kN$ and $P = 17.0kN$. To get more accurate results, further experiments must be done.

As mentioned above, it is point A, not B, which dominates the shakedown behaviour. Due to the stress concentration, the elastic stress at point A should be higher than that at B. Hence the correct shakedown limit of the system must be smaller than that measuered for B. The theoretical shakedown limit of this system is $P^{'} = 16.885kN$. The difference between theoretical and experimental results is about 10%. In our opinion, one of the reasons for this error is that, in the previous shakedown theorem, the creep deformation is not considered. This creep can reduce the shakedown limit of a system.

**Due to liminitation of space, not all loading–deformation diagrams are shown here

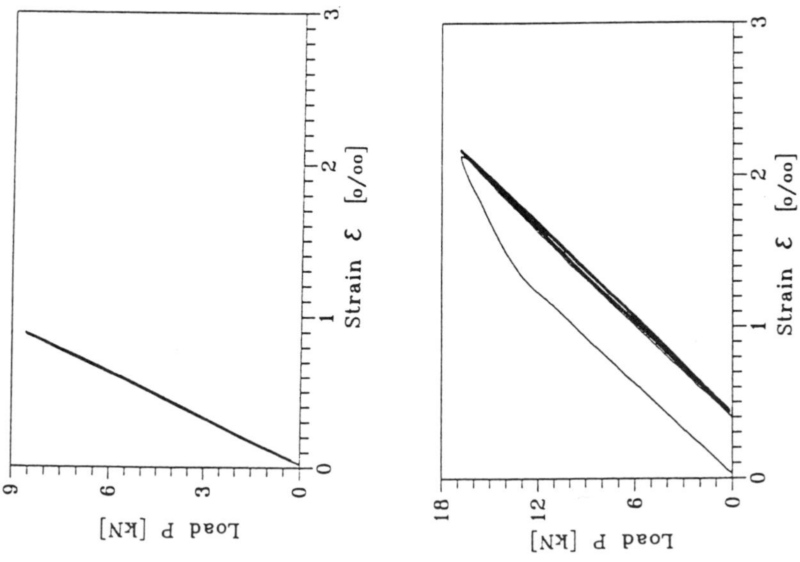

Fig. 4: The loading–deformation diagrams of point B for the first experiment

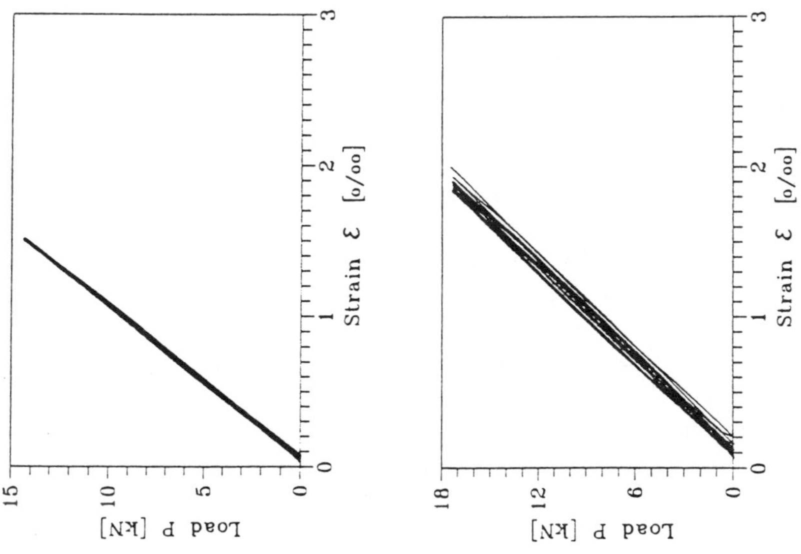

Fig. 5: The loading–deformation diagrams of point B for the second experiment

7 Conclusions

In this paper, a 3-d overlay-model for describing the nonlinear-kinematical hardening behaviour of ductile cristalline materials and a static shakedown theorem for this material model were treated. Analytical and numerical methods were formulated for determing the shakedown limits of systems with kinematical hardening materials. Two examples have been treated by using these methods. Experiments were carried out with CT-specimens. The experimental results verify the theoretical and numerical achievments in a satisfactory manner.

It is noteworthy that the material models considered in this paper are capable of describing the kinematic hardening behaviour of an elastic-plastic material, but creep, cyclic hardening and cyclic softening have not been considered. The CT-specimen has a special character. It fails only locally, so its shakedown load limit is rather independent of special deformation properties. To investigate the shakedown behaviour of systems with ductile materials in a general manner, more complicated material models should be considered, and experiments with other types of specimens should be carried out. These tasks are subjects of our current research.

Acknowledgement: This work was supported by Deutsche Forschungsgemeinschaft within the special research area SFB 319 "Stoffgesetz für das inelastische Verhalten metallischer Werkstoffe" under project B6.
The authors thank Dipl.-Ing. M. REININGHAUS *from the Technical University of Braunschweig for his valuable support of the experimental work.*

References

1924 MASING, G., "Zur Heyn'schen Theorie der Vestigung der Metalle durch verborgen elastische Spannungen", Wissenschaftliche Veröffentlichungen aus dem Siemens–Konzern, 3, pp. 231–239.

1932 BLEICH, H., "Über die Bemessung statisch unbestimmter Stahltragwerke unter Berücksichtigung des elastisch-plastischen Verhalten des Baustoffes", Der Bauingenieur 19/20, pp. 261–267.

1938a MELAN, E., "Der Spannungszustand eines Mises–Henckyschen Kontinuums bei veränderlicher Belastung", Sitzber. Akad. Wiss. Wien, IIa 147, pp. 73–78.

1938b MELAN, E., "Zur Plastizität des räumlichen Kontinuums", Ing.–Arch. 8, pp. 116–126.

1950 NEAL, B.G., "Plastic collapse and shake–down theorems for structures of strain–hardening material", J. Aero. Sci., 17, pp. 297–306.

1953 BESSELING, J. F., "A theory of plastic flow for anisotropic hardening in plastic deformation of an initially isotropic material", National aeronautical research institute, Amsterdam, Report pp. 410.

1956 KOITER, W.T., "A new general theorem on shake–down of elastic–plastic structures", Proc. Koninkl. Akad. Wet. B 59, pp. 24–34.

1956a PRAGER, W., "A new method of analyzing stresses and strains in work-hardening plastic solids", Journal of Applied Mechanics, pp. 493–496.

1956b PRAGER, W., "Shakedown in elastic–plastic media subjected to cycles of load and temperature", Proc. Symp. Plasticita nella Scienza delle Costruzioni, Bologna, pp. 239–244.

1957 ROZENBLUM, V.I., "On shakedown of uneven heated elastic–plastic bodies" (in Russian), Izw. Akad. Nauk SSSR. OTN, Mekh. Mash., 7, pp. 136–138.

1959 COFFIN, L. F.0 and TAVERNELLI, J. F., "The cycle strain and fatigue of metals", Transactions of society of AIME. 215, pp. 794-807.

1959 ZIEGLER, H., "A modification of Prager's hardening rule". Quart. Appl. Math., 17, pp. 55–65 (1959).

1965 PARIS, P. C. and SIH, G. C, "Fracture toughness testing and its applications", ASTM STP 381, American Society for testing and materials, pp. 30–83.

1966 FREDERICK, C. O. and ARMSTRONG, P. J., "Convergent internal stresses and steady cyclic states of stress", J. Srain Anal., 2, pp. 154–160.

1967 IWAN, W. D., "On a class of models for the yielding behavior of continuous and composite systems", Journal of Applied Mechanics, pp. 612–617.

1967 MRÓZ, Z., "On the description of anisotropic workhardening", J. Mech. Phys. Solids 15, pp. 163-175.

1969 CERADINI, G., "Sull' adattamento dei corpi elasto–plastici soggetti ad azioni dinamiche", Giorn. Genio Civile 106, No. 4/5, pp. 239–250.

1969 KÖNIG, J.A., "A shakedown theorem for temperature dependent elastic moduli", Bull. Acad. Polon. Sci. Sér. Sci. Tech. 17, pp. 161–165.

1972 MAIER, G., "A shakedown matrix theory allowing for workhardening and second–order geometric effects", Proc. Foundations of Plasticity, Warsaw.

1972 ZIENKIEWICZ, O. C, NAJAK, G. C. and OWEN, D. R. J., "Composite and 'overlay' models in numerical analysis of elasto-plastic continua", Pro. of Int. Symp. on Foundations of plasticity, ed. A. Sawczuk, Noorhoff Publishing Leyden.

1973 CORRADI, L. and MAIER, G., "Inadaption theorems in the dynamics of elastic–workhardening structures", Ing.–Arch. 43, pp. 44–57.

1975 DAFALIAS, Y.F. and POPOV, E.P., "A model of nonlinearly hardening materials for complex loading", Acta Mechanica 21, pp. 173-192.

1981 SCHITTKOWSKI, K., "The nonlinear programming method of Wilson, Han and Powell with an augmented Lagrangian type line search function", Numer. Math., 38, pp. 83–127.

1986 WEICHERT, D., "On the influence of geometrical nonlinearities on the shakedown of elastic–plastic structlures", Int. J. Plasticity, 2, pp. 135–148.

1990 SCHEER, J., SCHEIBE, H. J. and KUCH, D., "Untersuchung von Trägerschwäschungen unter wiederholter Belastung bis in den plastischen Bereich", Bericht Nr. 6099, Institut für Stahlbau, TU Braunschweig.

1990 SCHEIBE, H. J., "Zum zyklischen Materialverhalten von Baustahl und dessen Berücksichtigung in Konstruktionsberechnung", Ph.D. thesis, Institut für Stahlbau, Universität Braunschweig.

1992 MAHNKEN, R., "Duale Methoden für nichtlineare Optimierungsprobleme in der Strukturmechanik", Ph.D. thesis, Institut für Baumechanik und Numerische Mechanik, Universität Hannover.

1992a STEIN, E., ZHANG G. and KÖNIG, J.A., "Shakedown with nonlinear hardening including structural computation using finite element method", Int. J. Plasticity, 8, pp. 1–31.

1992b STEIN, E. and ZHANG, G., "Theoretical and numerical shakedown analysis for kinematic hardening materials", in Proc. 3rd Conf. on Computational Plasticity, Barcelona, pp. 1–25.

1992 ZHANG, G., "Einspielen und dessen numerische Behandlung von Flächentragwerken aus ideal plastischem bzw. kinematisch verfestigendem Material", Ph.D. thesis, Institut für Baumechanik und Numerische Mechanik, Universität Hannover.

1993a STEIN, E., ZHANG, G. and HUANG, Y., "Modeling and computation of shakedown problems for nonlinear hardening materials", Computer Methods in Mechanics and Engineering, 1-2, pp. 247-272.

1993b STEIN, E., ZHANG, G. and MAHNKEN, R., "Shake-down analysis for perfectly plastic and kinematic hardening materials, in Stein, E. (ed.) Progress in computational analysis of inelastic structures", Springer–Verlag, Wien–New York, CISM courses and lecture, 321, pp. 175–244.

1994 STEIN, E. and HUANG, Y., "An analytical method to solve shakedown problems with linear kinematic hardening materials", Int. J. of solids and structures, 31, No. 18, pp. 2433-2444.

RELATED MATHEMATICAL MODELS FOR SOLIDS AND STRUCTURES DEFORMATION AND FAILURE PROCESSES UNDER REPEATED LOADING

D.A.GOKHFELD, O.S.SADAKOV, O.F.CHERNIAVSKY
Cheliabinsk State Technical University
Lenin Ave., 76, 454080 Cheliabinsk, Russia

ABSTRACT

The survey reflects the development of investigations carried out in the research laboratory of ChSTU during recent decades. In the main, the results can be reduced to a branching hierarchy of related rheological models; its first level (perfect plasticity, perfect viscosity) is followed by a set of models based on the concept of micro-inhomogeneity of real materials. The latter approach leads to more adequate description of plasticity and creep including some specific regularities inherent to materials which belong to different groups. On the base of general analysis of rheology of structures made of real materials, a set of new effective structural design methods has been suggested. The study is completed by damage accumulation and crack growth models related to the rheological model by common state variables .

INTRODUCTION

To exhibit clearly the most relevant and characteristic features of any complicated process it is usually necessary to ignore some less essential details. Afterwards, the discrepancies which can result from the schematic character of the adopted model must be studied. Hence the idea follows, that it is reasonable to dispose of a set of models which differ one from another by the level of compromise between adequacy and complexity. On the other hand, the requirements for any model are related to the adopted concept of unsafe states of structures. In particular, supposing that the failure-free performance of any structure under low-cycle loading conditions is ensured only when there is no unlimited evolution of inelastic strains we naturally come to the shakedown theory. The main advantage of the latter is its extreme clarity: the limit (i.e. unsafe) states of a structure - alternating plasticity and strain accumulation - are distinguished qualitatively. This was the reason by which at the initial stage of the investigations our attention was focused on this approach. The main results are reflected in the book (GOKHFELD & CHERNIAVSKY, 1980).
However the advantages of limit analysis do not remove the necessity of more adequate description of plastic and viscous deformation processes. This is particularely so because sometimes (e.g. for thermo-stressed structures) it is either impossible or unreasonable to

Z. Mróz et al. (eds.), Inelastic Behaviour of Structures under Variable Loads, 51–81.

prevent these processes. In such situations the problem reduces to lifetime evaluation. Hence it follows that inelastic deformation analysis of structures should be completed by methods of low-cycle failure prediction.

Different trends in the development of rheological models are not considered herein, and we refer the specialists to the recently published excellent surveys by CHABOCHE [1989] and OHNO [1990] and to our former survey (GOKHFELD et al. [1978]). In this survey we emphasise the important role which micro-inhomogeneity plays in the deformation behavior of real materials.

The simplest way to model the inhomogeneity is to use the Masing scheme: The actual physical inhomogeneity of materials is simulated by a hypothetical structure which consists of components characterized by some distribution of their rheological parameters. Models belonging to this type are called "the structural models" due to ZARUBIN [1970] who associated the term with RABOTNOV's [1969] structural parameters. In English terminology they are usually called "composite" or "layers" models.

Among the structural models (SM) the basic version (the basic model, BM) occupies the central position. It is concerned with materials which are in a cyclically stable state, before or after previous cyclic loading. The subelements (SEs) are supposed to be perfectly viscous (viscous SEs were first employed by BESSELING [1974]); as has been shown (GOKHFELD & SADAKOV [1984,1991]) in the SM framework, creep and plasticity are given a common description.

For proportional loading conditions, the BM constitutive equations can be reduced to a state equation directly formulated in macro-terms. The latter reveals rather clear and simple regularities of deformation behavior under variable-repeated loading and hence the term "similarity principle" (SP) appears. Further development of the SM is related with simulation of specific rheological properties such as cyclic hardening, physical anisotropy, different resistance against tension and compression, inverse influence of micro-failures on rheological characteristics, temperature shape memory and others which are outside the framework of anisotropic hardening. These derived models form a branching hierarchy: any of the models can be reduced to the BM and then to the model of perfect viscosity (or plasticity) by successive simplifications.

However, the possibilities which appear for the rheology of structures analysis are the most impressive ones. Until now, such analysis was available for perfectly plastic structures only (KOITER [1960]). According to the BM, any structural component made of real material is assumed to consist ultimately of SEs which are endowed with perfect viscosity. Thus the problem is dealt with as the generalization of the traditional analysis to a system of high dimensionality of the corresponding matrices. The results have promoted the development of new rational design methods and calculation procedures.

The lifetime of a structure may be limited either by inadmissible geometry changes of some components, or by crack initiation and growth. To predict the latter situation, damage accumulation and crack growth models have been suggested. Their main peculiarity consists in the use of new state variables which are common to those of the basic rheological model.

Thus the suggested system of models embraces all the main stages of the problem of structure lifetime evaluation. The interconnections are illustrated by the scheme (Fig. 1).

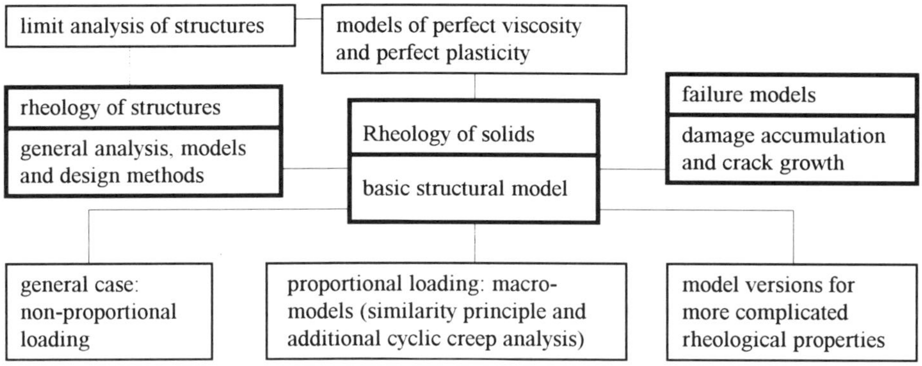

Fig. 1. Related models for rheology and failure of solids and structures description

It should be stressed that much attention has been paid to experimental investigations, due to the necessity of identifying and verifying the models. This is not considered here.
From the cited books and papers the reader can get some additional information. Unfortunately, our previous publications in English concerning the structural model are rather fragmentary. This survey attempts to give a more systematic explanation of the state-of art of the approach.

SIMPLEST REOLOGICAL MODELS AND THE LIMIT ANALYSIS OF STRUCTURES

This term is used to describe models with perfect plasticity and perfect viscosity. Due to the simplicity of their models the limit analysis of structures has become a convenient engineering tool: in some important branches of industry and civil engineering the corresponding methods have been introduced into the national engineering codes.
Development of shakedown theory is considered in many specialized books and surveys (KOITER [1960], SAWCZUK [1974], GOKHFELD & CHERNIAVSKY [1978, 1980], KÖNIG & MAIER [1981], KÖNIG [1987], and others).
It is generally acknowledged that the main problem of the theory is the determination of loading conditions at which strain accumulation begins; prediction of alternating plasticity conditions is rather trivial. By restatement of the fundamental theorems (GOKHFELD & CHERNIAVSKY [1980]) the necessity of integration in current time was excluded from consideration offering a wide field of applications. The problem is reduced to the use of the non-classical variational methods (including Pontriagin's maximum principle (CHERNIAVSKY [1970])). In many cases of practical importance some approximate methods, especially those which are based on the kinematical theorem, prove to be the most effective.

Structures subject to non-stationary heat fluxes (either with or without mechanical forces) present one of the most important fields of application of shakedown theory. Moreover, some specific limiting conditions have been found which are extremely favorable for appearance of ratchetting; we can approach the conditions in extended bodies of homogeneous shape (such as plates, cylindrical shells and some others) subjected to repeated moving heat fluxes. These give rise to quasi-stationary temperature fields which move with the heat source. Hence the idea of a forming technology without contact between the raw piece and the tool follows (GOKHFELD [1970], GOKHFELD & CHERNIAVSKY [1980]).

Analogous situations may appear for moving mechanical loads, e.g. contact forces (GOKHFELD & CHERNIAVSKY [1980]).

This book gives a detailed consideration of various aspects of the shakedown theory. However, some problems concerned with new applications of the theory and its further development are worth attention.

Progress in modern technologies leads to some processes which are accompanied by dilatation effects; these may be due to diffusion of active gases (such as hydrogen or nitrogen), or chemical interaction between metals and aggressive liquids or radiation fluxes. Diffusion separators used for the production of pure hydrogen contain membranes of plane or cylindrical shape made of palladium or other alloys. They are subject to variable dilatation caused by non-stationary operating regimes of the plant. The hydrogen atoms are the only ones which can pass such membranes. Calculations show that elastic stresses associated with dilatancy caused by hydrogen can be several orders of magnitude larger than actual thermal or mechanical (i.e. produced by pressure) stresses. In actual operating conditions this causes cracks and/or strain accumulation.

The corresponding shakedown analysis is analogous to that at cyclic thermal processes. However, the influence of hydrogen saturation on cyclic stress-strain diagram of the metal should be taken into account. We recall that in the framework of the shakedown theory, mechanical properties of materials related to temperature, time, previous loading, radiation and other factors can be approximately modelled by using the corresponding experimental stress-strain diagram represented according to the Prandtl scheme. Moreover, in some cases it is possible to use several different diagrams for an object according to the peculiarities of different failure mechanisms. An example is illustrated in Fig. 2. Here the shakedown diagram (chart of states) is shown for a cylindrical shell subjected to constant internal pressure and cyclic temperature changes. The loading cycle includes hold-time period at maximum temperature.

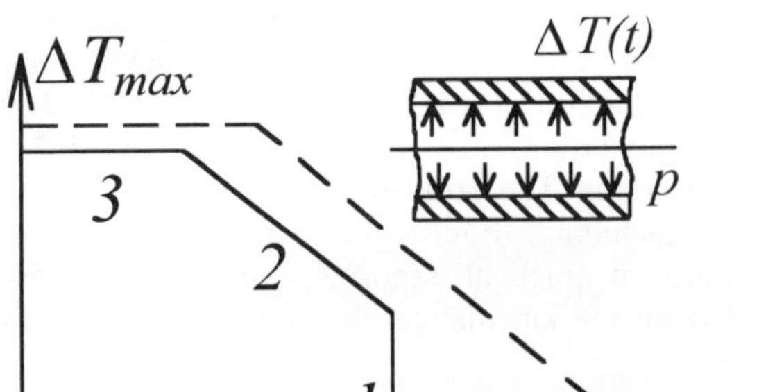

Fig.2. Shakedown of a creeping structure

Line 1 corresponds to a state of "limit equilibrium" caused by creep (strain grows with time beyond all bounds); determination of the corresponding conditions is to be based on the isochronic curve (RABOTNOV [1969]) for total operation time. Cyclic strain accumulation (line 2) in this case is due to both creep and plasticity which occur at hold-time periods and unloading stages, respectively. Thus, the corresponding two stress-strain diagrams should be used for calculations. Finally, unlike the considered limit states, the determination of alternating inelasticity conditions (line 1) should be based on the stabilized stress-strain diagram (i.e. corresponding to "saturation" of isotropic hardening of the material).

Deformation behavior of structures which suffer cyclic external actions beyond their shakedown limits is a problem of theoretical and practical interest. According developments in the framework of shakedown theory were started by FREDERICK & ARMSTRONG [1966], MROZ [1971] and other authors (see the references in GOKHFELD & CHERNIAVSKY [1980], GOKHFELD & SADAKOV, in this volume). Various types of shaping production processes (including the above mentioned without-contact technology) may be considered as one of the actual fields of applications for the methods of direct SCS determination.

However, the corresponding analysis often requires additional investigations concerning stability of deformation processes. It can be shown, that shakedown theory can determine only the parameters of a branching point of the process: often there is an infinite set of kinematically possible fields of strain increments per cycle corresponding to a given set of parameters. The type of the branching point, as well as supercritical deformation behavior of structures should be studied by the traditional methods of stability theory for inelastic deformation processes. If the corresponding analysis is not carried out, there may be essential discrepancies between the predicted and real deformation mechanisms.

Completing the paragraph, we note that the problems which are traditionally considered within the framework of perfect plasticity can be generalized to perfect viscosity conditions. In addition to the example considered above, we refer to the book of BIRGER at al. [1975] where carrying capacity of a turbine disk (BIRGER) and SCS of a tube (SHORR) have been considered as problems of creep.

BASIC STRUCTURAL MODEL

According to the structural ("composite" or "layer") model, any body element is assumed to consist of subelements (SEs) which have equal total strains ($\tilde{\varepsilon} = \varepsilon$), temperature ($\tilde{T} = T$), elastic moduli ($\tilde{E} = E$) and thermal expansion coefficients ($\tilde{\alpha} = \alpha$), but different rheological functions (RFs) determining the dependence of the inelastic strain rate on individual stress and temperature. An important simplifying assumption is the mutual similarity of all the RFs; this minimizes number of the SEs:

$$\dot{p} \equiv \tilde{\Phi}(\tilde{r}, T) = \Phi(\tilde{r} / z, T) \tag{1}$$

Here Φ is the common RF of a solid; it gives its secondary creep strain rate. The parameters z which identify any SE, give the inhomogeneity function (IF). It should be noted that henceforth elastic strain $\tilde{r}$ is used as an argument instead of stress $\tilde{\sigma}$; this simplifies the non-isothermal deformation analysis since the temperature dependence of elastic modulus is terminally excluded from analysis.

As we stated before, the basic version of the structural model (BM) is intended to describe deformation behavior of materials which are in their cyclically stable states, i.e. the BM does not reflect isotropic hardening processes. Hence, only two specific functions - the RF and the IF, must be identified (excluding the common elastic and thermal characteristics). Determination of the IF is associated with the stress-strain diagram.

To generalize the model for arbitrary stress states, the corresponding tensors are introduced. As usual, stress and strain tensors are divided into spherical and deviatoric parts

$$\varepsilon_{ij} = \rho_{ij} + p_{ij} + \vartheta_{ij} = e_{ij} + \varepsilon_o \delta_{ij}, \qquad \varepsilon_o = \frac{1}{3}\varepsilon_{ii}$$

$$\rho_{ij} = r_{ij} + \rho_o \delta_{ij}, \qquad \rho_o = \frac{1}{3}\rho_{ij} \tag{2}$$

$$\vartheta_{ij} = \vartheta_o \delta_{ij}, \quad p_{ii} = 0$$

Hence,

$$e_{ij} = r_{ij} + p_{ij}, \quad \varepsilon_o = \rho_o + \vartheta_o$$

$$r_{ij} = s_{ij}/(2G), \qquad \rho_o = \sigma_o/(3K) \qquad \vartheta_o = \alpha \Delta T \tag{3}$$

$$s_{ij} = \sigma_{ij} - \sigma_o \delta_{ij}$$

The notation is obvious; r_{ij} denotes the components of elastic strain deviator (unlike the one-dimensional version where it is elastic strain, i.e. a component of the corresponding tensor).

As in the Masing model, it is assumed that $\tilde{\varepsilon}_{ij} = \varepsilon_{ij}$ for any SE. This implies $\tilde{\varepsilon}_o = \varepsilon_o$ and

$$\tilde{e}_{ij} = \tilde{r}_{ij} + \tilde{p}_{ij} = e_{ij} = r_{ij} + p_{ij} \tag{4}$$

then,

$$\tilde{r}_{ij} = \tilde{s}_{ij}/(2G), \qquad r_{ij} = s_{ij}/(2G) \tag{5}$$

$$r_{ij} = \langle \tilde{r}_{ij} \rangle, \qquad \mathrm{p}_{ij} = \langle \tilde{\mathrm{p}}_{ij} \rangle, \qquad s_{ij} = \langle \tilde{s}_{ij} \rangle;$$

the brackets $\langle \ \rangle$ signify ensemble averaging.

It is supposed that creep rates of perfectly viscous SEs do not depend on the spherical stress tensor; due to isotropy of any SE inelastic properties, vectors $\dot{\tilde{\mathbf{p}}}$ and $\tilde{\mathbf{r}}$ are collinear in deviatoric space . The assumption of similarity of the SEs rheological functions implies

$$\dot{\tilde{p}}_{ij} = \frac{\tilde{r}_{ij}}{\tilde{r}_e} \Phi(\tilde{r}_e / z, T) \tag{6}$$

where

$$\tilde{r}_e = \left(\frac{1}{2}\tilde{r}_{ij}\tilde{r}_{ij}\right)^{1/2} \qquad (7)$$

is the elastic strain intensity.

Eqs (4) - (7) should be completed by defining the parameters of the SEs rheological properties (z) (i.e. by definition of the IF). Using the BM constitutive equations, we can obtain the response of the model for any type of loading program. In particular, numerical analysis shows that the model predicts all the known deformation effects which are characteristic for non-proportional loading: evolution of yield surface; scalar and vector delay effects,; specific plastic strain accumulation under cyclic loading superimposed on initial stress in conditions when their vectors are not collinear, see GOKHFELD & SADAKOV [1984, 1990].

In engineering applications, the number of SEs, n, should usually not exceed 5 - 7; this is enough to provide the necessary accuracy. Accordingly, index k, $k=1,2,...n$, is used instead of the tilde in relationships.

Returning to the one-dimensional case we have:

$$\begin{aligned} &\varepsilon_k = r_k + p_k + \vartheta_k = \varepsilon = r + p + \vartheta, \\ &r_k = \sigma_k / E, \quad r = \sigma / E, \quad \vartheta_k = \vartheta = \alpha T \\ &\dot{p}_k = \Phi(r_k / z_k, T) \\ &\sigma = \langle \sigma_k \rangle, \quad p = \langle p_k \rangle \end{aligned} \qquad (8)$$

The IF here is presented in the form of discrete relative weights g_k of the SEs. Both the identification parameters g_k, z_k are standardized by equations

$$\langle z_k \rangle = 1, \qquad \Sigma g_k = 1. \qquad (9)$$

Here

$$z_k = \sigma_b^k / \sigma_b = r_b^k / r_b \qquad (10)$$

where σ_b^k is the asymptotic ultimate stress of any perfectly viscous SE at given strain rate $\dot{\varepsilon}$ and temperature T, while σ_b is ultimate strength of a material at the corresponding conditions.

It is known that, if the notions of true stress and natural strain are used, the tangent modulus of the stress-strain diagram does not reach zero. On the other hand, it is often asssumed that the strain rate is constant at secondary creep. This actual situation can be modelled by assuming there is at least one SE which has an infinitely large parameter z (at finite relative weight). The advantages of the corresponding version of BM are not in contrast with the true description of actual deformation properties: besides, this enhances the identification procedure. Finally, such an approach simplifies the direct determination of the SCS of structure (GOKHFELD & SADAKOV [1984 and in this volume]).It should be stressed that in spite of the form of the given equations (e.g. eqn (6)), which are typical for creep, the model embraces both rheonomic and scleronomic behavior, depending on the character of the RF . Scleronomic behavior can be moedelled by an RF of singular type

$$\dot{p} = 0 \text{ at } |r| < r_s, \qquad \dot{p} = 0/0 \text{ at } r = r_s \ (r_s = \sigma_s/E) \tag{11}$$

Here σ_s is the yield limit and $\sigma_s^k = z\sigma_s$.
As experiments show, actual RFs of structural alloys within operating temperature ranges of the latter tend to approach function (11). Usually, actual RFs do not pass through the coordinate origin, i.e. there is some stress interval $0 \le \sigma < \sigma_o$ which corresponds to negligible creep rate (σ_o is the creep limit). Then, creep rate increases in stress slowly and begins to grow rapidly only when approaching the ultimate stress.

SIMILARITY PRINCIPLE

If we limit ourselves to proportional loading conditions, and introduce assumptions based on the above mentioned peculiarities concerning RFs of structural alloys, we can reduce the BM constitutive equations to a state equation formulated in macro-terms directly (GOKHFELD & SADAKOV [1984, 1989, 1990]). This can be presented in the form

$$\dot{p} = \Phi(sr_b, T)[1 - K(C)] \tag{12}$$

and is valid for any stage of variable-repeated deformation if a new system of macroscopic parameters is used. The system is related to current values r, ε and their magnitudes r_i, ε_i at the turning points of the loading trajectory (conditions which determine their appearance or erasure from the "memory" of a material will be considered below). The new coordinates are

$$r_* = r - r_\nu, \qquad \varepsilon_* = \varepsilon - \varepsilon_\nu \tag{13}$$

where index ν indicates coordinates of the last (i=ν) turning point; its parameters play a governing role in the next deformation stage.
Eqn (12) contains the following new state parameters: current tangent modulus $K = K(C)$ which is a function of secant modulus C -

$$C = r_*/\varepsilon_*, K(\varepsilon) = d\varphi(\varepsilon)/d\varepsilon, \ K(C) \equiv K(\varepsilon(C)), \ \textit{where } C = \varphi(\varepsilon)/\varepsilon; \tag{14}$$

The similarity parameter κ defined by the parametric equation

$$r_* = \kappa \cdot \varphi(\varepsilon_*/\kappa) \tag{15}$$

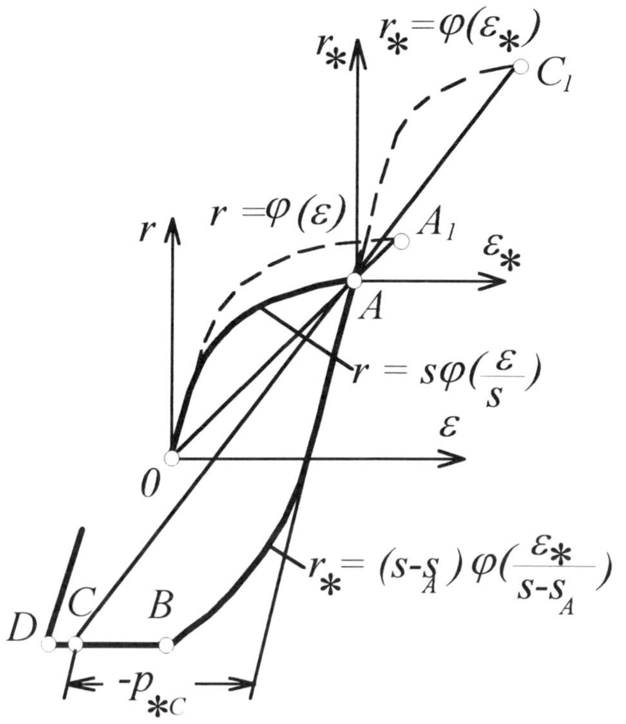

Fig. 3. On the definition of state parameters

can be presented as a ratio $\kappa = OA/OA_1$ (Fig. 3). Here the point A is determined by the current coordinates r_*, ε_*, while point A_1 is characterized by the same tangent modulus C and situated on the basic diagram, i,e. the diagram obtained at values of strain rate $\dot{\varepsilon}$ and temperature T adopted as the basic values. State parameter s (see eqn (12) is determined by the equation

$$\kappa = s - s_v \qquad (16)$$

Suppose the loading cycle include a hold-time period. The creep curve can be plotted by integration of eqn (12). In this case state variables C and s depend on the current position of the state point (eqs (13-16)). In its turn, this position is determined by actual creep rate and time. The solution can be obtained if eqn (12) is complemented with some type of statical or kinematical condition which makes the process definite: e.g. $\dot{\sigma} = 0$ (pure creep), $\dot{\varepsilon} = 0$ (pure relaxation).

At a loading stage characterized by constant strain rate $\dot{\varepsilon}$ and temperature T, integration of eqn (12) leads to *s=const*, equation (16) then gives *κ=const*. In such situations eqn (15) can be used directly instead of integration of eqn (12). This time parameter κ is constant and should be determined from eqn (16), where s is equal to

$$s = \Phi^o(\dot{\varepsilon}, T). \qquad (17)$$

Here Φ^o is the inverse function with respect to the RF. It should be noted that eqn (17) is valid only at constant values of $\dot{\varepsilon}$, T during the considered deformation stage, i.e. for variable $\dot{\varepsilon}$, T the rigorous solution requires the integration of eqn (12)).

New turning points emerge when reversals (sign changes) of either strain rate $\dot{\varepsilon}$ or secant modulus rate $\dot{C} = dC/dt$ occur. The latter case is connected with a rapid temperature fall (or strain rate increase).

Accordingly, there are situations which lead to erasure of the last turning point from the memory (or the last ν and last but one μ, simultaneously). In such cases, for instance when the secant modulus C reaches the value it had at the previous turning point, the specific role of the last turning point passes to the last among the points remaining in the memory. The formalized memory rules are given in publications (GOKHFELD & SADAKOV [1984, 1989, 1990]).

Thus, the complex of relationships (12) - (16) completed by the mentioned memory rules permits one to construct stress-strain diagrams and creep curves for any actual loading program. There are clear regularities which can be characterized by similarity laws for stress-strain diagrams at any given values of σ and ε and at any stage of variable-repeated loading (eqn (14)), as well as for the corresponding creep rate fields (eqn (12)) with respect to coordinate axes {r , ε }. The regularities exhibited by the developed system of equations and definitions has been termed "the similarity principle" (SP).

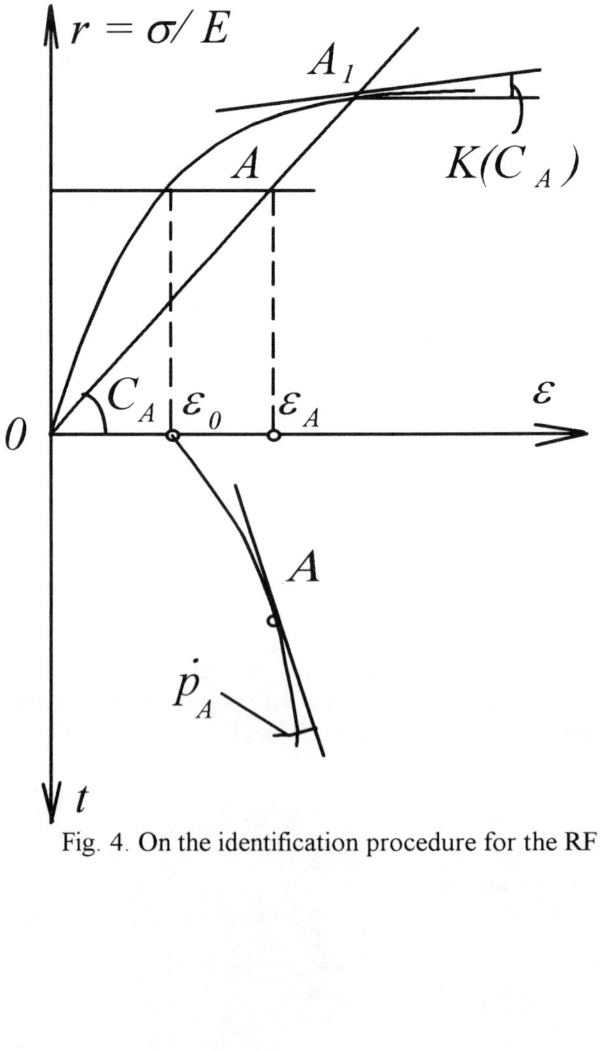

Fig. 4. On the identification procedure for the RF

It should be added that the suggested state parameters C, s can be interpreted in terms of crystal physics, C as a number of activated slip planes in a body element, and s as their loading state.

A more rational identification procedure for the RF is based on state equation (12). The latter can be rewritten in the form

$$\Phi(sr_b, T) = \frac{\dot{p}}{1 - K(C)}$$

The necessary values can be determined from the stress-strain diagram and creep curve of a material (Fig. 4).

It should be noted that the agreement between the results obtained by this procedures and those based on alternative utilization of rheonomic and scleronomic characteristics additionally confirm the consistency of the derived common description for both types of inelastic deformation processes.

However, there are some limitations for the SP: the assumptions which have been made require closed hysteresis loops, i.e. cyclic creep and relaxation are excluded. The asymptotic position of the loop can be determined by equations (formulated in macro-terms) which follow from the analysis of the SEs deformation behavior under cyclic loading (GOKHFELD & SADAKOV,[1984] and in this volume).

ISOTROPIC HARDENING

Simplicity of the basic model (BM) is due first of all to the disregard of cyclic isotropic hardening. In reality, deformation processes accompanied by a variation of the stress-strain diagram are usually relatively short compared with the lifetime of a structure, and they end with the formation of a new stress-strain diagram which remains almost constant under further loading. But any general rule may have exceptions; in some cases the stabilization process is extended or the stress-strain diagram does not reach a stabilized state at all. On the other hand, with changes in the loading program (e.g. inclusion of hold-time periods and even change of strain amplitudes), the stabilization process may lead to additional hardening or to some recovery, especially at elevated temperatures. It should be added that for rheonomic materials, these processes lead to variation of both characteristics - stress-strain diagram and creep curve.

To model isotropic hardening by the SM, any SE should be endowed with such property (MADUDIN & SADAKOV [1977], GOKHFELD & SADAKOV [1984]). Thus, expression (1) should be changed to:

$$\tilde{p}_{ij} = \frac{\tilde{r}_{ij}}{\tilde{r}_e}\Phi\left(\frac{\tilde{r}_e}{\tilde{\alpha}z}, T\right) \qquad (18)$$

Proceeding from the simplest assumption that the hardening parameter $\tilde{\alpha}$ depends only on Odquist's parameter and that the functions by which it is determined are similar for any SE, we shall have

$$\tilde{\alpha} = \tilde{\psi}(\tilde{\lambda}, T) = \psi(\tilde{\lambda}/z, T) \qquad (19)$$

where

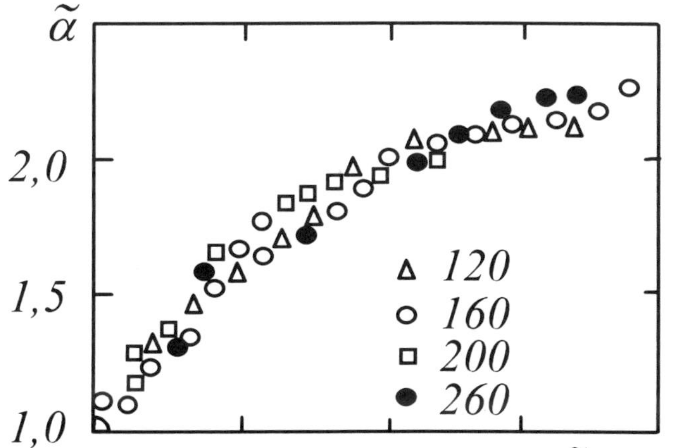

Fig. 5. Hardening of SEs against Odquist's parameter

$$\tilde{\lambda} = \left(\frac{1}{2} d\tilde{p}_{ij} d\tilde{p}_{ij} \right)^{1/2}; \tag{20}$$

as before, z is the identifying coefficient.

Hardening at cyclic loading of structural alloys often decreases monotonically until the stabilized state ($\dot{\tilde{\alpha}} = 0$, $\tilde{\alpha} = \tilde{\alpha}^*$) is reached. This corresponds to a "saturation" in SEs which were involved in inelastic deformation. The hardening process resumes if the strain amplitude increases.

Hardening functions (19) can be determined by tests. For this purpose, the following property should be used: points of stress-strain curves of successive halfcycles which are characterized by equal tangent modulus practically correspond to equal amounts of SEs involved in inelastic deformation. In this way we can trace (halfcycle-by-halfcycle) the variation of the hardening function of any SE. The results obtained for a stainless steel (ε = 2,5 %, T = 450°C) are illustrated in Fig. 5; the numbers specify the initial yield stresses of the corresponding SEs $\tilde{\sigma}_s = Ezr_b$; r_b is the ultimate strength of the material.

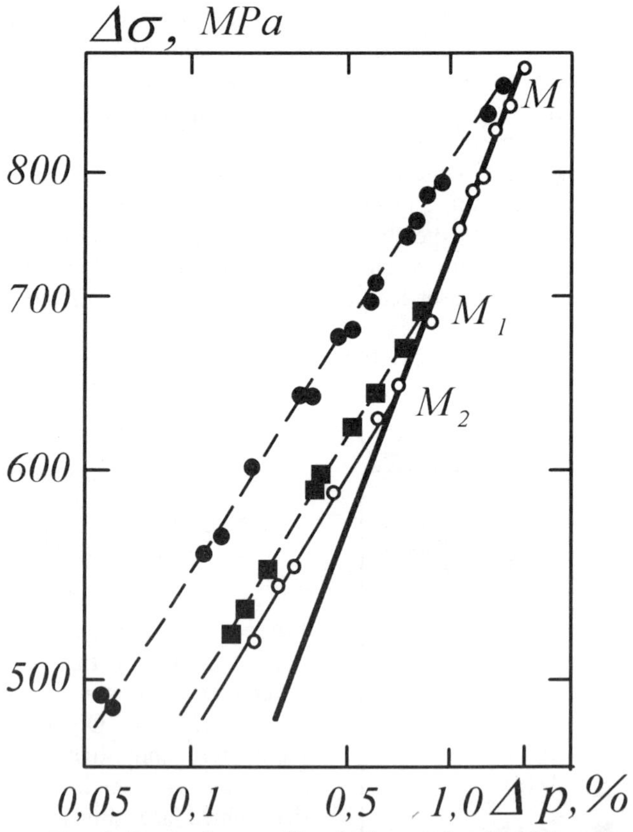

Fig. 6. Dependence of hardening on loading history

Reversibility of isotropic hardening is important, especially at elevated temperatures. This property can be modelled by introducing an additional member into eqn(19) after it is differentiated:

$$d\tilde{\alpha} = \tilde{\psi}'(\tilde{\alpha}) d\tilde{\lambda} - R(\tilde{\alpha}) dt \tag{21}$$

Thus, deformation is accompanied by two processes which exert antithetical influence: hardening which is caused by inelastic strain, and time-dependent recovery. Stabilization of the state ($\dot{\tilde{\alpha}} = 0$) can be interpreted as a kind of "dynamic equilibrium":

$$\psi'(\tilde{\alpha})\dot{\tilde{\lambda}} - R(\tilde{\alpha}) = 0 \tag{22}$$

Hence it follows that

$$\dot{\tilde{\lambda}} = R(\tilde{\alpha})/\psi'(\tilde{\alpha}) \tag{23}$$

This signifies that the equilibrium state can be disturbed by change of strain rate; usually its fall leads to discovery (softening) while its increase leads to additional hardening. In this way we can analyse the expected influence of a hold-time period inclusion or of temperature rise.

As the experimental results show, the parameters in the stabilized stress-strain diagrams depend on loading prehistory (Fig. 6, GOKHFELD et al. [1979]). The stabilization process was studied at stage-by-stage increasing or decreasing strain amplitudes. As was found, the second case is characterized by somewhat higher level of hardening. It should be noted that the effect takes place even after a few (at least one) overloading cycles.

Under non-proportional loading conditions, the level of hardening depends on the type of strain trajectory. Apparently LAMBA & SIDEBOTTOM [1978] were the first who studied this phenomenon. Further it was shown that the latter embraces both plasticity and creep (MURAKAMI, KAWAI & YAMADA [1990]).

Some results of investigations in this field (APAJCHEV, IVANOV & PONKIN [1991]) are illustrated in Fig. 7 and 8.

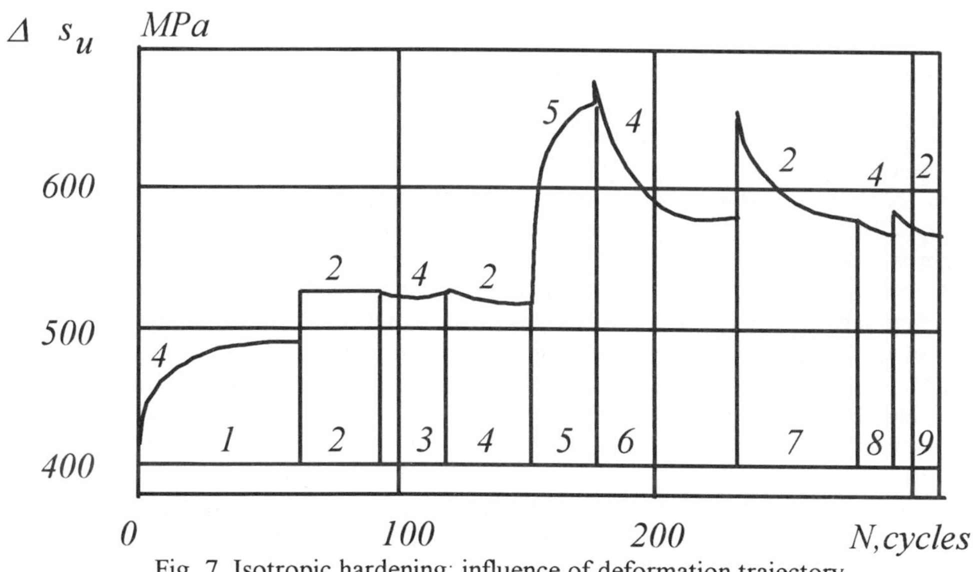

Fig. 7. Isotropic hardening: influence of deformation trajectory

The first one shows the variation of the stress amplitude intensity at successive stages characterized by constant intensity of strain amplitude, but different strain trajectories (types of trajectories are marked by figures besides the curve, in both figures.

To describe the considered peculiarities a special version of the SM was derived. The latter differs from the BM by introduction of modified memory surfaces of any SE in inelastic strain space. The center of any surface may shift, and its dimension (radius) may change up or down, depending on strain intensity and deformation trajectory. Hardening as well as recovery of any SE is modelled by the equation

$$\tilde{\sigma}_s = \tilde{\sigma}_{so} + \tilde{L}\tilde{\rho} \qquad (24)$$

Here $\tilde{\sigma}_s$, $\tilde{\sigma}_{so}$ are current and initial yield stress of any SE, respectively (they are equal to radii of the corresponding yield surfaces in the stress space); $\tilde{\rho}$ is the radius of its memory space, $\tilde{L}$ is a constant.

The considered version of the SM can be generalized for rheonomic materials, by making use of surfaces of equal creep potential instead of the yield surfaces.

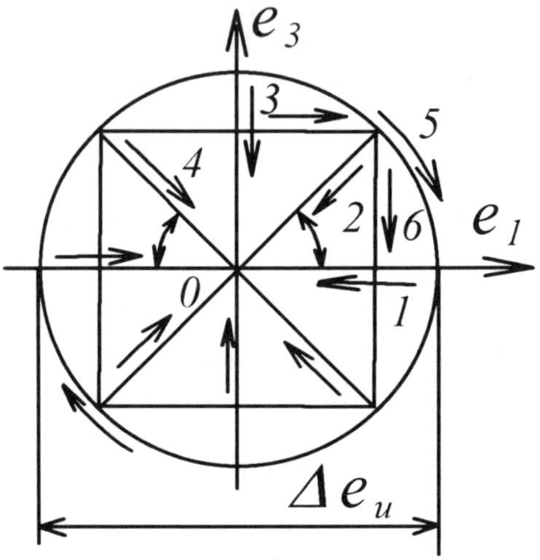

Fig. 8. Types of deformation trajectories

STRUCTURAL MODEL FOR ANISOTROPIC SOLIDS

Some possibilities which follow from the concept of micro-inhomogeneity of solids may be considered perhaps as unique: this concept leads to a unified description of both types of anisotropy (physical and deformation). To model physical anisotropy of some kind any SE should be endowed with the same property. However, deformation anisotropy is a property which is due to common deformation of inhomogeneous SEs, i.e. is modelled automatically. As usually, the SEs are assumed to be perfectly viscous; their strain rates are determined by corresponding creep potentials. In the general case of anisotropy, surfaces of equal potential are represented by ellipsoids in stress space.

A suitable method for constructing the ellipsoids has been developed. It consists in determining of an orthogonal tensor basis in the second rank tensor's space under condition that symmetry of the basis corresponds to that of deformation properties of the considered material.

The mentioned linear tensor space is defined by the following condition: there should be a one-to-one correspondance between any element $\mathbf{x}$ of tensor space and a symmetrical second rank tensor X_{ij} of the physical space; it is required that $\mathbf{x} \cdot \mathbf{y} = x_{ij} y_{ij}$. In particular, Hooke's law in the tensor space is represented as $\boldsymbol{\sigma} = \mathbf{C} \cdot \boldsymbol{\varepsilon}$ with a symmetrical second rank tensor of the elasticity constants $\mathbf{C}$. The tensor basis which corresponds to any type of symmetry can be found without essential difficulties by eqn $C_{ijhk} = k_m A^{(m)}_{ijhk}$ $(\mathbf{C} = k_m \mathbf{A}_m)$. For instance, for an isotropic solid (spherical symmetry) we have two such tensors: $A^{(1)}_{ijkl} = \delta_{ij}\delta_{kl}$ and $A^{(2)}_{ijkl} = \delta_{ik}\delta_{jl} + \delta_{kj}\delta_{il}$ (δ_{ij} are Kronecker's symbols). In addition to the space symmetry, the tensor should satisfy the conditions of tensor symmetry which are characteristic for the elastic constants: $C_{ijkl} = C_{jikl} = C_{ijlk} = C_{klij}$.

It should be noted that the index form is employed here only to discriminate the physical tensors from the tensors belonging to tensor space. In reality, the invariant, i.e. the without-index form of presentation has an exceptional importance in the problem of a tensor basis determination. This is just the approach by which the necessary basic tensors can be easily found.

For cubic symmetry (such type of symmetry is characteristic of steels manufactured by the oriented crystallization technology) there are three basic tensors: two of them remain such as in the previous case while the additional tensor is $A^{(3)}_{ijkl} = \delta_{ij}\delta_{kl}\delta_{il}$; the latter has the indicated coordinates when the space basis is related to the crystallographic axes (it should be remembered that coordinates of the first two tensors are independent of the adopted orthogonal- standardized space basis).

For transversal isotropy there are five different tensors which have the necessary symmetry, and so on.

But this is only the first step in the necessary direction. For any type of symmetry an orthogonal basis should be constructed which satisfies the following requirement: the sum of the basic tensors is to be equal to tensor $A^{(2)}_{ijkl}/2$; the latter will be denoted U_{ijkl} and it has the following properties: $U_{ijkl}\sigma_{kl} = \sigma_{ij}$; $\sigma_{ij}U_{ijkl} = \sigma_{kl}$. Such basis may be defined as the orthogonal-standardized one, it satisfies the condition $B^{(m)}_{ijkl}B^{(n)}_{klrs} = \delta_{mn}B^{(m)}_{ijrs}$. For isotropic solid , $\boldsymbol{B^{(1)}} = \boldsymbol{A^{(1)}}/3$, $\boldsymbol{B^{(2)}} = \boldsymbol{U} - \boldsymbol{B^{(1)}}$. By these tensors the spherical part of the stress tensor ($B^{(1)}_{ijkl}\sigma_{kl} = \frac{1}{3}\sigma_{kk}\delta_{ij}$) and its deviatoric part $s_{ij} = B^{(2)}_{ijkl}\sigma_{kl}$ are separated. The corresponding tensors of tensor space $\mathbf{B}_1$, $\mathbf{B}_2$ are the projecting ones: $\mathbf{B}_1 \cdot \boldsymbol{\sigma} = \boldsymbol{\sigma}_o$, $\mathbf{B}_2 \cdot \boldsymbol{\sigma} = \mathbf{s}$, $\boldsymbol{\sigma} = \boldsymbol{\sigma}_o + \mathbf{s}$, $\boldsymbol{\sigma}_o \cdot \mathbf{s} = 0$. Moreover, vector $\boldsymbol{\sigma}_0$ belongs to the one-dimensional space, while vector $\mathbf{s}$ - to the five-dimensional one. Similarly, it can be written for strains: $\boldsymbol{\varepsilon} = \boldsymbol{\varepsilon}_o + \mathbf{e}$, $\boldsymbol{\varepsilon}_o = \mathbf{B}_1 \cdot \boldsymbol{\varepsilon}$, $\mathbf{e} = \mathbf{B}_2 \cdot \boldsymbol{\varepsilon}$. Then Hooke's law obtains its simplest form:

$$\boldsymbol{\sigma}_o = k_1 \boldsymbol{\varepsilon}_o, \quad \mathbf{s} = k_2 \mathbf{e} \tag{25}$$

(this follows from decomposition: $\mathbf{C} = k_1\mathbf{B}_1 + k_2\mathbf{B}_2$).

Thus, the tensor space is divided into two subspaces which are, respectively, one- and five- dimensional. The elastic properties in each subspace are characterized by isotropy. This applies also to the rheological properties: the surfaces of equal creep potentials are represented by ellipsoids which have the only specific axis which is in the spherical sub-

space. Usually, this ellipsoid degenerates into a cylinder; in this case its dimension along the specific axis tends to infinity. A cross-section of the six-dimensional cylinder by a hyperplane which is orthogonal to this direction presents a five-dimensional sphere.
For cubic symmetry, tensor $B^{(1)}_{ijkl}$ proves to be the same as for isotropic solid ($A_{ijkl}/3$); but $B^{(2)}_{ijkl} = A^{(3)}_{ijkl} - A^{(1)}_{ijkl}/3$ when $B^{(3)}_{ijkl} = U_{ijkl} - A^{(3)}_{ijkl}$. Thus, each of the stress and strain tensors is divided into three orthogonal components ($\boldsymbol{\sigma}_o = \mathbf{B}_1 \cdot \boldsymbol{\sigma}$, $\mathbf{s}_1 = \mathbf{B}_2 \cdot \boldsymbol{\sigma}$, $\mathbf{s}_2 = \mathbf{B}_3 \cdot \boldsymbol{\sigma}$; similarly $\boldsymbol{\varepsilon}_o$, $\boldsymbol{\varepsilon}_1$, $\boldsymbol{\varepsilon}_2$, $\boldsymbol{\varepsilon} = \boldsymbol{\varepsilon}_o + \boldsymbol{\varepsilon}_1 + \boldsymbol{\varepsilon}_2$) which are respectively situated in one-dimensional ($\boldsymbol{\sigma}_o$, $\boldsymbol{\varepsilon}_o$), two-dimensional ($\mathbf{s}_1$, $\mathbf{e}_1$) and three-dimensional ($\mathbf{s}_2$, $\mathbf{e}_2$) subspaces. Now, Hooke's law includes three independent relationships which include, respectively, three elastic moduli:

$$\boldsymbol{\sigma}_o = k_1 \boldsymbol{\varepsilon}_o, \quad \mathbf{s}_1 = k_2 \mathbf{e}_1, \quad \mathbf{s}_2 = k_3 \mathbf{e}_2 \tag{26}$$

Ellipsoids of equal creep potentials are presented by circles in the two-dimensional subspace ($\mathbf{s}_1$, $\mathbf{e}_1$) and by three-dimensional spheres ($\mathbf{s}_2$, $\mathbf{e}_2$). The surface of equal creep potential can be introduced by a function $\psi(\sigma_{ef})$ where

$$\sigma_{ef}^2 = \boldsymbol{\sigma} \cdot \mathbf{B}_2 \cdot \boldsymbol{\sigma} + \varphi(\boldsymbol{\sigma} \cdot \mathbf{B}_3 \cdot \boldsymbol{\sigma}) \boldsymbol{\sigma} \cdot \mathbf{B}_3 \cdot \boldsymbol{\sigma} \tag{27}$$

or in physical space

$$\sigma_u^2 = B^{(2)}_{ijkl} \sigma_{ij} \sigma_{kl} + \varphi\left(B^{(3)}_{ijkl} \sigma_{ij} \sigma_{kl}\right) B^{(3)}_{ijkl} \sigma_{ij} \sigma_{kl} \tag{28}$$

The function (28) maintains the necessary symmetry properties, and this corresponds to the assumption that the spherical component of the stress tensor does not influence creep.The case of transversal isotropy (cylindrical symmetry) corresponds to five tensor components reflecting different types of symmetry; it is characterized by definite difficulties associated with construction of the orthogonal-standardized tensor basis. Two basic tensors can be found easily: they determine two orthogonal two-dimensional subspaces in the six-dimensional tensor space. From the remaining three tensors of the initial basis one more two-dimensional subspace is determined. We cannot discuss here the problem in detail, and limit ourselves by the following remark: in the latter subspace the principal directions can be determined by which elastic properties are characterized. In other words, some angular constant, specific for any material, can be found and by this way the necessary orthogonal-standardized pair of basic tensors can be distinguished among an infinite set of such pairs.
Thus, the five elastic constants include four elastic moduli (for the corresponding subspace of six-dimensional tensor space: two of them are one-dimensional, the other two two-dimensional) and a dimensionless angular constant . Then, Hooke's law takes the form:

$$\sigma^{(m)}_{ij} = k_m \varepsilon^{(m)}_{ij} \tag{29}$$

where

$$\sigma_{ij}^{(m)} = B_{ijkl}^{(m)}\sigma_{kl}, \quad \varepsilon_{ij}^{(m)} = B_{ijkl}^{(m)}\varepsilon_{kl}, \quad m = 1...4;$$
$$B_{ijkl}^{(1)} = B_{ijkl}^{(1)}(\chi), \quad B_{ijkl}^{(2)} = B_{ijkl}^{(2)}(\chi) \tag{30}$$

In the latter case, determination of surfaces of equal creep potential encounters additional difficulties. To meet them, some new concepts are necessary. Corresponding versions of such concepts which are in accordance with the general principle of physical consistency and simplicity can be suggested. Unfortunately, there are no experimental results at our disposal yet, to guide the choice of the most rational concept.
By development of the approach, the constitutive equations of the SM intended to model rheological properties of anisotropic bodies can be written in the following form:

$$\varepsilon_{ij} = \tilde{\varepsilon}_{ij} = r_{ij} + p_{ij} = \tilde{r}_{ij} + \tilde{p}_{ij} \tag{31}$$

$$\dot{\tilde{p}}_{ij} = \frac{R_{ijkl}\tilde{\sigma}_{kl}}{\tilde{\sigma}_u}\Phi\left(\frac{\tilde{\sigma}_e}{z}, T\right) \tag{32}$$

$$\tilde{\sigma}_e^2 = \tilde{\sigma}_{ij}\tilde{\sigma}_{kl}R_{ijkl}, \quad R_{ijkl} = \sum_m \varphi_m^2 B_{ijkl}^{(m)}, \quad \varphi_m = \varphi_m\left(\tilde{\sigma}_{ij}\tilde{\sigma}_{kl}B_{ijkl}^{(m)}\right) \tag{33}$$

$$\tilde{r}_{ij} = D_{ijkl}\tilde{\sigma}_{kl}; \quad D_{ijkl} = \sum_m k_m^{-1}B_{ijkl}^{(m)} \tag{34}$$

$$\sigma_{ij} = \left\langle \tilde{\sigma}_{ij} \right\rangle \equiv \int_0^\infty \tilde{\sigma}_{ij} y(z)dz \tag{35}$$

The functions which are determinative and should be found by tests for any material, are: the rheological function $\Phi(r, T)$; the inhomogeneity function which should be presented as a distribution density function for the SE parameters; characteristics of elasticity k_m and of rheological anisotropy φ_m . For sake of simplicity the latter characteristics can be assumed to be constant (if this does not contradict experimental data); by this way, similarity of the ellipsoids which corresponds to different values of creep potential $\psi(\sigma_{ef})$ is postulated. Then, to identify the model it is necessary to obtain by tests the RF and stress-strain diagrams at the corresponding number of different stress states (m=1,2,...,n). Equations of the BM can be presented as the simplest particular case corresponding to isotropic solids (φ_m=1).

GRAPHITE AS ONE OF LOW -PLASTIC MACRO-INHOMOGENEOUS MATERIALS

Artificial structural graphites are used in the nuclear power industry, metallurgy, and other branches of industry. This is stipulated by their thermal and radiation stability, high-temperature strength and some additional useful properties. Adequate mathematical description of deformation behavior and fracture of graphite is a rather complicated problem.

This is conditioned by a complex of specific mechanical and physical properties such as anisotropy, considerable difference between tension and compression resistance, variation of cyclic stress-strain curves from the very beginning of the cyclic deformation process, essential influence of progressive micro-failures on deformation characteristics, and other peculiarities. Formation of micro-cracks can be revealed by the acoustic emission method; the latter confirms that the process takes place from the very beginning of loading.

To model these properties in the framework of the models of structural type, rather essential complications are required. It should be mentioned that the first attempt to construct a version of the structural model for graphite is due to JENKINS [1965].

The SEs are divided into two groups: the viscous subelements (SEV) and the brittle perfectly elastic (SEB). Behavior of the SEV is determined by the RF which is presented here (for the one-dimensional case) in the form

$$\dot{p}_k = \Phi\left(r_k / \left(\kappa_k \varphi(r_k) z_k\right), T\right) \tag{36}$$

Here $\varphi(r_k)$ is a function by which difference of resistance against tension and compression is allowed:

$$\varphi(x) = H(x) + \psi H(-x) \tag{37}$$

where ψ is a constant of the material, H - the Heavyside function ($H(x) = 0$ at $x < 0$, $H(x) = 1$ at $x \geq 0$).

The parameter κ_k models isotropic softening of the SEs (SEV group); this type of behavior is the most typical for graphite under cyclic loading. Dependence of the parameter on loading prehistory is modelled like isotropic hardening (see (19), (20)), and this is in agreement with the experimental data obtained for graphites.

$$\kappa_k = \exp\left(-B\lambda_k / z_k\right) \tag{38}$$

$$\lambda_k = \int_0^t \left|\dot{p}_k\right| dt \tag{39}$$

Micro-crack development and "play" of the forming clearances in the group of brittle SEs (SEB) at cyclic loading leads to specific variations of the hysteresis loop shape (crescent loops and formation of "beaks") and in "loosening", which in turn leads to progressive decreasing of resistance against deformation.

The SEB group can be presented by one SE of this type which has corresponding relative weight g_e. Thus, instead of eqn (9) we have

$$\sum_{k=1}^{n-1} g_k + g_e = 1 \tag{40}$$

Accordingly, initial stress-strain curves at tension and compression halfcycles are:

$$\sigma = f(\varepsilon) + g_e f_e(\varepsilon)$$
$$\sigma = \psi f(\varepsilon/\psi) + g_e f_e(\varepsilon) \tag{41}$$

The function f_e is determined by the perfectly elastic SE (SEB) which undergoes progressive failure. As a first approximation it is assumed that this function can be presented by a two-member broken line. Its first section (at tension) is determined by decreasing (cycle-by-cycle) modulus accordingly to the crack evolution:

$$E_F = E(1 - D \ln N) \tag{42}$$

Here E is the initial value of elastic modulus, N - number of loading cycles, D - characteristic of a material (as the experiments show its dependence on strain amplitude is rather weak within moderate values of strain amplitudes).
The clearances caused by cracks grow at tension, but decrease at compression, and within the strain interval.

$$\varepsilon_{\min} \le \varepsilon \le \chi\varepsilon_{\min}, \qquad \varepsilon_{\min} < 0 \tag{43}$$

(where approximately $\chi \cong 0.6$) the modulus returns to its initial value E.
It is worth noting that stress-strain diagrams of graphite are characterized by rather small changes of tangent modulus; accordingly, relative weight of SEB proves to be equal to 0.25 - 0.5 for graphites, while its value is only 0.01 - 0.05 for structural alloys.
Development of micro-failures, which can be interpreted as damage accumulation (the "loosening" effect), leads to gradual fall of mechanical properties. This can be illustrated by successive variation of stress-strain diagrams and creep curves obtained by (GOKHFELD et al. [1989]).
Fig. 9 shows the dependence of cyclic creep rates on durability of hold-time periods at two levels of maximum stress ($R = -\infty$, T=700° C). We emphasize that at equal total time (e.g. determined by the line AA) the more the accumulated strain, the less is the hold-time period and, accordingly, the more is the number of loading cycles which occur during the corresponding time. Thus, unlike structural alloys, creep rates in graphite depends on loading frequency, and can reach magnitudes which essentially exceed those at statical loading at equal stress levels.
It is known that the statical strength of graphite increases with temperature. However, the opposite tendency has been revealed by tests at cyclic loading conditions (Fig. 10). Apparently this effect is due to interaction between fatigue (which is responsible for the "loosening" of graphite) and creep. Final fracture of a specimen at the stress-cycle loading occurs when some limit value of strain is reached. Experiments show that the magnitude of the latter at any temperature can be assumed as constant (at least, this is true up to the spread in experimental data) independently of the type of loading:

$$\varepsilon_{\max} = \varepsilon_F(\operatorname{sign} \varepsilon) \tag{44}$$

Here ε_F is the ultimate strain at given temperature under statical loading. Thus, durability (number of cycles to failure) can be predicted by employing this rheological model (Fig. 10, solid lines). Experimental results are marked by the dots.

Dependence of temperature expansion on load-temperature history is one of the characteristic features of graphite. To model this phenomenon it may be assumed that the SEs have different thermal expansion coefficients . Hence it follows that temperature changes lead to thermal micro-stresses which interact with those which are caused by mechanical loads. The corresponding analysis is not considered here as its results practically coincide with those which are discussed in the separate paper of CHERNIAVSKY published in this volume. It is worth noting that the version of the model suggested in the cited paper presents an attempt to connect directly physical characteristics of the components of graphite (grains, binder) with those of the conglomerate. By contrast, the model which is here considered is more formalized, its identification is completely based on data from macro-experiments.

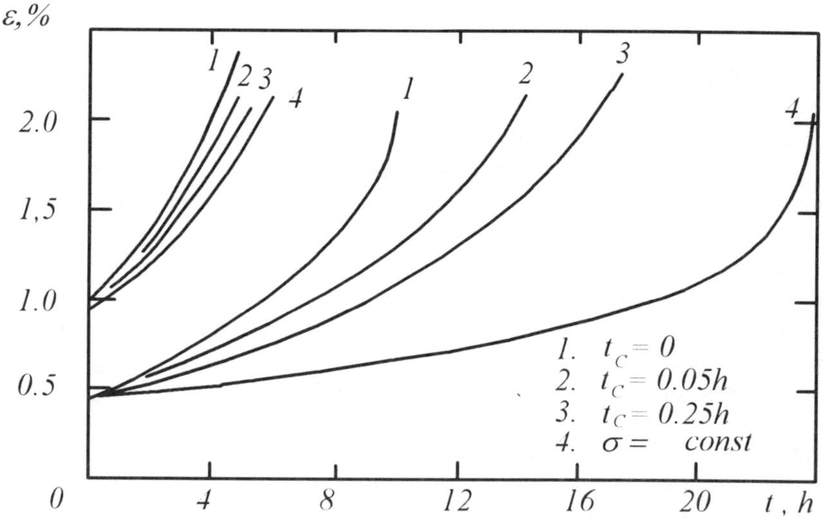

Fig. 9. On influence of hold-time periods upon creep rate in graphite

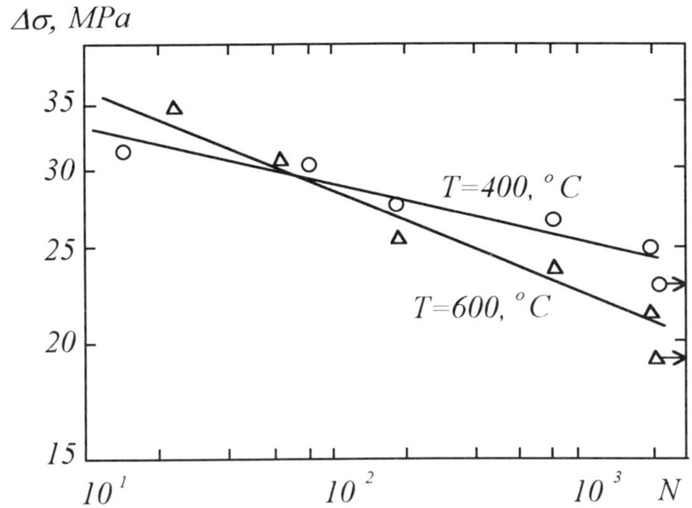

Fig. 10. Temperature dependence of low-cycle failure of graphite

Generalization of the suggested one-dimensional model to arbitrary stress states encounter some additional difficulties related to physical anisotropy of graphite. The latter is caused by anisotropy of grains and by ordering in their orientation due to the manufacturing process. The resulting properties are often close to transverse isotropy and can be characterized by a specific direction determined by unit vector **a** (the latter is oriented along the third Cartesian axis). For linear elasticity of a transversely isotropic solid

$$\sigma_{ij} = C_{ijkl}\varepsilon_{kl} \tag{45}$$

the tensor of elasticity constants C_{ijkl} can be decomposed into five linearly independent tensors :

$$C_{ijkl} = \mu_{\alpha} C^{\alpha}_{ijkl}, \qquad \alpha=1,\ldots 5 \tag{46}$$

In particular, they may have the following form

$$\left.\begin{aligned} C^{1}_{ijkl} &= A^{(1)}_{ijkl} = \delta_{ij}\delta_{kl} \\ C^{2}_{ijkl} &= U_{ijkl} = \left(\delta_{kj}\delta_{il} + \delta_{lj}\delta_{ki}\right)/2 \end{aligned}\right\} i,j,k,l = 1..3 \tag{47}$$

$$\left.\begin{aligned} C^{3}_{ijkl} &= \tilde{A}_{ijkl} = \delta_{ij}\delta_{kl} \\ C^{4}_{ijkl} &= \tilde{U}_{ijkl} = \left(\delta_{kj}\delta_{il} + \delta_{lj}\delta_{ki}\right)/2 \end{aligned}\right\} i,j,k,l = 1,2$$

$$C^{5}_{ijkl} = F_{ijkl} = \delta_{i3}\delta_{j3}\delta_{k3}\delta_{l3}$$

As it has been found, the orthogonal-standardized basis consists of four tensors only:

$$C_{ijkl} = k_\beta B^\beta_{ijkl} \tag{48}$$
$$\mathbf{B}^1 = \left(\tilde{\mathbf{A}}/2 + k\tilde{\mathbf{K}} + 2k^2\mathbf{F}\right)/\left(1+2k^2\right)$$
$$\mathbf{B}^2 = \left(k^2\tilde{\mathbf{A}} - k\tilde{\mathbf{K}} + \mathbf{F}\right)/\left(1+2k^2\right)$$
$$\mathbf{B}^3 = \tilde{\mathbf{U}} - \tilde{\mathbf{A}}/2$$
$$\mathbf{B}^4 = \mathbf{U} - \tilde{\mathbf{U}} - \mathbf{F}$$

where

$$\tilde{\mathbf{K}} = \mathbf{A}^{(1)} - \tilde{\mathbf{A}} - \mathbf{B}.$$

There are five elastic constants; four of them are elastic moduli (k_β) while the fifth is a dimensionless constant (k) by which two tensors of the basis are determined. Orthogonal normality signifies the following relation:

$$D^\alpha_{ijkl} D^\beta_{lkmn} = D^\alpha_{ijkl} D^\beta_{lkmn} = D^\beta_{ijkl} D^\alpha_{lkmn} = \delta_{\alpha\beta} D^\alpha_{ijmn}\ D^\alpha_{ijkl} D^\beta_{lkmn} = D^\beta_{ijkl} D^\alpha_{lkmn} = \delta_{\alpha\beta} D^\alpha_{ijmn}.$$

Besides that, sum of all the basic tensors is equal to tensor $\mathbf{U}$ which has the property: $U_{klij}\sigma_{jl} = \sigma_{ij}U_{jikl} = \sigma_{kl}$. After tensors $\boldsymbol{\sigma}, \boldsymbol{\varepsilon}$ are partitioned, the orthogonal components $\boldsymbol{\sigma}^\alpha, \boldsymbol{\varepsilon}^\alpha$

$$\sigma^\alpha_{ij} = D^\alpha_{ijkl}\sigma_{kl}, \qquad \varepsilon^\alpha_{ij} = D^\alpha_{ijkl}\varepsilon_{kl} \tag{49}$$

can be connected by Hooke's law in its simplest form:

$$\sigma^\alpha_{ij} = K_\alpha \varepsilon^\alpha_{ij}, \quad \alpha=1..4,\ i,j=1..3 \tag{50}$$

To describe plastic behavior of solids the orthogonal-standardized tensor basis should be constructed, and then the corresponding tensor function (yield surface) determined. If incompressibility of materials at inelastic deformation is taken into account (their "loosening" is to be modelled by the SEB) it follows that $k = 1/2$ and hence $\mathbf{D}^1 = \mathbf{A}/3$, $\mathbf{D}^2 = \tilde{\mathbf{A}}/2 + \mathbf{B} - \mathbf{A}/3$, $\varepsilon^1_{ij} = D^1_{ijhk}\varepsilon_{kl} = \theta/3$ where θ is elastic cubic strain ($p^1_{ij} = 0$).
Experiments show that resistance of graphite against shears caused by stress components σ^3_{ij} (corresponding surface elements are parallel and orthogonal to vector $\mathbf{a}$) and σ^4_{ij} (mutually orthogonal surface elements are parallel to this vector) are practically equal. Thus, the mentioned components may be united:

$$\sigma^\tau_{ij} = \sigma^3_{ij} + \sigma^4_{ij}, \qquad \varepsilon^\tau_{ij} = \varepsilon^3_{ij} + \varepsilon^4_{ij} \tag{51}$$

The simplest yield surface which can be adopted for any SEV is:

$$\sigma_e^k = \sqrt{\sigma_{ij}^k R_{ijkl} \sigma_{kl}^k / 2} = \sigma_s$$
$$\mathbf{R} = \left(\tilde{\mathbf{A}}/2 + \mathbf{B} - \mathbf{A}/3\right) / \lambda_k^2 + \mathbf{U} - \mathbf{B} - \tilde{\mathbf{A}}/2 \tag{52}$$

Here σ_s is the yield stress; λ_k^2 indicates how many times yield stress increases if deviator $\boldsymbol{\sigma}^\tau$ is substituted by $\boldsymbol{\sigma}^2$.
Difference in resistance against tension and compression can be reflected by a corresponding shift of the yield surfaces:

$$\sigma_e^k = \sqrt{\sigma_{ij}^k R_{ijkl} \sigma_{kl}^k / 2} - h_{mn} \sigma_{mn}^k$$
$$h_{mn} = h_k \left(\delta_{mn} - 3\delta_{m3}\delta_{n3}\right) \tag{53}$$
$$\sigma_e^k - \sigma_s^k\left(\sigma_k^o\right) \leq 0 \tag{54}$$

As a first approximation, characteristics of anisotropy λ_k and relative shifts of the individual SE yield surfaces h_k might be equal. For isotropic solids $\lambda_k = h_k = 1$.
Generalization for rheonomic materials is taken to be straightforward: it is assumed that there exist creep potential surfaces characterized by a scalar function $\psi_k\left(\sigma_{ij}^k\right)$, and creep rate is determined by its gradient. This leads to the following equation:

$$\dot{p}_{ij}^k = \Phi\left(\sigma_e^k / z_k\right)\left(\frac{R_{ijkl}}{2\sigma_e^k} - h_{ij}^k\right) \tag{55}$$

where $\Phi(x) = d\psi(x)/dx$.

THERMOSOFTENING PLASTIC MATERIALS. SHAPE MEMORY PHENOMENON

The main regularities of deformation behavior of polymers can be modelled quite satisfactory by the BM after the proper identification procedure is carried out. However, simulation of shape memory effects specific for thermoplastic materials requires revision of some concepts in the BM. In particular, the hypothetical supposition concerning with similarity of the RFs of all the SEs with respect to temperature (see eqn. (1)) must be withdrawn.
To model the mentioned peculiarities of thermoplastic materials, the functions by which creep rates of the SEs are determined may be presented in the following form:

$$\dot{p}_{kl} = A_k(T)\Phi(\sigma_{kl}/z_l), \qquad A_k(T) = A(T - T_{\nu k}) \tag{56}$$

where $T_{\nu k}$ is individual vitrification temperature of any SE which bounds its purely elastic behavior. Thus, inhomogeneity is characterized by the two independent distributions: $T_{\nu k}$ and z_l, where k, l are individual numbers identifying any SE. The corresponding identification procedure has been developed.
Here we limit ourselves to qualitative description of the shape memory effects using the proper simplification of the suggested formulation:

$$\dot{p}_k = 0 \quad \text{at} \quad T < T_{\nu k}, \quad \dot{p}_k = 0/0 \quad \text{at} \quad T > T_{\nu k} \tag{57}$$

In accordance with the latter, any SE behaves as an elastic one until temperature is below its vitrification temperature, while at $T > T_{\nu k}$ its resistance against deformation vanishes.
The elastic modulus of any SE is supposed to be constant, however, as it follows from eqn(57), the ensemble elastic modulus proves to be temperature dependent: this is caused by variation of relative weight of the SEs, which remain their elastic properties.
Thus, for increasing temperature, the deformation process will follow the corresponding thermomechanical surface (TMS) determined by equation $\sigma = E(T)\cdot\varepsilon$ independently of previous loading history, even if it includes unloadings. However, if the next loading stage is preceded by cooling (now inequality $T < T_{\nu k}$ becomes valid for some additional part of the SEs) deformation will follow another TMS:

$$\sigma_* = E(T)\varepsilon_*, \quad \sigma_* = \sigma - \sigma_\nu, \quad \varepsilon_* = \varepsilon - \varepsilon_\nu \tag{58}$$

(the notations remain the same as in previous parts of the paper; the index ν denotes the last turning point, which corresponds to the initial moment of cooling). The unloading leads to residual strain and related self-equilibrated stresses which can be easily determined.
Shape restoration is illustrated by Fig. 11. Let a bar characterized by some curvature in its initial state (Fig. 11a, dash line) be subjected to a bending load by which the sign of its curvature is changed and heated (Fig. 11a). At the next stage of the process, the bar is cooled at constant external load and then unloaded, i.e. its strains which are due to difference between the two mentioned TMS become frozen (Fig.11b). Now, repeated heating up to the same temperature will lead to vanishing of the residual stresses as the corresponding SEs loss their resistance to deformation. Then the bar is restored to its initial (bent, i.e. natural) shape.

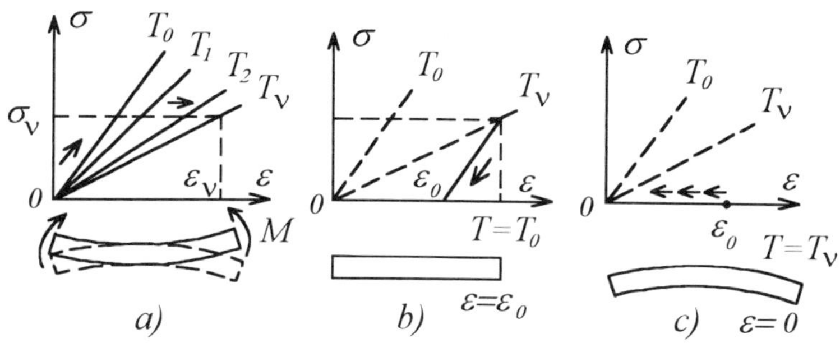

Fig. 11. On the shape memory effect

RHEOLOGY OF STRUCTURES

These SM versions illustrate "flexibility" of the approach with respect to simulation of specific rheological properties. However, the most remarkable peculiarity of the SM is related to rheology of structures. If any volume element of a solid is treated as a structure consisting of SEs, any structure made of real material can be considered as some "superstructure" which consists of such SEs. The general deformation analysis which becomes possible in this way, illuminates the main regularities of rheological processes in structures subject to non-stationary (e.g. cyclic) loading. In its turn, the analysis has lead to a set of macro-models, design methods and procedures which can be employed for lifetime prediction of structures.

This analysis is realized by making use of vector representations (only the linear formulation of the problem is considered here): as it is known, vector properties are independent of space dimension. On the other hand, perfect viscosity of the SEs leads to use of surfaces of equal creep potentials (reduced to yield surfaces for perfect plasticity) and this permits one to use the kinematical analogies of Prager's type.

As it is impossible to dwell here on the analysis procedure (GOKHFELD & SADAKOV [1984]) let us limit ourselves by some peculiarities which are most essential for applications.

1) The adopted interpretation of memory effects in solids and structures suggests an idea which is similar to the Ritz method: to simulate deformation behavior of structures, models characterized by low degree of freedom (i.e. based on proper kinematical hypothesis) can be used.

2) Response of any structure subjected to cyclic loading can be resolved into "cyclical" and "statical" parts which often can be determined separately, in succession (here some analogy with the limit analysis approach is seen).

3) Asymptotic stabilization of deformation behavior is the main regularity for structures subjected to regular loading. Hence the problem of the direct determination of steady cyclic state (SCS) follows. If the BM is used, SCS is reduced to alternating inelasticity superposed on constant accumulated strain, independently on cyclic loading conditions.
4) At local inelastic zones (e.g. stress concentrations) external actions are represented by the elastic stress in the zone. The iteration procedure embraces the mentioned zone only. Hence appear some new possibilities in macro-modeling; the effects are rather impressive, especially in the case of non-isothermal loading. The approach can be considered as a wide generalization of Neuber's method.
The general matrix model of a structure suggests the following four ways of realization of the corresponding calculation procedures.
a) Decreasing the number of degree of freedom:
- by combination of the ideas of FEM and the Ritz method (this approach has been termed "the vector model");
- if the actual number of degrees of freedom is not evident, it can be revealed by direct FEM application to the first loading cycle; reconstruction excludes the less significant degrees of freedom ("modification of FEM").
b) Decreasing of the number of representative points in a structure:
- the case of local inelastic zones (see above);
- use of generalized variables ("the generalized models of structures").
c) Decreasing the number of representative time-moments . This is the principal problem, as the traditional step-by-step procedure is often too time-consuming. Now there are possibilities for saving time.
i) The extrapolation method. Since stabilization processes in structures can be approximated by a piecewise-exponential law, the following procedure can be used: after the computations are performed for three successive cycles, and parameters of the exponent have been determined, extrapolation to about 40 - 50 cycles may be carried out. Then the procedure is repeated until the stress-strain state is stabilized to the given accuracy. Compared to the usual step-by-step procedures, this takes an order of magnitude less computing time.
ii) The SCS direct determination. A whole set of methods has been developed for various types of loading conditions (GOKHFELD & SADAKOV, in this volume). Some actual situations which have not been considered in the cited paper should be mentioned:
- inelastic bending of a rotating specimen (the "running wave"); there is a phase lag of the plane of bending relatively the bending moment plane (MARTYNENKO [1991]);
- two-frequency loading: apparently the problem cannot be solved by traditional methods (the results have not been published yet);
- proportional (or close to proportional) cyclic loading: the cyclic part of the response of a structure can be determined directly by using the generalized similarity principle (GSP) formulated in generalized variables (GOKHFELD & SADAKOV [1991b]). The GSP can be also used for specific loading conditions when cyclic loading is superimposed on a constant stress (or strain) which is not collinear to the cyclic load. The analysis is reduced to a state equation; to solve the problem, two functions only should be known: the RF of

a material and the deformation diagram of the considered structural member formulated in generalized variables.

As the loading diagrams obtained for various structural members under a certain loading program may not differ very much, a unique possibility appears: universal nomographs can be constructed, any of which is valid for a group of structural members under some fixed loading program (the RFs of the materials should differ only by a multiplier; in particular, they must have equal indices if the RFs are described by Norton's law).

Finally, it should be noted that employment of GSP leads to some new possibilities in the identification problem: now the basic experiments may be carried out on the structural components directly, i.e. special specimens are not required. This allows to evaluate the influence of the production processes on the rheological properties of materials.

LOW-CYCLE FAILURE

Failure of a structure subjected to low-cycle action of mechanical forces and temperature fields may be caused either by initiation and growth of cracks or by inadmissible geometry changes resulting from cyclic strain accumulation. The first situation often occurs in machinery. In this case lifetime evaluation needs some damage accumulation and crack propagation models.

Investigations concerning with LCF originate in the well known works of Coffin and Manson. However, to model the influence of different kinds of loading cycles (e.g. non-isothermal cycles, cycles with hold-time periods, etc.) the models of kinematical type seem to be preferable. As it is usually supposed, the LCF damage is divided into two parts

$$\omega = \omega_a + \omega_s \tag{59}$$

which represent cyclical and statical damages respectively. Only the first one is considered below. Component ω_a should be divided into two parts corresponding to damages caused by plastic and creep deformation processes: this is due to a specific property of creep at compression which is usually interpreted as damage "curability" effect. Thus,

$$\omega_a = \omega_p + \omega_c \tag{60}$$

Bearing in mind the close relation between inelastic deformation and failure processes, it seems reasonable to construct the damage function employing the state parameters K and s which was used before in the SP formulation. It should be remembered that these parameters can be interpreted in terms of crystal physics (see paragraph "Similarity principle"). However, the Odquist parameter $\lambda_* = \lambda - \lambda_v$ (asterisk indicates that the value should be counted off from the previous turning point of the stress-strain diagram) instead of $K = K(C)$ is more convenient; there is a close correspondence between the two parameters. At tension-compression loading cycles $\lambda_* = |p_*|$. Thus, damage rates are defined by equations (GOKHFELD et al. [1991]):

$$\begin{aligned} \dot{\omega}_p &= H(|s| - s_c) D(s,T) \lambda_*^{\alpha(T)} \dot{\lambda}_* \\ \dot{\omega}_c &= H(\omega_c) H(s_c - |s|) \lambda_*^{\alpha(T)} \dot{\lambda}_* \end{aligned} \tag{61}$$

Here H is Heavyside's function; damage function $D(s,T)$ and power index $\alpha(T)$ should be identified from the necessary experimental data; s_c demarcates areas of viscous ($|s| < s_c$) and plastic ($|s| > s_c$) damage mechanisms.

Identification of the model can be simplified by adopting piecewise-constant approximation for the damage function. $D(s,T)$ is equal to:

$$\begin{aligned} &D_p^+(T) \text{ at } s > s_c(T), \quad D_c^+(T) \text{ at } 0 < s < s_c(T) \\ &D_p^-(T) \text{ at } -s_c(T) < s < 0, \quad D_c^-(T) \text{ at } s < -s_c(T) \end{aligned} \tag{62}$$

The necessary data can be obtained by cyclic tests of the four characteristics types symbolized by Manson as *pp, cp, pc, cc*. According to the adopted approximation, the damage per cycle is equal to:

$$\Delta\omega = \frac{(\Delta p)^m}{m} D^*, \quad D^* = D_p^+ \rho_+^m + D_c^+\left(1 - \rho_+^m\right) + D_p^- \rho_-^m + D_c^-\left(1 - \rho_-^m\right) \tag{63}$$

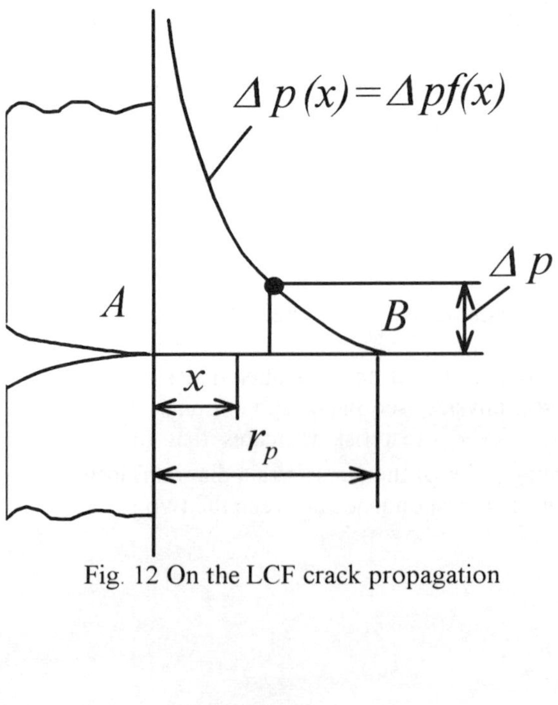

Fig. 12 On the LCF crack propagation

where

$m = \alpha + 1$; $\rho_+ = \lambda_*^+/\Delta p$, $\rho_- = \lambda_*^-/\Delta p$ are relative plastic parts of inelastic strain amplitudes in tension (+) and compression (-) halfcycles.

It has been confirmed by experiments that the model reflects the main LCF regularities with quite satisfactory accuracy, and the predicted numbers of loading cycles to failure at various loading programs show somewhat better correspondence compared to other known LCF damage models.

The next problem consists in prediction of crack propagation. This process of damage accumulation in a point of a body can be considered as a quasi-stationary one with respect to moving coordinate axes with origin at the

crack tip. Crack growth can be interpreted as a conveyor: any body element which is situated on the future crack trajectory passes in succession through elastic and inelastic deformation zones until it fails approaching the crack tip, where its damage reaches the critical magnitude ($\omega^* = 1$). At this moment the crack length grows by increment.
Let inelastic strain peak-to-peak amplitudes $\Delta p(x)$ at the points along the crack trajectory be determined by a function (Fig. 12)

$$\Delta p(x) = \Delta p \cdot f(x) \tag{64}$$

where Δp denotes some characteristic value while $f(x)$ is a function increasing as the point approaches the crack tip; r_p determines the extent of the inelastic zone (damage caused by elastic strains is neglected).
Since inelastic zone extent is usually small, it is assumed that rate $l' = dl/dN$ is constant within an interval which is comparable with r_p/l'. Hence,

$$x = r_p - l'N \tag{65}$$

where N is the number of loading cycles after the volume element has passed into the inelastic zone.
On the other hand, this number of loading cycles corresponds to the lifetime of the body element ($x = 0$, $N = N_f$). The damage accumulation model (eqs (62), (63)) leads to the following expression for crack growth rate

$$l' = \frac{D^*}{m} \Delta p^m \int_0^{r_p} f^m(x) dx \tag{66}$$

Peak-to-peak amplitude is in a certain conformity with the double amplitude of the stress intensity factor (MAKHUTOV [1981]). If dependence Δp is approximated as a power function $\Delta p = C(\Delta K_I)^n$, for a loading cycle of the simplest type (*pp*), the results obtained from equation (66) coincide with the known Paris' formula. However, it should be stressed that this equation predicts crack growth for more complicated cycles (*pc, cp, cc*) also.

This study has been carried out in the framework of a research project sponsored by the Russian Fund for Fundamental Research (grant 93-013-16779)

REFERENCES

1991 APAJCHEV, M.V., IVANOV, I.A. & PONKIN, A.V., "Modelling of isotropic hardening effects at nonproportional cyclic loading", in Russian, J. Strength Problems, No. 7, 47.

1974 BESSELING,J.F., "Plasticity and creep theory in engineering mechanics", Top. Appl. Contin. Mech., Wien-New York, p.115-136.

1975 BIRGER, I.A., SHORR, B.F., DEMIANUSHKO, I.V., DULNEV, R.A. & SIZOVA, R.N., "Thermostrength of machine parts", in Russian, Mashinostrojenije, Moscow.

1989 CHABOCHE, J.L., "Constitutive equations for cyclic plasticity and cyclic visco-plasticity", Int. J.Plasticity, 5, 247.

1970 CHERNIAVSKY, O.F., "On solving limit equilibrium and shakedown problems with the help of Pontragin's maximum principle", in Russian, Izvestiya AN SSSR, Mechanics of solids, No.4, 99.

1966 FREDERICK, C.O. & ARMSTRONG, P.J., "Convergent internal stress and steady cyclic states of stress", J. Strain Anal. 1, No. 2, 154.

1970 GOKHFELD, D.A., "Carrying capacity of structures at temperature changes", in Russian, Mashinostrojenije Publ., Moscow.

1973 GOKHFELD, D.A. & CHERNIAVSKY, O.F., " Shakedown of elastoplastic structures" (a survey), in Russian, in: Ishlinsky, A., Chernyi, G. (eds), Mechanics,Recent Developments in Forein Science, 18, Problems in Theory of Plasticity and Creep, Mir Publ., pp. 7-53.

1980 GOKHFELD, D.A. & CHERNIAVSKY, O.F., "Limit analysis of structures at thermal cycling", Sijthoff and Noordhoff, Alphen aan den Rijn-Rockville.

1979 GOKHFELD, D.A., GORSKY, S.V., KONONOV, K.M. & SADAKOV, O.S., "Influence of loading prehistory on resistance of materials against cyclic elasto-plastic loading", in Russian, J.Strength Problems, No. 1, 6.

1980 GOKHFELD, D.A., KONONOV, K.M., POROSHIN, V.B. & SADAKOV, O.S., "Coupled mathematical models for cyclic inelastic deformation and damage accumulation processes", Trans. 10th. Int. Conf. on SMiRT, Anaheim (U.S.A.), pp.19-24.

1978 GOKHFELD, D.A., KONONOV, K.M., SADAKOV, O.S. & CHERNIAVSKY, O.F., "Problems of strength of thermostressed structures", in Russian, in: Sedov,L.I. (Chief ed.), Advances in science and technology, ser. Mechanics of deformable solid, 12, VINITI Publ., Moscow, pp. 91-194.

1989 GOKHFELD, D.A., LEZHNEV, S.V. & SHVETSOV, A.G., "Deformation and fracture in graphite materials under cyclic loading", in Russian, J. Works laboratory, 55, No. 7, 55.

1991 GOKHFELD, D.A., POROSHIN, V.B. & SADAKOV, O.S., "Plasticity and creep, LCF and crack propagation processes: related state equations", Trans. 11th Int. Conf. on SMiRT, Tokyo, 1991, L 12/5

1984 GOKHFELD, D.A.& SADAKOV, O.S., "Plasticity and creep of structural elements under repeated loading", in Russin, Mashinostrojenije, Moscow.

1990 GOKHFELD, D.A.& SADAKOV, O.S., "On the main approaches to strength and lifetime evaluation of structures under variable repeated load and temperature", in: Kleiber, M., Konig, J.-A. (eds), Inelastic Solids and Structures (Antoni Sawczuk Memorial Volume), Pineridge Press, Swansea, U.K., pp,405-432.

1991a GOKHFELD, D.A.& SADAKOV, O.S., "A unified mathematical model for plasticity and creep under variable-repeated loading", in: Zyczkowski, M. (ed), Creep in Structures, IUTAM Sympos., Cracow/Poland, 1990,Springer-Verlag, Berlin, pp. 23-28.
1991b GOKHFELD, D.A.& SADAKOV, O.S., "Generalized similarity principle for plasticity and creep analysis of structural elements", Trans. 11th Int. Conf.on SMiRT, Tokyo, pp. 581-592.
1965 JENKINS, G.M., "The stress-strain relationships of polycrystalline graphites under compression up to fracture", Carbon, 3, No. 1.
1960 KOITER, W.T., "General theorems for elastic-plastic solids", ser. Progress in solid mechanics, 1 (eds. I.N. Sneddon & R. Hill), North-Holland Publ., Amsterdam.
1978 KÖNIG J.A. "Shakedown of elastic-plastic structures", PWN-Elsevier, Amsterdam.
1981 KÖNIG J.A., MAIER, G. "Shakedown analysis of elastoplastic structures, a review of recent developments", Nucl. Eng. Des. 66*, 81.
1978 LAMBA, H.S. & SIDEBOTTOM, O.M., "Cyclic plastisity for nonproportional paths", Trans. ASME, J. Eng. Mater. and Technol., 100, No. 1, 96.
1981 MAKHUTOV, N.A., "Deformation criteria of destruction and design of structural members", in Russian, Mashinostrojenije Publ., Moscow.
1991 MARTYNENKO, M.E., "Direct determination of steady inelastic deformations under action of the type of running wave", in Russian, in: Berezin (ed.), Strength of Machines and Apparatus, Proc. Chel. State Tech. Univ., Cheliabinsk, pp. 33-37.
1971 MROZ, Z., "On the theory of steady plastic cycles in structures", Prep.1st Int. Conf. on SMiRT, 5, Part 1, Berlin.
1990 MURAKAMI, S., KAWAI, M. & Jamada J., "Creep after cyclic plasticity under multiaxial conditions for type 316 stainless steel at elevated temperature", Trans. ASME, J. Eng. Mater. and Technol., 112, No. 3 , 346.
1990 OHNO, N., "Recent topics in constitutive modelling of cyclic plasticity and viscoplasticity", Appl. Mech. Review, 43, No. 11.
1988 POROSHIN, V.B., " Influence of the deformation cycle shape on damage accumulation at various types of low-cycle loading with hold_time periods", in Russian, J. Strength Problems, No. 1, 38.
1969 RABOTNOV, Yu.N., "Creep problems in structural members", North-Holland Publ., Amsterdam.
1991 SADAKOV, O.S., MADUDIN, V.N. & APAJCHEV, M.V., "On the state equations for anisotropic elasto-viscoplastic loadings", in: Zyczkowski, M. (ed.), Creep in Structures, IUTAM Symp., Crakow/Poland, 1990, Springer Verlag, Berlin, pp. 179-185.
1974 SAWCZUK, A., "Shakedown analysis of elastic-plastic structures", Nucl. Eng. Des., 28, 121.
1974 ZARUBIN, V.S., "Models of non-isothermal plasticity and creep", in Russian, in: Proc. of All-Union Symp. on LCF at Elevated Temp., Cheliabinsk, 1, pp. 58-78.

NONLINEAR BEHAVIOR OF LOW-PLASTIC STRUCTURES

A. CHERNIAVSKY
Cheliabinsk State Technical University
Lenin Ave. 76, Cheliabinsk, 454080, Russia

ABSTRACT

Structural-inhomogeneous materials like artificial graphite, ceramics of some sorts, cast iron, are often used for structural members subjected to intensive thermal loading that leads to considerable thermal stresses and hence to low-cycle fatigue at repeated loading. Experience reveals that accuracy of the known mathematical models is insufficient for life-time predictions of structural fracture under such conditions. A version of a structural model is proposed for the description of stress-strain behavior of such materials. Adequacy of the model has been confirmed by results of tests carried out for various loading programs.

Plastic strains of low-plastic materials are often insufficient to prevent crack forming. In thermostressed structures it often leads to crack mesh formation and growth. A method for analyzing the crack kinetic is suggested.

INTRODUCTION

Many thermo-stressed structural members are made of low-plastic materials such as artificial graphite, ceramics, cast iron and others. Relatively low values of Young's modulus and thermal expansion coefficient allows the employment such materials at high temperature gradients. However, shakedown at cyclic thermal loading, caused by residual strain and stress formation, cannot be achieved, because of low ultimate plastic strain of these materials. Cyclic loading is accompanied by low-cycle fatigue crack formation and growth. In more typical loading conditions, structural members are subjected to intensive surface heating. When temperatures of different surface points are almost equal and temperature gradients along the surface normal are sufficiently high, a typical fracture mechanism is the surface crack mesh formation rather then growth of a single crack. Formation of a crack mesh often leads to subsequent surface crumbling that makes further exploitation of the structure impossible. Such a situation appears, for example, in metallurgical equipment such as large-scale electrodes for arc furnaces, moulds; body bricks of high-temperature nuclear reactors, etc. Experience reveals, that thermal crack meshes can either grow with the number of loading cycles or stop after a few of them. The latter situation arises from unloading due to formation of discontinuities with respect to thermal or other types of self-equilibrated stress. Thus we can have a specific type of shakedown, in addition to the well-known ones resulting from residual stress and energy dissipation.

Z. Mróz et al. (eds.), Inelastic Behaviour of Structures under Variable Loads, 83–91.

Additional complications of the task are caused by stress-strain behavior peculiarities of materials which are used for heat-stressed structural members. It was assumed for a long time that low-cycle failure of such materials as artificial graphite is impossible. However, recent investigations reveal that, at high temperature, low-cycle phenomenon are significant for such materials. Furthermore, deformation properties of these materials are more complicated then those of traditional plastic alloys. Mechanical and physical properties of materials like artificial graphite are characterized by the following principal peculiarities:
- physical anisotropy of a body which is due to any crystalline grain anisotropy and non-chaotic distribution of the crystalline axes orientation;
- non-linearity of loading and unloading stress-strain curves practically over their whole extent, and hence hysteresis loop formation;
- essential differences between tension and compression resistance, which is characteristic for the whole class of low plasticity materials;
- softening effect which embraces the greater part of cyclic deformation processes;
- interaction between deformation and fracture processes which is reflected, in particular, in specific changes of the hysteresis loop form due to micro-cracks formation and their play;
- intensive creep rate increasing when creep at constant stress is periodically interrupted by a few loading cycles;
- cyclic creep and cyclic relaxation processes at stress and strain loading cycles, respectively;
- dependence of thermal expansion on load-temperature history (the observed coefficient of thermal expansion can differ by a factor of 3-4 depending on the preceding loading).

These particularities are joined with the non-uniform structure of material; existence of anisotropic grains with different orientation; adhesive (binding) with properties different from grain ones; various defects (pores and cracks of different size). Strength calculation in this case requires proper (sometimes rather complex) mathematical models.

STRESS-STRAIN BEHAVIOR OF LOW-PLASTIC MATERIALS. MATHEMATICAL MODELING

Several approaches to the construction of material deformation models are known at present time. One of them - phenomenological - assumes fairly simple determinative equations. Their parameters are calculated on the base of specimens' testing results, when specimens are considerably larger than the average grain size of the material. The second approach includes use of information about material structure and its structural members' properties. Such an approach may ensure the adequate description of material behavior at various loading programs, but there are some difficulties in representing the complex structure of a real material. The combination of these two approaches is possible; it consists of choosing a simple model scheme and determining the model parameters with the aid of both accessible physical information about real material structural members' properties and mechanical testing data. These data assure the proper estimation of the model's parameters for those parameters which can not be determined directly (or such determination would be a rather awkward task).

In conformity with such a combined approach, a structural model of graphite is proposed for modeling material element deformation and fracture. As in Masing's model, it consists of subelements (SE) with different properties. Distinctions from Masing's model are: use of physical information about material structure along with mechanical testing data for SE properties; consideration of SE fracture capability. Two types of SEs are supposed: brittle, linear elastic (up to the fracture), which corresponds to adhesion in artificial graphite; plastic ones, which reflect rheological behavior of a single crystallite under uniaxial loading in some direction. As far as possible, the characteristics of an element are determined by an approach which is similar to the slide theory for material of anisotropy grains with non-uniform orientation. Distribution of crystalline axis orientations [Sosedov 1975], defined with the X-ray method, and characteristics of a single graphite crystalline in its specific directions [Viatkin 1967] are used. Graphite monocrystals are essentially anisotropic, this leads to considerable differences in the characteristics (elastic modulus, yield stresses, c.t.e) of SEs which model the various orientations. This anisotropy results in a high level of stresses in SE during mechanical loading or heating, and thus to a high probability of plastic deformation and (or) fracture of SE's. It is assumed that cracks, formed at tensile stress in SE, can be closed by compressive stress in the SE (i.e. resistance to deformation can be restored) while cracks caused by compression completely destroy SE, leading to loss of its resistance.

For the computer-based realization of the model, a large number (about 100) of SE can be admitted, with such a number, the model stress-strain diagram, even at crack formation and closure remains smooth enough. By using brittle SEs, we can model the diffuse fracture of material (excluding micro-crack interaction). It is necessary to note that there are many microcracks in artificial graphite caused by cooling in the manufacturing process (so-called Mrozovsky' cracks); these have no influence on deformation processes at temperatures less than 1100K.

The considered model is formulated by the following equations:

$$\varepsilon_i = \varepsilon \qquad \sigma = \frac{1}{n}\sum_i \sigma_i$$

$$i = 1..m \qquad \sigma_i = E_i\left(\varepsilon - \alpha\Delta T - p_i\right) \qquad \dot{p}_i = \Phi\left(\frac{\sigma_i}{\sigma_i^y}, T, \mathbf{n}_i\right)$$

$$E_i = \mathbf{n}_i\mathbf{n}_i..\mathbf{E}..\mathbf{n}_i\mathbf{n}_i \qquad \alpha_i = \mathbf{a}..\mathbf{n}_i\mathbf{n}_i$$

$$\sigma_i^y = \sigma_i^y\left(\tau^y, \mathbf{n}_i, \lambda_i\right) \qquad \lambda_i = \int\left|dp_i\right|$$

$$\mathbf{n}_i: \quad \tilde{f}\left(\mathbf{n}_i\right) = f\left(\mathbf{n}\right)$$

$$i = m+1..n \qquad \sigma_i = E'r_i\left\{1 + \left[H\left(-E'r_i\right) - 1\right]H\left(\max\frac{r_i}{\varepsilon_i^{ut}} - 1\right)\right\}H\left(\min\frac{r_i}{\varepsilon_i^{ut}} - k\right)$$

$$E' = E'(\mu) \qquad \mu = \int |d\varepsilon|$$

$$r_i = \varepsilon - \alpha T \qquad \varepsilon_i^{ut}: \quad \tilde{\gamma}(\varepsilon_i^{ut}) = \gamma(\varepsilon)$$

Here ε, σ - the model strain and corresponding stress quantities with subscripts refer to the respective SE (at $i=1...m$ - to plastic, $i=m+1...n$ - to brittle ones); T - temperature; p - inelastic strain; Φ - rheological function; E, α - tensors of elastic constants and thermal expansion coefficients of graphite crystalline; **n** - vectors, normal to crystallite's slide planes, empirical distribution of their directions $f(\mathbf{n})$ coincide with distribution f, determined by the X-ray method; E', α' - elastic modulus and thermal expansion coefficient (c.t.e.) of brittle subelements, ε_i^{ut} - ultimate tensile strain of SE, $\tilde{\gamma}$ and γ - empirical (for the model' SE) and defined from specimens testing distributions of those strains, k - the ratio of the ultimate SE tensile strain to its ultimate compressive strain (assumed to be equal for all brittle SE's); H - Heavyside's function ($H(x)=1$ for $x>0$ or $x=0$ and $H(x)=0$ for $x<0$). Heavyside's function is used for brittle SE fracture modeling and simulation of cracked SE deformation: factor $H\left(\min \frac{r_i}{\varepsilon_i^{ut}} - k\right)$ describes fracture at compression, $H\left(\max \frac{r_i}{\varepsilon_i^{ut}} - 1\right)$ - at tension, and $\left[H(-E' r_i) - 1\right]$ - breakdown cracks shutting at compression (minimum and maximum corresponds to whole loading program).

Basic tests for determination of the model characteristics for a given material are: determination of crystalline orientation distribution (by the X-ray method), short-time and durable (creep) test at non-repeated loading, cyclic tests with at least two different strain levels.

The model is designed as a program for personal computers and allows one to do interactive modeling in cojunction with existing programs for strength calculation. In this case it is necessary to store the set of data (by which SEs state is described) for every reference point of the structure (or for every finite element). The relatively small volume of this set (5K) allows one to store information for a large number of reference points.

Experimental verification of the model was carried out for loading programs which differ from the basic on used for the model identification, including isothermal cyclic loading with constant minimum and maximum stresses; cyclic loading with hold-time periods during loading cycles; non-isothermal cyclic loading. The model allows one to describe material deformation well. So, at stress-cycle loading until $N=(0.7...0.8)N_f$ (where N_f - number of cycles to fracture) at 300..1100K

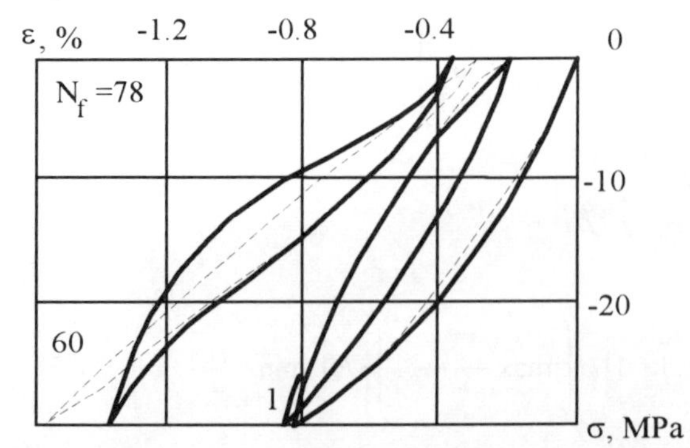

Fig. 1. Stress vs strain curves of artificial graphite (bold line - experiment, thin dotted line - the model)

stress-strain diagrams predicted by the model spread in the experimental data' scatter band (Fig. 1).

For numbers of cycles exceeding approximately $0.7N_f$,experimental stress-strain diagrams take C- and S-like form. These effects are apparently related to the interaction of cracks formed in the specimen and cannot be described by the model.

A typical loading program for thermo-stressed structural members is the non-isothermal one with different stress (strain) alteration rates. It is necessary to note that in some cases thermal and mechanical loading for graphite happens to be decidedly non-independent. For example, at non-isothermal loading, a considerable part of residual strain (produced during loading and unloading) has relaxed during heating and cooling, even at temperatures when creep at constant stress is negligible small (420K).

For non-isothermal loading, "strain vs loading history" relations obtained by experiment were compared with those calculated with the aid of the model and calculated by the model, where c.t.e. of all SE were equal (as is assumed in structural models of metals and alloys). Mechanical loading and subsequent unloading cause formation of plastic strains and residual stresses in SE's. During heating, because of SE's c.t.e. differences, additional stresses appear. This leads to changes of plastic strains and creep rates of SE's and displayed as changing of outward model (specimen) c.t.e. and residual strain changes during "heating-cooling" cycle. For a model with equal SE's c.t.e., heating causes changes to residual strains only in consequence of SE residual stress relaxation speed increase. In this case, as in experiment, specimen (model) residual strain decreased, but this decrease appeared far less than the experimental one (if rheology of the model's SE is determined through creep). The most significant difference was achieved from the program "loading - heating - cooling - unloading". Strains, calculated with the aid of the model described above, are within the experimental data scatter band, while if c.t.e. of SEs are assumed equal the difference may reach 25%, i.e. 0.2% of strain which is approximately the ultimate tensile strain of graphite used in this study.

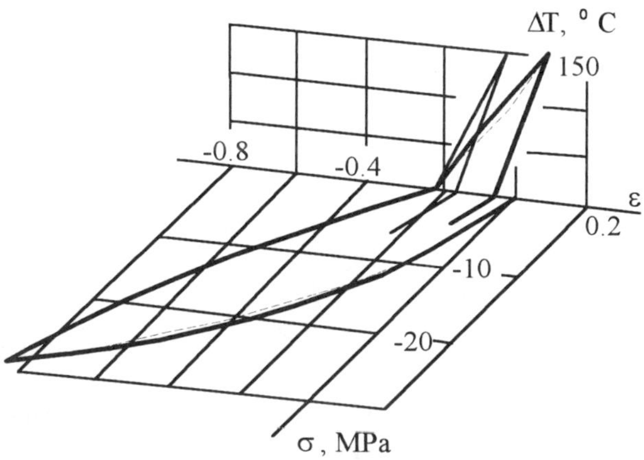

Fig. 2. Nonisothermal loading of graphite (bold line - experiment, dotted line - the model, thin line - model with equal SE' c.t.e.)

Destruction of the SEs by which existing diffuse failure processes in the material are simulated, can finally lead to total model "fracture". If the sum of yield stresses of plastic SEs (changes in a result of SE softening) and ultimate stresses of brittle SE (for SE, which are not destroyed to the moment in consideration) proves to be lower than applied stress, the model strain will increase in unconstrained manner for infinitely small load increase. The fracture criterion in this case appears to be close to the constant deformation one (some differences caused by differences in SEs plastic strains and residual stresses after different loading programs). A comparison of results predicted by the simulation of the deformation and fracture process with experimental data is illustrated by Fig. 3.

Analogous results have been obtained for loading cycles with hold-time periods. Here the number of cycles to fracture depends on both hold-time length and extremal cycle stresses. Fracture time at stresses more than 0.7 of ultimate value are close to fracture time at constant stress creep. At less stress levels, on account of the greater number of cycles, softening effects become conspicious, and fracture time at cyclic loading with endurances become less than time to fracture at creep. Fracture time at high stresses and large duration of cycle (i.e., at low numbers of cycle to failure) may be insignificantly greater than fracture time at constant stress creep. This may be explained by the decrease of overall strain as the result of loading and subsequent unloading (fracture, as noted above, described by criteria, not far from constant strain) while the cyclic part of the according damage and creep acceleration is small.

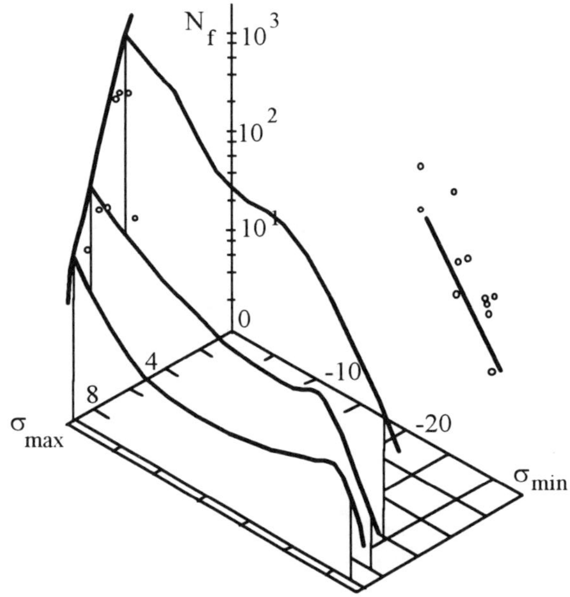

Fig. 3. Lifetime of graphite specimens under cyclic loading (points - experiment, line - the model)

Results of calculation of the structural member stress-strain state under nonisothermal cyclic loading, obtained by the structural rheological model and visco-elastic one are essentially different. Comparison of the results with structure operating experience shows that results obtained with the aid of the structural model seem more realistic.

CRACK MESH GROWTH ANALYSIS

Plastic strains of low-plastic materials like graphite are often insufficient to prevent crack formation. When the crack grows in a self-equilibrated (e.g. thermal) stress field, the stress intensity factor decreases, so intended cyclic loading can cause either crack stoppage (a kind of shakedown) or unlimited growth with a number of cycles. Uniformity of conditions on a considerable part of the structure's surface leads to restricted influence of cracks on the structural member's stress-strain state, and single crack appearance at the structure's surface cannot cause stress decrease in rather outlying points and thus cannot prevent the formation of other cracks. Under such conditions not a single crack, but a mesh of interacting cracks, forms at the structural member's surface. From the structural design point of view it is necessary to estimate both length and distance between neighboring cracks because of their influence on the danger of crumbling. Existing calculation methods are insufficient to resolve this task: analytical methods are applicable only for simple loading cases and structure geometry while numerical methods align for isolated cracks and happen to be very time-consuming for crack mesh. An energetic approach of fracture mechanics is proposed for the analysis of crack mesh growth. The particularity of the application of the approach in the given case is the approximation method for the elastic energy calculation for a non-uniformly heated body with a large number of cracks. Only periodical crack meshes are investigated (fig. 4).

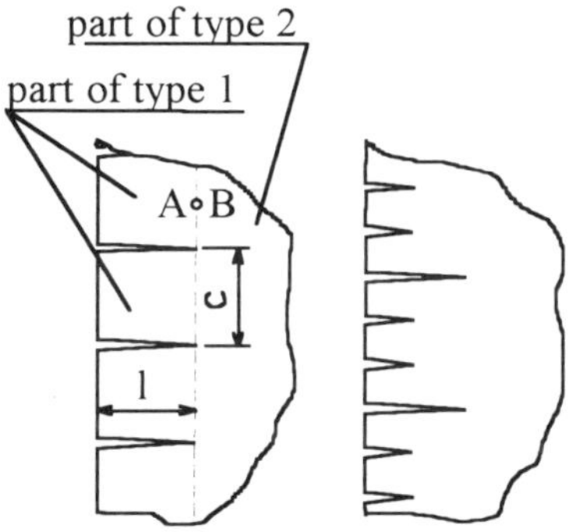

Fig. 4. Crack mesh

For the energy calculation, a body is divided into several parts by the plane which passes through crack tips. Stress fields in each part can be divided into two components: stresses σ_t, caused by nonuniform heating of a given part; and stresses σ_f, caused by forces that should be applied to part of the boundaries to satisfy strain compatibility conditions. These stresses are orthogonal from energetic point of view, i.e. $\int_V \sigma_t \cdot\cdot\, \varepsilon_f dV = \int_V \sigma_f \cdot\cdot\, \varepsilon_t dV = 0$ (this can be proved using results obtained in [Gokhfeld at al. 1984]). So, total strain energy can be calculated as a sum $U = U_t + U_f$ where U_t is energy related to σ_t while U_f is related to σ_f. The calculation of U_t is not very complicated as the parts, obtained by the member's division, do not contain cracks, and hence no singularities in the stress fields. Calculation of U_f is more difficult because just the corresponding part of stresses are caused by the interaction of cracks.

The calculation of U_f is based on Lagrange' variational principle, and displacements minimizing the respective functional are determined approximately by Ritz' method. In this way a rather simple expression for energy calculation has been obtained:

$$U_f = E\left(1-\mu^2\right)\Delta\varepsilon^2 \frac{c}{\xi}\frac{\psi}{1+\psi} \qquad \psi = \frac{e^{4\xi l}-1}{(e^{2\xi l}+1)^2} \qquad \xi = \frac{1}{c}\sqrt{12(1+\mu)(1-\mu)^2}$$

Here E, μ -are Young modulus and Poisson ratio, respectively; l and c - length of crack and distance between them; ε_A ε_B - strains in points A and B (see fig. 4) calculated for each body's part separately. The accuracy of the method was evaluated by comparison of the stress intensity coefficient K_I calculated for some simple cases with known analytical solutions. The differences in K_I are between 7 and 10% .

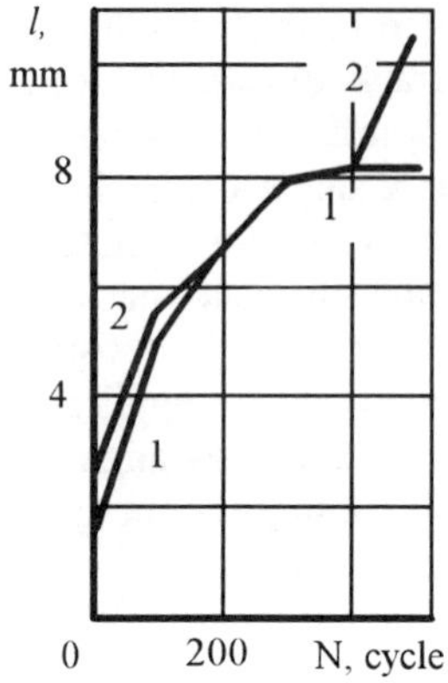

Fig. 5. Cracks in graphite brick

With the aid of the described method of elastic energy calculation (and, respectively, stress intensity coefficients determination) and obvious constraints for crack mesh growth (crack length cannot decrease during loading process and distance between cracks cannot be changed) crack kinetics has been examined, i.e. crack mesh change during loading process has been traced. Fig. 5 illustrates results obtained for a graphite block of prismatic shape which is subjected to cyclic surface thermoshocks. Here l - length of the cracks, N - number of cycles, curves 1 and 2 correspond to different distances between cracks (3.2 and 6.4 mm respectively). As is seen, according to the energy minimum principle, the number of growing cracks can decrease after some number of loading cycles, i.e., instead of a mesh of short cracks, a wider mesh of deep cracks will develop.

In the next figure (fig. 6), the crack mesh in a cast-iron wall of a mould for nickel alloys is shown. In this case crack mesh propagation stopped after a number of cycles. Comparison of predicted crack parameters (triangles) with the observed ones (points) show that the accuracy of the suggested method is sufficient for engineering design purposes.

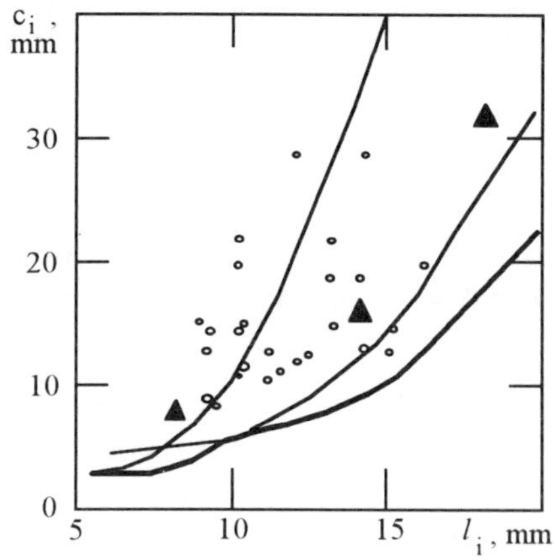

Fig. 6. Cracks in cast-iron mould

Any dilatation effects, cyclic (joined with cyclic temperature change) or monotonic (material swelling), can be considered.

However, it should be noted that such a method, as with any kinetic calculation, is fairly time consuming. At least in many important practical cases, it may be enough to know if cracks stop after some loading cycles and to know the region in the "length of cracks - distance between cracks"-plane, where all the points corresponding to progressed (or not progressed at given loading) cracks are situated. An estimation of the region's boundary can be obtained by the following method: define regions on the "length-distance" plane, where progressed cracks are situated at specific instants; the final region is obtained as these regions merge. In such an approach, the strain energy calculation can be simplified by elimination of some items (this elimination leads, as it is proved, only to increased evaluation of crack length). Besides that, contrary to the kinetic calculation, it is not necessary to account for the number of impending instants - it is enough to examine only some specific ones. These simplifications allow to reduce the amount of calculations more than ten times. Herewith, only the estimation of the not growing crack region can be obtained (in fact an over-estimation).Figure 6 shows the results of the kinetic crack calculation (circles) for the already-mentioned mould, compared with the boundary of a region, which consist of cracks that can grow (bold lines; thin lines correspond to specific instants of time). The boundary of the approximately calculated region is not very far from results of the kinetic calculation.

Calculations show that in some structures it is possible to control crack growth with the aid of defined deep (and distant) notches, creating for example a mesh of few but longer cracks, instead of a fine mesh on the smooth surface. The proposed method allows the prediction of the notch influence on the crack propagation and thus the choice of a notch system that increases structure lifetime.

CONCLUSION

Low-plastic materials are characterized by rather complicated stress-strain behavior, in particular, by cyclic softening, creep acceleration when constant stress is periodically interrupted by loading cycles, strains changes because of microstresses' relaxation at non-isothermal loading. The mathematical model for artificial graphite proposed here allows the description of all these processes. Stress-strain calculations with this model are in better agreement with structure testing results than calculations using traditional visco-elastic material descriptions.

If the material's plasticity is insufficient and thermal - or any other dilatation - loading is almost uniform on a considerable part of the structure's surface, then a crack mesh can form. When the cracks grow, they can deteriorate the surface by crumbling or can cause macro-crack formation and structure breakdown. The proposed method allows to follow the evolution of crack growth in a body with a large number of interacting cracks and so to analyze crack mesh growth by means of a mathematical modeling. The relatively short calculation time make it possible to study variants and to choose technological or design options which reduce the danger of crack mesh formation and and so to increase the lifetime of the structure.

REFERENCES

1975 SOSEDOV V.P. Properties of carbon-based structural materials - Metallurgia (in Russian)

1967 VIATKIN S.E. Nuclear graphite - Atomizdat (in Russian)

1984 GOKHFELD D.A. SADAKOV O.S. Plasticity and creep of structures under repeated loading. Moscow, Mashinostroenie (in Russian)

GENERAL THEORY OF ACCOMODATED INELASTIC STRUCTURES UNDER VARIABLE LOADS

ELASTIC-VISCOPLASTIC SOLIDS SUBJECTED TO THERMAL AND LOADING CYCLES

Castrenze POLIZZOTTO
Università di Palermo
Dipartimento di Ingegneria Strutturale & Geotecnica, DISEG
Viale delle Scienze, I-90128 Palermo, Italy

ABSTRACT — A class of elastic-viscoplastic materials with dual internal variables, thermodynamic potential and temperature-dependent plastic and creep data is considered. For solids (or structures) of such materials, subjected to cyclic loads and temperature variations, the existence of a steady-state response is ascertained and its periodicity characteristics established. Particular steady-state responses, like, elastic and inelastic shakedown, are addressed. By means of a sensitivity analysis of the steady cycle with respect to the load parameter changes, a number of basic features of inelastic shakedown (the viscoplastic counterpart of plastic shakedown) are also addressed.

INTRODUCTION

Elastic-viscoplastic solids subjected to periodic loads and temperature variations are considered here within the framework of small displacements and strains, in order to investigate the consequent steady-state (or stabilized) response, along with other aspects of the structural behaviour. In the case of material data independent of temperature variations, Frederick and Armstrong proved the existence of a steady-state response for elastic-perfectly plastic and elastic-perfectly viscoplastic materials, showing that the stresses and plastic strain rates eventually become periodic with the same period as the loads, (cf. FREDERICK & ARMSTRONG [1966]; see also GOKHFELD & CHERNIAVSKY [1980], MARTIN [1975], PONTER [1972]). These results were extended to a class of elastic-viscoplastic kinematically hardening materials (cf. MRÓZ [1972], AINSWORTH [1979]), and to a class of generalized standard elastic-plastic materials (cf. POLIZZOTTO [1993 d]). Halphen considered standard elastic-plastic and elastic-viscoplastic materials with linear hardening and temperature-dependent material data (cf. HALPHEN [1979]).

Z. Mróz et al. (eds.), Inelastic Behaviour of Structures under Variable Loads, 95–128.

For a structure showing such a steady-state response under cyclic loads and temperature variations, two different phases can be distinguished, namely

1. – The short term transient response (lasting, in general, only few cycles) which depends on the initial conditions and exhibits no periodicity features, and

2. – The long term stabilized response exhibiting periodicity features independent of the initial conditions (*steady cycle*).

Since the steady-state response phase in general covers almost the whole working life of the structure, methods for direct evaluation (i.e., without a step-by-step analysis) of the steady cycle represent a basic research issue in structural mechanics and engineering.

The material model considered here is a combination of one obeying a time--independent plasticity law of associative type, and one obeying a time-dependent plasticity law of Perzyna type (cf. PERZYNA [1974]), with the interaction between these two simple behaviours being controlled through suitable internal variables. Such a material model is able to cope with a wide class of realistic materials. It is proved here that – under certain restrictive hypotheses – a solid of such a material, if subjected to cyclic loads and temperature variations, exhibits a steady cycle in which the stresses, the internal variables, as well as the plastic and creep strain rates, become periodic in accordance with the loads. Besides, some related questions are addressed, like the conditions under which elastic shakedown occurs, the equations governing the steady cycle, and the basic properties of inelastic shakedown (the viscoplastic counterpart of plastic shakedown).

The plan of the paper is the following. After this introduction, the material constitutive equations and certain related central inequalities are presented, then the pertinent initial boundary value problem is addressed and the existence and uniqueness features of the steady-state response established. Then, the ad-hoc equation set governing the steady cycle is provided. Subsequently, the concept of elastic shakedown within the context of elastic-viscoplastic materials is discussed, and the sensitivity of the steady cycle to the load changes together with some basic features of inelastic shakedown are studied.

Notation. A compact notation is used throughout the paper. Vectors and tensors are denoted by bold-face symbols. The scalar product is marked by a dot between two vectors, and by a colon between two second-order tensors, e.g. $\mathbf{u}\cdot\mathbf{v} = u_i v_i$, $\boldsymbol{\sigma} : \boldsymbol{\varepsilon} = \sigma_{ij}\varepsilon_{ij}$, where the repeated index summation rule holds. Index saturation between vectors and tensors of different orders obeys the following rules: $(\boldsymbol{\sigma}\cdot\mathbf{n})_i = \sigma_{ij}n_j$, $(\mathbf{A} : \boldsymbol{\sigma})_{ij} = A_{ijhk}\sigma_{hk}$, $\boldsymbol{\sigma} : \mathbf{A} : \boldsymbol{\sigma} = \sigma_{ij}A_{ijhk}\sigma_{hk}$, etc. Derivative with respect to time is denoted by an upper dot, for instance $\dot{\boldsymbol{\varepsilon}} = \partial \boldsymbol{\varepsilon}/\partial t$. The symbol $:=$ denotes equality by definition. Other symbols will be defined where they first appear.

CONSTITUTIVE EQUATIONS AND CENTRAL INEQUALITIES

The elastic-viscoplastic material model considered here is characterized by a yield function $f = f(\boldsymbol{\sigma}, \boldsymbol{\chi}, \theta)$ and by a viscoplastic potential $\Omega = \Omega(\boldsymbol{\sigma}, \boldsymbol{\chi}, \theta)$, where $\boldsymbol{\sigma}$ denotes the stress tensor, $\boldsymbol{\chi}$ the internal variables (here called micro-stresses – scalar, vector or tensor, but here $\boldsymbol{\chi}$ is formally treated as a second order tensor), and θ is the temperature variation. The potential Ω is taken as one of Perzyna type (cf. PERZYNA [1974]), i.e.

$$\Omega := \gamma \Phi(< g >) \tag{1}$$

where γ is a temperature-dependent coefficient of fluidity (i.e. inverse viscosity η), and $\Phi(x)$ is a convex function increasing as the argument $x \geq 0$ increases and such that $\Phi(0) = d\Phi/dx = 0$ for $x = 0$; also, the operator $< \cdot >$ is the Macauley operator, i.e. $< x >= (x + |x|)/2$, and $g = g(\boldsymbol{\sigma}, \boldsymbol{\chi}, \theta)$ is the creep function, satisfying the inequality

$$g(\boldsymbol{\sigma}, \boldsymbol{\chi}, \theta) \geq f(\boldsymbol{\sigma}, \boldsymbol{\chi}, \theta) \tag{2}$$

for all values of the arguments. In other words, the surface $f(\boldsymbol{\sigma}, \boldsymbol{\chi}, \theta) = 0$ in the $(\boldsymbol{\sigma}, \boldsymbol{\chi})$-space contains the analogous surface $g(\boldsymbol{\sigma}, \boldsymbol{\chi}, \theta) = 0$. One possible choice for g is $g = f + k$ where $k \geq 0$ is a scalar parameter. Both f and g are assumed smooth and convex in the $(\boldsymbol{\sigma}, \boldsymbol{\chi})$-space for all θ values, and furthermore they increase as θ increases, namely $f(\boldsymbol{\sigma}, \boldsymbol{\chi}, \theta_1) > f(\boldsymbol{\sigma}, \boldsymbol{\chi}, \theta)$ for $\theta_1 > \theta$, and analogously for g.

The material constitutive equations read as follows:

$$f(\boldsymbol{\sigma}, \boldsymbol{\chi}, \theta) \leq 0, \quad \dot{\lambda} \geq 0, \quad \dot{\lambda} f(\boldsymbol{\sigma}, \boldsymbol{\chi}, \theta) = 0 \tag{3a}$$

$$\dot{\boldsymbol{\varepsilon}}^p = \dot{\lambda} \frac{\partial f}{\partial \boldsymbol{\sigma}}, \qquad \dot{\boldsymbol{\xi}}^p = -\dot{\lambda} \frac{\partial f}{\partial \boldsymbol{\chi}} \tag{3b}$$

$$\dot{\boldsymbol{\varepsilon}}^c = \frac{\partial \Omega}{\partial \boldsymbol{\sigma}} = \gamma \frac{\partial \Phi}{\partial g} \frac{\partial g}{\partial \boldsymbol{\sigma}}, \quad -\dot{\boldsymbol{\xi}}^c = \frac{\partial \Omega}{\partial \boldsymbol{\chi}} = \gamma \frac{\partial \Phi}{\partial g} \frac{\partial g}{\partial \boldsymbol{\chi}} \tag{4}$$

$$\xi = \xi^p + \xi^c, \qquad \boldsymbol{\chi} = \frac{\partial \Psi(\boldsymbol{\xi})}{\partial \boldsymbol{\xi}} \tag{5}$$

$$\boldsymbol{\varepsilon}^{in} = \boldsymbol{\varepsilon}^p + \boldsymbol{\varepsilon}^c, \qquad \boldsymbol{\varepsilon} = \mathbf{A} : \boldsymbol{\sigma} + \mathbf{a}\theta + \boldsymbol{\varepsilon}^{in}. \tag{6}$$

Here, ε^{in} denotes inelastic strain, decomposed into plastic and creep parts; $\boldsymbol{\xi}$ denotes another set of internal variables (called micro-strains, dual to $\boldsymbol{\chi}$), split into two parts, that is, $\boldsymbol{\xi}^p$ related to time-independent mechanisms of the material micro-structure, and $\boldsymbol{\xi}^c$ related to time-dependent ones, whereas $\Psi(\boldsymbol{\xi})$ is the pertinent thermodynamic potential, by hypothesis convex and such that $\Psi(\mathbf{0}) = 0$. Moreover, $\mathbf{A}$ is the compliance elasticity fourth-order tensor and $\mathbf{a}$ is the thermal expansion tensor. By hypotesis, both elastic and thermal expansion material data are independent of temperature variations. This material model can be described by the rheological scheme of Fig. 1 in case of uniaxial stress states.

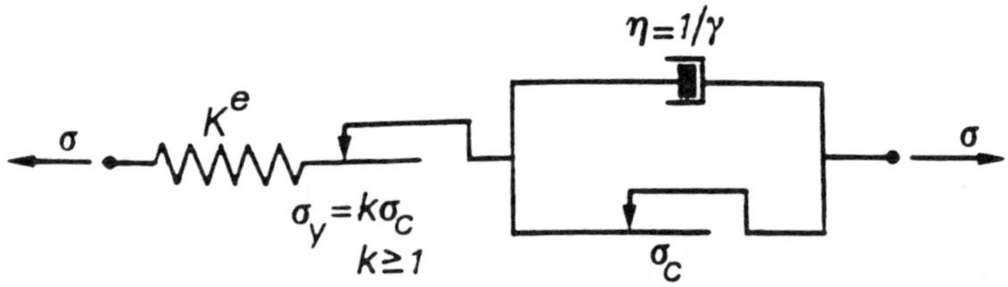

Fig. 1. Rheological model of an elastic-viscoplastic material in uniaxial stress states.

A material model obeying the above constitutive equations covers a wide range of material models (cf. LEMAITRE & CHABOCHE [1985], LUBLINER [1990]). Namely, a time-independent elastic-plastic material model is obtained by setting $\gamma = 0$ (i.e. infinite viscosity), whereas a time-dependent one is obtained by assuming an infinite yield stress. Though a time-independent elastic-plastic material model can be generated as a limit behaviour of the time-dependent one as $\gamma \to \infty$, here the combined constitutive model has been preferred for the sake of convenience. The introduced internal variable constitutive model enables one to describe different types of hardening laws, e.g. isotropic, kinematic, etc., (but with some important limitations inherent to the internal variable description combined with the hypothesis of associated plasticity). On taking

$$g := [(\boldsymbol{\sigma}^d - \boldsymbol{\chi}^d) : (\boldsymbol{\sigma}^d - \boldsymbol{\chi}^d)]^{\frac{1}{2}} - \frac{1}{\sqrt{3}}\chi_0 - k_0 \tag{7}$$

where $(\cdot)^d :=$ deviatoric part of $(\cdot)$, the Chaboche viscoplastic material model is obtained (cf. CHABOCHE [1977]). Also, if Φ is taken as a positively homogeneous function of $N+1$ degree, a viscoplastic model of Norton-Odqvist type is generated.

Finally, on taking $g \equiv f$, a unified viscoplasticity model is generated, namely a model in which inelastic strain is no longer decomposed into plastic and creep parts (cf. LUBLINER [1990]).

For the material model described by Eqs. (3-6), certain inequalities hold. These inequalities are crucial for the purposes of this work. Let the pair $(\bar{\boldsymbol{\sigma}}, \bar{\boldsymbol{\chi}})$ be plastically admissible at the themperature θ, i.e. $f(\bar{\boldsymbol{\sigma}}, \bar{\boldsymbol{\chi}}, \theta) \leq 0$, and let $(\boldsymbol{\sigma}, \boldsymbol{\chi})$ and $(\dot{\boldsymbol{\varepsilon}}^p, \dot{\boldsymbol{\xi}}^p)$ correspond to each other through Eqs. (3a, 3b) at the same of value θ. Then, in virtue of Drucker's postulate (cf. MARTIN [1975], LUBLINER [1990]), one can write

$$(\boldsymbol{\sigma} - \bar{\boldsymbol{\sigma}}) : \dot{\boldsymbol{\varepsilon}}^p - (\boldsymbol{\chi} - \bar{\boldsymbol{\chi}}) : \dot{\boldsymbol{\xi}}^p \geq 0 \tag{8}$$

which holds for any $(\bar{\boldsymbol{\sigma}}, \bar{\boldsymbol{\chi}})$, that is plastically admissible. Furthermore, let $(\boldsymbol{\sigma}, \boldsymbol{\chi})$ and $(\dot{\boldsymbol{\varepsilon}}^c, \dot{\boldsymbol{\xi}}^c)$ correspond to each other through Eqs.(4) with the same value of θ; then, due to the convexity of g, hence of Ω, in the $(\boldsymbol{\sigma}, \boldsymbol{\chi})$-space, the following inequality can be easily shown to hold

$$(\boldsymbol{\sigma} - \bar{\boldsymbol{\sigma}}) : \dot{\boldsymbol{\varepsilon}}^c - (\boldsymbol{\chi} - \bar{\boldsymbol{\chi}}) : \dot{\boldsymbol{\xi}}^c \geq \Omega(\boldsymbol{\sigma}, \boldsymbol{\chi}, \theta) - \Omega(\bar{\boldsymbol{\sigma}}, \bar{\boldsymbol{\chi}}, \theta) \tag{9}$$

which is valid for arbitrary pairs $(\bar{\boldsymbol{\sigma}}, \bar{\boldsymbol{\chi}})$. Thus, summing Eqs. (8, 9) with each other gives

$$(\boldsymbol{\sigma} - \bar{\boldsymbol{\sigma}}) : \dot{\boldsymbol{\varepsilon}}^{in} - (\boldsymbol{\chi} - \bar{\boldsymbol{\chi}}) : \dot{\boldsymbol{\xi}} \geq \Omega(\boldsymbol{\sigma}, \boldsymbol{\chi}, \theta) - \Omega(\bar{\boldsymbol{\sigma}}, \bar{\boldsymbol{\chi}}, \theta) \tag{10}$$

which holds for any plastically admissible pair $(\bar{\boldsymbol{\sigma}}, \bar{\boldsymbol{\chi}})$ at the temperature θ, i.e. $f(\bar{\boldsymbol{\sigma}}, \bar{\boldsymbol{\chi}}, \theta) \leq 0$.

With the aid of the second Eq. (5), one can write:

$$f(\boldsymbol{\sigma}, \boldsymbol{\chi}, \theta) = f(\boldsymbol{\sigma}, \frac{\partial \Psi}{\partial \boldsymbol{\xi}}, \theta) := \tilde{f}(\boldsymbol{\sigma}, \boldsymbol{\xi}, \theta) \tag{11a}$$

$$g(\boldsymbol{\sigma}, \boldsymbol{\chi}, \theta) = g(\boldsymbol{\sigma}, \frac{\partial \Psi}{\partial \boldsymbol{\xi}}, \theta) := \tilde{g}(\boldsymbol{\sigma}, \boldsymbol{\xi}, \theta); \tag{11b}$$

and since

$$\frac{\partial \tilde{f}}{\partial \boldsymbol{\sigma}} = \frac{\partial f}{\partial \boldsymbol{\sigma}}, \quad \frac{\partial \tilde{f}}{\partial \boldsymbol{\xi}} = \frac{\partial^2 \Psi}{\partial \boldsymbol{\xi} \partial \boldsymbol{\xi}} : \frac{\partial f}{\partial \boldsymbol{\chi}} \quad \text{and} \tag{12a}$$

$$\frac{\partial \tilde{\Omega}}{\partial \boldsymbol{\sigma}} = \frac{\partial \Omega}{\partial \boldsymbol{\sigma}}, \quad \frac{\partial \tilde{\Omega}}{\partial \boldsymbol{\xi}} = \frac{\partial^2 \Psi}{\partial \boldsymbol{\xi} \partial \boldsymbol{\xi}} : \frac{\partial \Omega}{\partial \boldsymbol{\chi}}, \tag{12b}$$

where $\tilde{\Omega} := \gamma \Phi(< \tilde{g} >)$, one has

$$\dot{\lambda} \frac{\partial \tilde{f}}{\partial \boldsymbol{\sigma}} = \dot{\boldsymbol{\varepsilon}}^p, \quad \frac{\partial \tilde{\Omega}}{\partial \boldsymbol{\sigma}} = \dot{\boldsymbol{\varepsilon}}^c, \quad \dot{\lambda} \frac{\partial \tilde{f}}{\partial \boldsymbol{\xi}} + \frac{\partial \tilde{\Omega}}{\partial \boldsymbol{\xi}} = -\frac{\partial^2 \Psi}{\partial \boldsymbol{\xi} \partial \boldsymbol{\xi}} : (\dot{\boldsymbol{\xi}}^p + \dot{\boldsymbol{\xi}}^c) = -\dot{\boldsymbol{\chi}}. \tag{13}$$

Thus, one easily realizes that Eqs. $(3a, 3b, 4)$ can be rewritten as follows

$$\tilde{f}(\boldsymbol{\sigma}, \boldsymbol{\xi}, \theta) \leq 0, \quad \dot{\lambda} \geq 0, \quad \dot{\lambda}\tilde{f}(\boldsymbol{\sigma}, \boldsymbol{\xi}, \theta) = 0 \tag{14a}$$

$$\dot{\boldsymbol{\varepsilon}}^p = \dot{\lambda}\frac{\partial \tilde{f}}{\partial \boldsymbol{\sigma}}, \qquad \dot{\boldsymbol{\chi}}^p = -\dot{\lambda}\frac{\partial \tilde{f}}{\partial \boldsymbol{\xi}} \tag{14b}$$

$$\dot{\boldsymbol{\varepsilon}}^c = \frac{\partial \tilde{\Omega}}{\partial \boldsymbol{\sigma}}, \qquad \dot{\boldsymbol{\chi}}^c = -\frac{\partial \tilde{\Omega}}{\partial \boldsymbol{\xi}} \tag{15}$$

where $\dot{\boldsymbol{\chi}}^p$ and $\dot{\boldsymbol{\chi}}^c$ constitute a partition of $\dot{\boldsymbol{\chi}}$, i.e. $\dot{\boldsymbol{\chi}} = \dot{\boldsymbol{\chi}}^p + \dot{\boldsymbol{\chi}}^c$.

A class of generalized standard materials is considered here, characterized by Ψ such that both $\tilde{f}$ and $\tilde{g}$ are convex in the $(\boldsymbol{\sigma}, \boldsymbol{\xi})$-space for every θ, (this class contains, at least, materials with a quadratic Ψ). For such materials, inequalities similar to Eqs. (8, 9) hold, that is

$$(\boldsymbol{\sigma} - \bar{\boldsymbol{\sigma}}) : \dot{\boldsymbol{\varepsilon}}^p - (\boldsymbol{\xi} - \bar{\boldsymbol{\xi}}) : \dot{\boldsymbol{\chi}}^p \geq 0 \tag{16}$$

$$(\boldsymbol{\sigma} - \bar{\boldsymbol{\sigma}}) : \dot{\boldsymbol{\varepsilon}}^c - (\boldsymbol{\xi} - \bar{\boldsymbol{\xi}}) : \dot{\boldsymbol{\chi}}^c \geq \tilde{\Omega}(\boldsymbol{\sigma}, \boldsymbol{\xi}, \theta) - \tilde{\Omega}(\bar{\boldsymbol{\sigma}}, \bar{\boldsymbol{\xi}}, \theta) \tag{17}$$

where $\bar{\boldsymbol{\xi}}$ corresponds to $\bar{\boldsymbol{\chi}}$, i.e. $\bar{\boldsymbol{\chi}} = \partial\Psi/\partial\boldsymbol{\xi}|_{\boldsymbol{\xi}=\bar{\boldsymbol{\xi}}}$. Thus, summing Eqs. (16, 17) with each other yields

$$(\boldsymbol{\sigma} - \bar{\boldsymbol{\sigma}}) : \dot{\boldsymbol{\varepsilon}}^{in} - (\boldsymbol{\xi} - \bar{\boldsymbol{\xi}}) : \dot{\boldsymbol{\chi}} \geq \tilde{\Omega}(\boldsymbol{\sigma}, \boldsymbol{\xi}, \theta) - \tilde{\Omega}(\bar{\boldsymbol{\sigma}}, \bar{\boldsymbol{\xi}}, \theta) \tag{18}$$

which holds for arbitrary pairs $(\bar{\boldsymbol{\sigma}}, \bar{\boldsymbol{\xi}})$, such that $\tilde{f}(\bar{\boldsymbol{\sigma}}, \bar{\boldsymbol{\xi}}, \theta) \leq 0$. Finally, when Eqs. (10, 18) are added to each other, and taking into account that the r.h.s. of these equations are equal to each other, one obtains:

$$(\boldsymbol{\sigma} - \bar{\boldsymbol{\sigma}}) : \dot{\boldsymbol{\varepsilon}}^{in} - \frac{1}{2}[(\boldsymbol{\chi} - \bar{\boldsymbol{\chi}}) : \dot{\boldsymbol{\xi}} + (\boldsymbol{\xi} - \bar{\boldsymbol{\xi}}) : \dot{\boldsymbol{\chi}}] \geq \Omega(\boldsymbol{\sigma}, \boldsymbol{\chi}, \theta) - \Omega(\bar{\boldsymbol{\sigma}}, \bar{\boldsymbol{\chi}}, \theta). \tag{19}$$

It is worth remarking that in Eq. (19) the equality sign holds if and only if either $\bar{\boldsymbol{\sigma}} = \boldsymbol{\sigma}$ and $\bar{\boldsymbol{\chi}} = \boldsymbol{\chi}$, hence $\boldsymbol{\xi} = \bar{\boldsymbol{\xi}}$, (in which case $\dot{\boldsymbol{\varepsilon}}^{in}$, $\dot{\boldsymbol{\xi}}$, $\dot{\boldsymbol{\chi}}$ are not vanishing), or $\dot{\boldsymbol{\varepsilon}}^{in} = \mathbf{0}$, $\dot{\boldsymbol{\xi}} = \dot{\boldsymbol{\chi}} = \mathbf{0}$ and $\Omega(\bar{\boldsymbol{\sigma}}, \bar{\boldsymbol{\chi}}, \theta) = 0$, hence $\Omega(\boldsymbol{\sigma}, \boldsymbol{\chi}, \theta) = 0$, (in which case $\boldsymbol{\sigma}$, $\boldsymbol{\chi}$, $\boldsymbol{\xi}$ may be different from $\bar{\boldsymbol{\sigma}}$, $\bar{\boldsymbol{\chi}}$, $\bar{\boldsymbol{\xi}}$, respectively).

Equation (19) can be further generalized by applying it to two different material states, say $(\boldsymbol{\sigma}_a, \boldsymbol{\chi}_a, \dot{\boldsymbol{\varepsilon}}_a^{in}, \dot{\boldsymbol{\xi}}_a)$ and $(\boldsymbol{\sigma}_b, \boldsymbol{\chi}_b, \dot{\boldsymbol{\varepsilon}}_b^{in}, \dot{\boldsymbol{\xi}}_b)$, both related to the same temperature θ. So, one obtains the inequality:

$$(\boldsymbol{\sigma}_a - \boldsymbol{\sigma}_b) : (\dot{\boldsymbol{\varepsilon}}_a^{in} - \dot{\boldsymbol{\varepsilon}}_b^{in}) - \frac{1}{2}\left[(\boldsymbol{\chi}_a - \boldsymbol{\chi}_b) : (\dot{\boldsymbol{\xi}}_a - \dot{\boldsymbol{\xi}}_b) + (\boldsymbol{\xi}_a - \boldsymbol{\xi}_b)(\dot{\boldsymbol{\chi}}_a - \dot{\boldsymbol{\chi}}_b)\right] \geq 0, \tag{20}$$

where the equality sign holds if and only if either $\boldsymbol{\sigma}_a = \boldsymbol{\sigma}_b$, $\boldsymbol{\chi}_a = \boldsymbol{\chi}_b$, $\boldsymbol{\xi}_a = \boldsymbol{\xi}_b$, in which case $\dot{\boldsymbol{\varepsilon}}_a^{in} = \dot{\boldsymbol{\varepsilon}}_b^{in} \neq \mathbf{0}$, $\dot{\boldsymbol{\xi}}_a = \dot{\boldsymbol{\xi}}_b \neq \mathbf{0}$, $\dot{\boldsymbol{\chi}}_a = \dot{\boldsymbol{\chi}}_b \neq \mathbf{0}$, or $\dot{\boldsymbol{\varepsilon}}_a^{in} = \dot{\boldsymbol{\varepsilon}}_b^{in} = \mathbf{0}$, $\dot{\boldsymbol{\xi}}_a = \dot{\boldsymbol{\xi}}_b = \mathbf{0}$, $\dot{\boldsymbol{\chi}}_a =$

$\dot{\boldsymbol{\chi}}_b = \mathbf{0}$, in which case $\boldsymbol{\sigma}_a$, $\boldsymbol{\chi}_a$, $\boldsymbol{\xi}_a$ may differ from $\boldsymbol{\sigma}_b$, $\boldsymbol{\chi}_b$, $\boldsymbol{\xi}_b$, respectively. Inequality (20) formally coincides with inequalities previously proposed for time-independent plasticity (cf. HALPHEN & NGUYEN [1975], POLIZZOTTO [1993d]), but here inelastic strain rates appear in place of the time-independent plastic strain rates. A material obeying inequalities (19) and (20) is referred to as generalized standard elastic-viscoplastic material in the following.

INITIAL/BOUNDARY VALUE PROBLEM AND STEADY CYCLE

A continuous solid body is considered here. In the undeformed state, the body occupies the domain V of the three-dimensional Euclidean space and is referred to a Cartesian orthogonal co-ordinate system $\mathbf{x} = (x_1, x_2, x_3)$. It is restrained upon the part $\partial_u V$ of its boundary ∂V. The material obeys the constitutive equations previously described, and has temperature-independent elastic and thermal espansion properties. The body is subjected to external actions like body forces in V, tractions on $\partial_t V = \partial V \backslash \partial_u V$, imposed displacements on $\partial_u V$ and temperature variations in V. By hypothesis, all these actions vary with time $t \geq 0$ in a quasi-thermostatic manner, and are all periodic with the *same* period Δt. By assumption, the body undergoes small displacements and strains, such that geometrical nonlinearities are ruled out. At time $t = 0$, the initial conditions are specified as given initial inelastic strains and consequent self-stresses.

For the subsequent developments, it is useful to consider the thermo-elastic response of the body to the above actions and initial conditions, that is, the body's response that would be obtained on disregarding its inelastic behaviour. Symbols like $(\cdot)^E$ will be employed henceforth to denote any quantity pertaining to the latter response.

The body's actual response, denoted with symbols $\boldsymbol{\sigma}$, $\boldsymbol{\varepsilon}$, $\mathbf{u}, \ldots$ for stresses, strains, displacements, etc., could be evaluated by a step-by-step procedure in conjuction with a discretization method (e.g. the finite element method, the boundary element method). Instead, our present purpose is to investigate whether the body's response to a given cyclic loading, after an initial transient phase, stabilizes into a steady-state response, and also to establish the related periodicity characteristics.

Before proceeding in this direction, it is useful to establish a preliminary result. For this purpose, let the body experience two distinct "isothermal" state evolutions, denoted with symbols $\boldsymbol{\sigma}_a, \boldsymbol{\varepsilon}_a, \mathbf{u}_a$, etc., and $\boldsymbol{\sigma}_b, \boldsymbol{\varepsilon}_b, \mathbf{u}_b$, etc., respectively, with $\theta_a(\mathbf{x}, t) = \theta_b(\mathbf{x}, t) = \theta(\mathbf{x}, t)$ for all $\mathbf{x} \in V$ and for all $t \geq 0$. By assumption, a particular condition occurs in the body, such that the difference fields defined as $\Delta(\cdot) := (\cdot)_a - (\cdot)_b$ satisfy the identity

$$\Delta\boldsymbol{\sigma} : \Delta\dot{\boldsymbol{\varepsilon}}^{in} - \frac{1}{2}\frac{d}{dt}(\Delta\boldsymbol{\chi} : \Delta\boldsymbol{\xi}) = 0 \text{ in } V, \text{ all } t \in J \tag{21}$$

where J is a specified (closed) time interval, say $J = \{t : t_0 \le t \le t_1, t_0 \ge 0\}$. It is worth remarking that the l.h.s. of Eq. (21) coincides with the l.h.s. of inequality (20), provided that the two state evolutions constitute time-continuous deformation processes, as it is assumed here.

In view of what was stated about the equality sign holding in inequality (20), the condition of Eq. (21) implies the following:

- $\Delta\dot{\boldsymbol{\varepsilon}}^p = \Delta\dot{\boldsymbol{\varepsilon}}^c = \mathbf{0}$ and $\Delta\dot{\boldsymbol{\chi}} = \Delta\dot{\boldsymbol{\xi}} = \mathbf{0}$ in $V \times J$;
- $\Delta\boldsymbol{\sigma} = \mathbf{0}$, $\Delta\boldsymbol{\chi} = \Delta\boldsymbol{\xi} = \mathbf{0}$ at points $\mathbf{x}$ in the region $V_p \subset V$ and at times $t = t_p(\mathbf{x}) \in J_p$ where the material yields, as well as at points $\mathbf{x}$ in the region $V_c \subset V$ and at times $t \in t_c(\mathbf{x}) \in J_c$ where the material creeps;
- $\Delta\boldsymbol{\sigma}$, $\Delta\boldsymbol{\chi}$, $\Delta\boldsymbol{\xi}$ may not vanish in $V_{p+c} := V_p \cup V_c$ and at times $t \in J_e := J \backslash J_{p+c}$, where $J_{p+c} := J_p \cup J_c$, as well as in the region $V_e := V \backslash V_{p+c}$ and for all $t \in J$.

Additionally, since $\Delta\boldsymbol{\chi}$ and $\Delta\boldsymbol{\xi}$ are time-independent fields in $V \times J$, it follows that $\Delta\boldsymbol{\chi} = \Delta\boldsymbol{\xi} = \mathbf{0}$ in $V_{p+c} \times J$ (because these fields vanish at certain times there). Also, the residual stress rate difference field, $\Delta\dot{\boldsymbol{\sigma}}^R$, vanishes in $V \times J$ like $\Delta\dot{\boldsymbol{\varepsilon}}^{in}$ (to which it is the elastic response), such that $\Delta\boldsymbol{\sigma}^R$ is time-independent in $V \times J$. Therefore, upon validity of Eq. (21), the following evolution uniqueness requisite can be stated

Evolution uniqueness requisite — For a generalized standard elastic-viscoplastic structure for which two different isothermal state evolutions and some time interval J can be envisaged such as to satisfy Eq. (21), the following properties hold:

i) The plastic and creep strain rate difference fields vanish identically, i.e. $\Delta\dot{\boldsymbol{\varepsilon}}^p = \Delta\dot{\boldsymbol{\varepsilon}}^c = \mathbf{0}$ in V for all $t \in J$;

ii) The internal variable difference fields, $\Delta\boldsymbol{\chi}$ and $\Delta\boldsymbol{\xi}$, are time-independent in $V \times J$ and in particular they vanish in the region where inelastic strain rates take place, i.e. $\Delta\boldsymbol{\chi} = \Delta\boldsymbol{\xi} = \mathbf{0}$ in $V_{p+c} \times J$;

iii) The residual stress difference field, $\Delta\boldsymbol{\sigma}^R$, is time-independent in $V \times J$;

iv) The stress difference field, $\Delta\boldsymbol{\sigma} = \Delta\boldsymbol{\sigma}^E + \Delta\boldsymbol{\sigma}^R$, vanishes in the region where inelastic strain rates take place, i.e. $\Delta\boldsymbol{\sigma} = \mathbf{0}$ in $V_{p+c} \times J_{p+c}$.

If the two considered state evolutions occur under the *same* load history, such that $\Delta\boldsymbol{\sigma}^E \equiv \mathbf{0}$, hence, $\Delta\boldsymbol{\sigma} = \Delta\boldsymbol{\sigma}^R$ identically, the properties i) and ii) above remain unaltered, whereas those under points iii) and iv) can be unified into the following point v), namely

v) The stress difference fields $\Delta\boldsymbol{\sigma}$ and $\Delta\boldsymbol{\sigma}^R$ are equal to each other and time--independent in $V \times J$, and in particular they vanish in $V_{p+c} \times J$.

The statement above is an extension to viscoplasticity of an analogous statement previously given for time-independent plasticity (cf. POLIZZOTTO [1993d]).

Next, let the actual response of the body introduced at the beginning of this Section be again considered and let the symbol $\Delta(\cdot)$ denotes the difference of values of a related variable $(\cdot)$ at times t and $t + \Delta t$, as for instance $\Delta\boldsymbol{\sigma}(\mathbf{x}, t) := \boldsymbol{\sigma}(\mathbf{x}, t + \Delta t) - \boldsymbol{\sigma}(\mathbf{x}, t)$. Since Eqs. (3-6) are continuously satisfied for all $t \geq 0$ and in the whole V, and since $\theta(\mathbf{x}, t + \Delta t) = \theta(\mathbf{x}, t)$ for all $t \geq 0$ in V, it follows that inequality (20) applies, that is, with the present notation,

$$\ell := \Delta\boldsymbol{\sigma} : \Delta\dot{\boldsymbol{\varepsilon}}^{in} - \frac{1}{2}\frac{d}{dt}(\Delta\boldsymbol{\chi} : \Delta\boldsymbol{\xi}) \geq 0 \text{ in } V, \text{ for all } t \geq 0. \tag{22}$$

From the second Eq. (6) written in the rate form, that is

$$\dot{\boldsymbol{\varepsilon}} = \mathbf{A} : \dot{\boldsymbol{\sigma}} + \mathbf{a}\dot{\theta} + \dot{\boldsymbol{\varepsilon}}^{in}, \tag{23}$$

and evaluated at times t and $t + \Delta t$, it follows that

$$\Delta\dot{\boldsymbol{\varepsilon}}^{in} = \Delta\dot{\boldsymbol{\varepsilon}} - \mathbf{A} : \Delta\dot{\boldsymbol{\sigma}} \text{ in } V \tag{24}$$

which is valid for all $t \geq 0$. Then, substituting Eq. (24) into Eq. (22) and with subsequent integration over V gives

$$\int_V \ell \, dV = \int_V \Delta\boldsymbol{\sigma} : \Delta\dot{\boldsymbol{\varepsilon}} \, dV - \int_V \Delta\boldsymbol{\sigma} : \mathbf{A} : \Delta\dot{\boldsymbol{\sigma}} \, dV - \frac{1}{2}\frac{d}{dt}\int_V \Delta\boldsymbol{\chi} : \Delta\boldsymbol{\xi} \, dV. \tag{25}$$

Since $\Delta\boldsymbol{\sigma}$, as the difference between two stress states in equilibrium with the same loads, is a self-stress field, and since $\Delta\dot{\boldsymbol{\varepsilon}}$, as the difference between two strain rate states compatible with the same displacement rates on $\partial_u V$, is a compatible strain rate field with zero displacement rates on $\partial_u V$, it follows that, by the virtual work principle, the first integral on the r.h.s. of Eq. (25) vanishes. Thus, on introducing the state functional

$$L := \frac{1}{2}\int_V \Delta\boldsymbol{\sigma} : \mathbf{A} : \Delta\boldsymbol{\sigma} \, dV + \frac{1}{2}\int_V \Delta\boldsymbol{\chi} : \Delta\boldsymbol{\xi} \, dV, \tag{26}$$

Eq. (25), in virtue of Eq. (22), reads

$$-\int_V \ell \, dV = \frac{dL}{dt} \leq 0. \tag{27}$$

This means that, during the loading process, L decreases monotonically as time elapses. Since L is positive definite and hence cannot take negative values, it necessarily must stop decreasing at a certain time t_s, such that $dL/dt = 0$ for all $t \geq t_s$, and as a consequence Eq. (22) is satisfied as an equality for $t \geq t_s$. Thus, in the light of the evolution uniqueness requisite, it can be stated that in the stabilized phase the body experiences, under the same load history, two distinct isothermal state evolutions satisfying Eq. (21) with $t_0 = t_s$ and $t_1 = +\infty$; and that the evolution uniqueness requisite applies with its properties i), ii) and v). Additionally, for the problem at hand it is $\Delta\boldsymbol{\chi} = \Delta\boldsymbol{\xi} = \mathbf{0}$ and $\Delta\dot{\boldsymbol{\varepsilon}}^{in} = \mathbf{0}$ in the body's region $V_e = V \backslash V_{p+c}$ where no inelastic strains occur after t_s, and thus also $\Delta\boldsymbol{\sigma}^R = \mathbf{0}$ in V for all $t \geq t_s$. Therefore, the following proposition can be stated.

Proposition 1 — In a structure of generalized standard elastic-viscoplastic material subjected to cyclic loads and temperature variations, there exists a stabilization time after which the response is characterized by stresses $\boldsymbol{\sigma}$, inelastic strain rates $\dot{\boldsymbol{\varepsilon}}^{in} = \dot{\boldsymbol{\varepsilon}}^p + \dot{\boldsymbol{\varepsilon}}^c$ and internal variables $\boldsymbol{\chi}$, $\boldsymbol{\xi}$ that are all periodic with the same period as the loads and temperature variation.

The restrictive hypothesis of smooth yield and creep functions, f and g, can be removed using classical arguments for perfect instantaneous plasticity (cf. MARTIN [1975]) such that the above result can be considered valid also for nonsmooth functions f and g, but this point is not addressed here for simplicity.

It is worth noticing that, by Proposition 1, for loads and temperature variations for which $V_{p+c} = \emptyset$, the body's stress and hardening states at any time $t \geq t_s$ are recovered after a complete cycle of time length Δt, whereas this is not the case for the body's inelastic strain state. In fact, the inelastic strain states at times $t \geq t_s$ and $t + \Delta t$ in general differ from each other by a time-independent strain field $\Delta\boldsymbol{\varepsilon}^{in}$ which is compatible in V with zero displacements on $\partial_u V$ — except when $\Delta\boldsymbol{\varepsilon}^{in}$ vanishes identically (*inelastic shakedown*).

It is of interest to establish which characteristics of the stabilized response are independent of the initial conditions. For this purpose, a procedure similar to the one employed for the derivation of Proposition 1 is followed. Let symbols as $(\cdot)_a$ and $(\cdot)_b$ label two different body's responses to the same loads and temperature variations, but with different initial conditions, and let the symbol $\Delta(\cdot)$ indicate the difference of values of a variable $(\cdot)$ in the two responses, but at the same time t, e.g. $\Delta\boldsymbol{\sigma}(\mathbf{x}, t) := \boldsymbol{\sigma}_a(\mathbf{x}, t) - \boldsymbol{\sigma}_b(\mathbf{x}, t)$. Since both responses meet Eqs. (3-6), inequality (20) applies and takes again the form of Eq. (22), but with the difference fields having the newly established meanings. Obviously, the same reasoning followed previously from Eq. (23) to Eq. (27) can be followed here, with the consequence that Eq. (22) is satisfied as an inequality for $t \geq t_s$. Thus, on the light of the uniqueness requisite, it can be stated that the body experiences, under the same load, two different isothermal state evolutions satisfying Eq. (21) with $t_0 = t_s$ and $t_1 = +\infty$, and that this requisite applies with its properties i), ii) and v). Moreover, as the situation when the two inelastic strain responses may differ from each other by a compatible time-independent strain field $\Delta\boldsymbol{\varepsilon}^{in}$ can be excluded, one can state that $\Delta\boldsymbol{\varepsilon}^{in} = \mathbf{0}$ in V_{p+c} for all $t \geq t_s$ and that in $V_e = V\backslash V_{p+c}$ some nonvanishing time-independent $\Delta\boldsymbol{\varepsilon}^{in}$, $\Delta\boldsymbol{\chi}$, $\Delta\boldsymbol{\xi}$ may exist as a consequence of the different initial conditions inducing different strain processes. Therefore, it can be concluded that, fot all $t \geq t_s$, $\dot{\boldsymbol{\varepsilon}}_a^{in} = \dot{\boldsymbol{\varepsilon}}_b^{in}$, $\dot{\boldsymbol{\chi}}_a = \dot{\boldsymbol{\chi}}_b$, $\dot{\boldsymbol{\xi}}_a = \dot{\boldsymbol{\xi}}_b$ in V, and $\boldsymbol{\sigma}_a = \boldsymbol{\sigma}_b$, $\boldsymbol{\varepsilon}_a^{in} = \boldsymbol{\varepsilon}_b^{in}$, $\boldsymbol{\chi}_a = \boldsymbol{\chi}_b$, $\boldsymbol{\xi}_a = \boldsymbol{\xi}_b$ in V_{p+c} (but not necessarily in V_e). The following proposition can thus be stated.

Proposition 2 — In a structure with a generalized standard elastic-viscoplastic material subjected to cyclic loads and temperature variations (but unspecified initial conditions), the long term response exhibits uniqueness for the inelastic strain rates $\dot{\boldsymbol{\varepsilon}}^{in} = \dot{\boldsymbol{\varepsilon}}^p + \dot{\boldsymbol{\varepsilon}}^c$ and the internal variable rates $\dot{\boldsymbol{\chi}}$, $\dot{\boldsymbol{\xi}}$ in the whole V, as well as for the stresses $\boldsymbol{\sigma}$, inelastic strains $\boldsymbol{\varepsilon}^{in}$ and internal variables $\boldsymbol{\chi}$, $\boldsymbol{\xi}$ in the region V_{p+c} where inelastic strain occurs in the long term, whereas in the elastic region $V_e = V\backslash V_{p+c}$ some time-independent $\boldsymbol{\varepsilon}^{in}$, $\boldsymbol{\chi}$, $\boldsymbol{\xi}$ may take place with values dependent upon the initial conditions and the ensuing transient straining process.

The synonymous expressions *steady-state response, steady-state cycle*, or simply *steady cycle*, allude to the long term stabilized response to given cyclic loads and temperature variations, but with unspecified initial conditions.

EQUATIONS GOVERNING THE STEADY CYCLE

For given cyclic loads and temperature variations, the steady cycle can be directly determined (that is, without any evolutionary analysis starting from $t = 0$), at least in principle. An *ad-hoc* equation set must be solved for this purpose. These equations account for the results previously obtained and read:

$$\boldsymbol{\sigma} = \boldsymbol{\sigma}^E + \boldsymbol{\sigma}^R, \quad \boldsymbol{\chi} = \frac{\partial \Psi}{\partial \boldsymbol{\xi}} \quad \text{in } V \times (0, \Delta t) \tag{28}$$

$$f(\boldsymbol{\sigma}, \boldsymbol{\chi}, \theta) \leq 0, \quad \dot{\lambda} \geq 0, \quad \dot{\lambda}\, f(\boldsymbol{\sigma}, \boldsymbol{\chi}, \theta) = 0 \quad \text{in } V \times (0, \Delta t) \tag{29}$$

$$\dot{\boldsymbol{\varepsilon}}^{in} = \dot{\lambda}\frac{\partial f}{\partial \boldsymbol{\sigma}} + \frac{\partial \Omega}{\partial \boldsymbol{\sigma}}, \quad -\dot{\boldsymbol{\xi}} = \dot{\lambda}\frac{\partial f}{\partial \boldsymbol{\chi}} + \frac{\partial \Omega}{\partial \boldsymbol{\chi}} \quad \text{in } V \times (0, \Delta t) \tag{30}$$

$$C^T \boldsymbol{\sigma}^R = \mathbf{0} \text{ in } V \times (0, \Delta t) \quad \boldsymbol{\sigma}^R \cdot \mathbf{n} = \mathbf{0} \text{ on } \partial_t V \times (0, \Delta t), \tag{31a}$$

$$C\dot{\mathbf{u}}^R = \mathbf{A} : \dot{\boldsymbol{\sigma}}^R + \dot{\boldsymbol{\varepsilon}}^{in} \text{ in } V \times (0, \Delta t), \quad \dot{\mathbf{u}}^R = \mathbf{0} \text{ on } \partial_u V \times (0, \Delta t) \tag{31b}$$

$$\Delta\boldsymbol{\varepsilon}^{in} := \int_0^{\Delta t} \dot{\boldsymbol{\varepsilon}}^{in}\, dt = C\mathbf{v} \text{ in } V, \quad \mathbf{v} = \mathbf{0} \text{ on } \partial_u V \tag{32}$$

$$\Delta\boldsymbol{\xi} := \int_0^{\Delta t} \dot{\boldsymbol{\xi}}\, dt = \mathbf{0} \text{ in } V, \tag{33}$$

where the same symbols as in the previous sections are employed.

Here, $\boldsymbol{\sigma}^E = \boldsymbol{\sigma}^E(\mathbf{x}, t)$, $0 \leq t \leq \Delta t$, denotes the known thermo-elastic response to the loads and temperature variations specified in $V \times (0, \Delta t)$. The stress $\boldsymbol{\sigma}$ is expressed as the superposition of $\boldsymbol{\sigma}^E$ with the residual stress field $\boldsymbol{\sigma}^R$, the latter being associated with the residual displacement field $\mathbf{u}^R$. C is the compatibility differential operator, C^T its adjoint, namely

$$C(\cdot) = \frac{1}{2}[\text{grad}(\cdot) + \text{grad}(\cdot)^T], \quad C^T(:) = \text{div}(:),$$

whereas $\mathbf{n}$ denotes the unit external normal to ∂V. Thus, Eq. (31*a*) states that $\boldsymbol{\sigma}^R$ constitutes a self-stress field, and Eq. (31*b*) that $\dot{\boldsymbol{\sigma}}^R$ and $\dot{\mathbf{u}}^R$ are the elastic responses to $\dot{\boldsymbol{\varepsilon}}^{in}$. Moreover, Eq. (32) states that the *inelastic strain ratchet* $\Delta\boldsymbol{\varepsilon}^{in}(\mathbf{x})$, that is, the net inelastic strain cumulated in a complete cycle, is compatible with the displacements $\mathbf{v}(\mathbf{x})$ vanishing on $\partial_u V$, whereas Eq. (33) states that the micro-strain ratchet $\Delta\boldsymbol{\xi}(\mathbf{x})$ is identically vanishing. Equations (32, 33) amount to stating that the body's stress and hardening states existing at $t = 0$ are fully recovered at $t = \Delta t$. Any strain rate history pair $(\dot{\boldsymbol{\varepsilon}}^{in}, \dot{\boldsymbol{\xi}})$ specified in $V \times (0, \Delta t)$ and satisfying, with some $\mathbf{v}$, Eqs. (32, 33) is referred to as a *Plastic Accumulation Mechanism* (PAM) (or *Ratchet Mechanism*), in the following. The concept of PAM, introduced previously in the framework of time-independent plasticity (cf. POLIZZOTTO et al. [1991]), generalizes the concept of "kinematically admissible plastic strain cycle" due to Koiter (cf. KOITER [1960]). The PAM pertaining to the solution to Eqs. (28-33) is characterized by $\mathbf{v} \equiv \Delta\mathbf{u}^R$, as it can be easily shown. Equations (28-33) are an extension to viscoplasticity of the analogous equations for time-independent plasticity (cf. POLIZZOTTO [1993a, 1993d]). They describe a problem similar to an evolutive one in viscoplasticity with internal variables, but with the initial conditions replaced by time-integral conditions, Eqs. (32, 33). Solving such a problem is in general quite difficult and for this reason direct solution methods are hoped for.

It can be proven that the solution to Eqs. (28-33) is unique for all, except for $\boldsymbol{\sigma}^R$, $\boldsymbol{\chi}$ and $\boldsymbol{\xi}$, which are uniquely determined within the region $V_{p+c} \subseteq V$ (if any), where inelastic strains occur. In order to accomplish this task, let us assume that two solutions do exist, respectively labeled with symbols $(\cdot)_a$ and $(\cdot)_b$. Since both solutions satisfy Eqs. (28-30), inequality (20) holds at every $\mathbf{x} \in V$ and every

$t \in (0, \Delta t)$, that is, on setting, e.g., $\Delta\boldsymbol{\sigma}(\mathbf{x}, t) := \boldsymbol{\sigma}_a(\mathbf{x}, t) - \boldsymbol{\sigma}_b(\mathbf{x}, t)$, an inequality similar to Eq. (22) can be written, namely

$$\ell := \Delta\boldsymbol{\sigma} : \Delta\dot{\boldsymbol{\varepsilon}}^{in} - \frac{1}{2}\frac{d}{dt}(\Delta\boldsymbol{\chi} : \Delta\boldsymbol{\xi}) \geq 0 \quad \text{in } V \times (0, \Delta t). \tag{34}$$

Again the reasoning from Eq. (23) to Eq. (27) can be adopted here, and then it can be stated that

$$\int_0^{\Delta t} \int_V \ell \, dV \, dt = L(0) - L(\Delta t) = 0 \tag{35}$$

because L (given by Eq. (26), but with the difference fields having the new meanings established here) takes equal values at $t = 0$ and $t = \Delta t$. As a consequence, ℓ vanishes identically. In the light of the evolution uniqueness requisite, this fact implies that the body experiences, under the same load history, two isothermal state evolutions satisfying Eq. (21) with $t = 0$, $t_1 = \Delta t$, such that the mentioned requisite again applies with its properties i), ii) and v). Therefore, it can be stated that:

- $\Delta\dot{\boldsymbol{\varepsilon}}^{in}$, $\Delta\dot{\boldsymbol{\chi}}$, $\Delta\dot{\boldsymbol{\xi}}$ vanish in $V \times (0, \Delta t)$, that is $\dot{\boldsymbol{\varepsilon}}^{in}$, $\dot{\boldsymbol{\chi}}$, $\dot{\boldsymbol{\xi}}$ are uniquely determined in $V \times (0, \Delta t)$;

- $\Delta\boldsymbol{\sigma}$, $\Delta\boldsymbol{\chi}$, $\Delta\boldsymbol{\xi}$ vanish in $V_{p+c} \times (0, \Delta t)$, that is $\boldsymbol{\sigma}$, $\boldsymbol{\chi}$, $\boldsymbol{\xi}$ are uniquely determined in $V_{p+c} \times (0, \Delta t)$;

- $\boldsymbol{\sigma}^R$, $\boldsymbol{\chi}$, $\boldsymbol{\xi}$ are determined in $V_e = V \backslash V_{p+c}$ to within some time-independent fields.

So, the uniqueness statement given above is proven. In the particular case in which $V_{p+c} = \emptyset$, it turns out that $\dot{\boldsymbol{\varepsilon}}^p = \dot{\boldsymbol{\varepsilon}}^c = \mathbf{0}$, $\dot{\boldsymbol{\chi}} = \dot{\boldsymbol{\xi}} = \mathbf{0}$ identically, whereas $\boldsymbol{\sigma}^R$, $\boldsymbol{\chi}$, $\boldsymbol{\xi}$ are time-independent nonuniquely determined fields in the whole V (*elastic shakedown*).

A question which arises at this point is whether the solution to Eqs. (28-33) coincides, at least in part, with the steady-state response described previously. This point can be clarified by letting symbols $(\cdot)_a$ be used for the solution to Eqs. (28-33), and symbols $(\cdot)_b$ for the steady-state response. Let a piece of the steady-state response of time lenght Δt be considered, and let a local time t, $0 \leq t \leq \Delta t$, be introduced in such a way that a- and b-labeled quantities turn out to be all functions of t and are related to the same load and temperature at every t in $(0, \Delta t)$. Since both types of solution satisfy the constitutive equations, an inequality formally identical to Eq. (34) can be written. Then, again following the reasoning from Eq. (23) to Eq. (27), one arrives at an equality like Eq. (35), with L given by Eq. (26), but the difference fields having the new meanings. Since these fields have equal values at $t = 0$ and $t = \Delta t$, it follows that ℓ vanishes identically and that, as a consequence, the state evolution requisite applies to the present case with $t_0 = 0$ and $t_1 = \Delta t$ and with its properties i), ii) and v). Thus, it can be stated that, for every $t \in (0, \Delta t)$, $\dot{\boldsymbol{\varepsilon}}_a^{in} = \dot{\boldsymbol{\varepsilon}}_b^{in}$, $\dot{\boldsymbol{\varepsilon}}_a^p = \dot{\boldsymbol{\varepsilon}}_b^p$, $\dot{\boldsymbol{\varepsilon}}_a^c = \dot{\boldsymbol{\varepsilon}}_b^c$, $\dot{\boldsymbol{\chi}}_a = \dot{\boldsymbol{\chi}}_b$, $\dot{\boldsymbol{\xi}}_a = \dot{\boldsymbol{\xi}}_b$ in V, and furthermore $\boldsymbol{\sigma}_a = \boldsymbol{\sigma}_b$, $\boldsymbol{\chi}_a = \boldsymbol{\chi}_b$, $\boldsymbol{\xi}_a = \boldsymbol{\xi}_b$ in V_{p+c} (where inelastic strain occurs), but $\boldsymbol{\sigma}_a$, $\boldsymbol{\chi}_a$, $\boldsymbol{\xi}_a$ may not coincide with $\boldsymbol{\sigma}_b$, $\boldsymbol{\chi}_b$, $\boldsymbol{\xi}_b$ in $V_e = V \backslash V_{p+c}$ (where $\boldsymbol{\sigma}_a$, $\boldsymbol{\chi}_a$, $\boldsymbol{\xi}_a$ are not uniquely determined). This proves that the uniquely determined part of the solution to Eqs. (28-33) coincides with that part of the steady-state response which is independent of the initial conditions. Thus, the solution to Eqs. (28-33) can be referred to as the steady cycle.

As a consequence of what has been previously established, the stress tensor $\boldsymbol{\sigma}^R$ pertaining to the steady-cycle can be given the representation

$$\boldsymbol{\sigma}^R(\mathbf{x}, t) = \boldsymbol{\rho}(\mathbf{x}) + \boldsymbol{\tau}(\mathbf{x}, t) \tag{36}$$

where $\boldsymbol{\rho}$ (*post-transient residual stress*) is the residual stress tensor associated with the inelastic strains $\dot{\boldsymbol{\varepsilon}}_s^{in}(\mathbf{x})$ and the internal variables $\boldsymbol{\chi}_s(\mathbf{x})$, $\boldsymbol{\xi}_s(\mathbf{x})$ occurring at the stabilization time, whereas $\boldsymbol{\tau}$ is the pure cyclic residual stress tensor, that is, the residual stresses associated with the additional inelastic strains of the steady cycle. This implies that $\boldsymbol{\sigma}^R(\mathbf{x}, t^*) = \boldsymbol{\rho}(\mathbf{x})$, $\boldsymbol{\tau}(\mathbf{x}, t^*) = \mathbf{0}$ in V for some $t^* \in (0, \Delta t)$. Since the steady cycle characteristics are independent of the load and temperature cycle origin, it is always possible, maybe with a suitable choice of this origin, to obtain $t^* = 0$, that is to make the stabilization time occur at the beginning of some subsequent cycle. Thus, one can write

$$\boldsymbol{\tau}(\mathbf{x}, t) = \int_0^t \dot{\boldsymbol{\tau}}(\mathbf{x}, \bar{t})\, d\bar{t}. \tag{37}$$

In this way, $\boldsymbol{\tau}$ is uniquely determined in $V \times (0, \Delta t)$, whereas $\boldsymbol{\rho}$ is uniquely determined (at least) within V_{p+c}. In the framework of time-independent perfect plasticity, variational methods have been provided for the direct determination of the post-transient residual stresses $\boldsymbol{\rho}$ (cf. POLIZZOTTO [1993a, 1993b, 1993c]).

ELASTIC SHAKEDOWN OF AN ELASTIC-VISCOPLASTIC BODY

A special case of steady-state response is, in a sense, a trivial one, occurring when the steady cycle is characterized by identically vanishing inelastic strain rates, namely $V_{p+c} = \emptyset$, and some time-independent $\boldsymbol{\sigma}^R = \bar{\boldsymbol{\sigma}}^R(\mathbf{x})$, $\boldsymbol{\chi} = \bar{\boldsymbol{\chi}}(\mathbf{x})$ and $\boldsymbol{\xi} = \bar{\boldsymbol{\xi}}(\mathbf{x})$ take place in $V_e = V$. Equations (28-33) lose their effectiveness, except for the equilibrium conditions of Eq. (31a) for $\bar{\boldsymbol{\sigma}}^R$, along with the condition $g(\boldsymbol{\sigma}^E + \bar{\boldsymbol{\sigma}}^R, \bar{\boldsymbol{\chi}}, \theta) < 0$ in $V \times (0, \Delta t)$, hence *a-fortiori* $f(\boldsymbol{\sigma}^E + \bar{\boldsymbol{\sigma}}^R, \bar{\boldsymbol{\chi}}, \theta) < 0$ in $V \times (0, \Delta t)$. The occurrence of such a steady cycle is possible as far as the viscoplastic material model exhibits a finite elastic domain $g \leq 0$, as is assumed here.

A few crucial questions arise at this point, namely:

i) How Melan's and Koiter's theorems of classical elastic shakedown theory can be employed within the present viscoplasticity context?

ii) What kind of limit state does occur in the structure under an elastic shakedown limit load?

The answer to the first question is intuitively clear. It is, in fact, sufficient to treat the creep function g as if it were a yield one, and accordingly formulate Melan's and Koiter's theorems. Melan's theorem (KOITER [1960], KÖNIG [1987], MARTIN [1978], MAIER [1987], MAIER & NOVATI [1987], POLIZZOTTO et al. [1991]) can be phrased as follows.

Melan theorem — For a given elastic-viscoplastic solid subjected to cyclic loads and temperature variations, a necessary and sufficient condition for elastic shakedown to occur is that there exist time-independent self-stress and micro-stress fields, $\bar{\boldsymbol{\sigma}}^R$ and $\bar{\boldsymbol{\chi}}$, satisfying the condition

$$g(\boldsymbol{\sigma}^E + \bar{\boldsymbol{\sigma}}^R, \bar{\boldsymbol{\chi}}, \theta) \leq 0 \quad \text{in } V \times (0, \Delta t) \tag{38}$$

where $\boldsymbol{\sigma}^E = \boldsymbol{\sigma}^E(\mathbf{x}, t)$ is the thermo-elastic stress response and $\theta = \theta(\mathbf{x}, t)$ is the assigned temperature variation history.

Proof. The necessity is quite obvious. As to the sufficiency, as usual for time--independent plasticity, the proof can be by achieved assuming Eq. (38) to be satisfied in a stronger form, that is

$$g(m\hat{\boldsymbol{\sigma}}, \bar{\boldsymbol{\chi}}, \theta) \leq 0 \quad \text{in } V \times (0, \Delta t) \tag{39}$$

where $m > 1$ is some scalar and,

$$\hat{\boldsymbol{\sigma}} = \hat{\boldsymbol{\sigma}}(\mathbf{x}, t) := \boldsymbol{\sigma}^E(\mathbf{x}, t) + \bar{\boldsymbol{\sigma}}^R(\mathbf{x}) \quad \text{in } V \times (0, \Delta t). \tag{40}$$

Denoting by $\boldsymbol{\sigma}$, $\boldsymbol{\chi}$, $\boldsymbol{\xi}$, $\dot{\boldsymbol{\varepsilon}}^{in} = \dot{\boldsymbol{\varepsilon}}^c$, etc., the actual response to the given loads and temperature variations (note that $\dot{\boldsymbol{\varepsilon}}^p \equiv \mathbf{0}$), and viewing this response as a function of the general time T, and using inequality (19), one can write

$$(\boldsymbol{\sigma} - m\hat{\boldsymbol{\sigma}}) : \dot{\boldsymbol{\varepsilon}}^c - \frac{1}{2}\frac{d}{dT}[(\boldsymbol{\chi} - \bar{\boldsymbol{\chi}}) : (\boldsymbol{\xi} - \bar{\boldsymbol{\xi}})] \geq \Omega(\boldsymbol{\sigma}, \boldsymbol{\chi}, \theta) \quad \text{in } V\text{, all } T \geq 0 \tag{41}$$

where $\hat{\boldsymbol{\sigma}}$ and θ are also treated as functions of T, $\bar{\boldsymbol{\xi}}$ is the micro-strain corresponding to $\bar{\boldsymbol{\chi}}$ and the identically vanishing of $\Omega(m\hat{\boldsymbol{\sigma}}, \bar{\boldsymbol{\chi}}, \theta)$ has been accounted for. Since

$$\boldsymbol{\sigma} - m\hat{\boldsymbol{\sigma}} = -(m-1)\boldsymbol{\sigma} + m(\boldsymbol{\sigma} - \hat{\boldsymbol{\sigma}}), \tag{42}$$

Eq. (41), after an integration over V, gives

$$(m-1)\int_V D^c(\dot{\boldsymbol{\varepsilon}}^c,\theta)dV \le m\int_V(\boldsymbol{\sigma}-\hat{\boldsymbol{\sigma}}):\dot{\boldsymbol{\varepsilon}}^c\,dV - \frac{1}{2}\frac{d}{dT}\int_V(\boldsymbol{\chi}-\bar{\boldsymbol{\chi}}):(\boldsymbol{\xi}-\bar{\boldsymbol{\xi}})\,dV - \int_V \Omega(\boldsymbol{\sigma},\boldsymbol{\chi},\theta)\,dV \tag{43}$$

where D^c is the creep dissipation function; as the last subtractive posivive integral on the r.h.s. of Eq. (43) can be dropped, which causes the enforcement of inequality, Eq. (43) becomes

$$(m-1)\int_V D^c(\dot{\boldsymbol{\varepsilon}}^c,\theta)dV \le m\int_V(\boldsymbol{\sigma}-\hat{\boldsymbol{\sigma}}):\dot{\boldsymbol{\varepsilon}}^c dV - \frac{1}{2}\frac{d}{dT}\int_V(\boldsymbol{\chi}-\bar{\boldsymbol{\chi}}):(\boldsymbol{\xi}-\bar{\boldsymbol{\xi}})dV \tag{44}$$

which holds for all $T \ge 0$. By the identity

$$\dot{\boldsymbol{\varepsilon}}^{in} = \dot{\boldsymbol{\varepsilon}}^c = \dot{\boldsymbol{\varepsilon}} - \dot{\boldsymbol{\varepsilon}}^E - \mathbf{A}:(\dot{\boldsymbol{\sigma}} - \dot{\hat{\boldsymbol{\sigma}}}), \tag{45}$$

and making use of the virtual work principle, the equality

$$\int_V(\boldsymbol{\sigma}-\hat{\boldsymbol{\sigma}}):\dot{\boldsymbol{\varepsilon}}^c\,dV = -\frac{1}{2}\frac{d}{dT}\int_V(\boldsymbol{\sigma}-\hat{\boldsymbol{\sigma}}):\mathbf{A}:(\boldsymbol{\sigma}-\hat{\boldsymbol{\sigma}})\,dV \tag{46}$$

is shown to hold for all $T \ge 0$. Therefore, after an integration over $(0, T_1)$ and with the introduction of the positive definite functional

$$Q := \frac{m}{2}\int_V(\boldsymbol{\sigma}-\hat{\boldsymbol{\sigma}}):\mathbf{A}:(\boldsymbol{\sigma}-\hat{\boldsymbol{\sigma}})\,dV + \frac{1}{2}\frac{d}{dT}\int_V(\boldsymbol{\chi}-\bar{\boldsymbol{\chi}}):(\boldsymbol{\xi}-\bar{\boldsymbol{\xi}})\,dV, \tag{47}$$

inequality (44) becomes

$$\int_0^{T_1}\int_V D^c(\dot{\boldsymbol{\varepsilon}}^c,\theta)\,dV\,dT \le \frac{1}{m-1}[Q(0)-Q(T_1)] \le \frac{1}{m-1}Q(0) \tag{48}$$

where the subtractive positive term $Q(T_1)$ has been legitimately dropped. Finally, letting T_1 tend to infinity, inequality (48) reads

$$\int_0^{\infty}\int_V D^c(\dot{\boldsymbol{\varepsilon}}^c,\theta)\,dV\,dT \le \frac{1}{m-1}Q(0). \tag{49}$$

Since $Q(0)$ is given by

$$Q(0) = \frac{m}{2}\int_V \bar{\boldsymbol{\sigma}}^R:\mathbf{A}:\bar{\boldsymbol{\sigma}}^R\,dV + \frac{1}{2}\int_V \bar{\boldsymbol{\chi}}:\bar{\boldsymbol{\xi}}\,dV \tag{50}$$

and is thus finite, Eq. (49) shows that the total dissipation work in the body is finite too and thus it must stop increasing at a certain time. This implies that elastic shakedown occurs. Using limit considerations, it can be stated that shakedown occurs also for $m = 1$, though in that case the body may find itself in a shakedown limit state. So the theorem is proven.

Koiter's theorem requires the concept of Plastic Accumulation Mechanism (PAM). In the present context, a PAM consists of a creep strain rate history $\dot{\boldsymbol{\varepsilon}}^{c*}(\mathbf{x}, t)$ and a micro-strain rate history $\dot{\boldsymbol{\xi}}^{*}(\mathbf{x}, t)$, both specified in $(0, \Delta t)$, satisfying the *ratchet compatibility conditions*, i.e.

$$\Delta\boldsymbol{\varepsilon}^{c*} := \int_0^{\Delta t} \dot{\boldsymbol{\varepsilon}}^{c*} dt = C\mathbf{v}^* \text{ in } V, \quad \mathbf{v}^* = \mathbf{0} \text{ on } \partial_u V \tag{51a}$$

$$\Delta\boldsymbol{\xi}^* := \int_0^{\Delta t} \dot{\boldsymbol{\xi}}^{*} dt = \mathbf{0} \text{ in } V \tag{51b}$$

where $\mathbf{v}^* = \mathbf{v}^*(\mathbf{x})$ is some displacement field. M will denote the set of all the PAMs. Another ingredient required for the theorem formulation is the (fictitious) intrinsic dissipation function, say $\mathcal{D}_g(\dot{\boldsymbol{\varepsilon}}^{c*}, \dot{\boldsymbol{\xi}}^{*}, \theta)$, related to $g \leq 0$ and viewed as a yield function. $\mathcal{D}_g$ is determined (cf. MARTIN [1975], LEMAITRE & CHABOCHE [1985], LUBLINER [1990]) through the maximum intrinsic dissipation work theorem applied to g, i.e.

$$\mathcal{D}_g(\dot{\boldsymbol{\varepsilon}}^{c*}, \dot{\boldsymbol{\xi}}^{*}, \theta) = \max_{(\boldsymbol{\sigma}, \boldsymbol{\chi})} \boldsymbol{\sigma} : \dot{\boldsymbol{\varepsilon}}^{c*} - \boldsymbol{\chi} : \dot{\boldsymbol{\xi}}^{*} \quad \text{s.t. } g(\boldsymbol{\sigma}, \boldsymbol{\chi}, \theta) \leq 0 \tag{52}$$

where "s.t." stands for "subject to" and $\dot{\boldsymbol{\varepsilon}}^{c*}$, $\dot{\boldsymbol{\xi}}^{*}$, θ are all considered fixed. At the optimum, it is

$$\mathcal{D}_g(\dot{\boldsymbol{\varepsilon}}^{c*}, \dot{\boldsymbol{\xi}}^{*}, \theta) = \boldsymbol{\sigma}(\dot{\boldsymbol{\varepsilon}}^{c*}, \dot{\boldsymbol{\xi}}^{*}, \theta) : \dot{\boldsymbol{\varepsilon}}^{c*} - \boldsymbol{\chi}(\dot{\boldsymbol{\varepsilon}}^{c*}, \dot{\boldsymbol{\xi}}^{*}, \theta) : \dot{\boldsymbol{\xi}}^{*}. \tag{53}$$

$\mathcal{D}_g$ is positively homogeneous of degree one and differentiable for any nonvanishing pair ($\dot{\boldsymbol{\varepsilon}}^{c*}$, $\dot{\boldsymbol{\xi}}^{*}$), that is

$$\boldsymbol{\sigma}^* = \frac{\partial \mathcal{D}_g}{\partial \dot{\boldsymbol{\varepsilon}}^{c*}}, \quad -\boldsymbol{\chi}^* = \frac{\partial \mathcal{D}_g}{\partial \dot{\boldsymbol{\xi}}^{*}}. \tag{54}$$

The pairs ($\boldsymbol{\sigma}^*$, $\boldsymbol{\chi}^*$) and ($\dot{\boldsymbol{\varepsilon}}^{c*}$, $\dot{\boldsymbol{\xi}}^{*}$) correspond to each other through the fictitious plasticity laws

$$\dot{\boldsymbol{\varepsilon}}^{c*} = \dot{\lambda}^* \frac{\partial g}{\partial \boldsymbol{\sigma}^*}, \quad \dot{\boldsymbol{\xi}}^{*} = -\dot{\lambda}^* \frac{\partial g}{\partial \boldsymbol{\chi}^*} \tag{55a}$$

$$g(\boldsymbol{\sigma}^*, \boldsymbol{\chi}^*, \theta) \leq 0, \quad \dot{\lambda}^* \geq 0, \quad \dot{\lambda}^* g(\boldsymbol{\sigma}^*, \boldsymbol{\chi}^*, \theta) = 0 \tag{55b}$$

which can be derived as the Kuhn-Tucker conditions of the problem (52). $\mathcal{D}_g$ is similar to the intrinsic dissipation function $\mathcal{D}^p(\dot{\boldsymbol{\varepsilon}}^p, \dot{\boldsymbol{\xi}}, \theta) = \boldsymbol{\sigma} : \dot{\boldsymbol{\varepsilon}}^p - \boldsymbol{\chi} : \dot{\boldsymbol{\xi}}$ associated with the yield function f and the related flow laws, obtained by simply replacing f with g and Eqs. ($3a, 3b$) with Eqs. ($55a, 55b$).

Assuming that $\mathcal{D}_g$ is known, Koiter's theorem (cf. KOITER [1960], KÖNIG [1987], MARTIN [1975], POLIZZOTTO et al. [1991]) can be formulated as follows.

Koiter theorem — For a given elastic-viscoplastic solid subjected to cyclic loads and temperature variations, a necessary and sufficient condition for elastic shakedown not to occur is that there exists some PAM ($\dot{\boldsymbol{\varepsilon}}^{c*}$, $\dot{\boldsymbol{\xi}}^*$) satisfying the inequality

$$E[\dot{\boldsymbol{\varepsilon}}^{c*}, \dot{\boldsymbol{\xi}}^*] := \int_0^{\Delta t} \int_V \mathcal{D}_g(\dot{\boldsymbol{\varepsilon}}^{c*}, \dot{\boldsymbol{\xi}}^*, \theta)\, dV dt - \int_0^{\Delta t} \int_V \boldsymbol{\sigma}^E : \dot{\boldsymbol{\varepsilon}}^{c*}\, dV dt < 0 \tag{56}$$

where $\boldsymbol{\sigma}^E = \boldsymbol{\sigma}^E(\mathbf{x}, t)$ is the thermo-elastic stress response to the loads and temperature variations.

Proof of the necessity. — By hypothesis, elastic shakedown occurs. Then, the assumption is made that $E[\dot{\boldsymbol{\varepsilon}}^{c*}, \dot{\boldsymbol{\xi}}^*] \geq 0$ for all PAMs. In other words, E has an absolute (vanishing) minimum in M, such that the minimization problem

$$\min E[\dot{\boldsymbol{\varepsilon}}^{c*}, \dot{\boldsymbol{\xi}}^*] \text{ s.t. } (\dot{\boldsymbol{\varepsilon}}^{c*}, \dot{\boldsymbol{\xi}}^*) \in M \tag{57}$$

admits a solution, hence a solution must satisfy the related set of Euler-Lagrange equations. The latter equations can be obtained by considering the Lagrangian

$$E_a = E[\dot{\boldsymbol{\varepsilon}}^{c*}, \dot{\boldsymbol{\xi}}^*] + \int_V \boldsymbol{\rho}^* : (C\mathbf{v}^* - \Delta\boldsymbol{\varepsilon}^{c*}) dV - \int_V \boldsymbol{\chi}^* : \Delta\boldsymbol{\xi}^* dV - \int_{\partial_u V} \mathbf{p}^* \cdot \mathbf{v}^* dS \tag{58}$$

where the integrals appended to E come from considering the constraints of Eqs. ($51a, 51b$), whereas $\boldsymbol{\rho}^*$, $\boldsymbol{\chi}^*$, $\mathbf{p}^*$ are the pertinent Lagrange multipliers. Taking the first variation of E_a and applying the divergence theorem where appropriate, gives:

$$\begin{aligned}\delta E_a =& \int_0^{\Delta t} \int_V \left(\frac{\partial \mathcal{D}_g}{\partial \dot{\boldsymbol{\varepsilon}}^{c*}} - \boldsymbol{\sigma}^E - \boldsymbol{\rho}^*\right) : \delta\dot{\boldsymbol{\varepsilon}}^{c*} dV dt - \int_0^{\Delta t} \int_V \left(\frac{\partial \mathcal{D}_g}{\partial \dot{\boldsymbol{\xi}}^*} + \boldsymbol{\chi}^*\right) : \delta\dot{\boldsymbol{\xi}}^* dV dt \\ & - \int_V \delta\mathbf{v}^* \cdot C^T \boldsymbol{\rho}^* dV + \int_{\partial_t V} \delta\mathbf{v}^* \cdot \boldsymbol{\rho} \cdot \mathbf{n}\, dS + \int_{\partial_u V} (\boldsymbol{\rho} \cdot \mathbf{n} - \mathbf{p}^*) \cdot \delta\mathbf{v}^*\, dS \\ & + \int_V (C\mathbf{v}^* - \Delta\boldsymbol{\varepsilon}^{c*}) \cdot \delta\boldsymbol{\rho}^* dV + \int_V \Delta\boldsymbol{\xi}^* : \delta\boldsymbol{\chi}^* dV - \int_{\partial_u V} \mathbf{v}^* \delta\mathbf{p}^* dS.\end{aligned} \tag{59}$$

Therefore, the Euler-Lagrange equations turn out to be:

$$\boldsymbol{\sigma}^* := \boldsymbol{\sigma}^E + \boldsymbol{\rho}^* = \frac{\partial \mathcal{D}_g}{\partial \dot{\boldsymbol{\varepsilon}}^{c*}}, \quad -\boldsymbol{\chi}^* := \frac{\partial \mathcal{D}_g}{\partial \dot{\boldsymbol{\xi}}^*} \quad \text{in } V \times (0, \Delta t) \tag{60}$$

$$C^T \boldsymbol{\rho}^* = \mathbf{0} \text{ in } V, \quad \boldsymbol{\rho}^* \cdot \mathbf{n} = \mathbf{0} \text{ on } \partial_t V, \quad \boldsymbol{\rho}^* \cdot \mathbf{n} = \mathbf{p}^* \text{ on } \partial_u V, \tag{61}$$

besides Eqs. ($51a, 51b$). Equations (60, 61) identify the meaning of the Lagrange multipliers and, in particular, show that $\boldsymbol{\rho}^*$ and $\boldsymbol{\chi}^*$ take the meanings of self-stresses and micro-stresses, respectively, and that the pair $(\boldsymbol{\sigma}^*, \boldsymbol{\chi}^*)$, being derived from $\mathcal{D}_g$, satisfies the condition

$$g(\boldsymbol{\sigma}^*, \boldsymbol{\chi}^*, \theta) \leq 0 \text{ in } V \times (0, \Delta t). \tag{62}$$

Thus, by Melan's theorem, shakedown occurs. Since this contradicts the starting hypothesis, the conclusion must be inferred that necessarily a PAM must exist which satisfies Eq. (56).

Proof of the sufficiency. — By hypothesis, there exists a PAM $(\dot{\boldsymbol{\varepsilon}}^{c*}, \dot{\boldsymbol{\xi}}^*)$ which satisfies Eq. (56). Correspondingly, the additional hypothesis is made that elastic shakedown occurs. Then, by Melan's theorem, there must exist some self-stress field $\bar{\boldsymbol{\sigma}}^R$ and micro-stress field $\bar{\boldsymbol{\chi}}$, both time-independent, satisfying Eq. (38). Applying Eq. (52) gives

$$\mathcal{D}_g(\dot{\boldsymbol{\varepsilon}}^{c*}, \dot{\boldsymbol{\xi}}^*, \theta) \geq (\boldsymbol{\sigma}^E + \bar{\boldsymbol{\sigma}}^R) : \dot{\boldsymbol{\varepsilon}}^{c*} - \bar{\boldsymbol{\chi}} : \dot{\boldsymbol{\xi}}^* \quad \text{in } V \times (0, \Delta t), \tag{63}$$

that is, with an integration over $V \times (0, \Delta t)$ and taking into account Eqs. $(51a, 51b)$,

$$\int_0^{\Delta t} \int_V \mathcal{D}_g(\dot{\boldsymbol{\varepsilon}}^{c*}, \dot{\boldsymbol{\xi}}^*, \theta)\, dV dt \geq \int_0^{\Delta t} \int_V \boldsymbol{\sigma}^E : \dot{\boldsymbol{\varepsilon}}^{c*}\, dV dt, \tag{64}$$

which contradicts the initial hypothesis. Thus, elastic shakedown cannot occur.

In the above proofs, a similar reasoning as in (POLIZZOTTO et al. [1991]) has been followed, but here temperature effects have been accounted for.

As regards the second question, it is worth mentioning that — within plasticity theory — the limit state which takes place in the body under a load at the elastic shakedown limit has received full explanation in (PANZECA & POLIZZOTTO [1988], FUSCHI & POLIZZOTTO [1993]). First of all, the equations which describe such a limit state can be obtained as the Euler-Lagrange equations related to the problem for the determination of the maximum elastic shakedown load multiplier, β say, as shown e.g. in (POLIZZOTTO et al. [1991]) for a generalized standard elastic-plastic body. With the proviso that the pertinent yield function is here g, the value of $\beta = \beta_g$ can be derived either as the optimal objective value of the maximization problem (65), i.e.

$$\beta_g = \max_{(\beta, \boldsymbol{\rho}, \mathbf{h})} \beta \text{ s.t.} \begin{cases} g(\beta\bar{\boldsymbol{\sigma}}^E + \boldsymbol{\rho}, \mathbf{h}, \theta) \leq 0 \text{ in } V \times (0, \Delta t) \\ C^T\boldsymbol{\rho} = \mathbf{0} \text{ in } V,\ \boldsymbol{\rho}\cdot\mathbf{n} = \mathbf{0} \text{ on } \partial_t V, \end{cases} \tag{65}$$

or as part of the solution to the following set of Euler-Lagrange equations:

$$g(\boldsymbol{\sigma}^*, \boldsymbol{\chi}^*, \theta) \leq 0, \quad \dot{\lambda}^* \geq 0, \quad \dot{\lambda}^* g(\boldsymbol{\sigma}^*, \boldsymbol{\chi}^*, \theta) \leq 0 \text{ in } V \times (0, \Delta t) \tag{66a}$$

$$\boldsymbol{\sigma}^* = \beta_g \bar{\boldsymbol{\sigma}}^E + \boldsymbol{\rho}^*,\ \dot{\boldsymbol{\varepsilon}}^{c*} = \dot{\lambda}^* \frac{\partial g}{\partial \boldsymbol{\sigma}^*},\ \dot{\boldsymbol{\xi}}^* = -\dot{\lambda}^* \frac{\partial g}{\partial \boldsymbol{\chi}^*} \text{ in } V \times (0, \Delta t) \tag{66b}$$

$$C^T \boldsymbol{\rho}^* = \mathbf{0} \text{ in } V, \quad \boldsymbol{\rho}^* \cdot \mathbf{n} = \mathbf{0} \text{ on } \partial_t V \tag{67}$$

$$\Delta \boldsymbol{\varepsilon}^{c*} := \int_0^{\Delta t} \dot{\boldsymbol{\varepsilon}}^{c*} dt = C \mathbf{v}^* \text{ in } V, \quad \mathbf{v}^* = \mathbf{0} \text{ on } \partial_u V \tag{68a}$$

$$\Delta \boldsymbol{\xi}^* := \int_0^{\Delta t} \dot{\boldsymbol{\xi}}^* dt = \mathbf{0} \text{ in } V \tag{68b}$$

$$\int_0^{\Delta t} \int_V \bar{\boldsymbol{\sigma}}^E : \dot{\boldsymbol{\varepsilon}}^{c*} dV dt = 1. \tag{69}$$

For g being a real yield function, the above equations describe the impending steady cycle which is going to be eventually produced in the structure subjected to a cyclic load and temperature variation sligtly above the elastic shakedown limit. But g is here just a creep function and the above equations describe only a fictitious time-independent plasticity straining process. Nevertheless, the above steady cycle can be shown to represent the actual impending steady-cycle which is going to eventually establish itselfin the elastic-viscoplastic body as a result of a cyclic load and temperature variation slightly exceeding the limit β_g. For this purpose, let the Perzyna function $\Phi(< g >)$ be considered and let it be developed in a MacLaurin series around $g = 0$, truncating the series at the lowest nonvanishing derivative, say the n-th one. So

$$\Phi(< g >) \simeq \frac{1}{n!} \frac{d^n \Phi}{dg^n}\bigg|_0 < g >^n \tag{70}$$

and thus

$$\frac{d\Phi}{dg} \simeq \frac{1}{(n-1)!} \frac{d^n \Phi}{dg^n}\bigg|_0 < g >^{n-1} . \tag{71}$$

For instance, if Φ is taken as a homogeneous function of degree n, that is $\Phi :=< g >^n$, since

$$\frac{d < g >}{dg} = \begin{cases} 1 \text{ for } g \geq 0 \\ 0 \text{ for } g < 0, \end{cases} \tag{72}$$

one obtains:

$$\frac{d^n \Phi}{dg^n} = \frac{d^n \Phi}{d < g >^n} \left(\frac{d < g >}{dg} \right)^n = \begin{cases} n! \text{ for } g \geq 0 \\ 0 \text{ for } g \leq 0, \end{cases} \tag{73}$$

as illustrated by Fig. 2.

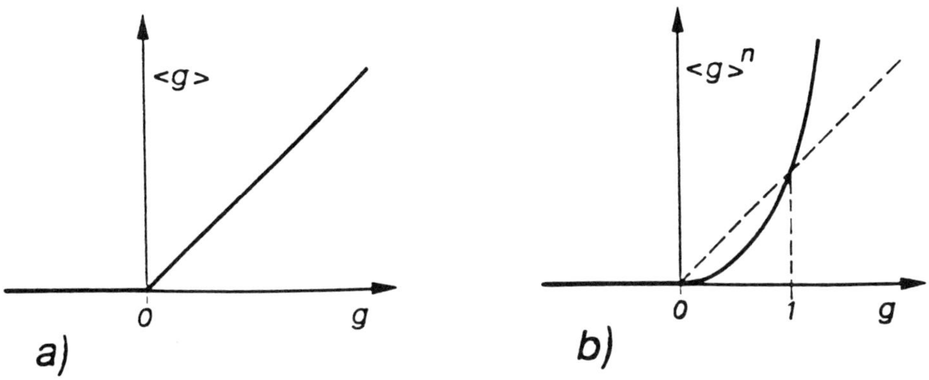

Fig. 2. Sketches representing: a) Macauley function $< g >$, and b) the power function $< g >^n$, $(n > 1)$.

Let $\boldsymbol{\sigma} = \boldsymbol{\sigma}^* + \mathbf{s}\zeta$ and $\boldsymbol{\chi} = \boldsymbol{\chi}^* + \mathbf{h}\zeta$ denote the body's stress and micro-stress states somewhere slightly above the elastic shakedown limit, with $\mathbf{s} = \mathbf{s}(\mathbf{x})$ and $\mathbf{h} = \mathbf{h}(\mathbf{x})$ being some fixed stress path from $(\boldsymbol{\sigma}^*, \boldsymbol{\chi}^*)$ over $g = 0$ (and externally oriented, e.g. along the normal to $g = 0$), whereas ζ is a small scalar parameter. The corresponding creep and micro-strain rates read

$$\dot{\boldsymbol{\varepsilon}}^c = \gamma \frac{d\Phi}{dg} \frac{\partial g}{\partial \boldsymbol{\sigma}} = \frac{\gamma}{(n-1)!} \left. \frac{d^n \Phi}{dg^n} \right|_0 < g >^{n-1} \frac{\partial g}{\partial \boldsymbol{\sigma}} \tag{74a}$$

$$-\dot{\boldsymbol{\xi}} = \gamma \frac{d\Phi}{dg} \frac{\partial g}{\partial \boldsymbol{\chi}} = \frac{\gamma}{(n-1)!} \left. \frac{d^n \Phi}{dg^n} \right|_0 < g >^{n-1} \frac{\partial g}{\partial \boldsymbol{\chi}}. \tag{74b}$$

Next, taking the limits of the ratios $\dot{\boldsymbol{\varepsilon}}^c / < g >^{n-1}$ and $\dot{\boldsymbol{\xi}} / < g >^{n-1}$ for $\zeta \to 0$, and on setting

$$\dot{\boldsymbol{\varepsilon}}^{c*} := \lim_{\zeta \to 0} \frac{\dot{\boldsymbol{\varepsilon}}^c}{< g >^{n-1}}, \quad \dot{\boldsymbol{\xi}}^* := \lim_{\zeta \to 0} \frac{\dot{\boldsymbol{\xi}}}{< g >^{n-1}} \tag{75}$$

and

$$\dot{\lambda}^* := \begin{cases} \dfrac{\gamma}{(n-1)!} \left. \dfrac{d^n \Phi}{d < g >^n} \right|_0 & \text{for } g = 0 \\ 0 & \text{for } g < 0, \end{cases} \tag{76}$$

one easily realizes that Eqs. $(74a, 74b)$, after dividing by $< g >^{n-1}$, taking the limits for $\zeta \to 0$ and accounting for Eqs. (75, 76), transforms in such away as to read like just the second and third components of Eq. (66*b*), with $\dot{\lambda}^*$ complying with the complementarity conditions in Eq. (66*a*). So the statement is proven. What precedes enables one to state the following proposition.

Proposition 3 – In a structure with a generalized standard elastic-viscoplastic material subjected to cyclic loads and temperature variations, Melan's and Koiter's theorems can be applied for the assessment of the body's elastic shakedown, provided that the relevant creep function be given the role of yield function; also, the limit state which takes place in the body under a load condition at the shakedown limit identifies with the impending steady cycle associated with a loading condition going to exceed the shakedown load limit.

SENSITIVITY OF THE STEADY CYCLE TO LOAD CHANGES

Let the external actions be considered as the superposition of time-variable actions specified to within a scalar load multiplier, say $\beta \geq 0$, and of a permanent (mechanical) load, that is, in an abridged notation,

$$P(t) = P_0 + P_C(t) = P_0 + \beta \bar{P}_C(t), \quad 0 \leq t \leq \Delta t \tag{77}$$

where P_0 denotes the permanent load and $\bar{P}_C(t)$ the reference cyclic loads and temperature variations. P^0 is in principle allowed to be any load distribution in $V \cup \partial_t V$, below the instantaneous plastic collapse value, but on occasion it may be a narrower set, for instance a one-parameter load family like $P_0 = \alpha \bar{P}_0$, $\alpha \geq 0$.

Different types of steady cycle are produced in a given structure depending on the load parameter values. In the space of these parameters, different regions can be envisaged, each of which is associated with the same type of steady cycle. As in time-independent plasticity (cf. MARTIN [1975], PONTER [1982], GOCHFELD & CHERNIAVSKY [1980], POLIZZOTTO et al. [1991], POLIZZOTTO [1993a, 1993d]), the typical steady cycles are, besides elastic shakedown:

— *Inelastic shakedown*, characterized by the identical vanishing of the inelastic strain ratchet, and thus by alternating inelastic strains (analogous to plastic shakedown of time-independent plasticity).

— *Ratcheting*, characterized by a nonvanishing inelastic strain ratchet, at least somewhere in V (also called "incremental collapse").

The steady cycles can be mapped into a convex domain B of the load parameter space, such that B can be decomposed into subdomains, namely $B = B_S \cup B_F \cup B_R$, where B_S, B_F and B_R collect load points producing, respectively, steady cycles as elastic shakedown, inelastic shakedown and ratcheting.

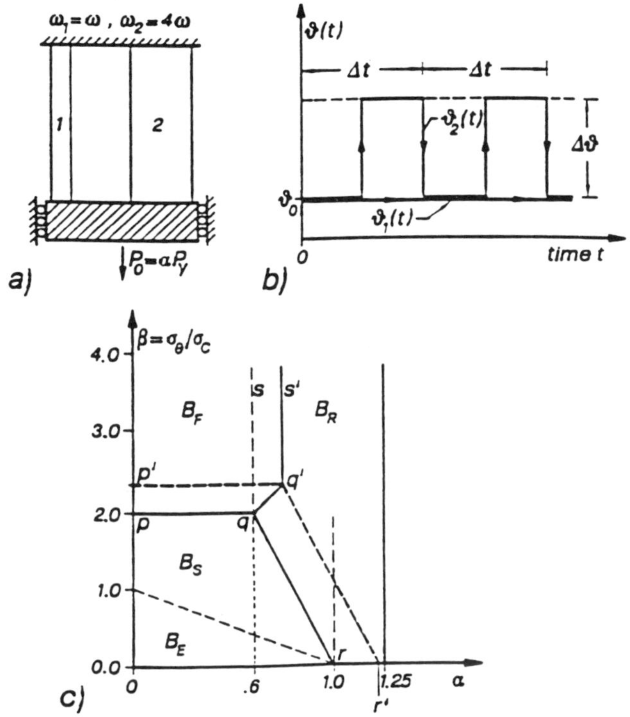

Fig. 3. Two-bar unsymmetric system under thermal cycles: a) Geometrical and loading scheme; b) Temperature variation histories; c) Interaction diagram in the (α, β)-plane, (σ_θ = maximal thermo-elastic stress).

A simple example is depicted in Figs. 3($a-c$) relating to a two-bar one-degree-of--freedom system. Bar 1 (of cross section ω) is taken at a constant temperature θ_0, whereas bar 2 (of cross section 4ω) suffers periodic temperature variations as depicted in Fig. 3(b), causing a maximum thermo-elastic stress of $\sigma_\theta = 4aE\Delta\theta/5$, where a = thermal expansion coefficient and E = Young modulus. The material is elastic-perfectly viscoplastic with the elastic limit σ_c and yield stress $\sigma_y = 1.25\sigma_c$, and these data are temperature independent. A permanent load $P_0 = \alpha P_y$, with $P_y = 5\omega\sigma_y$, is applied upon the rigid block, Fig. 3(a). With $\sigma_\theta = \sigma_c\beta$, the interaction diagram in the $(\alpha - \beta)$-plane is depicted in Fig. 3(c), where the line $p - q - r$ is the boundary of B_S, whereas the line $q - q' - s'$ is the border between B_F and B_R. The regions of B_F and B_R which are located between $p - q - r$ and $p' - q' - r'$ collect points related to the steady cycles with only time-dependent plasticity. The location of the bilateral line $p' - q' - r'$ (hence of $q - q' - s'$) depends upon the particular creep law adopted, as well as upon the ratio σ_y/σ_c and the time period Δt. In the case of only time-independent plastic behaviour, with $\sigma_y = \sigma_c$, line $p' - q' - r'$ coincides with $p - q - r$, and line $q - q' - s'$ with $q - s$; also, line $\alpha = 1.25$ (which marks the system instantaneous plastic collapse) transforms into $\alpha = 1.00$.

It is of interest to investigate the sensitivity of the steady cycle to the load parameters changes. This task is here achieved with methods like that in (POLIZZOTTO et al. [1990], POLIZZOTTO [1993a, 1993]); see also (MRÓZ [1972], AINSWORTH et al. [1980]). To this purpose, the thermo-elastic stress response to the loads as in Eq. (77) is cast in the form:

$$\boldsymbol{\sigma}^E(\mathbf{x}, t) = \boldsymbol{\sigma}_0^E(\mathbf{x}) + \beta \bar{\boldsymbol{\sigma}}_c^E(\mathbf{x}, t) \tag{78}$$

where $\boldsymbol{\sigma}_0^E$ is the elastic stress response to P_0, whereas $\bar{\boldsymbol{\sigma}}_c^E$ denotes the thermo-elastic stress response to the reference cyclic loads and temperature variations. Denoting by $\boldsymbol{\sigma}$, $\boldsymbol{\chi}$, $\dot{\boldsymbol{\varepsilon}}^{in}$, $\dot{\boldsymbol{\xi}}$, etc., the actual steady-state response of the body to the loads, the related inelastic intrinsic dissipation density $\mathcal{D}$ reads:

$$\mathcal{D} := \boldsymbol{\sigma} : \dot{\boldsymbol{\varepsilon}}^{in} - \boldsymbol{\chi} : \dot{\boldsymbol{\xi}} \geq 0 \tag{79}$$

and thus the total inelastic intrinsic dissipation in the steady cycle is

$$W := \int_0^{\Delta t} \int_V \mathcal{D}(\dot{\boldsymbol{\varepsilon}}^{in}, \dot{\boldsymbol{\xi}}, \theta)\, dV dt \geq 0. \tag{80}$$

By the first Eq. (28) and by Eq. (36), the stress $\boldsymbol{\sigma}$ can be cast as

$$\boldsymbol{\sigma} = \boldsymbol{\sigma}^E(\mathbf{x}, t) + \boldsymbol{\rho}(\mathbf{x}) + \boldsymbol{\tau}(\mathbf{x}, t). \tag{81}$$

Then, accounting for the identities,

$$\int_V \boldsymbol{\rho} : \Delta \boldsymbol{\varepsilon}^{in} dV = 0 \quad \text{(by the virtual work principle)} \tag{82}$$

$$\begin{aligned} \int_0^{\Delta t} \int_V \boldsymbol{\tau} : \dot{\boldsymbol{\varepsilon}}^{in} dV &= \int_0^{\Delta t} \int_V \boldsymbol{\tau} : \dot{\boldsymbol{\varepsilon}}^{R} dV - \int_0^{\Delta t} \int_V \boldsymbol{\tau} : \mathbf{A} : \dot{\boldsymbol{\tau}}\, dV \\ &= -\frac{1}{2} \int_V \boldsymbol{\tau} : \mathbf{A} : \boldsymbol{\tau}\, dV \Big|_0^{\Delta t} = 0 \quad \text{(by the periodicity of } \boldsymbol{\tau}) \end{aligned} \tag{83}$$

$$\int_0^{\Delta t} \int_V \boldsymbol{\chi} : \dot{\boldsymbol{\xi}}\, dV dt = \int \Psi(\boldsymbol{\xi})\, dV \Big|_0^{\Delta t} = 0 \quad \text{(by the periodicity of } \boldsymbol{\xi}), \tag{84}$$

it follows that the expression for W is:

$$W = A_0 + \beta \bar{A}_C \geq 0, \tag{85}$$

where

$$A_0 := \int_V \boldsymbol{\sigma}_0^E : \Delta\boldsymbol{\varepsilon}^{in} dV \;(Primary\ accumulation\ parameter) \tag{86a}$$

$$\bar{B}_C := \int_0^{\Delta t}\int_V \bar{\boldsymbol{\sigma}}_C^E : \dot{\boldsymbol{\varepsilon}}^{in} dV dt \;(Secondary\ net\ accumulation\ parameter). \tag{86b}$$

Notice that A_0 is vanishing in case of either elastic or inelastic shakedown, whereas $\bar{B}_C$ is vanishing only in case of elastic shakedown.

Obviously, W turns out to be a function of β, as well as of the steady load parameters P_0. The sensitivity of the steady cycle to the load changes is here investigated through the analogous sensitivity of W. Let (P_0, β) and $(P_0+\delta P_0, \beta+\delta\beta)$ be two neighbouring load points, with $(P_0, \beta) \notin B_S$. Denoting by $\boldsymbol{\sigma}$, $\boldsymbol{\chi}$, $\mathbf{u}$, $\dot{\boldsymbol{\varepsilon}}^{in}$, etc., the steady cycle related to (P_0, β), and by $\delta\boldsymbol{\sigma}$, $\delta\boldsymbol{\chi}$, $\delta\mathbf{u}$, $\delta\dot{\boldsymbol{\varepsilon}}^{in}$, etc., the steady cycle variation due to the load change, the hypothesis of stability and continuity is made, namely that the steady cycle variation vanishes as the load change $(\delta\beta, \delta P_0)$ is removed. Then, inequality (20) can be applied here to write

$$\delta^2\mathcal{D} = \delta\boldsymbol{\sigma} : \delta\dot{\boldsymbol{\varepsilon}}^{in} - \frac{1}{2}\frac{d}{dt}(\delta\boldsymbol{\chi} : \delta\boldsymbol{\xi}) \geq 0 \quad \text{in } V \times (0, \Delta t). \tag{87}$$

Let one note that the identity

$$\int_0^{\Delta t}\int_V \frac{d}{dt}(\delta\boldsymbol{\chi} : \delta\boldsymbol{\xi})\, dV dt = \int_V \delta\boldsymbol{\xi} : \frac{\partial^2\Psi}{\partial\boldsymbol{\xi}\partial\boldsymbol{\xi}} : \partial\boldsymbol{\xi}\, dV\Big|_0^{\Delta t} = 0 \tag{88}$$

holds due to the periodicity of $\boldsymbol{\xi}$, hence of $\delta\boldsymbol{\xi}$, in the steady cycle, and that $\delta\boldsymbol{\sigma}$ in Eq. (87) has the following form:

$$\delta\boldsymbol{\sigma} = \delta\boldsymbol{\sigma}_0^E + \delta\beta\bar{\boldsymbol{\sigma}}_C^E + \delta\boldsymbol{\rho} + \delta\boldsymbol{\tau}. \tag{89}$$

An integration of Eq. (87) over $V \times (0, \Delta t)$ and accounting for Eqs. (88, 89) gives

$$\begin{aligned} \delta^2 W &= \int_0^{\Delta t}\int_V \delta^2\mathcal{D}\, dV dt = \int_0^{\Delta t}\int_V \delta\boldsymbol{\sigma} : \delta\dot{\boldsymbol{\varepsilon}}^{in}\, dV dt \\ &= \delta^2 A_0 + \delta\beta\delta\bar{A}_C + \int_V \delta\boldsymbol{\rho} : \Delta\delta\dot{\boldsymbol{\varepsilon}}^{in}\, dV + \int_0^{\Delta t}\int_V \delta\boldsymbol{\tau} : \delta\dot{\boldsymbol{\varepsilon}}^{in}\, dV dt. \end{aligned} \tag{90}$$

Since the last two integrals on the r.h.s. of Eq. (90) can be easily shown to vanish, tha second variation of W finally reads

$$\delta^2 W = \delta^2 A_0 + \delta\beta\delta\bar{A}_C \geq 0 \tag{91}$$

where $\delta^2 A_0$ and $\delta \bar{A}_c$ are variations of A_0 and $\bar{A}_C$, i.e.

$$\delta^2 A_0 = \int_V \delta\boldsymbol{\sigma}_0^E : \Delta\delta\dot{\boldsymbol{\varepsilon}}^{in} dV, \quad \delta\bar{A}_C = \int_0^{\Delta t}\int_V \bar{\boldsymbol{\sigma}}_c^E : \delta\dot{\boldsymbol{\varepsilon}}^{in} dV\, dt\ . \tag{92}$$

Equation (91) is an extension onto the present context of an analogous equation holding for perfect plasticity (cf. POLIZZOTTO et al. [1990], POLIZZOTTO [1993a]) and for generalized standard elastic-plastic solids (cf. POLIZZOTTO [1993d]). As in the latter papers, Eq. (91) can be interpreted as an assessment of a stability principle for the steady cycle. The value of the second variation of W as a consequence of the load variation cannot be negative and provides a measure of the steady cycle sensitivity to the load changes.

STEADY CYCLE INSENSITIVITY TO LOAD INCREMENTS

The steady cycle related to any load $(P_0, \beta) \in B_S$ is fully insensitive to the load changes (it is always characterized by identically vanishing inelastic strain rates). Let $(P_0, \beta) \notin B_S$, and let $(\delta P_0, \delta\beta)$ be a small change of it, such that correspondingly $\delta^2 W = 0$ and, as a consequence,

$$\delta^2 \mathcal{D} = \delta\boldsymbol{\sigma} : \delta\dot{\boldsymbol{\varepsilon}}^{in} - \frac{1}{2}\frac{d}{dt}(\delta\boldsymbol{\chi} : \delta\boldsymbol{\xi}) = 0 \text{ in } V \times (0, \Delta t). \tag{93}$$

This equality is formally identical to Eq. (21), but with the variation field $\delta(\cdot)$ in place of the difference fields $\Delta(\cdot)$, and with $J = (0, \Delta t)$. This means that the evolution uniqueness requisite applies here in its general format, that is, with its properties i), ii) and iii), such that one can state that $\delta\dot{\boldsymbol{\varepsilon}}^{in} = \mathbf{0}$, $\delta\dot{\boldsymbol{\chi}} = \delta\dot{\boldsymbol{\xi}} = \mathbf{0}$ in $V \times (0, \Delta t)$, and that $\delta\boldsymbol{\chi}$ and $\delta\boldsymbol{\xi}$ are time-independent fields in V, both vanishing in V_{p+c}. Additionally, as $\delta\dot{\boldsymbol{\sigma}}^R \equiv \delta\dot{\boldsymbol{\tau}} \equiv \mathbf{0}$, hence $\delta\boldsymbol{\tau} \equiv \mathbf{0}$, Eq. (89) can be rewritten as

$$\delta\boldsymbol{\sigma} = \delta\boldsymbol{\sigma}_0^E(\mathbf{x}) + \delta\beta\bar{\boldsymbol{\sigma}}_C^E(\mathbf{x}, t) + \delta\boldsymbol{\rho}(\mathbf{x}) \text{ in } V_{p+c} \times J_{p+c}, \tag{94}$$

where J_{p+c} collects all times at which inelastic strain occurs in the steady cycle. J_{p+c} may consist of a single instant only in the case of elastic shakedown, or even in the condition of instantaneous plastic collapse, but both such events are to be excluded here. Thus, Eq. (94) implies that

$$\delta\beta = 0, \quad \delta\boldsymbol{\sigma} = \delta\boldsymbol{\sigma}_0^E(\mathbf{x}) + \delta\boldsymbol{\rho}(\mathbf{x}) = \mathbf{0} \text{ in } V_{p+c} \times J. \tag{95}$$

Since $\delta\boldsymbol{\sigma}_0^E$ constitutes a self-stress field within V_{p+c} by Eq. (95), it follows that the region V_{p+c}, where inelastic strains take place in the steady cycle, cannot include the application points of the (mechanical) load δP_0.

The above results can be summarized as follows:

a) Under a load $(P_0, \beta) \notin B_S$, the only increments such that $\delta^2 W = 0$ are increments of the permanent load, with $\delta\beta = 0$.

b) The steady cycle related to (P_0, β) is insensitive to those increments (if any) of the permanent load for which $\delta^2 W = 0$, in the sense that, upon such a load increment, the related PAM $(\dot{\boldsymbol{\varepsilon}}^{in}, \dot{\boldsymbol{\xi}})$ remains unaltered, as do the state variables $\boldsymbol{\sigma}$, $\boldsymbol{\chi}$, $\boldsymbol{\xi}$ in $V_{p+c} \times (0, \Delta t)$, along with the region V_{p+c} where inelastic strains occur.

c) The load increment δP_0 for which $\delta^2 W = 0$, if any, must have its application points within the region $V_e = V \backslash V_{p+c}$, where $\dot{\boldsymbol{\varepsilon}}^{in} \equiv \mathbf{0}$, and some stress increments $\delta\boldsymbol{\sigma}$ there take place such as to equilibrate δP_0.

A further implication of Eq. (93) is discovered by considering the vanishing integral

$$\int_0^{\Delta t} \int_V \delta\boldsymbol{\sigma} : \dot{\boldsymbol{\varepsilon}}^{in} dV dt = \int_V \delta\boldsymbol{\sigma}_0^E : \Delta\boldsymbol{\varepsilon}^{in} dV + \int_V \delta\boldsymbol{\rho} : \Delta\boldsymbol{\varepsilon}^{in} dV = 0 \tag{96}$$

in which $\dot{\boldsymbol{\varepsilon}}^{in}$ is related to the steady cycle of the load (P_0, β). The latter integral is vanishing because $\dot{\boldsymbol{\varepsilon}}^{in} \equiv \mathbf{0}$ in V_e (where $\delta\boldsymbol{\sigma} \not\equiv \mathbf{0}$), and $\delta\boldsymbol{\sigma} \equiv \mathbf{0}$ in V_{p+c} (where $\dot{\boldsymbol{\varepsilon}}^{in} \not\equiv \mathbf{0}$). As the last integral on the r.h.s. of Eq. (96) vanishes by the virtual work principle (namely, $\delta\boldsymbol{\rho}$ is a self-stress field, whereas $\Delta\boldsymbol{\varepsilon}^{in}$ is compatible with zero displacements on $\partial_u V$), Eq. (96) yields

$$\int_V \delta\boldsymbol{\sigma}_0^E : \Delta\boldsymbol{\varepsilon}^{in} dV = 0. \tag{97}$$

Since this equation holds for arbitrary choices of δP_0, hence for the field $\delta\boldsymbol{\sigma}_0^E$ in V, from Eq. (97) it follows that, necessarily, $\Delta\boldsymbol{\varepsilon}^{in} = \mathbf{0}$ in V, that is, inelastic shakedown occurs under the load (P_0, β). Furthermore, as the inelastic strain rate history $\dot{\boldsymbol{\varepsilon}}^{in}(\mathbf{x}, t)$ remains unaltered on application of δP_0, it can be stated that inelastic shakedown also occurs under the loads $(P_0 + \delta P_0, \beta)$ with arbitrary small δP_0 (applied at points $\mathbf{x} \in V_e$).

The converse is also true. In fact, let $(P_0, \beta) \in B_F$ and let $(P_0 + \delta P_0, \beta) \in B_F$ too for arbitrarily small δP_0 and β being constant. Since, by hypothesis, $\Delta\boldsymbol{\varepsilon}^{in} = \mathbf{0}$ and $\Delta\delta\boldsymbol{\varepsilon}^{in} = \mathbf{0}$ in V, by Eq. (91) and the first Eq. (92) one has $\delta^2 W = 0$, hence Eq. (93)

is satisfied, and consequently the above properties a) and b) hold. In conclusion, the following propositions can be stated.

Proposition 4 — For a structure of generalized standard elastic-viscoplastic material subjected to cyclic loads and temperature variations of the type (P_0, β), a necessary and sufficient condition for inelastic shakedown to occur under a loading condition which exceeds the elastic shakedown limit, is that the steady cycle be insensitive to the permanent (mechanical) load changes, that is, the second variation $\delta^2 W$ of the total inelastic intrinsic dissipation in a complete cycle, caused by δP_0, vanishes for arbitrarily small δP_0.

Proposition 5 — In a structure of generalized standard elastic-viscoplastic material subjected to cyclic loads and temperature variations of the type (P_0, β) and such as to cause inelastic shakedown, the relevant steady cycle is insensitive to (small) increments δP_0 of the permanent (mechanical) load, namely, on application of δP_0 the PAM $(\dot{\boldsymbol{\varepsilon}}^{in}, \dot{\boldsymbol{\xi}})$ remains unaltered together with the stresses $\boldsymbol{\sigma}$ and the internal variables $\boldsymbol{\chi}, \boldsymbol{\xi}$ within the region V_{p+c} where alternating inelastic strains take place, and additionally the region V_{p+c} — which cannot include the application points of δP_0 — remains fixed too.

It can also be stated that the steady cycle pertaining to a load $(P_0, \beta) \in B_F$ is sensitive only to increments $\delta\beta$ of the time-dependent load and temperature variation, and that the steady cycle pertaining to a load $(P_0, \beta) \in B_R$ is fully sensitive to the load changes $(\delta P_0, \delta\beta)$.

THE ELASTIC REGION V_e IN THE STEADY CYCLE

By definition, the elastic region V_e is that part of the body where no inelastic strains occur in the steady cycle, such that the inelastic strain $\boldsymbol{\varepsilon}^{in}$ and the internal variables $\boldsymbol{\chi}$, $\boldsymbol{\xi}$, if nonzero, are there time-independent (see Proposition 2). As for time-independent plasticity (cf. POLIZZOTTO [1993a, 1993d]), the following proposition can be stated:

Proposition 6 — In the steady cycle of a structure composed of generalized standard elastic-viscoplastic material, an elastic region V_e can exist if and only if the structure finds itself in a condition of shakedown, either inelastic shakedown (in which case $V_e \subset V$), or elastic shakedown (in which case $V_e = V$).

In order to prove the above statement, a load $(P_0, \beta) \notin B_S$ is considered together with the related steady cycle, the latter being described by the state and evolution variables $\boldsymbol{\sigma}$, $\boldsymbol{\varepsilon}$, $\mathbf{u}$, $\dot{\boldsymbol{\varepsilon}}^{in}$, etc.. By hypothesis, there exists a nonempty region $V_e \subset V$

where no inelastic strains occur in the steady cycle. Equations (28-33) are all satisfied by the latter variables, but

$$g(\boldsymbol{\sigma}, \boldsymbol{\chi}, \theta) < 0 \text{ in } V_e \times (0, \Delta t) \tag{98}$$

and *a-fortiori*

$$f(\boldsymbol{\sigma}, \boldsymbol{\chi}, \theta) < 0, \ \dot{\lambda} = 0 \text{ in } V_e \times (0, \Delta t). \tag{99}$$

Next, let a (small) permanent load δP_0 be applied on the body, with its application points in V_e, and let $\delta\boldsymbol{\sigma}_0^E$ be the corresponding elastic stress response. Let $\delta\boldsymbol{\rho}$, $\delta\boldsymbol{\chi}$ be some time-independent self-stresses and micro-stresses, defined as follows:

$$\delta\boldsymbol{\sigma} := \delta\boldsymbol{\sigma}_0^E + \delta\boldsymbol{\rho} = \mathbf{0}, \quad \delta\boldsymbol{\chi} = \mathbf{0} \text{ in } V_{p+c} := V \backslash V_e \tag{100a}$$

$$g(\boldsymbol{\sigma} + \delta\boldsymbol{\sigma}, \boldsymbol{\chi} + \delta\boldsymbol{\chi}, \theta) < 0 \text{ in } V_e \times (0, \Delta t). \tag{100b}$$

As a consequence of Eq.(100*b*), we have

$$f(\boldsymbol{\sigma} + \delta\boldsymbol{\sigma}, \boldsymbol{\chi} + \delta\boldsymbol{\chi}, \theta) < 0 \text{ in } V_e \times (0, \Delta t). \tag{101}$$

These choices are made possible by δP_0 being sufficiently small. Then, it can be recognized that Eqs. (28-33) remain satisfied if the same rate variables $\dot{\lambda}$, $\dot{\boldsymbol{\varepsilon}}^{in}$, $\dot{\boldsymbol{\xi}}$, $\dot{\mathbf{u}}^R$ and $\dot{\boldsymbol{\sigma}}^R$ are coupled with new state variables $\boldsymbol{\sigma}^*, \boldsymbol{\rho}^*, \boldsymbol{\chi}^*$, defined as

$$\boldsymbol{\sigma}^* := \boldsymbol{\sigma} + \delta\boldsymbol{\sigma} = \boldsymbol{\sigma}^{E*} + \boldsymbol{\rho}^* + \boldsymbol{\tau}, \ \boldsymbol{\chi}^* = \boldsymbol{\chi} + \delta\boldsymbol{\chi} \tag{102a}$$

$$\boldsymbol{\sigma}^{E*} := \boldsymbol{\sigma}^E + \delta\boldsymbol{\sigma}_0^E, \ \boldsymbol{\rho}^* = \boldsymbol{\rho} + \delta\boldsymbol{\rho} \tag{102b}$$

but with the stresses $\boldsymbol{\tau}(\mathbf{x}, t)$ unchanged. The new equation set so restated provides the steady cycle related to the load $(P_0 + \delta P_0, \beta)$. As a consequence of the uniqueness properties of the solution to Eqs. (28-33), it is possible to state that the steady cycles related to (P_0, β) and $(P_0 + \delta P_0, \beta)$ are characterized by the same rate variable histories, along with the same stresses $\boldsymbol{\sigma}$ and internal variables $\boldsymbol{\chi}$, $\boldsymbol{\xi}$ in V_{p+c}. Since this result remains valid whatever δP_0 is (but sufficiently small), it can be assessed that the steady cycle pertaining to (P_0, β) exhibits insensitivity to the permanent load changes. This means, by Proposition 4, that $\delta^2 W = 0$ for arbitrary small δP_0 and thus inelastic shakedown occurs under the load (P_0, β), with $V_{p+c} = V \backslash V_e$ being the region of the body where alternating inelastic strains occur. And that concludes the proof.

In the case of existence of a region V_e (and thus occurrence of inelastic shakedown), the following proposition can also be stated.

Proposition 7 — In a structure of generalized standard elastic-viscoplastic material which experiences a condition of inelastic shakedown, the pertinent elastic region V_e finds itself in a condition of elastic shakedown, where it remains on application upon it of arbitrary steady loads δP^0 within some range.

This property, firstly announced by Ponter for perfectly plastic materials (cf. PONTER [1983]), was previously proven for perfect plasticity (cf. POLIZZOTTO [1993a]) and then extended to generalized standard elastic-plastic materials (cf. POLIZZOTTO [1993d], see also MRÓZ [1972]). The extension to elastic-viscoplastic materials announced in Proposition 6 proceeds as in (POLIZZOTTO [1993d]), with the only difference that the Melan shakedown theorem (to be invoked there) must now be taken as specified here for elastic-viscoplastic materials. The consequent developments are quite straighforward and are thus not reported here for simplicity.

In reference to Proposition 7, it can be stated that the inelastic shakedown condition experienced by the body can persist with different permanent loads as long as the isolated body V_e (obtained after removal from V) has the capacity to shake down in the elastic regime (in the sense specified previously) under the action of additional mechanical loads δP_0. The transition from inelastic shakedown to ratcheting for V occurs when the isolated body V_e finds itself at the elastic shakedown limit under the additional loads δP_0, and this makes it possible to specify the loads corresponding to the limit state of the structure in which the transition from inelastic shakedown to ratcheting occurs, (cf. PONTER & KARADENIZ [1985a, 1985b], POLIZZOTTO [1993a, 1993b]).

CONCLUSIONS

A class of elastic-viscoplastic materials with temperature-dependent plastic and creep data has been considered. For such materials, the existence of a steady-state response in a continuous body subjected to cyclic loads and temperature variations has been proven and discussed under the hypothesis of small displacements and strains. Additionally, the equations governing this response (or steady cycle) have been provided and the peridicity characteristics of the steady cycle studied. Essentially, in the steady cycle, the stresses $\boldsymbol{\sigma}$ and the internal variables $\boldsymbol{\chi}$, $\boldsymbol{\xi}$, along with the inelastic strain rates $\dot{\boldsymbol{\varepsilon}}^{in} = \dot{\boldsymbol{\varepsilon}}^{p} + \dot{\boldsymbol{\varepsilon}}^{c}$, turn out to be periodic with the same period as the loads. The concept of elastic shakedown has been extended to the present viscoplastic model, showing how Melan's and Koiter's theorems can be rephrased for this purpose, and also establishing the physical nature of the elastic shakedown limit state in the light of the material viscoplastic behaviour. Also, the sensitivity of the steady cycle to the load parameter changes has been studied, showing that a number of properties, previously ascertained for time-independent plasticity models, hold true also in the present time-dependent plasticity context.

The results given in the present paper are of interest for structural design purposes, in particular for nuclear energy production plants applications. Further developments are hoped for, in particular in the following directions:

– Methods for the direct determination of the so-called interaction diagram, and in particular of the boundary between the inelastic shakedown load domain B_F and the ratcheting load domain B_R. To this purpose procedures like those envisaged for time-independent plasticity can be considered and developed (cf. PONTER [1983], PONTER & KARADENIZ [1985a, 1985b], PONTER et. al. [1990] and POLIZZOTTO [1993a, 1993b]).

– Methods for the direct determination of the so-called post-transient residual stresses, that is the residual stresses which develop in the body at the stabilization time. For this purpose, the so-called Minimum Total Plastic Over-Potential principle (cf. POLIZZOTTO [1993b, 1993c]), suitably extended to the present context, may be useful.

– Criteria for predicting the type of steady cycle produced by a given load. Melan's and Koiter's theorems are applicable for elastic shakedown for this purpose, whereas for inelastic shakedown, theorems given in (POLIZZOTTO [1993b]) for plastic shakedown in perfect plasticity may be suitably extended to the present context.

Acknowledgement – This paper is a part of a research project supported by the Italian Governement, Ministero dell'Università e della Ricerca Scientifica e Tecnologica (MURST).

REFERENCES

1960 KOITER, W.T., "General Theorems of Elastic-Plastic Solids", in SNEDDON, J.N., and HILL, R. (eds.),Progress in Solid Mechanics, Vol. 1, North Holland, Amsterdam, pp. 167-221.

1966 FREDERICK, C.O. and ARMSTRONG, P.J., "Convergent Internal Stresses and Steady Cyclic State of Stress", J. Strain Analysis **1**, 154.

1972 MRÓZ, Z., "On the Theory of Steady Plastic Cycles in Structures", In: Proc.

of the First Int. Conf. on Structural Mechanics in Reactor Technology, Vol. 6, Part L, pp. 489-501.

1972 PONTER, A.R.S., "Deformation, Displacement and Work Bounds for Structure in a State of Creep and Subjected to Variable Loading", J. Appl. Mech. **39**, 953.

1974 PERZYNA, P., "Internal Variable Description of Plasticity", in SAWCZUK, A. (ed.), Problems of Plasticity, Noordhoff Int. Pub., Leyden, The Nederlands, pp. 145-170.

1975 MARTIN, J.B., Plasticity: Fundamentals and General Results, The MIT Press, Cambridge, Ma.

1975 HALPHEN, B. and NGUYEN, Q.S., "Sur les Matériaux Standard Géneralisées", J. de Mécanique **14**, 641.

1977 AINWHORTH, R.A., "Bounding Solutions for Creeping Structures Subjected to Load Variations Above the Shakedown Limit", Int. J. Solids Structures **3**, 971.

1977 CHABOCHE, J.L., "Viscoplastic Constitutive Equations for the Description of Cyclic and Anisotropic Behaviour of Metals", Bull. of the Polish Academy of Sciences, Series Sc and Techn., Vol 25, no.1, pp. 33-42.

1979 HALPHEN, B., "Steady Cycles and Shakedown in Elastic-Viscoplastic and Plastic Structures", in Matériaux et Structures Sous Chargement Cyclique, Association Amicale des Ingenieurs Anciens Eléves de E.N.P.C., Paris, pp. 203-230.

1980 AINSWORTH, R.A., GOODALL, I.W. and WALTERS, D.J., "Cyclic Loading in the Creep Range", Proc. Int. Conf. on Engineering Aspects of Creep, I. Mech. E/ASME, Sheffield, C215/80, pp. 153-160.

1980 GOKHFELD, D.A. and CHERNIAVSKY, D.F., "Limit Analysis of Structures at Thermal Cycling", Sijthoff & Noordhoff, Alphen aan der Rijn, The Netherlands.

1983 PONTER, A.R.S., Shakedown and Ratchetting Below the Creep Range, Report EUR 8702 EN, Commission of European Communities, Brussels.

1985 LEMAITRE J. and CHABOCHE, J.L., Mécanique des Matériaux Solides, Dunod, Paris.

1985a PONTER, A.R.S. and KARADENIZ, S., "An Extended Shakedown Theory for Structures that Suffer Cyclic Thermal Loadings, Part 1: Theory", J. Appl. Mech. **52**, 877.

1985b PONTER, A.R.S., and KARADENIZ, S., "An Extended Shakedown Theory for Structures that Suffer Cyclic Thermal Toadings, Part 2: Applications", J. Appl. Mech. **52**, 883.

1987 KÖNIG, J.A., Shakedown of Elastic-Plastic Structures, PWN-Polish Scientific Publishers, Warsaw, and Elsevier, Amsterdam.

1987 MAIER, G., "A Generalization to Nonlinear Hardening of the First Shakedown Theorem for Discrete Elastic-Plastic Models", Atti Acc. Lincei, Rend. Fis. **8**, LXXX, 161.

1987 MAIER, G., and NOVATI, G., "Deformation Bounds for Elastic-Plastic Structures with Piecewise Linear Yield Laws and Nonlinear Hardening", in OWEN, D.R.J., HINTON, E. and ONATE, E., (eds.), Computational Plasticity, Pineridge Press, Swansea, UK., pp. 17-31.

1988 PANZECA, T. and POLIZZOTTO, C., "On Shakedown of Elastic-Plastic Solids", Meccanica **23**, 94.

1990 LUBLINER, J., Plasticity Theory, Macmillan Publishing Co., New York.

1990 POLIZZOTTO, C., BORINO, G., and FUSCHI, P., "On the Steady-State Response of Elastic Perfectly Plastic Solids to Cyclic Loads", in KLEIBER, M. and KÖNIG, J.A.(eds.), Inelastic Solids and Structures, Pineridge Press, Swansea, U.K., pp. 473-488.

1990 PONTER, A.R.S., KARADENIZ, S. and CARTER, K.F., "Extended Upper Bound Shakedown Theory and Finite Element Method for Axisymmetric Thin Shells", in KLEIBER, M. and KÖNIG, J.A.(eds.), Inelastic Solids and Structures, Pineridge Press, Swansea, U.K., pp. 433-449.

1993a POLIZZOTTO, C., "A Study on Plastic Shakedown of Structures. Part I: Basic Properties", J. Appl. Mech. **60**, 318.

1993b POLIZZOTTO, C., "A Study on Plastic Shakedown of Structures. Part II: Theorems", J. Appl. Mech. **60**, 324.

1993c POLIZZOTTO, C., "The Assessment of the Residual Post-Transient Stresses", Rend. Mat. Acc. Lincei S.9, **4**, 133.

1993d POLIZZOTTO, C., "Steady States and Sensitivity Analysis in Elastic-Plastic Structures Subjected to Cyclic Loads", J. Solids Struct. (being printed).

1993 FUSCHI, P. and POLIZZOTTO, C., "The Shakedown Load Boundary of an Elastic-Plastic Structure Under Combined Cyclic/Steady Loads", Meccanica, (submitted for publication).

SOME BASIC ELEMENTS OF THE SHAKEDOWN THEORY

B. NAYROLES[1]
Laboratoire Sols-Solides-Structures
Grenoble, France.

INTRODUCTION

The aim of the present exposition is to summarize the main algebraic properties of classical plasticity, and thus help the understanding of the mathematical context of shakedown studies. Hence nothing new must be expected, except some few improvements concerning presentation and hypothesis enlargement ; for instance the new concept of "elastic sanctuary" has recently been introduced by the author and D. Weichert [1993] in order to give a maximal extension to the assumptions which ensure the classical proofs. We shall stay within the linear framework generated by the hypothesis that displacements and strains remain small in the vicinity of some strained configuration.
It is well known that such a theory was in a first state of achievement in a famous paper by KOITER [1964], completed by CERADINI in the dynamic case [1969]. The great progress of the seventies consisted of the first existence theorems in elastoplasticity, by J. J. MOREAU [1973], JOHNSON [1976], SUQUET [1979] and STRANG [1978]. Besides, some other improvements in the shakedown theory were proposed, for instance by KÖNIG [1987] or DEBORDES and NAYROLES [1976]. This last paper exposed the theory for a body with a finite number of degrees of freedom, together with the new theory of "shakedown domains" mainly devoted to the structural decomposition techniques. Later on O. DEBORDES [1981] established that static and kinematical shakedown formulations effectively define dual problems.
No functional analysis framework will be chosen here, for two good reasons : first, we wish to give a general exposition of the algebraical basis of the theory, independently of any particular kind of structure. This structure may be composed of different pieces, with different modelizations : beams, shells, 3D continuum etc. If some functional analysis results exist for this last piece, with the space BD(Ω) introduced by SUQUET [1979] and STRANG [1978], almost everything remains to be done for other sorts of solids. The second reason is that, even in the case of a 3D continuum, any exposition of functional analysis would need a full article ; reader interested in such questions should read SUQUET[1988].

---

[1] Directeur de Recherches CNRS, laboratoire Sols-Solides-Structures, Grenoble

Z. Mróz et al. (eds.), Inelastic Behaviour of Structures under Variable Loads, 129–152.

CONSTRAINTS AND EQUILIBRIUM EQUATIONS

A basic framework of structural mechanics is constituted by the duality between displacements and loads on the one hand, and strains and stresses on the other. As usual in structural mechanics all these terms are understood in a generalized sense. For instance, a displacement may be the deflection of a beam, the rotation of the normal to a shell etc. ; associated loads are defined as the generalized forces associated with these generalized displacements, i.e. by their virtual work (or "virtual power", if preferred...).
Generalized displacements, loads, strains and stresses are also understood as fields defined on the material finite domain Ω of $\mathbb{R}^p$ occupied by the structure in its reference geometry[1]. At the generic point M of the material domain Ω these fields take values which may be called "local displacement", "local strain", etc.
In the finite dimensional case, all these quantities are usually represented by column vectors in numerical computations. The local value may be either an entry of such a column, or, more generally, a sub-column. Most often, the word "generalized" will be omitted.

Displacements and forces

A possible configuration is defined by a displacement field $\mathbf{u}$:

$$P \in \Omega \mapsto \mathbf{u}(P) \in \mathbb{R}^m, \ m : \text{positive integer} \tag{1}$$

submitted to a set of geometric constraints which are supposed to be frictionless and linear. Hence the set of fields which are compatible with these constraints is a vector space[2], say $\mathbb{U}$. An easy generalization consists in assuming the constraints to be affine, which means that possible displacements fields belong to an affine manifold $\mathbf{u}^o + \mathbb{U}$, where $\mathbb{U}$ is a linear space and $\mathbf{u}^o$ a given displacement field, possibly not belonging to $\mathbb{U}$, possibly depending on time t.
If a discretization has been made, the dimension of $\mathbb{U}$ is finite and its elements are columns, the m×n entries of which are the m local displacement components at the n nodes. In the general situation, however, the dimension of $\mathbb{U}$ is infinite and some regularity conditions about the fields should be added in order to ensure the validity of our findings. A skillful choice of these regularity conditions mainly depends on the kind of the problem considered ; for instance Hilbert Sobolev's spaces perfectly fit problems involving linear constitutive equations, as is well known, and do not fit elastoplasticity problems. The nature of constitutive equations for loads and stresses is the most significant fact to take into account in choosing functional spaces. Hence, the choice of regularity conditions must be either delayed or simplified when a new particular problem is studied or when, as now, a large family of problems is being considered. A worthwhile simplified choice is the following : all considered fields are indefinitely differentiable[3] with respect to P. This choice will legitimize most of the operations we need to do, allowing the properties of possible solutions to be studied ; of course it will rarely ensure the existence of a solution.
The generalized forces $\mathbf{f}$ are defined by their virtual work, i.e. as linear functionals defined on $\mathbb{U}$:

---

[1] In order to shorten the exposition we shall restrict it to Lagrangian kinematics.

[2] If the implicit constraints are not linear, the set of permitted displacements is no longer a vector space ; in good cases it is a differentiable variety of infinite dimension, the use of which would need some mathematical developments. The virtual displacements then belong to the linear tangent space etc.

[3] Remark : the structure is very often composed of several solid bodies, say Ω_ι, assembled by frictionless constraints taken into account in the definition of the displacement space : in this case regularity conditions are defined for each body.

$$\mathbf{u} \in \mathbb{U} \mapsto \langle\langle \mathbf{u}, \mathbf{f} \rangle\rangle \in \mathbb{R}$$

In the finite dimensional case, generalized forces are columns **f**, and their virtual work reduces to the product $\mathbf{f}^T\mathbf{u}$, the upper index "T" denoting matrix transposition. More generally **f** is a vector of $\mathbb{F}$, a space of linear forms defined on $\mathbb{U}$. Anyhow the virtual work is now expressed as a bilinear form which places this pair of spaces in duality :

$$(\mathbf{u}, \mathbf{f}) \in \mathbb{U} \times \mathbb{F} \quad \mapsto \quad \langle\langle \mathbf{u}, \mathbf{f} \rangle\rangle \in \mathbb{R} \tag{2}$$

Usually arguments are replaced by points and this bilinear form is denoted by << . , . >>. As generalized forces are defined by their virtual work we have the first of the following equivalencies ; moreover we shall assume that F has been chosen rich enough to ensure the second :

$$\begin{cases} \mathbf{f} = 0 \Leftrightarrow \forall\ \mathbf{u} \in \mathbb{U} & : \langle\langle \mathbf{u}, \mathbf{f} \rangle\rangle = 0 \\ \mathbf{u} = 0 \Leftrightarrow \forall\ \mathbf{f} \in \mathbb{U} & : \langle\langle \mathbf{u}, \mathbf{f} \rangle\rangle = 0 \end{cases} \tag{3}$$

For instance this latter equivalence holds if all regular fields of force densities belong to $\mathbb{F}$. The term "load" is usually devoted to generalized external forces ; the internal generalized forces are also represented by vectors of $\mathbb{F}$. The equilibrium is then expressed by the equation :

$$\mathbf{f}_e + \mathbf{f}_i = \mathbf{0} \tag{4}$$

where $\mathbf{f}_e$ denotes the sum of external forces (or loads), and $\mathbf{f}_i$ the sum of internal forces. Most often internal forces are defined in terms of stresses, and this vector $\mathbf{f}_i$ plays the same role as the divergence of the stress field in the classical theory of continuous media. We now introduce the (generalized) strains and stresses.

Strains and stresses

At point P of the material domain Ω a tensor space $\mathbb{T}^n$ is defined , with dimension n and inner product denoted by a single dot :

$$(\varepsilon,\sigma) \in \mathbb{T}^n \times \mathbb{T}^n \mapsto \varepsilon.\sigma \in \mathbb{R} \tag{5}$$

The possible local strain ε and possible local stress σ both belong to $\mathbb{T}^n$, and the product $-\varepsilon.\sigma$ is the local virtual work of the stress σ in the virtual strain ε. Here again, local stresses and strains are represented by columns of n entries in numerical computations ; nevertheless it may be worthwhile that notation expresses their tensorial nature in so far as numerical computations are not undertaken.
The tensor space $\mathbb{T}^n$ may vary from one part of the structure to another, depending on the model used for this part : beam, plate, 3D-continuous medium etc.
Now consider a strain field ε and a stress field σ defined over the material domain Ω ; the global virtual work of σ is now defined as the integral of the former local virtual work (5) over Ω :

$$- < \varepsilon , \sigma > = - \int_{\Omega} \varepsilon(P).\sigma(P)\, d\mu(P) \tag{6}$$

The integral must be understood with respect to a measure dm which, very often, is the usual Lebesgue measure dM for a volume, a surface or a line. This measure is concentrated at Gauss points when a finite element model is used. The sign "-", in both sides of equation (6), is chosen in such a way that, in the finite dimensional case, the transposition of the linear operator **D**, which will be introduced soon, identifies with matrix transposition.
A pair of functional spaces must be chosen, say $\mathbb{E}$ for strain fields ε and $\mathbb{S}$ for stress fields σ, equipped with the following three properties :

$$(\varepsilon, \sigma) \in \mathbb{E}\times\mathbb{S} \mapsto < \varepsilon , \sigma > = \int_{\Omega} \varepsilon(P).\sigma(P)\, d\mu(P) \in \mathbb{R} \tag{7}$$

All regular strain fields belong to $\mathbb{E}$ and to $\mathbb{S}$. (8)

$$\begin{cases} \varepsilon = 0 \iff \forall\, \sigma \in \mathbb{S} \quad < \varepsilon , \sigma > = 0 \\ \sigma = 0 \iff \forall\, \varepsilon \in \mathbb{E} \quad < \varepsilon , \sigma > = 0 \end{cases} \tag{9}$$

Condition (7) expresses that integral (6) does exist for every pair (ε, σ) in $\mathbb{E}\times\mathbb{S}$. Condition (8) means that no algebraic constraint is taken into account in the choice of these two spaces. Condition (9) is expressed in the followings terms by mathematicians : "the bilinear form (7) places the pair of spaces $(\mathbb{E}, \mathbb{S})$ in separating duality". Besides, it is a consequence of (8), provided the elements of $\mathbb{E}$ and $\mathbb{S}$ are defined only "almost everywhere" with respect to the measure $d\mu$.

The deformation operator $\mathcal{D}$ and its derivative D

When modelling a structure one defines a so called "deformation operator" $\mathcal{D}$, which expresses the strain field ε generated by a displacement **u** :

$$\mathbf{u} \in \mathbb{U} \mapsto \varepsilon = \mathcal{D}(\mathbf{u}) + \varepsilon^{\circ} \in \mathbb{E} \tag{10}$$

where ε° denotes some imposed strain, due, for instance, to a thermal dilatation[1], and to a possible imposed displacement $\mathbf{u}^{\circ}$. Formally $\mathcal{D}$ is a differential operator. When a 3D continuum in finite strain is concerned, $\mathcal{D}$ is :

$$\mathcal{D}(\mathbf{u})_{ij} = \frac{1}{2}\left(\frac{\partial u_j}{\partial x_j} + \frac{\partial u_i}{\partial x_j} + \frac{\partial u_k}{\partial x_i}\frac{\partial u_k}{\partial x_j}\right) \tag{11}$$

The same notation $\mathcal{D}$ will also[2] denote the mapping, defined by this formal operator, of a vector space of displacement $\mathbb{U}$ into a space of strain $\mathbb{E}$.

---

[1] Let the temperature be θ, the thermal dilatation tensor χ, then $\varepsilon^{o} = - \chi\, \theta$.

[2] It would be more precise to use another letter to define this mapping.

Generally $\mathbb{E}$ is bigger than $\mathcal{D}(\mathbb{U})$: indeed, there exist strain fields which are not the strains generated by a displacement **u** : usually some integrability conditions are required. For a simply connected 3D continuum in small strains and displacements, this condition is well known :

$$\mathbf{rot}\,(\mathbf{rot}^{T}\varepsilon) = 0$$

For the sake of simplicity, we shall assume the null set of $\mathcal{D}$ to reduce to $\{\mathbf{0}\}$:

$$\mathbf{u} = \mathbf{0} \Longleftrightarrow \mathcal{D}(\mathbf{u}) = \mathbf{0}$$

Let us now introduce $\mathbf{D} = \mathcal{D}'_{\mathbf{u}}$, the linear operator tangent to $\mathcal{D}$ at **u** :

$$\delta\mathbf{u} \in \mathbb{U} \mapsto \mathbf{D}\,\delta\mathbf{u} \in \mathbb{E} \tag{12}$$

which may be defined, in a weak sense, by :

$$\forall\,\sigma \in \mathbb{S} : \quad \lim_{\lambda \to 0} < \frac{\mathcal{D}(\mathbf{u}+\lambda\delta\mathbf{u}) - \mathcal{D}(\mathbf{u})}{\lambda}, \sigma > = < \mathbf{D}\,\delta\mathbf{u}\,, \sigma > \tag{13}$$

Let us explain this definition. First it expresses that, for any given $\delta\mathbf{u}$ in $\mathbb{U}$, the differential quotient (written in the left hand side of (13)) weakly converges to some element $\mathcal{D}'_{\mathbf{u}}(\delta\mathbf{u}) \in \mathbb{E}$ and thus is a linear form on $\mathbb{S}$. Now (13) implies that $\mathcal{D}'_{\mathbf{u}}(\delta\mathbf{u})$ is a homogenous function of $\delta\mathbf{u}$, i.e. :

$$\forall\,\mu \in \mathbb{R} \qquad \mathcal{D}'_{\mathbf{u}}(\mu\,\delta\mathbf{u}) = \mu\,\mathcal{D}'_{\mathbf{u}}(\delta\mathbf{u})$$

Finally if $\mathcal{D}'_{\mathbf{u}}(\delta\mathbf{u})$ linearly depends on $\delta\mathbf{u}$ and is more simply written $\mathbf{D}\,\delta\mathbf{u}$, $\mathbf{D} = \mathcal{D}'_{\mathbf{u}}$ is the tangent linear operator, which might be also called the "virtual deformation operator".
In the 3D continuum case $\mathcal{D}$ is given by (11) ; **D** is easily derived as :

$$(\mathbf{D}\,\delta\mathbf{u})_{ij} = \frac{1}{2}\left(\frac{\partial du_j}{\partial x_j} + \frac{\partial du_i}{\partial x_j} + \frac{\partial du_k}{\partial x_i}\frac{\partial u_k}{\partial x_j} + \frac{\partial du_k}{\partial x_j}\frac{\partial u_k}{\partial x_i}\right) \tag{14}$$

As is the case of $\mathcal{D}$ the same notation **D** is used for the formal differential operator and the linear mapping of $\mathbb{U}$ into $\mathbb{E}$.
Note that we omit parenthesis when an operator is linear ; for instance $\mathbf{D}\,\delta\mathbf{u}$ stands for $\mathbf{D}(\delta\mathbf{u})$.

Virtual work principle and equilibrium equation

The virtual work principle is now expressed in the following terms :
A stress field σ and a load **f** are in equilibrium in the configuration **u** of the structure if and only if :

$$\forall\, \delta\mathbf{u} \in \mathbb{U} : \langle\langle \delta\mathbf{u}\,,\ \mathbf{f} \rangle\rangle - \langle \mathbf{D}\,\delta\mathbf{u}\,, \sigma \rangle = 0 \tag{15}$$

which, due to the definition of the transposed operator, is equivalent to :

$$\mathbf{D}^{\mathrm{T}}\sigma = \mathbf{f} \tag{16}$$

$\mathbf{D}^{\mathrm{T}}$ may be called the "equilibrium operator".
As an example consider the 3D continuum case ; the equilibrium operator $\mathbf{D}^{\mathrm{T}}$ can be computed after the load space $\mathbb{F}$ has been specified. We may define loads as pairs :

$$\mathbf{f} = \{\varphi, \Phi\}$$

where φ is a field of volume force in Ω and Φ a field of surface force on the boundary $\partial\Omega$ (the outward unit vector is denoted by **n**). The virtual work is then defined by :

$$\langle\langle \delta\mathbf{u}\,, \mathbf{f} \rangle\rangle = \int_{\Omega} \varphi.\delta\mathbf{u}\ \ dV + \int_{\partial\Omega} \Phi.\delta\mathbf{u}\ \ dS \tag{17}$$

and the transposed operator is now computed classically :

$$\mathbf{D}^{\mathrm{T}}\sigma = \mathbf{f} \iff \begin{cases} -\dfrac{\partial}{\partial x_i}\left(\sigma_{ik} + \sigma_{ij}\dfrac{\partial u_k}{\partial x_j}\right) = \varphi_k & \text{in } \Omega \\ n_i\left(\sigma_{ik} + \sigma_{ij}\dfrac{\partial u_k}{\partial x_j}\right) = \Phi_k & \text{on } \partial\Omega \end{cases} \tag{18}$$

It can be seen that $\mathbf{D}^{\mathrm{T}}$ is not a simple differential operator since it also involves boundary terms.
Independently of constitutive equations, constraint equation (10) and equilibrium equation (16) are valid for any problem, and are relationships, first between the geometric variables **u** and ε, secondly between the strength variables **f** and σ. Two more equations are needed : first a constitutive equation for the load **f**, i.e. a relationship between **u** and **f** (possibly **f** is given independently of **u**), secondly a constitutive equation for the stress σ, i.e. a relationship between ε and σ.

The schema of duality with four linear spaces

The present situation, with four spaces in duality, may be summarized up by the next diagram :

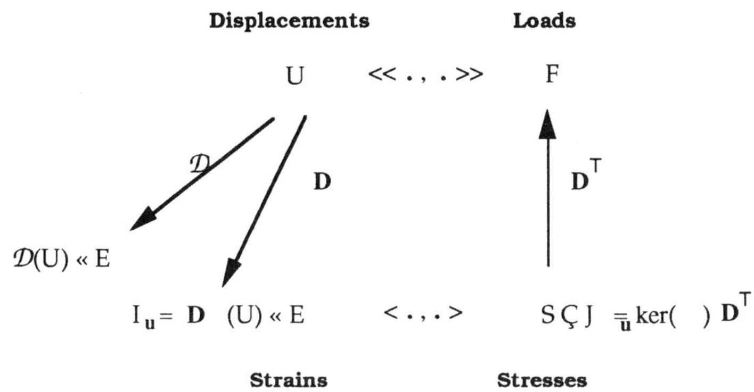

Figure 1

Generally neither $\mathcal{D}$ nor the image space $\mathcal{D}(\mathbb{U})$ are linear. On the other hand $\mathbf{D}$ is linear and the image $\mathbb{I}_{\mathbf{u}} = \mathbf{D}(\mathbb{U})$ is a vector subspace of $\mathbb{E}$. Equations (15) or (16) imply that the null space of $(\mathbf{D})^T$, denoted either by $\ker(\mathbf{D}^T)$ or by $\mathbb{J}_{\mathbf{u}}$, is the orthogonal space of $\mathbb{I}_{\mathbf{u}}$:

$$\mathbb{J}_{\mathbf{u}} = \ker(\mathbf{D}^T) = \left\{ \sigma \in \mathbb{S} \quad \forall\ \varepsilon \in \mathbb{I}_{\mathbf{u}} \ : < \sigma , \varepsilon > = 0 \right\} \tag{19}$$

$\mathbb{I}_{\mathbf{u}}$ is the vector space of integrable strain rate fields, $\mathbb{J}_{\mathbf{u}}$ the vector space of self-equilibrated stress fields. We shall assume the reciprocal statement to hold :

$$\mathbb{I}_{\mathbf{u}} = \left\{ \varepsilon \in \mathbb{E} \quad \forall\ \sigma \in \mathbb{J}_{\mathbf{u}} \ : < \sigma , \varepsilon > = 0 \right\} \tag{20}$$

as it always does in finite dimensional cases and in most other situations for a convenient choice of functional spaces.
As far as shakedown theory is concerned, the main fact is :
In the general framework of finite displacements, the orthogonal vector spaces $\mathbb{I}_{\mathbf{u}}$ and $\mathbb{J}_{\mathbf{u}}$ depend on the configuration u, and thus are moving subspaces of $\mathbb{E}$ and $\mathbb{S}$. This is one of the major obstacles to be cleared in extending the shakedown theory to finite displacements.
In the rest of this paper we shall generally omit the subscript "$_{\mathbf{u}}$" and remember that $\mathbb{I}$ and $\mathbb{J}$ depend on the configuration $\mathbf{u}$.

CONSTITUTIVE EQUATIONS

We now introduce the elastoplasticity constitutive laws. As far as shakedown only is concerned, we need not define them completely, as we should do to set up the evolution problem completely : we only want to outline the properties which possibly ensure shake-down.

When a structure is submitted to time-variable loads, shake-down consists, roughly speaking, in the ability of this structure to take a permanent plastic configuration, in the neighbourhood of which its response to variable loads is purely elastic. Of course this plastic configuration may be only a limit, reached at $t = +\infty$. Two main ideas lead shake-down theory and computation :

• Loads are not precisely known : they may vary within a given domain Δ_F of the load space $\mathbb{F}$.

• Around the limit, permanent plastic configuration elastic strains remain small, since they are elastic. Hence the basic shake-down theory is concerned with small strains, possibly around a configuration obtained after large deformations of the reference configuration. Moreover we do not care how this limit plastic configuration is reached.

Local constitutive equations

We now denote by ε and σ the local values of generalized strain and stress at a generic point P of the structure :

$$\varepsilon = \varepsilon(P) \in \mathbb{T}^n \quad ; \quad \sigma = \sigma(P) \in \mathbb{T}^n$$

It is assumed that ε is the sum of an elastic term x and of a plastic term p :

$$\varepsilon = x + p \tag{21}$$

each of them related to the stress s in its own way.

Elasticity

There exists a symmetric positive linear mapping $K : \mathbb{T}^n \mapsto \mathbb{T}^n$, depending on P only, such that :

$$\sigma = K.x \tag{22}$$

which means, using the indicial notation :

$$\sigma_i = K_{ij}\, x_j$$

Normal plasticity : Hill's inequality holds

The "normality law" of plastic rate may be written in the following way.

There exists in $\mathbb{T}^n$ a so called "elasticity domain", which is a closed convex set Γ, generally depending on the history of strain at point P, such that :

$$\begin{cases} \sigma \in \Gamma \\ \forall \sigma' \in \Gamma : \dot{p}.(\sigma - \sigma') \geq 0 \end{cases} \tag{23}$$

a statement which may be completely expressed by the subdifferential notation :

$$\dot{p} \in \partial\Psi_{\Gamma}(\sigma) \tag{24}$$

Inequality (23) is well known as "Hill's inequality". It expresses that plastic rate $\dot{p}$ is an outward normal to the convex Γ at the point σ. In particular $\dot{p}$ vanishes when σ belongs to the interior of Γ.

Perfect plasticity : the elasticity domain Γ is constant

When Γ is constant we have the so called "perfect plasticity". Then :

$$\dot{p}.\dot{\sigma} = 0 \tag{25}$$

is easily derived from (23) by replacing s' first by $\sigma(t+dt)$, then by $\sigma(t-dt)$.

Other plastic behaviours

In the general case Γ varies with time, for many possible reasons : work-hardening, soft-hardening, variations in temperature, material ageing etc. Hence Γ will rather be denoted by Γ_t. Different models have been constructed for many years in order to describe Γ_t evolution. Classical plasticity models assume that Γ_t depends only on the trajectory followed by the plastic strain p.

By definition, it is said one that "work-hardening" occurs at time t if $\dot{p}\,.\,\dot{\sigma}$, the left-hand side of (25), is positive ; conversely "work-softening" occurs when $\dot{p}\,.\,\dot{\sigma}$ is negative.

The theory of "generalized standard materials"[1] assumes the apparent work-hardening to reduce to a perfectly plastic behaviour, provided suitable hidden variables are exhibited. Such models imply Drucker's inequality $\dot{p}\,.\,\dot{\sigma} \geq 0$, i.e. no work-softening. Their great theoretical interest comes from the following property: any structure built of a generalized standard material reduces to an equivalent structure built of a perfectly plastic material ; in particular the classical theory of shake-down and limit loads in perfect plasticity applies to generalized standard materials.

As a matter of fact we do not know any shakedown theorem except when perfect elastoplasticity is assumed, either directly, or indirectly behind the machinery of generalized standard materials. Now we know at least two practical reasons why these models are not general enough in the present context of structural mechanics :

- Many materials, such as soils, do not obey a normality law of plastic rate ;
- Work-softening may occur.

When evoking a general plasticity law we shall simply write :

$$\dot{p} \in \Pi_t(\sigma) \tag{26}$$

[1] introduced by Nguyen Q.S. and others around J. Mandel in the seventies, and by the author in a similar way under the "postulat de plasticité parfaite sous-jacente". Another more sophisticated version was proposed by P. Suquet

where the set Π_t depends on the present stress σ, as indicated by the notation, and also on the history, at the point P, of the plastic strain p until the present time t. Writing (26) is somewhat empty of information when Π_t is not known ; at least it recalls that plastic flow is given by a multivalued mapping.

After Euromech 298 was held the author and D. Weichert searched for some maximal enlargement of the usual hypotheses which are the basis of shakedown theorems, provided the heart of the classical proofs could be retained. This led them to :

A new concept : the elasticity sanctuary

Let us come back to normal plasticity and consider the family $\{\Gamma_i\}$ of all possible elasticity domains, obtained for all practically possible histories of strains, temperature etc., and consider the interior of their intersection. It follows from (23) that

$$\mathcal{S} \subset \mathrm{Int}(\bigcap_i \Gamma_i) \Rightarrow \begin{cases} \sigma \in \mathcal{S} \Rightarrow \dot{p} = 0 \\ \forall \sigma' \in \mathcal{S} : \dot{p}.(\sigma - \sigma') \geq 0 \end{cases} \quad (27)$$

So, $\dot{p}$ verifies Hill's inequality for any subset $\mathcal{S}$ of the interior of this intersection. Conversely if Hill's inequality is verified by $\mathcal{S}$ and $\dot{p}$, the latter does not need to be normal to the present elasticity domain Γ_i : it only needs to be normal to the smallest cone with vertex σ and embedding $\mathcal{S}$, as illustrated by figure 2. This property allows hypothesis enlargement. With no more reference to normal plasticity we propose the next definition, NAYROLES & WEICHERT [1993].

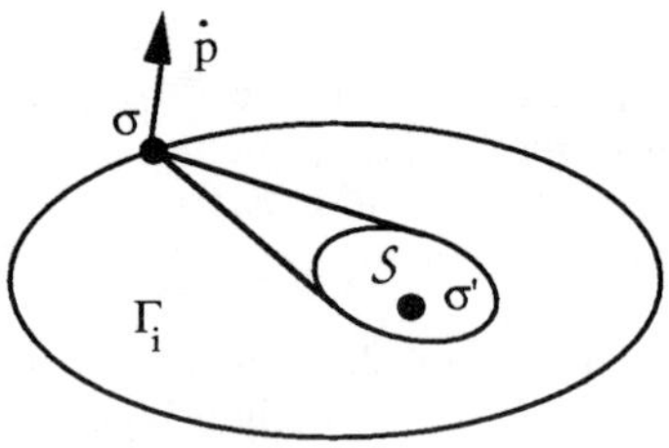

Figure 2 Elasticity sanctuary

Definition

A (constant) subset $\mathcal{S}$ of the stress space is called an "elasticity sanctuary" if it possesses the following properties :

$$\begin{cases} 0 \in \mathcal{S} \\ \sigma \in \mathcal{S} \Rightarrow \dot{p} = 0 \\ \forall \sigma' \in \mathcal{S} : \dot{p}.(\sigma - \sigma') \geq 0 \end{cases} \quad (28)$$

The family $\{\mathcal{S}_i\}$ of all elasticity sanctuaries has some immediate properties :

- $\{\mathcal{S}_i\}$ is stable with respect to the reunion and to the section of sets.
- For perfect plasticity, the greatest elasticity sanctuary is the interior of the elasticity domain.
- For normal plasticity, the interior of the intersection of the elasticity domain is an elasticity sanctuary, provided O belongs to it.

Hill's inequality (28) is the weakest hypothesis we need in adapting the classical proofs of the static shake-down theorems in perfect plasticity : the elasticity sanctuary now plays the role of the elasticity domain in perfect plasticity. As previously said, the classical normality law is no more assumed, nor work-hardening. Note that elasticity sanctuaries are more or less analogous to the domain used by some authors in defining a "plastic potential" when the normality law is not verified with respect to the elasticity domain.

Global form of constitutive equations

The basic decomposition of the strain into an elastic part and a plastic one is now understood in terms of fields :

$$\boldsymbol{\varepsilon} = \mathbf{x} + \mathbf{p} \tag{29}$$

where bold characters are used, as before, to denote global quantities such as fields, functional sets etc.

Elasticity

The elastic part $\mathbf{x}$ is linearly related to the stress field $\boldsymbol{\sigma}$ by the compliance $\mathbf{K}^{-1}$:

$$\mathbf{x} \in \mathbb{S} \mapsto \mathbf{x} = \mathbf{K}^{-1}\boldsymbol{\sigma} \in \mathbb{E}_K \subset \mathbb{E} \; ; \mathbb{E}_K = \mathbf{K}^{-1}(\mathbb{E}) \tag{30}$$

which, means, by definition :

$$\forall\, P \in \Omega : \mathbf{x}(P) = \mathbf{K}^{-1}(P)\, \boldsymbol{\sigma}(P)$$

Compliance is used rather than stiffness because stress fields are generally more regular than strain fields ; indeed plastic strain fields may not be regular and, however, must belong to $\mathbb{E}$. Hence $\mathbb{E}$ is generally bigger than the image $\mathbb{E}_K$ on which the stiffness $\mathbf{K}$ is defined.

Remark that, provided remarkably supple materials are ignored, the elastic part $\mathbf{x}$ is small ; large strains are mainly plastic.

Plastic flow

In the general case we may simply write (26) in bold letters

$$\dot{\mathbf{p}} \in \Pi_t(\boldsymbol{\sigma}) \tag{31}$$

which simply means that (26) is verified almost everywhere in Ω. The situation must be clear when normal plasticity is concerned. As usual, we define the elasticity domain of an elastoplastic structure as the set:

$$\Gamma = \left\{ \sigma \in \mathbb{S} \,/\, \forall\, P \in \Omega : \sigma(P) \in \Gamma_{\mathbf{P}} \right\} \tag{32}$$

and now the question is whether the next equivalence holds or not:

$$\left.\begin{array}{c} \sigma \in \Gamma \\ \forall\, \sigma' \in \Gamma : < \dot{\mathbf{p}} , (\sigma - \sigma') > \, \geq 0 \end{array}\right\} \Leftrightarrow \forall\, P \in \Omega : \left\{\begin{array}{c} \sigma(P) \in \Gamma_{\mathbf{P}} \\ \forall\, \sigma' \in \Gamma_{\mathbf{P}} \; : \; \dot{\mathbf{p}}(P) . (\sigma(P) - \sigma') \geq 0 \end{array}\right. \tag{33}$$

The right hand side obviously implies the left; the reverse may be false unless complementary assumptions are made: for instance functions $\dot{\mathbf{p}}$, σ, σ' are continuous and the elasticity domain Γ_P continuously depends on P. Other kinds of hypothesis are possible; anyhow we shall assume that equivalence (33) holds.
Hence, when studying a structure with normal elastoplasticity, the global form of constitutive equations will be (29)+(30) joined to :

$$\left\{\begin{array}{c} \sigma \in \Gamma \\ \forall\, \sigma' \in \Gamma : < \dot{\mathbf{p}} , (\sigma - \sigma') > \, \geq 0 \end{array}\right. \tag{34}$$

About elasticity sanctuaries consider a mapping Σ :

$$P \in \Omega \mapsto \Sigma(P) \subset \mathbb{T}^n \tag{35}$$

such that, for (almost) every P in Ω, Σ(P) is a elasticity sanctuary. Then Σ may be associated with the following subset $\mathcal{S}$ of $\mathbb{S}$:

$$\mathcal{S} = \left\{ \sigma \in \mathbb{S} \,/\, \forall\, P \in \Omega : \sigma(P) \in \Sigma(P) \right\} \tag{36}$$

Now it can be derived from (28) that $\mathcal{S}$ is an elastic sanctuary in the following sense:

$$\left\{\begin{array}{c} \mathbf{0} \in \mathcal{S} \\ \sigma \in \mathcal{S} \Rightarrow \dot{\mathbf{p}} = \mathbf{0} \\ \forall\, \sigma' \in \mathcal{S} : < \dot{\mathbf{p}} , (\sigma - \sigma') > \, \geq 0 \end{array}\right. \tag{37}$$

and, as in the case of (33), we shall assume that, conversely, (37) implies (28).
As usualy, constitutive equations take the same form at both levels, global and local, provided suitable regularity assumptions are made.

PLASTICITY AND SHAKEDOWN THEORY FOR QUASI STATIC EVOLUTIONS IN SMALL TRANSFORMATION

Introduction. General equations

In this paragraph we expose the classical, geometrically linear, theory of shakedown. We assume all operations are valid, which is always true in the finite dimensional case. The next paragraph will expose some difficulties encountered when setting the functional frameworks of more general cases.

As already quoted in the paragraph "Constitutive equations" the classical proofs of shakedown theorems are based on the perfectly elastoplastic behaviour, assumed directly or indirectly with the help of a generalized standard behaviour hypothesis. It is useful to recall the outlines of this structural behaviour.

When more general cases are concerned we shall write Hill's inequality for a possible sanctuary of elasticity $\mathcal{S}$.

In order to open later studies in large transformations it may be worthwhile to set this geometrically linear theory in the more general framework of finite transformations. So, forthcoming difficulties may already be observed.

Last, but not least, we consider dynamic evolutions : inertia loads are taken into account. Then evolution equations of an elastoplastic structure are :

$$\left\{\begin{array}{l} \varepsilon = \mathcal{D}(\mathbf{u}) + \varepsilon^{o} \\ \mathcal{D}^{T}\sigma = \mathbf{f} - \mathbf{M}\ddot{\mathbf{u}} \\ \sigma = \mathbf{K}\,(\varepsilon - \mathbf{p}) \\ \dot{\mathbf{p}} \in \Pi_t(\sigma) \end{array}\right. \tag{38}$$

where $\mathbf{M}$ is the mass operator, a non negative[1], symmetric linear mapping of $\mathbb{U}$ into $\mathbb{F}$.

When perfect elastoplasticity is concerned, Π_t reduces to the sub-differential cone :

$$\Pi_t = \partial\Psi_{\Gamma}(\sigma) \tag{39}$$

and, anyhow, (37) holds for any sanctuary of elasticity $\mathcal{S}$.

Small transformations and tangent linear elasticity

We now consider small transformations in the vicinity of some state whose configuration is $\mathbf{u}_o$, plastic strain $\mathbf{p}_o$. In studying these transformations we use the tangent deformation operator $\mathbf{D}$ and consider the associated tangent linear elasticity. We also introduce a relaxed configuration[2] $\bar{\mathbf{u}}$ which plays the role of a new reference configuration ; $\bar{\mathbf{u}}$ is deduced from $\mathbf{u}_o$ by removing elastically the load $\mathbf{f}$ and the imposed strain ε^o. Let us introduce the new variables $\mathbf{v}$ and $\mathbf{q}$:

$$\left\{\begin{array}{l} \mathbf{u} = \bar{\mathbf{u}} + \mathbf{v} \\ \mathbf{p} = \mathbf{p}_o + \mathbf{q} \end{array}\right. \tag{40}$$

---

[1] vanishing for quasi-static studies.

[2] This configuration may be virtual if the real removal of $\mathbf{f}$ and ε^{o} would imply some new plastic strain.

and assume that $\mathbf{v}$ is "small" ; then :

$$\mathcal{D}(\mathbf{u}) = \mathbf{D}\,\mathbf{v} + \mathcal{D}(\bar{\mathbf{u}})$$

and

$$\varepsilon = \mathbf{D}\,\mathbf{v} + \varepsilon^{o} + \mathcal{D}(\bar{\mathbf{u}}) \tag{41}$$

replaces (38.1) in small transformations studies.

The equations of tangent linear elasticity are derived from (38) when $\dot{\mathbf{p}} = 0$, i.e. when $\Pi_t = \{0\}$, whereas (38.1) is replaced by (41). When ε has been eliminated one obtains :

$$\begin{cases} \mathbf{D}^{T}\sigma^{e} = \mathbf{f} - \mathbf{M}\ddot{\mathbf{v}}^{e} \\ \sigma^{e} = \mathbf{K}\left[\mathbf{D}\mathbf{v}^{e} + \varepsilon^{o}\right] \end{cases} \tag{42}$$

The structural stiffness is :

$$\mathbf{L} = \mathbf{D}^{T}\mathbf{K}\,\mathbf{D} \tag{43}$$

and the existence of $\mathbf{L}^{-1}$ is henceforth assumed ; then we may introduce the operators :

$$\begin{cases} \mathbf{A} = \mathbf{K}\;\mathbf{D}\mathbf{L}^{-1} \\ \mathbf{B} = \mathbf{K} - \mathbf{A}\,\mathbf{D}^{T}\mathbf{K} \end{cases} \tag{44}$$

and the "tangent elastic response" (v^e, ε^e, σ^e) is given by :

$$\begin{cases} \sigma^{e} = \mathbf{A}\;(\mathbf{f} - \mathbf{M}\ddot{\mathbf{v}}^{e}) + \mathbf{B}\,\varepsilon^{o} \\ \mathbf{v}^{e} = \mathbf{L}^{-1}(\mathbf{f} - \mathbf{M}\ddot{\mathbf{v}}^{e} - \mathbf{D}^{T}\mathbf{K}\,\varepsilon^{o}) \\ \varepsilon^{e} = \mathbf{D}\;\mathbf{v}^{e} + \varepsilon^{o} \end{cases} \tag{45}$$

Note that :

$$\sigma = \sigma^{e} + \bar{\sigma}\;;\; \sigma^{e} = \mathbf{K}\,\varepsilon^{e}\;;\; \bar{\sigma} = \mathbf{K}\left[\mathcal{D}(\bar{\mathbf{u}}) - \mathbf{p}_{o}\right] \tag{46}$$

The field $\bar{\sigma}$ is the residual stress in the relaxed configuration $\bar{\mathbf{u}}$.
In (45) the elastic response σ^e is expressed as the sum of a self equilibrated field :

$$\mathbf{B}\,\varepsilon^{o} \in \mathbb{J}$$

and of the image, by the stiffness, of the integrable strain field $\mathbf{D}\mathbf{L}^{-1}(\mathbf{f} - \mathbf{M}\ddot{\mathbf{v}}^{e})$:

$$\mathbf{A}\,(\mathbf{f} - \mathbf{M}\ddot{\mathbf{v}}) \in \mathbf{K}(\mathbb{I})$$

Every stress field σ may be expressed in the same way, with the help of two projectors:

$$\begin{cases} \sigma = \sigma_I + \sigma_J \\ \sigma_I = \mathbf{P_I}\sigma = \mathbf{A}\,\mathbf{D}^T\sigma \\ \sigma_J = \mathbf{P_J}\sigma = (\mathbf{I} - \mathbf{P_I})\sigma \end{cases} \tag{47}$$

These two projectors are orthogonal with respect to the compliance :

$$\forall\, \sigma \in \mathbb{S} : < \mathbf{K}^{-1}\mathbf{P}_{\mathbf{I_u}}(\sigma)\,,\, \mathbf{P}_{\mathbf{J_u}}(\sigma) > = 0 \tag{48}$$

since the former factor belongs to $\mathbb{I}$ and the latter to $\mathbb{J}$.
The situation may be observed in the space $\mathbb{S}$ of stress fields: $\mathbb{J}$ and $\mathbf{K}(\mathbb{I})$ are orthogonal supplementary subspaces with respect to the scalar product defined by $\mathbf{K}^{-1}$:

$$(\sigma, \sigma') \in \mathbb{S}\times\mathbb{S} \mapsto (\sigma | \sigma')_{\mathbf{K}} = <\mathbf{K}^{-1}\sigma,\ \sigma'> \tag{49}$$

The subspace space $\mathbf{K}(\mathbb{I})$ is isomorphic to the load space $\mathbf{F}$. Hence its elements may also be called loads, by a language short cut.
The schematic figure illustrates this orthogonality and the operators defined above.

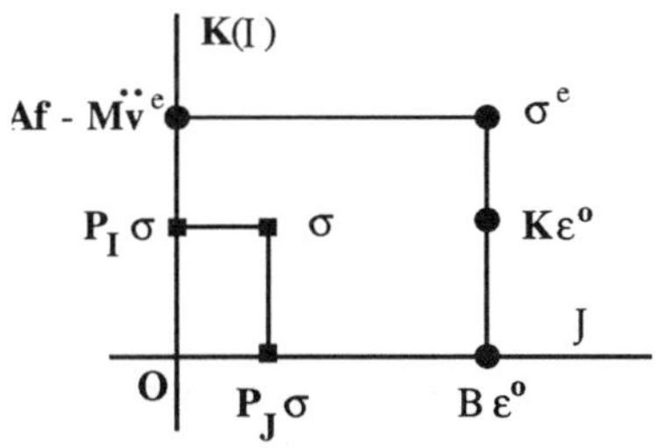

Figure 3

Perfect elastoplasticity in quasi-static small transformations

The evolution of a perfectly elastoplastic structure in small strains is a reference model: it helps us to understand the plastic properties of more general behaviours.
Evolution equations are already written down in (38) in the general case ; using (39), (41) and the quasi-static elastic solution gives:

$$\begin{cases} \sigma = \sigma^e + \tau \\ \sigma = \mathbf{K}\left[\mathbf{D}\mathbf{v} + \varepsilon^o - \mathbf{q} + \mathcal{D}(\bar{\mathbf{u}}) - \mathbf{p}_o\right] \\ \sigma \in \Gamma \;\; ; \;\; \forall\, \sigma' \in \Gamma : <\dot{\mathbf{q}}\,, (\sigma - \sigma')> \geq 0 \end{cases} \tag{50}$$

Unknowns are $\mathbf{v} \in \mathbb{U}$, $\mathbf{q} \in \mathbb{E}$, $\tau \in \mathbb{J}$ and $\sigma \in \mathbb{S}$. A well defined evolution problem would need some more conditions, such as initial conditions, or periodic conditions etc.
Figure 4 shows, in the stress space, the roles played by various quantities.

As in figure 3, the orthogonal subspaces $\mathbf{K}(\mathbb{I})$ and $\mathbb{J}$ are represented as axes of coordinates. The global elasticity domain of the structure, denoted by Γ, is the tall ellipse leaning over to the right.

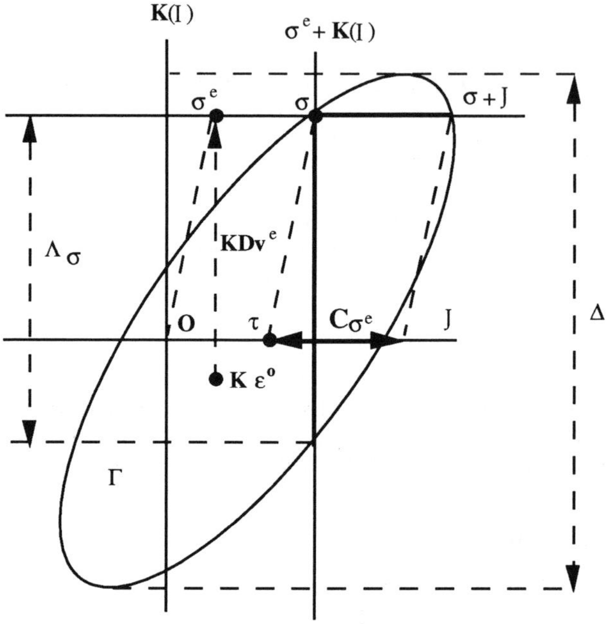

Figure 4

The orthogonal projection Δ of Γ on $\mathbf{K}(\mathbb{I})$ is the set of loads which can be equilibrated by a stress field, called "safe loads". Its boundary is constitued of limit loads.
Λ_σ, the projection on $\mathbb{I}$ of $(\sigma+\mathbb{I}) \cap \Gamma$, is the present domain of elasticity for loads.
$(\sigma+\mathbb{J}) \cap \Gamma$ is the set of plastically admissible stress fields which are in equilibrium with the given load **f**.
Let us now split the plastic strain rate $\dot{\mathbf{p}}$ into two orthogonal components $\dot{\mathbf{p}}_\mathbf{I}$ and $\dot{\mathbf{p}}_\mathbf{J}$:

$$\dot{\mathbf{p}} = \dot{\mathbf{p}}_\mathbf{I} + \dot{\mathbf{p}}_\mathbf{J} \quad \text{with :} \begin{cases} \dot{\mathbf{p}}_\mathbf{I} = \mathbf{K}^{-1}\,\mathbf{P}_\mathbf{I}\,\mathbf{K}\,\dot{\mathbf{p}} \\ \dot{\mathbf{p}}_\mathbf{J} = \mathbf{K}^{-1}\,\mathbf{P}_\mathbf{J}\,\mathbf{K}\,\dot{\mathbf{p}} = \mathbf{K}^{-1}\,\mathbf{B}\,\dot{\mathbf{p}} \end{cases}$$

The former is the integrable part of the plastic strain whereas the latter is derived from self-equilibrated stress field. Indeed we derive from (42.2) and (50.2) :

$$\tau = \sigma - \sigma^e = \mathbf{K}\big[\mathbf{D}(\mathbf{v} - \mathbf{v}^e) - \mathbf{q} + \mathcal{D}(\bar{\mathbf{u}}) - \mathbf{p}_o\big] \tag{51}$$

Then, as τ and $\mathbf{K}[\mathcal{D}(\bar{\mathbf{u}}) - \bar{\mathbf{p}}]$ belong to $\mathbb{J}$:

$$\mathbf{q}_\mathbf{I} = \mathbf{D}(\mathbf{v} - \mathbf{v}^e) \tag{52}$$

$$\tau = \mathbf{K}\left[- \mathbf{q_J} + \mathcal{D}(\overline{\mathbf{u}}) - \mathbf{p}_o\right] \tag{53}$$

which yields, as we have the equality $\dot{\mathbf{p}} = \dot{\mathbf{q}}$:

$$\dot{\tau} + \mathbf{K}\,\dot{\mathbf{p}}_{\mathbf{J}} = 0 \tag{54}$$

which is the announced result.
Now consider the convex set :

$$\mathbf{C}_{\sigma^e} = \mathbb{J} \cap (\Gamma - \sigma^e) = \left[(\sigma^e + \mathbb{J}) \cap \Gamma\right] - \sigma^e \tag{55}$$

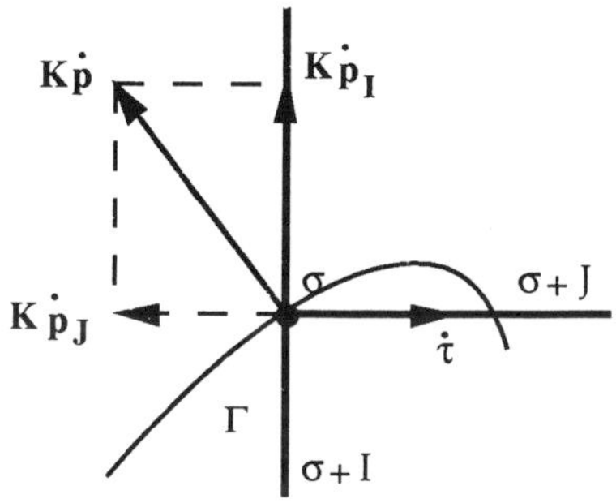

Figure 5 : Plastic strain rate components.

then:

$$\tau' \in \mathbf{C}_{\sigma^e} / \sigma' = \sigma^e + \tau' \in \Gamma / < \dot{\mathbf{p}}\,, \tau - \tau' > = < \dot{\mathbf{p}}_{\mathbf{J}}\,, \tau - \tau' > \geq 0$$

and finally we derive from (54) :

$$\tau \in \mathbf{C}_{\sigma^e} \;\; ; \;\; \forall\, \tau' \in \mathbf{C}_{\sigma^e} \; : \; < \mathbf{K}^{-1}\dot{\tau} \;, \tau - \tau' > \leq 0 \tag{56}$$

This variational inequality played an important role for the first proof of stress field existence, by J.J. Moreau [1973]. Note that $\mathbf{C}_{\sigma^e}$ is a given moving closed convex set in the subspace $\mathbb{J}_{\mathbf{u}}$; equation (56) defines what Moreau called the problem of "sweeping process by a moving convex set".

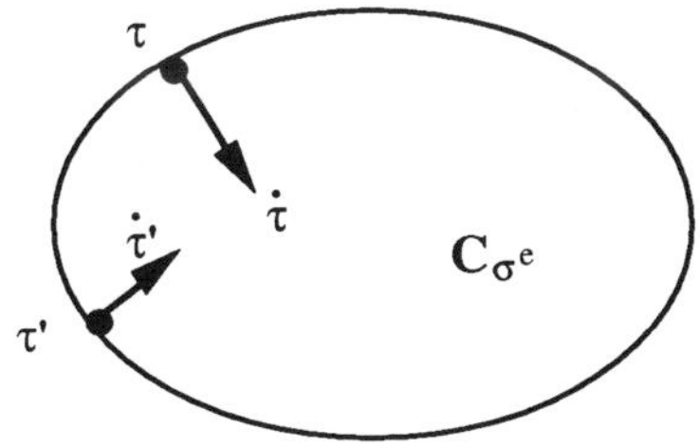

Figure 6 : sweeping process by a moving convex set.

The unknown moving point τ is constrained to lie in the convex $\mathbf{C}_{\sigma^e}$, as by a frictionless constraint: its celerity $\dot{\tau}$ is an inward normal to the convex, vanishing if τ is in the interior. Note that (56) defines the stress evolution, regardless of other unknowns; it may be written in the equivalent subdifferential form :

$$-\mathbf{K}^{-1}\dot{\tau} \in \partial\Psi_{\mathbf{C}_{\sigma^e}}(\tau) \tag{57}$$

An important property is "monotony" (in Minty's sense):

$$(\mathbf{y}_1 \in \partial\Psi_{\mathbf{C}_{\sigma^e}}(\tau_1) \text{ and } \mathbf{y}_2 \in \partial\Psi_{\mathbf{C}_{\sigma^e}}(\tau_2) / < \mathbf{y}_1-\mathbf{y}_2, \tau_1-\tau_2 > \geq 0 \tag{58}$$

Now consider the energetic norm :

$$\|\tau\|_{\mathbf{K}}^2 = \frac{1}{2} < \mathbf{K}^{-1}\tau, \tau > \tag{59}$$

which is the elastic energy of stress field **t**, and choose in (58) $\mathbf{y}_1 = -\dot{\tau}_1$, $\mathbf{y}_2 = -\dot{\tau}_2$; it yields :

$$\frac{d}{dt}\|\tau_1-\tau_2\|_{\mathbf{K}} \leq 0 \tag{60}$$

We may thus assert:

Contraction property

Let τ_1 and τ_2 be any two solutions of (57) ; the elastic energy of their difference is a non-increasing function of time.

Shakedown ability is due to this contractivity when a constant solution $\bar{\tau}$ has been exhibited : every other solution will get nearer to $\bar{\tau}$. This is precisely Melan's result. For the same reason accommodation properties under periodic loads are encountered when a periodic solution τ_p has been found.
Let us now look at stronger results.

Shakedown in small transformations

Basic ideas

A unique solution to system (38) or (50) should exist[1] for given initial conditions, for dynamic evolutions as well as for quasi-static ones (provided loads remain safe, i.e. smaller than limit loads) . This solution can be computed numerically and we could see a huge improvement in numerical methods during the eighties. For instance, metal forming is now concerned with accurate numerical simulations. Most often, however, data are not well known from any point of view : constitutive equations, initial conditions, loads and thermal

[1] Once again remember that we do not want to specify what kind of structure is concerned : so we do not have such a general existence theorem.

conditions. This ignorance is particularly blatant when civil engineering structures are concerned, for many reasons.

Generalizing the concept of limit analysis, the main idea of shake-down theory is as follows: first it is assumed that bounding domains Θ_f and Θ_e are known for **f** and ε^o : whatever the real history of the structure may be, the memberships:

$$\mathbf{f} \in \Theta_f \ ; \ \varepsilon^o \in \Theta_e \tag{61}$$

hold ; hence :

$$\sigma^e \in \Theta_s = \mathbf{A_u}(\Theta_f) + \mathbf{B_u}(\Theta_e) \tag{62}$$

also holds.
Then assume some self equilibrated stress field $\bar{\tau}$ can be exhibited, such that:

$$\bar{\tau} + \Theta_\sigma \subset \Gamma \tag{63}$$

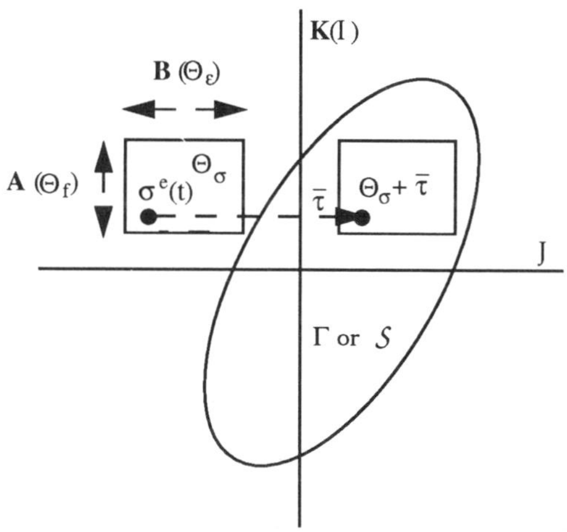

Figure 7 : shakedown hypotheses

then a sensible conjecture is that, eventually, the real self equilibrated stress field $\tau(t)$ will "shake down" to some τ_∞ such that:

$$\tau_\infty + \Theta_\sigma \subset \Gamma$$

and the plastic strain rate will progressively vanish. Note that $\bar{\tau}$ is a particular constant solution of (56) and $\bar{\tau} + \sigma^e$ a particular solution of (50) for suitable **v** and **q**.
The term "shake down" probably may be understood as "converge". In fact this assumption (63) is correct for engineering, since some safety factor is always used, a bit too weak for a mathematical proof.

THE SHAKEDOWN THEOREMS (Koiter, Ceradini)

The correct assumption, due to KOITER [1964], is precisely the existence of a safety factor m for which (63) remains true :

Theorem for quasi static evolutions (W.T. Koiter)

Let $\sigma^e(t)$ denote the stress field of the quasi-static elastic response of the structure to loads and imposed displacements; let $\sigma(t) = \sigma^e(t) + \tau(t)$ denote the quasi-static elastoplastic response.

If there exists a constant self equilibrated stress field $\overline{\tau}$, a real number m >1, a time t_o and a sanctuary of elasticity $\mathcal{S}$ such that :

$$\forall t > t_o \quad m(\overline{\tau} + \sigma^e(t)) \subset \mathcal{S} \tag{64}$$

then the plastic work is bounded :

$$\int_{t_o}^{\infty} < \dot{\mathbf{p}}(t), \sigma(t) > dt \leq \frac{m}{m-1} \| \sigma(t_o) - \sigma^e(t_o) - \overline{\tau} \|_{\mathbf{K}}^2 \tag{65}$$

Note that the upper bound is the elastic energy of the initial difference between the real solution $\sigma(t)$ and the particular solution which has been exhibited : $\sigma^e(t) + \overline{\tau}$.
We shall give a general proof, including the dynamic case studied by CERADINI [1969], which we are to study now. Let us compare the equations of the elastic response (already written in (42) and written again here for the reader's comfort) :

$$\left\{ \begin{aligned} \mathbf{D}^T \sigma^e &= \mathbf{f} - \mathbf{M}\ddot{\mathbf{v}}^e \\ \sigma^e &= \mathbf{K}\left[\mathbf{D}\mathbf{v}^e + \varepsilon^o\right] \end{aligned} \right. \tag{66}$$

with equations satisfied by the elastoplastic response ($\mathcal{S}$ is a sanctuary of elasticity):

$$\left\{ \begin{aligned} &\mathbf{D}^T \sigma = \mathbf{f} - \mathbf{M}\ddot{\mathbf{v}} \\ &\sigma = \mathbf{K}\left[\mathbf{D}\mathbf{v} + \varepsilon^o - \mathbf{q} + \mathcal{D}(\overline{\mathbf{u}}) - \overline{\mathbf{p}}\right] \\ &\forall\, \sigma' \in \mathcal{S} : < \dot{\mathbf{q}}, \sigma - \sigma' > \; \geq 0 \end{aligned} \right. \tag{67}$$

Let us introduce the kinetic energy norm $\| . \|_{\mathbf{M}}$:

$$\dot{\mathbf{u}} \in \mathbb{U} \mapsto \|\dot{\mathbf{u}}\|_{\mathbf{M}}^2 = \frac{1}{2} << \dot{\mathbf{u}}, \mathbf{M}\dot{\mathbf{u}} >> \tag{68}$$

and state Ceradini's theorem :

Extension to dynamics (G. Ceradini)

Let $\sigma^e(t)$ denote the stress field of the dynamic elastic response of the structure to loads and imposed displacements ; if there exists a constant self equilibrated stress field $\bar{\tau}$, a real number m >1, a time t_o and a sanctuary of elasticity $\mathcal{S}$ such that :

$$\forall t > t_o \quad m(\bar{\tau} + \sigma^e(t)) \subset \mathcal{S} \tag{69}$$

then the plastic work is bounded :

$$\int_{t_o}^{\infty} < \dot{\mathbf{p}}(t), \sigma(t) > dt \leq \frac{m}{m-1} \left(\| \sigma(t_o) - \sigma^e(t_o) - \bar{\tau} \|_{\mathbf{K}}^2 + \| \dot{\mathbf{v}}(t_o) - \dot{\mathbf{v}}^e(t_o) \|_{\mathbf{M}}^2 \right) \tag{70}$$

The result is completely analogous to the static one, but the kinetic energy of the initial difference has been added to its elastic energy.

Proof

Remember that, in the case of perfect plasticity, the interior of Γ is an elasticity sanctuary and that, in general plasticity, Γ embeds $\mathcal{S}$. Hence, the next proof is a simple extension of the classical.

Let σ_1 denote the stress :

$$\sigma_1 = \sigma^e + \bar{\tau}$$

Combining equilibrium equations in (66.1) and (67.1) yields :

$$\mathbf{D}^T(\sigma - \sigma_1) = \mathbf{M}(\ddot{\mathbf{v}}^e - \ddot{\mathbf{v}})$$

which implies :

$$< \dot{\mathbf{v}}^e - \dot{\mathbf{v}} \, , \mathbf{D}^T(\sigma - \sigma_1) > \; = \frac{d}{dt} \left[\| \dot{\mathbf{v}}^e - \dot{\mathbf{v}} \|_{\mathbf{M}}^2 \right] \tag{71}$$

Now using constitutive equations (66.2) and (67.2) provides :

$$\mathbf{D}(\dot{\mathbf{v}}^e - \dot{\mathbf{v}}) = -\dot{\mathbf{q}} + \mathbf{K}^{-1}(\dot{\sigma}_1 - \dot{\sigma}) \tag{72}$$

By the definition of transposition :

$$<< \dot{\mathbf{v}}^e - \dot{\mathbf{v}} \, , \mathbf{D}^T(\sigma - \sigma_1) >> \; = \; < \mathbf{D}(\dot{\mathbf{v}}^e - \dot{\mathbf{v}}) \, , \sigma - \sigma_1 >$$

which gives, with the help of (72) :

$$<< \dot{\mathbf{v}}^e - \dot{\mathbf{v}}\ ,\ \mathbf{D}^T(\sigma - \sigma_1) >> = - < \dot{\mathbf{q}}, \sigma - \sigma_1 > - \frac{d}{dt}\left[\|\sigma_1 - \sigma\|_{\mathbf{K}}^2\right]$$

which is now combined with (71) to give the equality:

$$-\frac{d}{dt}\left[\|\sigma_1 - \sigma\|_{\mathbf{K}}^2\right] - \frac{d}{dt}\left[\|\dot{\mathbf{v}}^e - \dot{\mathbf{v}}\|_{\mathbf{M}}^2\right] = < \dot{\mathbf{q}}, \sigma - \sigma_1 > \tag{73}$$

Note that the right hand side of (73) is not negative since $\dot{\mathbf{q}}$ is the plastic strain rate associated with σ and σ_1 is in the sanctuary. Now remember that $\sigma' = m\sigma_1$ is also in the sanctuary ; then, by Hill's inequality (67.3):

$$0 \leq <\dot{\mathbf{q}}, \sigma - m\sigma_1> = < \dot{\mathbf{q}}, \sigma - m(\sigma + \sigma_1 - \sigma) > = -(m-1) < \dot{\mathbf{q}}, \sigma > + m <\dot{\mathbf{q}}, \sigma - \sigma_1>$$

Using (73) we obtain:

$$< \dot{\mathbf{q}}, \sigma > \leq - \frac{m}{m-1}\frac{d}{dt}\left\{\|\sigma_1 - \sigma\|_{\mathbf{K}}^2 + \|\dot{\mathbf{v}}^e - \dot{\mathbf{v}}\|_{\mathbf{M}}^2\right\}$$

and, after integrating (remember that $\dot{\mathbf{p}} = \dot{\mathbf{q}}$):

$$\int_{t_o}^{\infty} < \dot{\mathbf{p}}(t), \sigma(t) > dt \leq \frac{m}{m-1}\left[\|\sigma_1 - \sigma\|_{\mathbf{K}}^2 + \|\dot{\mathbf{v}}^e - \dot{\mathbf{v}}\|_{\mathbf{M}}^2\right]_{\infty}^{t_o}$$

Finally, since the contribution of ∞ is not positive, we obtain:

$$\int_{t_o}^{\infty} < \dot{\mathbf{p}}(t), \sigma(t) > dt \leq \frac{m}{m-1}\left(\|\sigma(t_o) - \sigma^e(t_o) - \bar{\tau}\|_{\mathbf{K}}^2 + \|\dot{\mathbf{v}}(t_o) - \dot{\mathbf{v}}^e(t_o)\|_{\mathbf{M}}^2\right)$$

which ends the proof.

ABOUT CONVERGENCE

Plastic work has been bounded from ; is there any corollary concerning stress field convergence, displacement convergence or strain upperbounds ? In the general framework we want to keep, the answer is "probably" for strain and displacement convergence, pessimistic for stress convergence, negative for local strain upperbounding. In the finite dimensional case however, the theoretical answer may be deceptively positive because the practical result quickly deteriorates when then dimension increases.
Suppose, for instance, that at every point P of the domain Ω, the elasticity domain Γ_P (or the sanctuary $\mathcal{S}_P$) is bounded from below by a constant bowl with center 0 and radius r:

$$\exists r>0\ \forall P \in \Omega : B(O, r) = \left\{\sigma \in \mathbb{T}^n / |\sigma| < r\right\} \subset \Gamma_p \tag{74}$$

Then, if $\dot{p} \neq 0$, Hill's inequality applied to $\sigma' = \dot{p}r/|\dot{p}|$ implies:

$\dot{p}.\sigma \geq |\dot{p}|\, r$

This inequality holds when $\dot{p} = 0$ and implies, after integrating over Ω:

$$\|\dot{\mathbf{p}}(t)\|_1 = \int_\Omega |\dot{\mathbf{p}}(P,t)|\, d\mu(P) \leq \frac{\dot{W}_p}{r} \tag{75}$$

The plastic power is denoted by $\dot{W}_p$ and the L^1 norm by $\|.\|_1$.
Note that, in the finite dimensional case, the space integral in (75) reduces to a finite sum over the set of Gauss points G_i:

$$\|\dot{\mathbf{p}}\|_1 = \sum_i |\dot{\mathbf{p}}(G_i,t)|\, \mu_i \tag{76}$$

Integrating (75) with respect to time provides:

$$\int_{t_o}^{\infty} \|\dot{\mathbf{p}}(t)\|_1\, dt \leq \frac{W_p}{r} \tag{77}$$

Hence, if the plastic work is bounded, so is the trajectory length of the plastic strain $\mathbf{p}$ in $[L^1(\Omega)]^n$, provided $\mathbf{p}$ exists as a time function with values in a space of measurable fields. Then, as $[L^1(\Omega)]^n$ is a complete space, $\mathbf{p}(t)$ strongly converges to some $\mathbf{p}(\infty)$. Unfortunately, it can be shown (cf. SUQUET [1988]) that plastic strains may not be space functions but only measures: concentrated strains may occur, even before a limit loading is reached, even when loads are defined by regular force densities; nevertheless, the convergence of $\mathbf{p}(t)$, in a suitable measure space, may probably be ensured by a similar argument. It will ensure also the convergence of the displacement $\mathbf{u}(t)$ in a suitable space of $\mathbf{BD}(\Omega)$ type.
In the finite dimensional case this kind of difficulty is avoided; (77) is an effective upper bound for the total time variation of $\mathbf{p}$, and $\mathbf{p}(t)$ converges to some $\mathbf{p}(\infty)$. An upper bound is:

$$\|\mathbf{p}(\infty) - \mathbf{p}(t_o)\|_1 \leq \frac{W_p}{r}$$

We must remark, however, that it is an upper bound of the $[L^1(\Omega)]^n$ norm: the difference may reach high values on small domains, or at some few particular Gauss points when the finite dimensional case is considered. Then, if one is interested in a local upper bound for strains (for some practical failure criterion, for instance) this upper bound quickly fails when the dimension increases.
Still more deceptive is the stress convergence in the finite dimensional case. Indeed all norms are equivalent in „^{n}: hence there exists some $\alpha > 0$ such that the Hlbert norm $\|.\|_2$ is bounded from above by $\alpha\|.\|_1$ and we may get an inequality similar to (75):

$$\|\dot{\mathbf{p}}(t)\|_2 = \Big[\sum_i |\dot{\mathbf{p}}(G_i,t)|^2 \mu_i\Big]^{\frac{1}{2}} \leq \alpha\, \|\dot{\mathbf{p}}(t)\|_1 \leq \alpha \frac{\dot{W}_p}{r} \tag{78}$$

For the same reason (finite dimension) there exists some $\beta > 0$ such that:

$$\|\dot{\mathbf{p}}\|_{\mathbf{K}} \leq \beta \|\dot{\mathbf{p}}\|_2$$

Using (54) we obtain :

$$\|\dot{\tau}\|_{\mathbf{K}} = \|\dot{\mathbf{p}}_J\|_{\mathbf{K}} \leq \|\dot{\mathbf{p}}\|_{\mathbf{K}} \leq \beta \|\dot{\mathbf{p}}\|_2 \leq \alpha\beta \frac{\dot{W}_p}{r}$$

Hence, if the plastic work is bounded, the trajectory length of τ is bounded; then $\tau(t)$ converges towards some $\tau(\infty)$. This line of proof fails completely in the infinite dimensional case because the norm $\|.\|_2$ cannot be bounded from above by the norm $\|.\|_1$.

REFERENCES

1964 KOITER, W. T,. "General theorems for elastic-plastic solids ", Progress in Solid Mechanics, 1, Sneddon & Hill, North Holland P. C.

1969 CERADINI, G,. "Sull'adattamento dei corpi elasto-plastici sogetti ad azioni dinamiche", Giornale del Genio Civile, 415.

1973 MOREAU, J. J., "On Unilateral Constraints, Friction and Plasticity" Lecture Notes, CIME June, Edizione Cremonese.

1976 DEBORDES, O., et NAYROLES, B., "Sur la théorie et le calcul à l'adaptation des structures élastoplastiques",. J. de Mécanique, 15, N° 1.

1976 JOHNSON, C., "Existence theorems for plasticity problems", J. Math. Pures et Appl., 55, pp. 431-444.

1978 STRANG, G,. "A family of model problems in plasticity", Proc. Symp. Comp. Meth. in Appl. Sci. Ed. R. Glowinski an J.L. Lions, Lecture Notes in Math., 704, pp. 292-308.

1979 SUQUET, P., "Un espace fonctionnel pour les équations de la plasticité", Ann. Fac. Sci. Toulouse, 1, pp. 77-87.

1981 DEBORDES, O., "Przystosowanie układow sprezysto-idealnie plastycznych", Metody analizy funkcjonalnej w plastycznosci, Ossolineum.

1981 NAYROLES, B,. "Elementy analizy wypukłejw mechanice ciała stałego", Metody analizy funkcjonalnej w plastycznosci, Ossolineum.

1987 KÖNIG, J.A., "Shakedown of Elastic-Plastic Structures", Elsevier, Amsterdam.

1988 SUQUET, P., "Discontinuities and Plasticity", Non Smooth Mechanics and Applications, Ed. J.J. Moreau and P.D. Panagiotopoulos, Int. Centre for Mech. Sci. Courses and Lectures, 302, pp. 280-340.

1993 NAYROLES, B., and WEICHERT, D., "La notion de sanctuaire d'élasticité et l'adaptation des structures ", C. R. Acad.Sci. Paris, 316, Série II, pp. 1493-1498.

A VARIATIONAL DEDUCTION OF THE UPPER AND LOWER BOUND SHAKEDOWN THEOREMS BY MARKOV AND HILL'S PRINCIPLES OVER A CYCLE

Géry DE SAXCE
Mechanics of Materials and Structures
Polytechnic Faculty of Mons
9, rue de Houdain, B-7000 Mons (Belgium)

INTRODUCTION

This work was part of the author's doctoral thesis and was briefly presented at the first Belgian Congress of Mechanics in Brussels (cf. DE SAXCE [1986,1987]). Some additional developments can be found in a European report (cf.SAVE et al. [1991]). The paper presents two dual variational principles governing the collapse behaviour of the elastic-perfectly plastic structures under variable loading. On this basis, an original deduction of the bound shakedown theorems is given, in a similar way to MANDEL's corresponding approach for the rigid plastic material under proportional loading. In the second part, the regularity of the solutions is discussed. Applying the theorems to the thick wall tube problem, we show that DIRAC's distribution is required to obtain the plastic fatigue kinematical solution.

ELASTIC-PERFECTLY PLASTIC MATERIAL

The first usual assumption concerns the existence of a convex set K, called the elastic domain, defined at any point of the structure Ω. It is supposed that the stress field is plastically admissible (P.A.) :

$$\sigma \in K \quad in \quad \Omega$$

An important additional hypothesis concerning the elastic domain is that

$$0 \in K \tag{1}$$

The second basic assumption is that the plastic yielding rule is a normality law. In other words, the plastic strain rate field $\dot{\varepsilon}^p$ associated with σ satisfies *HILL's maximum*

Z. Mróz et al. (eds.), Inelastic Behaviour of Structures under Variable Loads, 153–167.

power principle :

$$\forall\ \sigma' \in K, \qquad (\sigma' - \sigma)\,.\,\dot{\varepsilon}^p \leq 0 \tag{2}$$

Then, the dissipation power function is defined by :

$$D(\dot{\varepsilon}^p) = \sup_{\sigma' \in K} \ \sigma'\,.\,\dot{\varepsilon}^p$$

Two fundamental properties follow immediately from the definition. First, one can state the so-called *maximum dissipation principle* :

$$\forall\ \sigma' \in K, \qquad D(\dot{\varepsilon}^p) \geq \sigma'\,.\,\dot{\varepsilon}^p \tag{3}$$

In particular, hypothesis (1) implies

$$D(\dot{\varepsilon}^p) \geq 0 \tag{4}$$

Comparing ineqs (2,3), if $\dot{\varepsilon}^p$ is associated with σ, one has

$$D(\dot{\varepsilon}^p) = \sigma\,.\,\dot{\varepsilon}^p \tag{5}$$

Secondly, D is a *convex* function, as superior envelope of linear functions of $\dot{\varepsilon}^p$ From ineq (3) and eqn (5), it is easy to deduce the following *convexity inequality* :

$$\forall \dot{\varepsilon}^{p'}, \qquad D(\dot{\varepsilon}^{p'}) - D(\dot{\varepsilon}^p) \geq \sigma\,.\,(\dot{\varepsilon}^{p'} - \dot{\varepsilon}^p) \tag{6}$$

FICTITIOUS ELASTIC FIELDS AND RESIDUAL FIELDS

Let a given structure occupy the volume Ω, bounded by the surface S, and subjected to the following external actions :

(*i*) *surface tractions* : $p(\sigma) = \sigma.n = \overline{p} = \alpha\ \overline{p}^0 \quad on\ S_1$

$$(ii)\quad displacements: \quad u = \overline{u} = 0 \quad on\ S_0 \tag{7}$$

The parts S_0 and S_1 are disjoint and $S = S_0 \cup S_1$. α is a positive real number called the *load factor*. The reference load $\overline{p}^0$ is supposed to belong to a convex reference load domain P^0 which, for instance, can be described by an arbitrary combination of elementary loads varying between prescribed limits. Then, the traction $\overline{p}$ belongs to a homothetic domain $P = \alpha\ P^0$. For actions (7), the elastic responses in displacement u^E, in strain ε^E and in stress σ^E in a corresponding fictitious perfectly elastic structure, satisfy Hooke's law. Because of the linearity of the elastic response, the stress field σ^E belongs to a convex polyedral load domain $\Sigma = \alpha\ \Sigma^0$, homothetic to the load domain Σ^0 corresponding to the reference load domain P^0. Then, the field defined by :

$$\rho = \sigma - \sigma^E \tag{8}$$

belongs to the set of the residual stress fields :

$$N = \{\ \rho\ |\ div\ \rho = 0\ in\ \Omega\ and\ p(\rho) = \rho . n = 0\ on\ S_1\ \}$$

Moreover, the field defined by :

$$\eta = \varepsilon - \varepsilon^E$$

belongs to the set of residual strain fields :

$$N^* = \{\ \eta\ |\ \exists v\ such\ that\ \eta = grad_s v\ in\ \Omega\ and\ v = 0\ on\ S_0\ \}$$

Of course, applying GREEN's formula, one has :

$$\forall\ \rho \in N,\ \forall\ \eta \in N^*, \quad \int_\Omega \rho . \eta\ d\Omega = 0 \tag{9}$$

STATIC AND KINEMATIC ADMISSIBLE FIELDS

The key idea of the static approach is to define *admissible* residual stress fields (in

MELAN's sense) $\overline{\rho}(x)$, corresponding to the load domain Σ, such that (cf. MELAN [1936]) :

$$\begin{aligned} &(i)\ \overline{\rho}\ \textit{is time independent,} \\ &(ii)\ \overline{\rho}\ \textit{is a residual stress field} : \overline{\rho} \in N, \\ &(iii)\ \overline{\rho}\ \textit{is P.A. in the sense that} : \\ &\qquad \sigma^E + \overline{\rho} \in K \textit{ in } \Omega \textit{ at any time.} \end{aligned} \tag{10}$$

with $\sigma^E \in \Sigma$. Indeed, the following proposition can be proved (cf. MELAN [1936]):

THEOREM 1 *(MELAN's theorem) : if an admissible residual stress field* $\overline{\rho}$ *can be found, the structure shakes down.*

The key idea of the dual approach is to characterize the collapse mode of the structure over a load cycle for a *periodic* action of period T. Then, the *admissible* plastic strain rates (in KOITER's sense) $\dot{\varepsilon}^p(x,t)$, corresponding to the load domain Σ, are defined by (cf. KOITER [1960]) :

$$\begin{aligned} &(i)\ \textit{for the plastic strain increment,} \quad \Delta\varepsilon^p = \oint \dot{\varepsilon}^p\, dt \in N^* \\ &(ii)\ \dot{\varepsilon}^p\ \textit{is P.A. in the sense that} : \quad \int_\Omega \oint \sigma^E . \dot{\varepsilon}^p\, dt\, d\Omega > 0 \end{aligned} \tag{11}$$

Hence, the following theorem, dual to MELAN's theorem, can be demonstrated (cf. HALPHEN [1978]) :

THEOREM 2 *(HALPHEN's theorem) : if the structure does not shake down, the plastic strain fields tends to an admissible field.*

MANDEL'S APPROACH FOR THE LIMIT AND SHAKEDOWN ANALYSIS

For the rigid-perfectly plastic material, the boundary value problem (B.V.P.) corresponding to given external actions $\overline{p}$ on S_1, and characterizing the collapse, is:

Find some couple $(\dot{u}, \sigma)$ *such that :*

(i) $\dot{u}$ *is admissible (kinematically admissible and P.A.).*

(ii) σ *is admissible (statically admissible and P.A.).*

(iii) $\dot{\varepsilon}(\dot{u})$ *and* σ *are associated by the normality law in* Ω. *(12)*

The corresponding variational formulation consists in stating the following two propositions (cf. MARKOV [1947], HILL [1948]) :

THEOREM 3 *(MARKOV's principle): if* $\dot{u}$ *is a solution of the B.V.P. (12), it is a solution to the following problem :*

Find some $\dot{u}$ *such that the functional*

$$\Phi(\dot{u}) = \int_{\Omega} D(\dot{\varepsilon}(\dot{u}))\ d\Omega - \int_{S_1} \overline{p}.\dot{u}\ dS$$

is minimum among the admissible velocity fields.

THEOREM 4 *(HILL's principle) : if* σ *is a solution to the B.V.P. (12), it is solution of the following problem :*
Find some σ *such that the functional*

$$\Pi(\sigma) = -\int_{S_0} p(\sigma).\overline{\dot{u}}\ dS$$

is minimum among the admissible stress fields.

The propositions are recalled without proof, but it can be remarked that the two variational principles are duals of each other in the sense that for any couple $(\dot{u}^k, \sigma^s)$ of admissible fields:

$$\Phi(\dot{u}^k) + \Pi(\sigma^s) \geq 0 \tag{13}$$

because of ineq (3) and GREEN's formula (cf. MROZ [1973]). In particular, for the solution $(\dot{u}, \sigma)$ of the B.V.P., the equality is achieved in ineq (13) because of eqn (5):

$$\Phi(\dot{u}) + \Pi(\sigma) = 0$$

Now, let us consider the usual assumption of the limit analysis theory :

(i) proportional loading : $\overline{p} = \alpha\, \overline{p}^0$

(ii) fixed supports : $\overline{u} = 0 \text{ on } S_0$

It can be remarked that because of the hypothesis (ii), the problem of theorem 4 degenerates into :

Find some σ among the admissible stress fields

where no variational functional has to be minimized. Only admissibility conditions have to be satisfied.

The aim of the limit analysis is to define the conditions to be satisfied by the loading in order that the B.V.P. problem (12) has a solution. Following an approach initially proposed by MANDEL [1966], one can deduce the so-called lower and upper bound theorems from MARKOV and HILL's principles.

Let us come back again to the shakedown analysis. The considerations developed in the previous section suggest the possibility of characterizing the collapse corresponding to given external actions $\overline{p} \in P$ by means of the following B.V.P. :

Find some couple $(\dot{\varepsilon}^p, \overline{\rho})$ *such that :*

(i) $\dot{\varepsilon}^p$ *is an admissible plastic strain rate field.*

(ii) $\overline{\rho}$ *is an admissible residual stress field.*

(iii) $\dot{\varepsilon}^p$ and $(\sigma^E + \overline{\rho})$ *are associated by the normality law in Ω at any time.* *(14)*

The goal of the next sections is to follow for the shakedown analysis the same approach as the one proposed by MANDEL for the limit analysis and, first of all, to state and prove new variational principles corresponding to MARKOV and HILL's ones.

MARKOV'S PRINCIPLE OVER A CYCLE

In order to construct the ad hoc functional, MARKOV's principle is modified in the following way :
- the history of the structure over the whole collapse cycle is considered,
- the admissible field is now $\dot{\varepsilon}^p$,
- the external actions are replaced by the equivalent stress field σ^E in the corresponding fictitious elastic body.

Then, the following proposition may be stated (cf. DE SAXCE [1986,1987]) :

THEOREM 5 *(MARKOV's principle over a cycle) : if* $\dot{\varepsilon}^p$ *is a solution to the B.V.P. (14) , it is a solution to the following problem :*

Find some $\dot{\varepsilon}^p$ *such that the functional*

$$\Phi_s(\dot{\varepsilon}^p) = \int_\Omega \oint D(\dot{\varepsilon}^p)\ dt\ d\Omega - \int_\Omega \oint \sigma^E . \dot{\varepsilon}^p\ dt\ d\Omega \tag{15}$$

is minimum among the admissible plastic strain rate fields.

Proof. Using eqn (8) and the convexity ineq (6), one has, for any admissible field $\dot{\varepsilon}^{p'}$:

$$\Phi_s(\dot{\varepsilon}^{p'}) - \Phi_s(\dot{\varepsilon}^p) \geq \int_\Omega \oint \overline{\rho} . (\dot{\varepsilon}^{p'} - \dot{\varepsilon}^p)\ dt\ d\Omega$$

$\overline{\rho}$ is admissible, therefore time independent, whence, one has

$$\Phi_s(\dot{\varepsilon}^{p'}) - \Phi_s(\dot{\varepsilon}^p) \geq \int_\Omega \overline{\rho} . (\Delta\varepsilon^{p'} - \Delta\varepsilon^p)\ d\Omega$$

As $\Delta\varepsilon^{p'}, \Delta\varepsilon^p \in N^*$ and $\overline{\rho} \in N$, the proof is achieved by applying GREEN's eqn (9):

$$\Phi_s(\dot{\varepsilon}^{p'}) \geq \Phi_s(\dot{\varepsilon}^p)$$

Remark : by using the classical Lagrangian multiplier technique, applied to the admissibility conditions (11) for $\dot{\varepsilon}^p$, it is easy to prove that EULER's variational equations are the admissibility conditions (10) for $\overline{\rho}$.

HILL'S PRINCIPLE OVER A CYCLE

Similar considerations can be developed for the static formulation, leading to the following proposition (cf. DE SAXCE [1986]) :

THEOREM 6 *(HILL's principle over a cycle) : if* $\overline{\rho}$ *is a solution to the B.V.P. (14), it is a solution to the following problem :*

Find some $\overline{\rho}$ *among the admissible residual stress fields.* *(16)*

The proposition is trivial. Let us remark that, as for HILL's principle restricted to limit analysis, the statical problem is degenerate in the sense that no variational functional has to be minimized. Conventionally, one may assume that the statical problem is a variational principle with admissibility constraints and a vanishing functional :

$$\Pi_s(\overline{\rho}) = 0 \tag{17}$$

As for the rigid-perfectly plastic material, it can be proved that the two variational principles (15) and (16) are dual of each other in the sense that for any couple $(\dot{\varepsilon}^{pk}, \overline{\rho}^s)$ of admissible fields, one has :

$$\Phi_s(\dot{\varepsilon}^{pk}) + \Pi_s(\overline{\rho}^s) \geq 0 \tag{18}$$

Indeed, taking into account ineq (3) and eqn (8,15,17), and that $\overline{\rho}^s$ is time independent, we have

$$\Phi_s(\dot{\varepsilon}^{pk}) + \Pi_s(\overline{\rho}^s) \geq \int_\Omega \overline{\rho}^s . \Delta\varepsilon^{pk} \, d\Omega$$

GREEN's eqn (9) implies that the ineq (18) holds. In particular, for the limit state

$(\dot{\varepsilon}^p, \overline{\rho})$, solution to the B.V.P. (14), eqn (5) replaces ineq (3) and the equality is achieved in ineq (18) :

$$\Phi_s(\dot{\varepsilon}^p) + \Pi_s(\overline{\rho}) = 0 \tag{19}$$

LOWER BOUND THEOREM

Taking into account eqns (17,19), it can be remarked that for the solution $\dot{\varepsilon}^p$ to the B.V.P. (14) corresponding to a load domain $\Sigma = \alpha^a \Sigma^0$, one has

$$\Phi_s^a(\dot{\varepsilon}^p) = \int_\Omega \oint D(\dot{\varepsilon}^p)\, dt\, d\Omega - \alpha^a \int_\Omega \oint \sigma^{E0} . \dot{\varepsilon}^p\, dt\, d\Omega = 0 \tag{20}$$

Similarly, following MARTIN [1975], we define the kinematical load factor α^k associated with an admissible plastic strain rate field $\dot{\varepsilon}^{pk}$ by the equality :

$$\int_\Omega \oint D(\dot{\varepsilon}^{pk})\, dt\, d\Omega = \alpha^k \int_\Omega \oint \sigma^{E0} . \dot{\varepsilon}^{pk}\, dt\, d\Omega \tag{21}$$

The relation has sense because of ineq (4) and $\dot{\varepsilon}^{pk}$ is P.A. (ineq (11)(ii)). Thus, the following proposition can be stated :

THEOREM 7 *(KOITER's lower bound theorem) : if a solution* $(\dot{\varepsilon}^p, \overline{\rho})$ *to the B.V.P. (14) exists for the load domain* $\Sigma = \alpha^a \Sigma^0$, α^a *is the lower bound of the kinematic factors* α^k

$$\alpha^a \le \alpha^k \tag{22}$$

Moreover, problem (15) has no solution for $\alpha^k > \alpha^a$.

Proof. First of all, let us prove that $\alpha^a \le \alpha^k$. Because of theorem 5 and eqn (20), for any admissible plastic strain rate field $\dot{\varepsilon}^{pk}$ and the load domain $\Sigma = \alpha^a \Sigma^0$:

$$\Phi_s^a(\dot{\varepsilon}^{pk}) = \int_\Omega \oint D(\dot{\varepsilon}^{pk})\, dt\, d\Omega - \alpha^a \int_\Omega \oint \sigma^{E0} . \dot{\varepsilon}^{pk}\, dt\, d\Omega \ge \Phi_s^a\,(\dot{\varepsilon}^p) = 0$$

Taking into account definition (21) of the kinematical factors, one has :

$$(\alpha^k - \alpha^a) \int_\Omega \oint \sigma^{E0} . \dot{\varepsilon}^{pk} \, dt \, d\Omega \geq 0$$

Because of the admissibility condition (11)(ii), the bound property (22) is proved. Finally, let us demonstrate that problem (15) has no solution for $\alpha^k > \alpha^a$. It is sufficient to prove that functional (15) is not coercive (cf. EKELAND & TEMAM [1975]). Indeed, one has for the load domain $\Sigma = \alpha^k \Sigma^0$:

$$\Phi_s^k(\dot{\varepsilon}^p) = \int_\Omega \oint D(\dot{\varepsilon}^p) \, dt \, d\Omega - \alpha^k \int_\Omega \oint \sigma^{E0} . \dot{\varepsilon}^p \, dt \, d\Omega$$

Taking into account eqn (20) and the admissibility condition (11)(ii), one has:

$$\Phi_s^k(\dot{\varepsilon}^p) = (\alpha^a - \alpha^k) \int_\Omega \oint \sigma^{E0} . \dot{\varepsilon}^p \, dt \, d\Omega < 0$$

Then, by choosing the minimizing sequence $(\dot{\varepsilon}_n^p)$ defined by : $\dot{\varepsilon}_n^p = n \, \dot{\varepsilon}^p$ (n=1,2,..), it is easy to prove that $\Phi_s^k(\dot{\varepsilon}_n^p) < 0$ tends to $-\infty$ when $\|\dot{\varepsilon}_n^p\|$ tends to $+\infty$.

UPPER BOUND THEOREM

The following proposition has to be proved :

THEOREM 8 *(KOITER's upper bound theorem) : if a solution* $(\dot{\varepsilon}^p, \overline{\rho})$ *to the B.V.P. (14) exists for the load domain* $\Sigma = \alpha^a \Sigma^0$ *, problem (16) has a solution* $\overline{\rho}^s$ *for* $\Sigma = \alpha^s \Sigma^0$ *if and only if* $\alpha^s \leq \alpha^a$ (23)
Moreover, if $\alpha^s < \alpha^a$*, the structure shakes down for the load factor* α^s.

Proof. Necessary condition. Introducing :

$$\sigma = \alpha^a \sigma^{E0} + \overline{\rho} \in K, \qquad \sigma^s = \alpha^s \sigma^{E0} + \overline{\rho}^s \in K \tag{24}$$

and taking into account that $\overline{\rho}$ *and* $\overline{\rho}^s$ are time independent, one has:

$$\int_\Omega \oint (\sigma^s - \sigma) \, . \, \dot{\varepsilon}^p \, dt \, d\Omega =$$
$$\int_\Omega (\overline{\rho}^s - \overline{\rho}) \, . \, \Delta\varepsilon^p \, d\Omega + (\alpha^s - \alpha^a) \int_\Omega \oint \sigma^{E0} . \, \dot{\varepsilon}^p \, dt \, d\Omega \qquad (25)$$

The first term of the right hand side vanishes because of GREEN's eqn (9). Owing to the convexity ineq (2), one has :

$$(\alpha^s - \alpha^a) \int_\Omega \oint \sigma^{E0} . \, \dot{\varepsilon}^p \, dt \, d\Omega \le 0$$

Then, it is concluded that ineq (23) holds because of the admissibility condition (11)(ii). Sufficient condition. Taking into account eqn (1), eqn (24)(i) and the convexity of the elastic domain K, one has

$$\forall\lambda \in [0,1], \quad \lambda \, (\overline{\rho} + \alpha^a \sigma^{E0}) \in K$$

Because of ineq (23), the following particular value $\lambda = \dfrac{\alpha^s}{\alpha^a}$ can be chosen. Hence, the residual stress field $\overline{\rho}^s = \lambda \, \overline{\rho}$ is plastically admissible ($\overline{\rho}^s + \alpha^s \sigma^{E0} \in K$) and consequently admissible in MELAN's sense. Then, $\overline{\rho}^s$ is a solution to the problem (16). Finally, one has to prove that the structure shakes down for the load factor $\alpha^s < \alpha^a$. Let us assume that there is a plastic strain rate field $\dot{\varepsilon}^{ps}$ satisfying (11)(i) and associated with $(\overline{\rho}^s + \alpha^s \sigma^{E0})$ by the normality law in Ω at any time. So, one may replace $\dot{\varepsilon}^p$ *by* $\dot{\varepsilon}^{ps}$ in eqn (25) and, by the same arguments, conclude that

$$(\alpha^a - \alpha^s) \int_\Omega \oint \sigma^{E0} . \, \dot{\varepsilon}^{ps} \, dt \, d\Omega \le 0 \qquad (26)$$

Moreover, because $\dot{\varepsilon}^{ps}$ is associated with $\overline{\rho}^s$ and eqn (5), one has :

$$D(\dot{\varepsilon}^{ps}) = \sigma^s . \dot{\varepsilon}^{ps}$$

Taking into account GREEN's eqn (9), one has :

$$\int_\Omega \oint D(\dot{\varepsilon}^{ps})\ dt\ d\Omega = \alpha^s \int_\Omega \oint \sigma^{E0} . \dot{\varepsilon}^{ps}\ dt\ d\Omega \tag{27}$$

Ineq (26) and eqn (27) imply:

$$\left(\frac{\alpha^a}{\alpha^s} - 1\right) \int_\Omega \oint D(\dot{\varepsilon}^{ps})\ dt\ d\Omega \leq 0$$

The first multiplicative factor is positive. Hence :

$$\int_\Omega \oint D(\dot{\varepsilon}^{ps})\ dt\ d\Omega = 0$$

Because of ineq (4), $\dot{\varepsilon}^{ps}$ vanishes and the structure shakes down for the load factor α^s.

Remarks : As was shown by DE SAXCE [1986], it is not necessary to consider a time-periodic loading, but the weaker assumption of a random loading. In this context, the load cycle integrals are replaced by average values :

$$\lim_{\tau \to \infty} \frac{1}{\tau} \int_\Omega \int_0^\tau D(\dot{\varepsilon}^p)\ dt\ d\Omega \quad and \quad \lim_{\tau \to \infty} \frac{1}{\tau} \int_\Omega \int_0^\tau \sigma . \dot{\varepsilon}^p\ dt\ d\Omega$$

APPLICATION : THE THICK-WALL TUBE

Let us consider the shakedown analysis of a thick-wall tube with internal radius a and external radius b, which is in plane strain state under the influence of an internal pressure varrying within the range $0 \leq p(t) \leq \overline{p}$. For sake of simplicity, the maximum value is taken as the load factor α and $0 \leq p^0(t) \leq 1$. For the TRESCA plastic yielding criterion , the static solution is well-known and was obtained first by

HODGE [1954]. For a large enough tube, the collapse occurs by plastic fatigue at r = a, and the statical load factor is twice the elastic pressure α^e. The aim of this section is to present the corresponding kinematic solution, using theorem 7. Let σ_Y be the plastic yield stress. If $\sigma_r \leq 0$ and $\sigma_\theta \geq 0$, the plastic yielding occurs for Tresca's condition when $|\sigma_\theta - \sigma_r| = \sigma_Y$. Using the expression for the fictitious elastic stress field (cf. HODGE [1954]), definition (21) of the kinematic load factor gives:

$$\alpha^k = \frac{b^2 - a^2}{2a^2 b^2}\sigma_Y \frac{\int_a^b \oint |\dot{\varepsilon}_\theta^p| dt . r\, dr}{\int_a^b \oint p^0 \dot{\varepsilon}_\theta^p \, dt . \frac{dr}{r}} \tag{28}$$

Now, following the indications given by the static solution, we take the following plastic strain rate field

$$\dot{\varepsilon}_\theta^p(r,t) = \dot{\varepsilon}_0^p(t)\ \delta(r - c) \tag{29}$$

where δ (r - c) denotes DIRAC's distribution at r = c and the plastic strain rate is alternating :

$$\begin{aligned} \dot{\varepsilon}_0^p(t) &= \varepsilon_0 > 0 && \text{when } p^0(t) = 1 \\ &= -\dot{\varepsilon}_0 && \text{when } p^0(t) = 0 \\ &= 0 && \text{otherwise} \end{aligned}$$

Of course, $\Delta\varepsilon^p = 0$ belongs to N*. Conversely, $\dot{\varepsilon}^p$ is admissible but does not belong to N* because of the presence of DIRAC's distribution. Introducing the field (29) into eqn (21), one has :

$$\alpha^k = \frac{b^2 - a^2}{2a^2 b^2}\sigma_Y \frac{\int_a^b 2\ \dot{\varepsilon}_0 \delta(r - c)\ dt . r\, dr}{\int_a^b \dot{\varepsilon}_0 \delta(r - c)\ dt . \frac{dr}{r}} = 2\alpha^e\left(\frac{c}{a}\right)^2$$

Applying theorem 7, one obtains the minimum value $\alpha^k = 2\ \alpha^e$ for c = a. Because the static and kinematic load factors are equal, the solution is exact and equal to the shakedown load factor $\alpha^a = 2\,\alpha^e$.

Next, the physical interpretation of DIRAC's distribution is given. As has been remarked by MARTIN [1975], the collapse actually occurs for a maximum value

$\bar{p} = \alpha^k > \alpha^a$. Besides, it may be shown by an elastic-plastic evolution analysis and considering the asymptotic behaviour for α^k next to α^a that the half increment of the hoop plastic strain is different from zero in a small ring (cf. DE SAXCE [1986]) :

$$a \leq r \leq c = \sqrt{\frac{\alpha^k}{\alpha^a}}\, a$$

and has the following distribution within this range :

$$\Delta_{1/2}\varepsilon_\theta^p = \Delta\varepsilon_0^p \left(\frac{r}{a}\right)^2$$

With the normalization condition :

$$\int_a^c \Delta_{1/2}\varepsilon_\theta^p \, 2\pi r \, dr = 1$$

$\Delta_{1/2}\varepsilon_\theta^p$ tends to DIRAC's distribution when

$\alpha^k \rightarrow \alpha^a$

In conclusion, the kinematic solutions given by the shakedown analysis are much less regular than those of limit analysis. The more appropriate functional frame seems to be that of the bounded functions of deformation used by TEMAM & STRANG [1980], and SUQUET [1978].

REFERENCES

1986 DE SAXCE G., "Sur quelques problèmes de mécanique des solides considérés comme matériaux à potentiels convexes", Doctor thesis, Faculty of applied sciences, University of Liège, Belgium.

1987 DE SAXCE G., "A Variational Theory of the Plastic Shakedown for Continua by Mandel's approach", Proc. of the First Belgian Congr. of Appl. and Theor. Mech., Brussels, ed. M. Deville, pp. 56-59.

1991 SAVE M., DE SAXCE G., BORKOWSKI A., "Computation of Shakedown Loads, Feasibility Study", Contract RA1-0100-B (GDF/DSF), Commission of the European Communities, Nuclear Sc. and Tech., Report EUR 13618 EN.

1975 MARTIN J.B., "Plasticity, Fundamentals and General Results", M.I.T. Press, Cambridge (Massachusetts) and London.

1936 MELAN E., "Theory statisch unbestimmter Systeme aus ideal plastischen Baustoff", Sitz Ber. Akad. Wiss. Wien IIa, 145, 195.

1978 HALPHEN B., "Periodic Solutions in Plasticity and Viscoplasticity", Proc. IUTAM Symp. in Var. Meth. in Mech. Solids, Evanston (Ilinois), S. Nemat-Nasser ed., Pergamon Press, Oxford, pp.273-277.

1960 KOITER W.T., "General Theorems for Elastic-Plastic Solids", Progress in Solid Mech., Vol. 1, I.N. Sneddon, R. Hill ed., North Holland, Amsterdam.

1966 MANDEL J., "Cours de mécanique des milieux continus", T.2, Mécanique des Solides, Gauthier-Villars, Paris.

1947 MARKOV A.A., "On Variational Principles in Theory of Plasticity", Prek. Math. Mech., 11, 339.

1948 HILL R., "A Variational Principle of Maximum Plastic Work in Classical Plasticity", Quart. J. Mech. Appl. Math., 1, 18.

1975 EKELAND I., TEMAM R., "Convex Analysis and Variational Problems", New York, North Holland.

1973 MROZ, Z., "Mathematical Models of Inelastic Material Behaviour", University of Warterloo, Waterloo, Canada.

1954 HODGE P.G., "Shakedown of Elastic-Plastic Structures", Residual Stresses in Metals and Metal Constructions, Reinhold Pub. Co, New York.

1980 TEMAM R., STRANG G., "Function of Bounded Deformation", Arch. Rat. Mech. Anal., 75, 7.

1978 SUQUET P.M., "Existence et régularité des solutions des équations de la plasticité", C.R. Acad. Sc. Paris, Série A, N°286, 1129 and 1201.

INFLUENCE OF CYCLIC CREEP ON THE UPPER BOUND TO SHAKEDOWN INELASTIC DEFLECTIONS

Stanisław DOROSZ
Polish Academy of Sciences
Institute of Fundamental Technological Research
ul. Świętokrzyska 21, 00-049 Warsaw, Poland

Abstract: For structures of linear elastic, perfectly plastic and creeping materials subjected to cyclic loading, a bounding principle for the maximum residual deflection at a given point is presented. The bound is valid for a simple non-interactive creep.

INTRODUCTION

The shakedown theorems do not provide information about the magnitude of displacements before shakedown occurs. Some procedures must be available for estimating the order of magnitude of displacement if a given structure is to shake down under a given load program. In the case when only limits of variation of the cyclic loading are given, checking the magnitudes of displacement by means of any step by step procedure fails to solve the problem. In this case only a global approach accounting for all the possible load paths may be accepted.

Some structures are subjected to cycles of temperature variation as well as to cycles of repeated loading. At elevated temperatures exceeding approximately $0.3T_m$ (where T_m is the melting temperature) and for higher stress magnitudes viscous strains are observed to develop. The presence of creep may cause not only a stress redistribution within a given structure but can change the whole picture of structural behaviour. Both the creep and instantaneous plastic deformation may increase indefinitely at loads less than shakedown limit calculated under the assumption that the creep does not arise. This fact requires essentially new formulation of the shakedown analysis problem.

Various approaches have already been employed. Until 1972 there were no methods capable of assessing the order of deformation prior to shakedown.

Z. Mróz et al. (eds.), Inelastic Behaviour of Structures under Variable Loads, 169–181.

The methods proposed for obtaining bounds on the residual deflections of a structure may be divided into two groups, as suggested, e.g., by KÖNIG and MAIER [1981].

In the first group the bounding methods are called "direct" methods. An upper bound on the residual component of a local quantity, e.g. the deflection, is here obtained by maximizing it with respect to certain parameters which govern the plastic strains over some domain where their post-shakedown values can be proved to be contained. Such methods have been suggested by: BRZEZIŃSKI and KÖNIG [1972], MAIER [1972], VITIELLO [1972], BRZEZIŃSKI and KÖNIG [1973], MAIER and VITIELLO [1974] and LOI and GRUNDY [1978].

In the second group the methods are called "indirect" and rest on the use of a dummy load, analogous to the use of the principle of virtual work to determine deflections in linear elastic structures, combined with some inequality derived from the constitutive laws, or with an upper bound for some work or energy function. The basis of the so-called "indirect" methods or "dummy force" methods is the consideration of the effect of an imaginary or virtual force on the structure. It is completely analogous to the well-known principle of virtual work in the analysis of structures.

PONTER [1972] derived a displacement bound, using a dummy or virtual surface traction or force at the point where the residual displacement is to be determined and using the virtual linear elastic stress field associated with this traction or force. CAPURSO [1974] found theorems of bounds on the complementary and the direct plastic work. DOROSZ [1976, 1978] provided an upper bound to the maximum residual deflection by deriving a lower bound to the plastic dissipation function and using this result to find an upper bound to the total plastic work performed in an arbitrary load program within the prescribed limits. He used the virtual force technique to obtain the maximum deflection bound. POLIZZOTTO [1978] generalized Ponter's method for the case where the yield surface is described by piecewise linear yield functions, allowing for strain hardening effects, if any. For elastic-perfectly plastic structures Polizzotto's formula yielded the same residual deflection bound as PONTER [1972] had found. DOROSZ and SAWCZUK [1981] published an extension of the theory to the case of beams and frames where account was taken of the fact that plastic strains are spread over finite zones rather than being concentrated at plastic hinges. LOI [1980] derived an upper bound on residual displacements in beams and frames. He used the virtual force method combined with CAPURSO'S [1974] upper bound on the direct plastic work. The methods for low temperature shakedown problems are reviewed by KÖNIG [1986].

Over recent years some papers on the shakedown displacement in the creep range have appeared: PONTER [1972], LECKIE [1974], POLIZZOTTO 1982], AINSWORTH [1981]. Analysis of such problems is, in general, a very difficult task. Therefore, in this paper a procedure allowing to bound the deformations and accounting for the creep strain of a given structure has been derived.

Special attention is given to developing the bounds on the plastic dissipation DOROSZ [1978] as well as on the dissipation of the creep strain energy by using the norm of the strain tensor.

It should be noted that this bound is valid for simple non-interactive creep and plastic laws in which effects such as creep recovery and cyclic hardening or softening are neglected. The bound obtained may provide a useful amendment to shakedown analysis and seems to be useful in the case when a safe bound is sufficient.

FUNDAMENTALS

Consider a body occupying volume V, bounded by a surface S. Points of the body are described by rectangular Cartesian coordinates (x_i). The body is subjected to the following external actions: a body force X_i, a given thermomechanical loading $P_i(x,t)$ over part S_p of the surface S, and zero surface velocities over the remainder S_u of S . The temperature T is a given function of position and time.

We assume that the model of material is elastic-plastic-creeping. The stress state σ_{ij} at the specified point $\mathbf{x}$ of the body is bounded by the convex yield condition

$$F = f\left(\sigma_{ij}(x,t)\right) - \tau(x,T) \leq 0 \ , \tag{1}$$

where τ is the yield stress.

Total strain ε_{ij} is additively decomposed into elastic ε^E, plastic ε^p_{ij}, creeping ε^c_{ij}, and thermal portions ε^T_{ij}

$$\varepsilon_{ij} = \varepsilon^E_{ij} + \varepsilon^p_{ij} + \varepsilon^c_{ij} + \varepsilon^T_{ij} \ . \tag{2}$$

The elastic strain is related to the total stress by Hooke's law via a positive definite elasticity tensor E_{ijkl} with its usual symmetry properties, thus

$$\sigma_{ij} = E_{ijkl}\varepsilon^E_{kl} \ . \tag{3}$$

The plastic strain rate tensor $\dot{\varepsilon}^p_{ij}$ is given by the associated flow rule which can be written in the following way

$$\dot{\varepsilon}^p_{ij} = \begin{cases} \dot{\lambda}\dfrac{\partial f}{\partial \sigma_{ij}} \ , \ \dot{\lambda} \geq 0 \ \text{ if } F = 0 & \dfrac{\partial f}{\partial \sigma_{kl}}\dot{\sigma}_{kl} + \dfrac{d\tau}{dT}\dot{T} = 0 \\ 0 \qquad\qquad\qquad \text{if } F < 0 \text{ or } F = 0 \ , & \dfrac{\partial f}{\partial \sigma_{kl}} + \dot{\sigma}_{kl}\dfrac{d\tau}{dT}\dot{T} < 0 \ . \end{cases} \tag{4}$$

The creep strain rate $\dot{\varepsilon}^c_{ij}$ is assumed to obey the general time hardening law of the form

$$\dot{\varepsilon}^{\,c}_{ij} = \Phi^n(\sigma_{ij})\frac{\partial \Phi(\sigma_{ij})}{\partial \sigma_{ij}}k(t) \ , \tag{5}$$

where Φ is a convex homogeneous function of degree one of σ_{ij}, n is the creep constant and $k(t)$ is a positive function. The function $k(t)$ is provided by a uniaxial test which follows the same time history of loading as the structural loading history. Therefore, we assume that for a particular history of structural loading we may find some function $k(t)$ which provides a reasonable approximation to the material behaviour. It may be noted that the theorem proved below does not explicitly involve the form of the $k(t)$. Therefore, the approximation is such that the same $k(t)$ is appropriate at all points of the body. The derivation of the theory does not require a more detailed specification of Φ which may represent Mises or Tresca conditions, or any other possible forms.

The thermal strain ε_{ij}^T is proportional to the temperature T with thermal coefficient α_{ij}^T, BOLEY and WEINER [1960]

$$\varepsilon_{ij}^T = \alpha_{ij}^T T \tag{6}$$

The boundary value problem for the body subjected to the loading specified above can be formulated as follows:

$$\begin{aligned} \sigma_{ij,j} + X_i(x,t) &= 0 && \text{in } V\,, \\ \sigma_{ij} n_j &= P_i(x,t) && \text{on } S_P\,, \\ \varepsilon_{ij}^T &= \alpha_{ij}^T T(x,t) && \text{in } V\,, \\ u_i &= 0 && \text{on } S_U\,. \end{aligned} \tag{7}$$

The total stress tensor σ_{ij} can be additively decomposed into the elastic stress σ_{ij}^E and the residual stress ρ_{ij}, thus

$$\sigma_{ij} = \sigma_{ij}^E + \rho_{ij} \tag{8}$$

$$\rho_{ij} = \rho_{ij}^p + \rho_{ij}^v \tag{9}$$

where σ_{ij}^E denotes the elastic stress resulting from external loading. The residual stress tensor ρ_{ij} results from plastic ε_{ij}^p and creep strains ε_{ij}^c. Thus the residual stress tensor can be divided into the corresponding parts ρ_{ij}^p and ρ_{ij}^c.

For the sake of simplicity, in the exposition of the theory the total deflection u_i is split into quantities u_i^E called the elastic deflection, u_i^{Rp} called the residual deflection due to plastic strains, and u_i^{Rc} called the residual deflection due to creep strains.

The corresponding boundary value problems can be written as follows

$$\left.\begin{array}{ll}
\sigma^E_{ij,j} + X_i(x,t) = 0 & \text{in } V\,, \\
\rho^p_{ij,j} = 0 \quad \rho^c_{ij,j} = 0 & \text{in } V\,, \\
\sigma^E_{ij} n_j = P_i(x,t) & \text{on } S_P\,, \\
\rho^p_{ij} n_j = 0 \quad \rho^c_{ij} n_j = 0 & \text{on } S_P\,, \\
\sigma^E_{ij} = E_{ijkl}\varepsilon^E_{kl} & \text{in } V\,, \\
\rho^p_{ij} = E_{ijkl}\varepsilon^{Rp}_{kl} \quad \rho^c_{ij} = e_{IJKL}\varepsilon^{Rc}_{kl} & \text{in } V\,,
\end{array}\right\} \quad (10.a)$$

$$\left.\begin{array}{ll}
\varepsilon^E_{kl} = (u^E_{k,l} + u^E_{l,k}) & \text{in } V\,, \\
\varepsilon^{Rp}_{kl} + \varepsilon^p_{kl} = (u^{Rp}_{k,l} + u^{Rp}_{k,l}) \quad \varepsilon^{Rc}_{k,l} + \varepsilon^c_{k,l} = (u^{Rc}_{k,l} + u^{Rc}_{l,k}) & \text{in } V\,, \\
u^E_i = 0 & \text{on } S_u\,, \\
u^{Rp}_i = 0 \qquad\qquad u^{Rc}_i = 0 & \text{on } S_u\,.
\end{array}\right\} \quad (10.b)$$

Since ε^E_{ij} is kinematically compatible with u^E_i, equations (10b) show that the strains $\varepsilon^{Rp}_{kl}+\varepsilon^p_{kl}$ and $\varepsilon^{Rc}_{kl}+\varepsilon^c_{kl}$ are kinematically compatible with the residual displacements u^{Rp}_i and u^{Rc}_i, respectively. Our interest is focussed here on obtaining an upper bound to the maximum of the residual deflection at the considered point and in the specified direction.

EVALUATION OF UPPER BOUND FORMULAE

In order to formulate the upper bound theorem we shall consider another load problem. Assume the fictitious linear elastic structure with the same geometrical properties as the actual structure, to be acted upon by a fictitious dummy load with specified tractions $\bar{P}_i(x_a)$ on the surface portion S_p of S. In special cases the fictitious load may be a concentrated load. Making use of the reciprocal relation of the elasticity extended to the case of non-compatible strains which are identified here with the creeping and plastic strains, we can derive the following expression:

$$\int_{S_p} \bar{P} u_i dS_p = \int_V \bar{\sigma}_{ij} \varepsilon_{ij} dV =$$

$$= \int_V \bar{\sigma}_{ij} (\varepsilon_{ij}^E + \varepsilon_{ij}^p + \varepsilon_{ij}^c) dV = \tag{11}$$

$$= \int_V \bar{\sigma}_{ij} \varepsilon_{ij}^E dV + \int_V \bar{\sigma}_{ij} \varepsilon_{ij}^p dV + \int_V \bar{\sigma}_{ij} \varepsilon_{ij}^c) dV$$

which provides a relation for displacement at the considered point provided $\bar{P}_i$ becomes a concentrated point force, where $\bar{\sigma}_{ij}(x, x_o)$ stands for the elastic stress field in a fictitious purely elastic structure having the same geometry and loaded by the concentrated force P_i at point x_o

The actual displacement $u_i(x_o)$ can be estimated by the extremal magnitudes of $u_i^E(x_o), u_i^{Rp}(x_o)$ and $u_i^{Rc}(x_o)$ at the considered point x_o .

$$u^E(x_o, t_o) = \int_V \bar{\sigma}_{ij}(x_o, x) \varepsilon_{ij}^E(x, t_o) dV \tag{12}$$

$$u^{R_p}(x_o, t_o) = \int_V \bar{\sigma}_{ij}(x_o, x) \varepsilon_{ij}^p(x, t_o) dV \tag{13}$$

$$u^{R_c}(x_o, t_o) = \int_V \bar{\sigma}_{ij}(x_o, x) \varepsilon_{ij}^c(x, t_o) dV \tag{14}$$

The structural deformation can be considered to consist of three components:
- an elastic component. Bounds to $u^E(x_o, t_o)$ are straightforward or mostly existing analytical solutions,
- a component due to plastic strain $u^{R_p}(x_o, t_o)$,
- a component due to creep $u^{R_c}(x_o, t_o)$.

If the component due to stress redistribution is neglected, the initial elastic deformation together with the stationary creep deformation provide a lower bound to the true deformation.

There are the four distinct bases for the derivations discussed in the next sections:
- the convexity of a function,
- the existence of a positive definite function,
- a geometrical property of a surface,
- the Schwartz and Hölder's inequalities.

EVALUATION OF THE DISPLACEMENT DUE TO PLASTIC STRAIN

To evaluate $u^{Rp}(x_o, t_o)$ which is represented by the second integral on the right hand side of Eq.(13), we consider the following procedure:

1. if at $t = 0$ there is no plastic strain $\varepsilon^p_{ij} = 0$ then

$$\left\| \varepsilon^p_{ij} \right\| = \sqrt{\varepsilon^p_{ij}\varepsilon^p_{ij}} = \int_0^t \frac{d}{dt}\sqrt{\varepsilon^p_{ij}\varepsilon^p_{ij}}dt \leq \int_0^t \left\| \dot{\varepsilon}^p_{ij} \right\| dt \tag{15}$$

2. In the case of the yield condition with temperature dependence of the yield stress (1) the rate of plastic dissipation depends not only on the plastic strain rate but also on the instantaneous temperature

$$D_p = \sigma_{ij}\dot{\varepsilon}^p_{ij} = D_0(\dot{\varepsilon}^p_{ij})g(T) \ , \tag{16}$$

where D_0 is the energy rate referred to room temperature T, and $g(T)$ specifies the temperature dependence of τ , ANDERBERG [1982].

$$\tau(T) = \tau_0 g(T), \quad \tau_0 = \tau(0), \quad g(0) = 1 \tag{17}$$

Most metals exhibit properties satisfying the inequality $dg/dT < 0$.
If the yield surface is bounded or is a hyper-cylinder with the axis defined by the vector n and its intersection with the hyperplane, then there exist two constants κ and η such that

$$\kappa g(T) \left\| \dot{\varepsilon}^p_{ij} \right\| < \sigma_{ij}\dot{\varepsilon}^p_{ij} < \eta g(\tau) \left\| \dot{\varepsilon}^p_{ij} \right\| \tag{18}$$

The constants are given explicitly by the formulae

$$\kappa = \inf \frac{D_0(\dot{\varepsilon}^p_{ij})}{\left\| \dot{\varepsilon}^p_{ij} \right\|} \ , \qquad \eta = \sup \frac{D_0(\dot{\varepsilon}^p_{ij})}{\left\| \dot{\varepsilon}^p_{ij} \right\|} \ , \tag{19}$$

where

$$\left\| \dot{\varepsilon}^p_{ij} \right\| = \sqrt{\dot{\varepsilon}^p_{ij}\dot{\varepsilon}^p_{ij}} \tag{20}$$

in which the optimization is to be performed over the whole hypersurface. This inequality allows the dissipation of the plastic energy to be estimated by means of the norm of the plastic strain rate.

Thus the inequalities (15), (17), (18) and (19) above permit the residual displacement due to plastic strain at considered point x_0 and time t_0 to be bounded as follows:

$$
\begin{aligned}
u^{R_p}(x_o,t_o) &= \int_V \bar{\sigma}_{ij}(x_0,x)\varepsilon^p_{ij}(x,t_o)dV = \\
&\leq \int_V \|\bar{\sigma}_{ij}(x_0,x)\| \left\|\varepsilon^p_{ij}(x,t_0)dV\right\| dV = \\
&\leq \int_V \|\bar{\sigma}_{ij}(x_0,x)\| \int_0^{t_0} \left\|\dot{\varepsilon}^p_{ij}\right\| dtdV = \\
&\leq \int_V \|\bar{\sigma}_{ij}(x_0,x)\| \int_0^{t_0} \frac{\sigma_{ij}\dot{\varepsilon}^p_{ij}}{\kappa g(T)} dtdV = \\
&\leq \int_V \frac{\|\bar{\sigma}_{ij}\|}{\kappa g(T_{\max})} \int_0^{t_0} \sigma_{ij}\dot{\varepsilon}^p_{ij} dtdV \\
&\leq \max \frac{\|\bar{\sigma}_{ij}\|}{\kappa g(T_{\max})} \int_V \int_0^{t_0} \sigma_{ij}\dot{\varepsilon}^p_{ij} dtdV \ .
\end{aligned}
\tag{21}
$$

We see that the integral on the right hand side of this inequality is the total plastic energy dissipated during the shakedown process. If we assume that all the possible load-temperature variations are than lying within the modified shakedown limit, PONTER [1972], POLIZZOTTO [1982]:

$$
P_i(x,t) < \frac{n}{n+1} P_{sh} \tag{22}
$$

The dissipation of the plastic energy can now be bounded using the KOITER's [1960] upper bound theorem

$$
\int_V \int_0^{t_0} \sigma_{ij}\dot{\varepsilon}^p_{ij} dtdV \leq \frac{s}{s-1} \int_V A_{ijkl} \overset{R}{\sigma}_{ij} \overset{R}{\sigma}_{kl} \, dV \tag{23}
$$

where σ^R_{ij} is any arbitrary constant residual state of stress such that

$$
f\left\{\frac{n+1}{n}(\sigma^E_{ij}(x,t) + \sigma^R_{ij}(x))\right\} \leq 0 \tag{24}
$$

where we assumed

$$
s = \frac{n+1}{n} \tag{25}
$$

The bound to the residual displacements due to plastic strain will now have the following form

$$
u^{R_p}(x_0,t_0) \leq \max \frac{\|\bar{\sigma}_{ij}\|}{\kappa g(T_{\max})} (n+1) \int_V A_{ijkl} \overset{R}{\sigma}_{ij} \overset{R}{\sigma}_{kl} \, dV \tag{26}
$$

EVALUATION OF THE DISPLACEMENT DUE TO CREEP STRAIN

To evaluate u_i^{Rc} which is represented by the last integral on the right hand side of Eq. (14), we consider the following:

1. If at $t = 0$ there is no creep strain, then

$$\left\|\varepsilon_{ij}^{c}\right\| = \sqrt{\varepsilon_{ij}^{c}\varepsilon_{ij}^{c}} = \int_0^t \frac{d}{dt}\sqrt{\varepsilon_{ij}^{c}\varepsilon_{ij}^{c}}dt \leq \int_0^t \left\|\dot{\varepsilon}_{ij}^{c}\right\| dt \tag{27}$$

2. The creep strain rate $\dot{\varepsilon}_{ij}^{c}$ is assumed to obey the general time-hardening law of the form

$$\dot{\varepsilon}_{ij}^{c} = \Phi^{n}(\sigma_{ij})\frac{\partial\Phi(\sigma_{ij})}{\partial\sigma_{ij}}k(t) \tag{28}$$

   where Φ is a convex homogeneous function of degree one of σ_{ij}, n is the creep constant and $k(t)$ is a positive function. The function $k(t)$ is provided by a uniaxial test which follows the same time history of loading as the structural loading history.

The rate of creep energy dissipation per unit volume is defined by

$$D_c(\dot{\varepsilon}_{ij}^{c}) = \Phi^{n+1}(\sigma_{ij})k(t,T) \ , \tag{29}$$

where we use Euler's theorem for homogeneous functions.

The assumption concerning the creep law (28) permits the dissipation of creep energy (29) to be expressed in terms of the norm of the creep strain rate in the following way

$$\alpha(t,T)\left\|\dot{\varepsilon}_{ij}^{c}\right\|^{1+1/n} = \sigma_{ij}\dot{\varepsilon}_{ij}^{c} \tag{30}$$

Let us define:
– the norm of the stress tensor

$$\|\sigma_{ij}\| = \sqrt{\frac{2}{3}}\sigma_{ef} \tag{31}$$

– the norm of the creep strain rate:

$$\left\|\dot{\varepsilon}_{ij}^{c}\right\| = \sqrt{\frac{3}{2}}\dot{\varepsilon}_{ef}^{c} \ . \tag{32}$$

For relation between the effective creep rate and the effective stresses the following creep law can be assumed

$$\dot{\varepsilon}_{ef}^{c} = k(t,T)\sigma_{ef}^{n} = k(t,T)\Phi^{n} \ . \tag{33}$$

The rate of the dissipation of the creep energy can be now expressed as

$$D_c(\dot{\varepsilon}^c_{ij}) = \Phi^{n+1}(\sigma_{ij})k(t,T) = k(t,T)^{-1/n}\left\{\sqrt{\frac{2}{3}}\right\}^{1+1/n} \left\|\dot{\varepsilon}^c_{ij}\right\|^{1+1/n} \tag{34}$$

Comparing with (30), it is found that

$$\alpha(t,T) = k(t,T)^{-1/n}\left\{\sqrt{\frac{2}{3}}\right\}^{1+1/n} \tag{35}$$

The norm of the creep strain rate can be now expressed as follows

$$\left\|\dot{\varepsilon}^c_{ij}\right\| = \left[\frac{\sigma_{ij}\dot{\varepsilon}^c_{ij}}{\alpha(t,T)}\right]^{\frac{n}{n+1}} \tag{36}$$

By introducing another time scale

$$\tau = E\sigma_0^{n-1}\alpha_0 \tag{37}$$

the parameter α_0 becomes now only a temperature dependent constant. Thus the inequalities (14), (34), and (36) permit the residual displacement due to creep strain at the considered point x_0 to be bounded as follows:

$$\begin{aligned} u^{R_c}(x_0,t_0) &= \int_V \bar{\sigma}_{ij}(x_0,x)\varepsilon^c(x,t_0)dV = \\ &< \int_V \|\bar{\sigma}_{ij}(x_0,x)\| \left\|\varepsilon^c_{ij}(x,t_o)dV\right\| dV = \\ &< \int_V \|\bar{\sigma}_{ij}(x_0,x)\| \int_0^{t_0} \left\|\dot{\varepsilon}^c_{ij}\right\| dtdV = \\ &< \int_V \|\bar{\sigma}_{ij}(x_0,x)\| \int_0^{t_0} \left\{\frac{\sigma_{ij}\dot{\varepsilon}^c_{ij}}{\alpha_0(T)}\right\}^m dtdV \end{aligned} \tag{38}$$

where $m = \frac{n}{n+1}$. Applying Hölder's inequality for integrals

$$\begin{aligned} &\int_V \bar{\sigma}_{ij}(x_0,x)\varepsilon^c_{ij}(x,t_o)dV = \\ &< \frac{1}{\alpha_0^m}\int_V \|\bar{\sigma}_{ij}(x_0,x)\|\, \tau^{nm}\int_0^{\tau} \sigma_{ij}(x)\cdot\varepsilon^c_{ij}(x,\tau)d\tau dV = \\ &< \max_v \frac{\|\bar{\sigma}_{ij}(x_0,x)\|}{\alpha^m}(\tau V)^{nm}\left\{\int_V\int_0^{\tau}\sigma_{ij}\dot{\varepsilon}^c_{ij}d\tau dV\right\}^m . \end{aligned} \tag{39}$$

We see that the integrals on the right-hand side of this inequality represent the creep energy dissipated in the process. If we assume that all the possible load-temperature variations are lying within the modified shakedown limit, PONTER [1972], POLIZZOTTO [1982]:

$$P_i(x,t) \le \frac{n}{n+1} P_{sh} \ , \tag{40}$$

the dissipation of the creep energy can be bounded using PONTER's [1972] or POLIZZOTTO's [1982] upper bound. If we restrict the problem to the case of a structure subjected to repeated cycles of loading we may obtain more specific formula, see DOROSZ [1994].

CONCLUSIONS

It should be noted that the bound is valid for simple non-interactive creep and plasticity laws in which effects such as creep recovery and cyclic hardening or softening are neglected. Norton's law is quite often used in the case of variable loads in view of its mathematical simplicity.
The bound obtained may provide a useful addition to shakedown analysis, and it seems that it could be applied in cases where a safe bound is sufficient. Problems related to bound optimization will be considered in a subsequent paper, together with some numerical applications.

ACKNOWLEDGEMENT

The work was supported by the National Research Council (KBN) under the Reserch Project 309789101.

REFERENCES

1960 KOITER, W.T.: General theorems for elastic-plastic solids, In: *Progress in Solid Mechanics*, North Holland, Amsterdam, p.165.

1960 BOLEY, B.A., WEINER, J.H.: Theory of Thermal Stress, John Wiley & Sons.

1972 MRÓZ, Z.: On the theory of steady plastic cycles in structures, Proc. of SMiRT 1, 6 part L, p. 489.

1972 PONTER, A.R.S.: Deformation, displacement, and work bounds for structures in a state of creep and subjected to variable loading, J. Appl. Mech. Trans. of the ASME, 39, p. 953.

1972 VITIELLO, E.: Upper bounds to plastic strains in shakedown of structures subjected to cyclic loads. Meccanica, Vol. 7, pp. 205-213.

1972 BRZEZIŃSKI, R. and KÖNIG, J.A.: Evaluation of shakedown deflections of framed structures by linear programming. In: Symp. Plastic Analysis of Structures, Jassy, pp. 101-116.

1972 PONTER, A.R.S.: An upper bound on the small displacements of elastic, perfectly plastic structures, J.Appl. Mech., Vol. 39, pp. 959-963.

1973 MAIER, G.: A shakedown matrix theory allowing for work-hardening and second-order geometric effects, Foundations of Plasticity, ed. A.Sawczuk, Nordhoff, Leyden, pp.417-433.

1973 BRZEZIŃSKI, R. and KÖNIG, J.A.: Deflection analysis of elastic-plastic frames at shakedown, J. Struct. Mech., vol. 2, pp. 211-228.

1974 CAPURSO, M.: A displacement bounding principle in shakedown of structures subjected to cyclic loads. Int. J. Solids Structures, Vol. 10, pp. 77-92.

1974 LECKIE, F.A.: A review of bounding techniques in shake down and ratchetting at elevated temperature, Welding Research Council Bull., No. 195, p. 1.

1974 MAIER, G. and VITIELLO E.: Bounds on plastic strains and displacements in dynamic shakedown of workhardening structures, J. Appl. Mech., Vol. 41, pp. 434-440.

1976 DOROSZ, S.: An upper bound to maximum residual deflections of elastic--plastic structures at shakedown. Bull. de l'Acad. Pol. Sci., Ser. Sci. Techn., Vol 24, pp. 167-174.

1978 DOROSZ, S.: An improved upper bound to maximum deflections of elastic--plastic structures at shakedown, J. Struct. Mech., 76, 3, p. 267.

1978 POLIZZOTTO, C.: A unified approach to quasi-elastic shakedown problems for elastic-plastic solids with piecewise linear yield surface. Meccanica, vol. 13, pp. 109-120

1978 LOI, F.T. and GRUNDY, P.: Deflection stability of workhardening structures, J. Struct. Mech., vol.6, pp. 331-347.

1980 LOI, F.T.: Deflection bounding at shakedown, J. Struct. Div., Proc. ASCE, vol. 106, pp. 1209-1215.

1981 DOROSZ, S. and SAWCZUK, A: Deflections of elastic-plastic beams at finite spread of plastic zones. In: Physical Non-linearities in Structural Analysis. Eds.: J. Hult and J. Lemaitre, Berlin, Springer, pp. 64-73.

1981 KÖNIG, J.A., MAIER, G.: Shakedown analysis of elastoplastic structures: A review of recent developments, Nuclear Engineering and Design, 66, 81-95.

1981 AINSWORTH, R.A.: Bounding structural deformation due to combined creep and plasticity. IUTAM Symposium Creep in Structures, Springer Verlag.

1981 PONTER, A.R.S.: On the creep modified shakedown limit, IUTAM Symposium, Leicester 1980, Creep in Structures ed. A.R.S. Ponter, D.P. Hayhurst, Springer Verlag, p. 264,.

1982 ANDERBERG, Y.: Behaviour of steel at high temperatures, RILEM-COMMITTEE 44-PHT, Nov. 1982.

1982 POLIZZOTTO, C.: Bounding principles for elastic-plastic-creeping solids loaded below and above the shakedown limit, Meccanica, 17, p. 143.

1986 KÖNIG J.A.: Shakedown of Elastic-Plastic Structures, Elsevier & Polish Scientific Publishers , Warsaw.

1994 DOROSZ, S.: Method of calculation for elastic-plastic-creeping structures subjected to variable loading, (in preparation).

ON SHAKEDOWN THEOREMS IN THE PRESENCE OF SIGNORINI CONDITIONS AND FRICTION

Józef Joachim TELEGA
Polish Academy of Sciences
Institute of Fundamental Techological Research
ul. Świętokrzyska 21, 00-049 Warsaw, Poland

Abstract: The aim of this contribution is to study shakedown of structures when contact conditions are imposed on a part of the boundary. Three cases have been investigated: 1. Signorini's conditions without friction, 2. Friction with prescribed distribution of normal stress, and 3. Signiorini's conditions with friction.

INTRODUCTION

The shakedown theory is a well established domain of the small displacement plasticity theory, cf. COMI and CORIGLIANI [1991], DEBORDES and NAYROLES [1976], HALPHEN and SALENÇON [1987], KÖNIG [1987], MERCIER [1977], NAYROLES and WEICHERT [1993], POLIZZOTTO [1993], WESFREID [1980]. Geometrically nonlinear effects can be included, though in a restricted sense, see WEICHERT [1990]. In fact, the geometrically nonlinear shakedown theory follows the developments of the linear theory.

Though in the last two decades significant progress has been made in the field of contact mechanics, it has not yet influenced the shakedown theory. The results presented by JOHNSON [1985], JOHNSON and SHERCLIFF [1992] and NORDGREN [1992] are obtained within the framework of classical considerations. RYDHOLM [1981] formulated the shakedown theorem in the presence of friction. The result of this author seems to be valid in the case when normal contact stresses are known. The frictional contact conditions are formulated in a manner analogous to the relationships describing elastic-perfectly plastic behaviour. Signorini's conditions were out of scope of the dissertation by RYDHOLM [1981].

The aim of our contribution is mainly twofold; first, in Section 2, we shall formulate the shakedown theorem for the body with a frictionless unilateral contact with a rigid

Z. Mróz et al. (eds.), Inelastic Behaviour of Structures under Variable Loads, 183–201.

foundation. This means that apart from the classic mixed boundary conditions, Signorini's conditions are placed on the part Γ_2 of the boundary of the body. Essential here is the choice of the elastic problem; our choice of the Signorini's conditions on Γ_2 for this problem seems to be a natural one. However, other variants are also plausible.

Section 3 of the paper discusses the shakedown theorems in the presence of friction; first when normal contact stresses are prescribed, secondly when the Signorini's conditions with friction are placed on Γ_2. In the latter case the shakedown theorem has rather the form of a conjecture.

The results presented in the paper are by no means exhaustive. Our aim was to make at least a small step towards the inclusion of unilateral boundary conditions into the shakedown theory. Shortly before his untimely death, Professor J.A. König asked me about such a possibility. The present paper provides the first answer, clearly exhibiting that even the small displacement shakedown theory can be significantly enriched by taking into account various contact conditions.

1 SOME BASIC RELATIONS

Let $\Omega \subset \mathbb{R}^3$ be a bounded sufficiently regular domain and $\Gamma = \partial\Omega$ its boundary. We assume that $\Gamma = \bar{\Gamma}_0 \cup \bar{\Gamma}_1 \cup \bar{\Gamma}_2$, $\Gamma_0 \cap \Gamma_1 \cap \Gamma_2 = \emptyset$. On $\Gamma_0 \cup \Gamma_1$, the classic mixed boundary conditions will be imposed; Γ_2 is the surface of a potential contact. The bar over a set denotes its closure. By $\boldsymbol{N} = (N_i)$ $(i = 1, 2, 3)$ we denote the outward unit vector normal to Γ. Let $\boldsymbol{u} = (u_i)$ be the displacement vector and $\boldsymbol{s} = (s_{ij})$ the stress tensor. Our considerations are restricted to the geometrically linear theory. Throughout this paper, the summation convention is applied.

We can perform the following decompositions on Γ:

$$\boldsymbol{u} = u_N \boldsymbol{N} + \boldsymbol{u}_t \, , u_N = u_i N_i \, , \ \boldsymbol{u}_t = \boldsymbol{u} - u_N \boldsymbol{N} \, , \tag{1.1}$$

$$s_{ij} N_j = s_N N_i + s_{ti} \, , \ s_N = s_{ij} N_i N_j \, , \ s_{ti} = s_{ij} N_j - s_N N_i \, . \tag{1.2}$$

In Section 3 we shall discuss the shakedown problem of the body represented by Ω in the presence of Signorini's conditions with friction on Γ_2. The description of friction follows that expounded by us in TELEGA [1985, 1987, 1988, 1990]. Let the friction condition be given by

$$g_I(x, s_N, \boldsymbol{s}_t) \leq 0 \, , \ s_N \leq 0 \, , \ x \in \Gamma_2 \, , \ I = 1, \ldots, M \, . \tag{1.3}$$

We set

$$\mathbf{C}(x) = \{(s_N, \boldsymbol{s}_t) | g_I(x, s_N, \boldsymbol{s}_t) \leq 0 \, , \ I = 1, \ldots, M\} \, , \ x \in \Gamma_2 \, . \tag{1.4}$$

We assume that the set $\mathbf{C}(x)$ of admissible normal and tangential stresses at a point $x \in \Gamma_2$ is closed but not necessarily convex. However, the set

$$K(x, s_N) := \{\boldsymbol{t} | g_I(x, s_N, \boldsymbol{t}) \leq 0 \, , \ \boldsymbol{t} \cdot \boldsymbol{N} = 0 \, , \ I = 1, \ldots, M\} \, , \tag{1.5}$$

is assumed to be closed, convex and bounded for $x \in \Gamma_2$, $s_N \leq 0$; moreover $0 \in int K(x, s_N)$. As usual, the indicator function of $K(x, s_N)$ is defined by (cf. ROCKAFELLAR [1970]):

$$I_{K(x,s_N)}(\boldsymbol{t}) = \begin{cases} 0\ , & \text{if } \boldsymbol{t} \in K(x, s_N)\ , \\ +\infty\ , & \text{otherwise.} \end{cases} \tag{1.6}$$

The friction law is assumed in the subdifferential form

$$-\ \dot{\boldsymbol{u}}_t \in \partial I_{K(x,s_N)}(\boldsymbol{s}_t)\ ,\ x \in \Gamma_2\ . \tag{1.7}$$

The adopted description of friction generalizes the classic Coulomb's law and allows for anisotropic friction. The latter has been confirmed experimentaly, see ARIKAN and MURPHY [1991]; FELDER [1988]; KADIJK and BROESSE VAN GROENOU [1990]; NAYEB–HASHEMI, BLUCHER and MIRAGEAS [1991], cf. also HE and CURNIER [1994], ZMITROWICZ [1989].

The support function of $I_{K(x,s_N)}$ is the frictional dissipation density, say $D_f(x, s_N, \cdot)$. Being the dual function it is calculated according to (cf. ROCKAFELLAR [1970]):

$$D_f(x, s_N, \boldsymbol{t}^*) = \sup\{\boldsymbol{t}^* \cdot \boldsymbol{t} | \boldsymbol{t} \in K(x, s_N)\}\ . \tag{1.8}$$

Consequently, the friction law (1.7) is *equivalent* to

$$\boldsymbol{s}_t \in \partial_3 D_f(x, s_N, -\ \dot{\boldsymbol{u}}_t)\ , \tag{1.9}$$

and to

$$D_f(x, s_N, -\ \dot{\boldsymbol{u}}_t) = \boldsymbol{s}_t \cdot (-\ \dot{\boldsymbol{u}}_t)\ . \tag{1.10}$$

In (1.9) ∂_3 denotes subdifferentiation with respect to the third argument.

Some properties of D_f.

i. $D_f(x, s_N, \cdot)$ is a convex and continuous function, positively homogeneous of degree 1.

ii. $\exists\ c_1(s_N) \geq c_0(s_N) > 0$ such that

$$c_0(s_N)|\boldsymbol{t}^*| \leq D_f(x, s_N, \boldsymbol{t}^*) \leq c_1(s_N)|\boldsymbol{t}^*|\ .$$

iii. $D_f(x, s_N, \mathbf{o}) = 0$, $D_f(x, s_N, \boldsymbol{t}^*) \geq 0$.

Let us pass now to the description of the plastic behaviour of the body itself. By $C(x)$ we denote the elasticity convex at a point $x \in \Omega$. It is a bounded, closed and convex set generated by the yield condition; $0 \in \operatorname{int} C(x)$. For the body made of a homogeneous material the elasticity convex does not depend on $x \in \Omega$. Denoting by $\boldsymbol{d}^p$ the plastic strain rate, the associated flow rule is written in the subdifferential form

$$\boldsymbol{d}^p \in \partial I_{C(x)}(\boldsymbol{s}) \ , x \in \Omega \ . \tag{1.11}$$

Notice that (1.11) does not preclude anisotropic materials.

The plastic dissipation density is given by

$$D(x, \boldsymbol{E}) = \sup\{s_{ij}E_{ij} | \boldsymbol{s} \in C(x)\} \ , \ \boldsymbol{E} \in \mathbb{E}^3_s \ . \tag{1.12}$$

Here $\mathbb{E}^3_s$ stands for the space of symmetric 3×3 matrices; in the two–dimensional case we write $\mathbb{E}^2_s$.

Some properties of D.

a. $\exists \ c_3 \geq c_2 > 0$ such that

$$c_2|\boldsymbol{E}| \leq D(x, \boldsymbol{E}) \leq c_3|\boldsymbol{E}| \ , \forall \boldsymbol{E} \in \mathbb{E}^3_s \ .$$

b. $D(x, \mathbf{o}) = 0$, $D(x, \boldsymbol{E}) \geq 0$, for all $\boldsymbol{E} \in \mathbb{E}^3_s$.

c. $D(x, \cdot)$ is a convex and continuous function, positively homogeneous of degree 1.

2 SHAKEDOWN IN THE PRESENCE OF SIGNORINI'S CONDITIONS WITHOUT FRICTION

2.1. The aim of this section is to extend Melan theorem so as to include unilateral contact conditions on Γ_2 . By $t \geq 0$, we denote time. Body forces $\boldsymbol{f}(x,t) = (f_i(x,t))$ $(x \in \Omega, t \geq 0)$ and surface tractions $\boldsymbol{F}(x,t) = (F_i(x,t))$ $(x \in \Gamma_1, t \geq 0)$ are periodic functions of time, otherwise arbitrary. By $\boldsymbol{a} = (a_{ijkl})$ we denote the compliance tensor. We assume that

$$a_{ijkl} \in L^\infty(\Omega) \ . \tag{2.1}$$

This tensor does not change with time and satisfies the usual symmetry and coercivity conditions.

The strong formulation of the boundary initial value problem for an elastic-perfectly plastic body in the presence of Signorini's conditions without friction is formulated as

Problem P_s
Find $\boldsymbol{s}(x,t)$ and $\boldsymbol{u}(x,t)$ $(x \in \Omega, t \geq 0)$ such that

$$s_{ij,j} + f_i = 0 \ , \ \text{in } \Omega \times [0, \infty) \tag{2.2}$$

$$e_{ij}(\dot{\boldsymbol{u}}) = \dot{u}_{(i,j)} = a_{ijkl} \ \dot{s}_{kl} + d^p_{ij} \ , \ \text{in } \Omega \times [0, \infty) \tag{2.3}$$

$$\boldsymbol{d}^p \in \partial I_{C(x)}(\boldsymbol{s}) \ , \ \text{in } \Omega \times [0,\infty) \tag{2.4}$$

$$\boldsymbol{u} = 0 \ , \ \text{on } \Gamma_0 \times [0,\infty) \tag{2.5}$$

$$s_{ij}N_j = F_i \ , \ \text{on } \Gamma_1 \times [0,\infty) \tag{2.6}$$

$$\boldsymbol{s}_t = 0 \ , \ s_N \leq 0 \ , \ u_N \leq 0 \ , s_N u_N = 0 \ , \ \text{on } \Gamma_2 \times [0,\infty) \tag{2.7}$$

$$\boldsymbol{u}(x,0) = \boldsymbol{u}_0(x) \ , \ \boldsymbol{s}(x,0) = \boldsymbol{s}_0(x) \ , \ \text{in } \Omega \ . \tag{2.8}$$

The prescribed initial conditions $\boldsymbol{u}_0$ and $\boldsymbol{s}_0$ should be propertly chosen; (2.7) expresses the Signorini's conditions without friction. In the case when an initial gap exists between the body and the rigid obstacle, (2.7) is to be replaced by

$$\boldsymbol{s}_t = 0 \ , \ s_N \leq 0 \ , \ u_N - g \leq 0 \ , s_N(u_N - g) = 0 \ . \tag{2.7a}$$

The function $g(x)(x \in \Gamma_2)$, describing the initial gap, has to be prescribed.

Let us pass now to the variational formulation. Towards this end we set

$$(\boldsymbol{\sigma}, \boldsymbol{\tau}) = \int_\Omega a_{ijkl}\sigma_{ij}\tau_{kl}dx \ , \tag{2.9}$$

$$< \boldsymbol{e}, \boldsymbol{\tau} >= \int_\Omega e_{ij}\tau_{ij}dx \ , \tag{2.9a}$$

$$\mathsf{L}(\boldsymbol{v}) = \int_\Omega f_i v_i dx + \int_{\Gamma_1} F_i v_i dx \ , \tag{2.10}$$

$$\mathsf{K} = \{\boldsymbol{\tau} \in H_s | \boldsymbol{\tau}(x) \in C(x) \ , \ \text{a.e. } x \in \Omega\} \ . \tag{2.11}$$

Here a.e. means almost everywhere and H_s is the stress space, see TEMAM [1985]:

$$H_s = \{\boldsymbol{\tau} \in L^\infty(\Omega, \mathbb{E}^3_s) | \tau_{ij,j} \in L^4(\Omega)\} \ , \tag{2.12}$$

or, in the case of incompressible materials

$$H_s = \{\boldsymbol{\tau} \in L^2(\Omega, \mathbb{E}^3_s) | \tau_{ij,j} \in L^4(\Omega) \ , \tau^d_{ij} \in L^\infty(\Omega)\} \ , \tag{2.12a}$$

where $\boldsymbol{\tau}^d$ stands for the deviatoric part of $\boldsymbol{\tau} = (\tau_{ij})$. Though we use the same symbol for the stress spaces in the case of compressible and incompressible materials, yet such a notation should not lead to ambiguities.

For bodies exhibiting perfectly plastic behaviour displacement (velocity) fields belong to (cf. SUQUET [1988], TEMAM [1985]):

$$\mathsf{V} = \{\boldsymbol{v} \in V | \boldsymbol{v} = 0 \text{ on } \Gamma_0, v_N \leq 0 \text{ on } \Gamma_2\} \ , \tag{2.13}$$

where

$$V = BD(\Omega) = \{\boldsymbol{v} \in L^1(\Omega)^3 | e_{ij}(\boldsymbol{v}) \in M^1(\Omega)\} \ , \tag{2.14}$$

or, in the case of incompressible materials

$$V = \{\boldsymbol{v} \in BD(\Omega) | \text{ div } \boldsymbol{v} \in L^2(\Omega)\} \ . \tag{2.14a}$$

Here div means divergence, and $M^1(\Omega)$ denotes the space of bounded measures on Ω.

In variational formulations of evolution problems it is convenient to treat a function, say $h(x,t)(x \in \Omega, t \geq 0)$ as a so-called vector function $h(t)$, where

$$h(t) = \{h(x,t) | x \in \Omega\}. \tag{2.15}$$

The variational formulation of the evolution problem P_s takes the form of

Problem P

Find $\{\boldsymbol{s}, \boldsymbol{u}\} : [0, \infty) \to \mathsf{K} \times \mathsf{V}$ such that

$$(\dot{\boldsymbol{s}}\ (t), \boldsymbol{\tau} - \boldsymbol{s}(t)) - < \boldsymbol{e}(\dot{\boldsymbol{u}}\ (t)), \boldsymbol{\tau} - \boldsymbol{s}(t) > \geq 0 \ , \ \forall \boldsymbol{\tau} \in \mathsf{K} \tag{2.16}$$

$$< \boldsymbol{e}(\boldsymbol{v} - \boldsymbol{u}(t)), \boldsymbol{s}(t) > \geq \mathsf{L}(\boldsymbol{v} - \boldsymbol{u}(t)) \ , \ \forall \boldsymbol{v} \in \mathsf{V} \tag{2.17}$$

$$\boldsymbol{u}(0) = \boldsymbol{u}_0 \ , \ \boldsymbol{s}(0) = \boldsymbol{s}_0 \ . \tag{2.18}$$

for a.e. $t \in [0, \infty)$. The existence of a solution $\{\boldsymbol{s}(t), \boldsymbol{u}(t)\} \in \mathsf{K} \times \mathsf{V}$ to the problem P remains open. Consequently, the same can be said about the asymptotic behaviour of $\boldsymbol{s}(t)$ and $\boldsymbol{u}(t)$ when $t \to \infty$. The particular case when $\Gamma_2 = \emptyset$, has been studied by MERCIER [1977], cf. also DEBORDES and NAYROLES [1976], HALPHEN and SALENÇON [1987], KÖNIG [1987], NAYROLES AND WEICHERT [1993], TELEGA [1972], WESFREID [1980]. MERCIER [1977] shows that for classic boundary conditions, in suitable topologies

$$\left.\begin{array}{l} \boldsymbol{s}(t) - \boldsymbol{s}^e(t) \to 0 \\ \boldsymbol{u}(t) - \boldsymbol{u}^e(t) \to 0 \end{array}\right\} \text{ when } t \to \infty \ ,$$

i.e. shakedown occurs.

In passing we note that, to our knowledge, the only existence results for unilateral boundary value problems for plastic solids were provided within the holonomic (Hencky) plasticity, cf. REDDY and TOMARELLI [1990], TELEGA [1987].

2.2. We shall now generalize Melan theorem to the case when the body is, on Γ_2, in the unilateral frictionless contact with a rigid foundation. If $\{\boldsymbol{s}(x,t), \boldsymbol{u}(x,t)\}, x \in \Omega, t \geq 0$ is the actual solution to the boundary initial value problem for the elastic-perfectly plastic body subjected to variable (periodic) loading, then we may write (cf. HALPHEN and SALENÇON [1987], KÖNIG [1987]):

$$\boldsymbol{s} = \boldsymbol{s}^e + \boldsymbol{\rho} \qquad , \qquad \boldsymbol{u} = \boldsymbol{u}^e + \boldsymbol{u}^r. \tag{2.19}$$

Here, as usual, $\boldsymbol{s}^e, \boldsymbol{\rho}, \boldsymbol{u}^e$ and $\boldsymbol{u}^r$ denote elastic stresses, residual stresses, elastic displacements and residual displacements, respectively. On account of the presence of Signorini's conditions (2.7), the notion of the elastic solution must be made precise. It is natural to assume

$$\boldsymbol{s}_t^e = 0 \ , \ s_N^e \leq 0 \ , \ u_N^e \leq 0 \ , s_N^e u_N^e = 0 \ , \ \text{on } \Gamma_2 \times [0, \infty) \ . \tag{2.20}$$

For each $t \geq 0$, the elastic solution $\{\boldsymbol{s}^e(t), \boldsymbol{u}^e(t)\}$ exists and is unique, cf. TELEGA [1987, 1988].

Since the contact is without friction, the shakedown criterion assumes the form

$$W_p(t) < +\infty \qquad \text{for each} \qquad t \geq 0 \ , \tag{2.21}$$

where $W_p(t)$ denotes the total plastic work at a time instant $t \geq 0$, given by

$$W_p(t) = \int_0^t \int_\Omega D[x, \boldsymbol{d}^p(x, \tau)] dx \, d\tau \ . \tag{2.22}$$

Here $W_p(t)$ includes also the plastic work done at surfaces of discontinuities, if any. In such a case, $D(x, \cdot)$ is the function of a measure in the sense of BOUCHITTÉ and VALADIER [1988].

Now we are formulating the generalized Melan theorem when Signorini's conditions (2.7) are taken into account on Γ_2.

Theorem 2.1. If there exists a field of residual stresses $\boldsymbol{\rho}^*(x)$ $(x \in \Omega)$ and $m > 1$ such that

$$\boldsymbol{\rho}^*(x) + m\boldsymbol{s}^e(x, t) \in C(x) \ , \ x \in \Omega \ , \ t \geq 0 \ , \tag{2.23}$$

$$\boldsymbol{\rho}_t^*(x) = 0 \ , \ \boldsymbol{\rho}_N^*(x) \leq 0 \ , \ x \in \Gamma_2 \ , \tag{2.24}$$

and (2.20) is satisfied, then shakedown occurs.

Proof. Taking account of $(2.7)_2$, $(2.19)_1$ and $(2.20)_3$ we have

$$(s_N^e + \rho_N) u_N^e \geq 0 \ , \ \text{on } \Gamma_2, t \geq 0 \ .$$

Hence, by virtue of $(2.20)_3$ and $(2.20)_4$

$$s_N^e u_N^e + \rho_N u_N^e = \rho_N u_N^e \geq 0 \Longrightarrow \rho_N \leq 0.$$

Similary, by using $(2.7)_3$, $(2.19)_2$, $(2.20)_2$ and $(2.20)_4$ we obtain, on Γ_2 for $t \geq 0$:

$$s_N^e (u_N^e + u_N^r) \geq 0 \Longrightarrow u_N^r \leq 0.$$

The condition $(2.7)_4$ can be written in the following way

$$(s_N^e + \rho_N)(u_N^e + u_N^r) = 0 \ , \ \text{on } \Gamma_2 \times [0, \infty) \ .$$

Hence, by virtue of $(2.20)_4$

$$\rho_N u_N^e + s_N^e u_N^r + \rho_N u_N^r = 0 \ , \ \text{on} \ \ \Gamma_2 \times [0, \infty) \ . \tag{2.25}$$

Each term in Eq. (2.25) is nonnegative. In summary, we conclude that (2.7) and (2.20) imply, on $\Gamma_2 \times [0, \infty)$:

$$\boldsymbol{\rho}_t = 0 \ , \ u_N^r \leq 0 \ , \ \rho_N \leq 0 \ , \ \rho_N u_N^r = 0 \ . \tag{2.26}$$

Additionaly, from (2.25) one obtains

$$\rho_N u_N^e = 0 \ , \ s_N^e u_N^r = 0 \ , \ \text{on} \ \ \Gamma_2 \times [0, \infty) \ . \tag{2.27}$$

As is well known, the principle of virtual work rate is of primal importance. We shall now derive its form suitable in our case.

We set

$$\boldsymbol{s}^*(x,t) = \boldsymbol{\rho}^*(x) + m\boldsymbol{s}^e(x,t) \ . \tag{2.28}$$

Taking account of (2.19), we thus have

$$\boldsymbol{s}(x,t) - \frac{1}{m}\boldsymbol{s}^*(x,t) = \boldsymbol{\rho}(x,t) - \frac{1}{m}\boldsymbol{\rho}^*(x) \ . \tag{2.29}$$

It is worth noting that though $\rho_N \leq 0$ and $\rho_N^* \leq 0$, on Γ_2, yet obviously $\rho_N - \frac{1}{m}\rho_N^*$ is not necessarily nonpositive.

We recall that, cf. HALPHEN and SALENÇON [1987], KÖNIG [1987]

$$e_{ij}(\dot{\boldsymbol{u}}^{\,r}) = \dot{u}^{\,r}_{(i,j)} = a_{ijkl}\,\dot{\rho}_{kl} + d_{ij}^p \ , \tag{2.30}$$

and find, by virtue of $(2.24)_1$ and (2.29):

$$\begin{aligned} &\int_\Omega [s_{ij}(x,t) - \frac{1}{m}s_{ij}^*(x,t)]e_{ij}(\dot{\boldsymbol{u}}^{\,r}(x,t))dx = \\ &= \int_\Omega [\rho_{ij}(x,t) - \frac{1}{m}\rho_{ij}^*(x)]e_{ij}(\dot{\boldsymbol{u}}^{\,r}(x,t))dx = \\ &= -\int_\Omega [\rho_{ij}(x,t) - \frac{1}{m}\rho_{ij}^*(x)]_{,j}\,\dot{u}_i^{\,r}(x,t)dx + \\ &+ \int_{\Gamma_2} [\rho_N(x,t) - \frac{1}{m}\rho_N^*(x)]\,\dot{u}_N^{\,r}(x,t)d\Gamma \ . \end{aligned} \tag{2.31}$$

Taking account of (2.30) we finally obtain

$$\begin{aligned} &\int_\Omega a_{ijkl}(x)[\rho_{ij}(x,t) - \frac{1}{m}\rho_{ij}^*(x)]\,\dot{\rho}_{kl}\,(x,t)dx + \\ &+ \int_\Omega [s_{ij}(x,t) - \frac{1}{m}s_{ij}^*(x,t)]d_{ij}^p(x,t)dx = \\ &= \int_{\Gamma_2} [\rho_N(x,t) - \frac{1}{m}\rho_N^*(x)]\,\dot{u}^{\,r}_N d\Gamma \ . \end{aligned} \tag{2.32}$$

Notice that though $u_N^r \leq 0$, on Γ_2, yet $\dot{u}_N^r$ can be either nonnegative or nonpositive, depending on the loading.

Since $\boldsymbol{s}^*(x,t) \in C(x)$ $(x \in \Omega, t \geq 0)$ therefore

$$[s_{ij}(x,t) - s_{ij}^*(x,t)]d_{ij}^p(x,t) \geq 0 \ , \tag{2.33}$$

and consequently $(m > 1)$

$$\frac{1}{m}[s_{ij}(x,t) - s_{ij}^*(x,t)]d_{ij}^p(x,t) \geq 0 \ . \tag{2.34}$$

Further we may write

$$\frac{1}{m}\boldsymbol{s} - \frac{1}{m}\boldsymbol{s}^* = \boldsymbol{s} - \frac{1}{m}\boldsymbol{s}^* + \frac{1-m}{m}\boldsymbol{s} \ . \tag{2.35}$$

Now (2.34) and (2.35) yield

$$[s_{ij}(x,t) - \frac{1}{m}s_{ij}^*(x,t)]d_{ij}^p(x,t) \geq \frac{m-1}{m}s_{ij}(x,t)d_{ij}^p(x,t) \ . \tag{2.36}$$

By integration one obtains

$$\frac{m-1}{m}\int_0^t\int_\Omega s_{ij}(x,\tau)d_{ij}^p(x,\tau)dx\,d\tau \leq$$
$$\leq \int_0^t\int_\Omega [s_{ij}(x,\tau) - \frac{1}{m}s_{ij}^*(x,\tau)]d_{ij}^p(x,\tau)dx\,d\tau \ .$$

Hence, by taking account of Eq. (2.32) one has

$$\begin{aligned} 0 &\leq \frac{m-1}{m}\int_0^t\int_\Omega s_{ij}(x,\tau)d_{ij}^p(x,\tau)dx\,d\tau \leq \\ &\leq \int_0^t \left[-\frac{1}{2}\frac{d}{d\tau} \parallel \boldsymbol{\rho}(\tau) - \frac{1}{m}\boldsymbol{\rho}^* \parallel_a^2\right] d\tau + \\ &+ \int_0^t\int_{\Gamma_2} [\rho_N(x,\tau) - \frac{1}{m}\rho_N^*(x)]\,\dot{u}_N^r(x,\tau)d\Gamma\ d\tau \ , \end{aligned} \tag{2.37}$$

where

$$\parallel \boldsymbol{\rho}(\tau) - \frac{1}{m}\boldsymbol{\rho}^* \parallel_a^2 = \int_\Omega a_{ijkl}(x)[\rho_{ij}(x,\tau) - \frac{1}{m}\rho_{ij}^*(x)][\rho_{kl}(x,\tau) - \frac{1}{m}\rho_{kl}^*(x)]dx \ .$$

Signorini's conditions (2.26) imply

$$\rho_N(x,\tau)\,\dot{u}_N{}^r(x,\tau) = 0 \ , x \in \Gamma_2 \ . \tag{2.38}$$

Consequently, from (2.36) and (2.37) we obtain

$$0 \leq \frac{m-1}{m} \int_0^t \int_\Omega s_{ij}(x,\tau) d_{ij}^p(x,\tau) dx\, d\tau \leq$$

$$\leq -\frac{1}{2} \parallel \boldsymbol{\rho}(\tau) - \frac{1}{m}\boldsymbol{\rho}^* \parallel_a^2 + \frac{1}{2} \parallel \boldsymbol{\rho}(0) - \frac{1}{m}\boldsymbol{\rho}^* \parallel_a^2 - \tag{2.39}$$

$$-\frac{1}{m}\int_{\Gamma_2} \rho_N^*(x) \int_0^t \dot{u}_N^r(x,\tau) d\Gamma\; d\tau \;.$$

We recall that

$$\dot{u}_N^r(x,\tau) = \dot{u}_i^r(x,\tau) N_i(x) \;, \quad x \in \Gamma_2 \;,$$

and thus

$$u_N^r(x,t) = \int_0^t \dot{u}_N^r(x,\tau) d\tau \;,$$

provided that $u_N^r(x,0) = 0$ for $x \in \Gamma_2$. The last term in (2.39) is equal to

$$-\frac{1}{m}\int_{\Gamma_2} \rho_N^*(x) u_N^r(x,t) d\Gamma \leq 0 \;, \tag{2.40}$$

since $\rho_N^* \leq 0$ and $u_N^r(x,t) \leq 0$ on Γ_2, $t \geq 0$. Thus, we finally arrive at the classic inequality, cf. HALPHEN and SALENÇON [1987], KÖNIG [1987]

$$\frac{m-1}{m} W_p(t) \leq \frac{1}{2}\int_\Omega a_{ijkl}(x)[\rho_{ij}(x,0) - \frac{1}{m}\rho_{ij}^*(x)][\rho_{kl}(x,0) - \frac{1}{m}\rho_{kl}^*(x)] dx \;, \tag{2.41}$$

where

$$W_p(t) = \int_0^t \int_\Omega D[x, d^p(x,\tau)] dx\, d\tau =$$
$$= \int_0^t \int_\Omega s_{ij}(x,\tau) d_{ij}^p(x,\tau) dx\, d\tau \;. \tag{2.42}$$

Remark 2.1

In our definition of the elastic problem the mixed conditions on $\Gamma_0 \cup \Gamma_1$ are the classic ones. The novelty lies in Signorini's conditions (2.20); then the conditions (2.26) are a straightforward consequence of (2.7) (Signorini's conditions for the real problem) and (2.20). However, as we know from the proof of Theorem 2.1, additional conditions (2.27) appear. The last conditions couple in some sense the elastic and residual stress problems.

One can probably envisage a different choice of conditions on Γ_2 for the elastic and residual stress problems, obviously still satisfying (2.7).

The hypothesis of Th. 2.1 can be weakened by assuming

$$\boldsymbol{\rho}^*(x) = \begin{cases} \boldsymbol{\rho}_1^*(x) & \text{if} \quad x \in \Omega \;, \\ \boldsymbol{\xi}(x) & \text{if} \quad x \in \Gamma_2 \;. \end{cases}$$

Here $\boldsymbol{\xi}$ is independent of $\boldsymbol{\rho}_1$. Then ρ_N^* should be replaced by ξ_N in (2.31).

3 SHAKEDOWN IN THE PRESENCE OF SIGNORINI'S CONDITIONS WITH FRICTION

3.1. Now the condition $(2.7)_1$ will be replaced by the friction law (1.7). Obviously, the loading is still variable (periodic) and the initial conditions must be properly chosen.
Problem P_s^f
Find $\boldsymbol{s}(x,t)$ and $\boldsymbol{u}(x,t)$ $(x \in \Gamma, t \geq 0)$ such that

$$s_{ij,j} + f_i = 0 \ , \ \text{in } \Omega \times [0,\infty) \tag{3.1}$$

$$e(\dot{\boldsymbol{u}}) \in (a_{ijkl}\, \dot{s}_{kl}) + \partial I_{C(x)}(\boldsymbol{s}) \ , \ \text{in } \Omega \times [0,\infty) \tag{3.2}$$

$$\boldsymbol{u} = 0 \ , \ \text{on } \Gamma_0 \times [0,\infty) \tag{3.3}$$

$$s_{ij}N_j = F_i \ , \ \text{on } \Gamma_1 \times [0,\infty) \tag{3.4}$$

$$s_N \leq 0 \ , \ u_N \leq 0 \ , \ s_N u_N = 0 \ , \ \text{on } \Gamma_2 \times [0,\infty) \tag{3.5}$$

$$-\ \dot{\boldsymbol{u}}_t \in \partial I_{K(x,s_N))}(\boldsymbol{s}_t) \tag{3.6}$$

$$\boldsymbol{u}(x,0) = u_0(x) \ , \ \boldsymbol{s}(x,0) = \boldsymbol{s}_0(x) \ . \tag{3.7}$$

Let us pass to the variational formulation of the above evolution problem. We set

$$V_0 = \{\boldsymbol{v} \in V | \boldsymbol{v} = 0 \ \text{ on } \ \Gamma_0\} \ , \tag{3.8}$$

where V is defined either by (2.14) or (2.14a).

Problem P^f

Find $\{\boldsymbol{s}, \boldsymbol{u}\} : [0,\infty) \to \mathsf{K} \times V_0$ such that

$$(\dot{\boldsymbol{s}}\ (t), \boldsymbol{\tau} - \boldsymbol{s}(t)) - < \boldsymbol{e}(\dot{\boldsymbol{u}}\ (t)), \boldsymbol{\tau} - \boldsymbol{s}(t) > \geq 0 \ , \ \forall \boldsymbol{\tau} \in \mathsf{K} \ , \tag{3.9}$$

$$< \boldsymbol{s}(t), \boldsymbol{e}(\boldsymbol{v} - \dot{\boldsymbol{u}}\ (t)) > + \int_{\Gamma_2} D_f(x, s_N(t), -\boldsymbol{v}_t) d\Gamma - \int_{\Gamma_2} D_f(x, s_N(t), -\ \dot{\boldsymbol{u}}_t\ (t)) d\Gamma -$$

$$- \int_{\Gamma_2} s_N(t)(v_N - \dot{\boldsymbol{u}}\ (t)) d\Gamma \geq \mathsf{L}(\boldsymbol{v} - \dot{\boldsymbol{u}}\ (t)) \ , \ \forall \boldsymbol{v} \in V_0 \ , \tag{3.10}$$

$$\int_{\Gamma_2} s_N(t)(z - u_N(t)) d\Gamma \geq 0 \qquad \forall z \in K_N = \{w \in L^1(\Gamma_2) | w \leq 0\} \tag{3.11}$$

$$\boldsymbol{s}(0) = \boldsymbol{s}_0 \ , \ \boldsymbol{u}(0) = \boldsymbol{u}_0 \ . \tag{3.12}$$

The variational inequality (3.9) is the global form of the constitutive relationship (3.2). The implicit variational inequality (3.10) is obtained by multiplying (3.1) by $\boldsymbol{e}(\boldsymbol{v} - \dot{\boldsymbol{u}}(t))$

and the integration over Ω, and then taking into account the relation equivalent to (3.6):

$$D_f(x, s_N(t), -\boldsymbol{v}_t) - D_f(x, s_N(t), -\dot{\boldsymbol{u}}_t(t)) \geq (-\boldsymbol{s}_t(t)) \cdot (\boldsymbol{v}_t - \dot{\boldsymbol{u}}_t(t)) . \tag{3.13}$$

The last inequality results from the subdifferentiability of the convex function $D_f(x, s_N, \cdot)$.

The variational inequality (3.11) represents the global form of Signorini's conditions (3.5).

The problem P^f seems to be completely intact in the available literature. Nevertheless, we shall now attempt to examine the shakedown problem in the presence of the contact with friction.

3.2 Prior to discussing Melan theorem in the presence of Signorini's conditions with friction, it is instructive to examine the simpler case when normal stresses are prescribed on Γ_2 . Assume that

$$s_N(x,t) = p(x,t) \ , \ x \in \Gamma_2 \ , \ t \geq 0 \ , \tag{3.14}$$

where the function p is prescribed.

The shakedown criterion is assumed in the form

$$W_p(t) + \int_0^t \int_{\Gamma_2} D_f[x, p(x,\tau), -\dot{\boldsymbol{u}}_t(x,\tau)] d\Gamma d\tau < +\infty \tag{3.15}$$

for each $t \geq 0$.

We can now extend Melan's theorem.

Theorem 3.1. If there exists a field of residual stresses $\boldsymbol{\rho}^*(x)(x \in \Omega)$ and $m > 1$ such that

$$\boldsymbol{\rho}^*(x) + m\boldsymbol{s}^e(x,t) \in C(x) \ , \ x \in \Omega \ , \ t \geq 0 \tag{3.16}$$

$$\boldsymbol{\rho}_t^*(x) + m\boldsymbol{s}_t^e(x,t) \in K(x, p(x,t))) \ , \ x \in \Gamma_2 \ , \ t \geq 0 \tag{3.17}$$

$$\boldsymbol{\rho}_N^*(x) = 0 \ , \ x \in \Gamma_2 \ , \tag{3.18}$$

then shakedown occurs.

Proof. Performing similar calculations as in subsection 2.2 we obtain the principle of virtual work rate:

$$\begin{aligned} &\int_\Omega [\rho_{ij}(x,t) - \frac{1}{m}\rho_{ij}^*(x)] a_{ijkl}(x) \dot{\rho}_{kl}(x,t) dx + \\ &+ \int_\Omega [s_{ij}(x,t) - \frac{1}{m} s_{ij}^*(x,t)] d_{ij}^p(x,t) dx = \\ &= \int_{\Gamma_2} [\boldsymbol{\rho}_t(x,t) - \frac{1}{m}\boldsymbol{\rho}_t^*(x)] \cdot \dot{\boldsymbol{u}}_t^r(x,t) d\Gamma \ , \end{aligned} \tag{3.19}$$

because $\rho_N = \rho_N^* = 0$, on Γ_2. The friction law (1.7) is a counterpart of the flow law of rigid perfectly plastic solids. A straightforward application of the principle (3.19) does not lead to the sought estimate. An elastic regularization of the law (1.7) is needed. Let $\varepsilon > 0$ be a small parameter. For a fixed ε we write

$$\boldsymbol{u}_t^\varepsilon = \boldsymbol{u}_t^{e\varepsilon} + \boldsymbol{u}_t^{r\varepsilon} = \hat{\boldsymbol{u}}_t^{e\varepsilon} + \boldsymbol{u}_t^{p\varepsilon} \ , \text{ on } \Gamma_2 \ , \tag{3.20}$$

and assume that

$$\boldsymbol{u}_t^{e\varepsilon} = -\varepsilon \boldsymbol{A} \boldsymbol{s}_t^{e\varepsilon} \ , \tag{3.21}$$

$$-\overset{\bullet}{\boldsymbol{u}}{}_t^{p\varepsilon} \in \partial I_{K(x,p)}(\boldsymbol{s}_t^\varepsilon) \ . \tag{3.22}$$

Here

$$\hat{\boldsymbol{u}}_t^{e\varepsilon} = \boldsymbol{u}_t^{e\varepsilon} - \varepsilon \boldsymbol{A} \boldsymbol{\rho}_t^\varepsilon \ , \tag{3.23}$$

and $\boldsymbol{A}$ is a positive definite matrix which does not depend on ε. Thus we have

$$\boldsymbol{u}_t^\varepsilon = -\varepsilon \boldsymbol{A} \boldsymbol{s}_t^{e\varepsilon} - \varepsilon \boldsymbol{A} \boldsymbol{\rho}_t^\varepsilon + \boldsymbol{u}_t^{p\varepsilon} \ . \tag{3.24}$$

Consequently

$$\left.\begin{array}{r} \varepsilon \boldsymbol{A} \boldsymbol{s}_t^{e\varepsilon} + \varepsilon \boldsymbol{A} \boldsymbol{\rho}_t^\varepsilon \to 0 \\ \\ \boldsymbol{u}_t^{p\varepsilon} \to \boldsymbol{u}_t \end{array}\right\} \text{ when } \varepsilon \to 0 \ , \tag{3.25}$$

provided that the sequences $\{\boldsymbol{s}_t^{e\varepsilon}\}_{\varepsilon>0}$, $\{\boldsymbol{\rho}_t^\varepsilon\}_{\varepsilon>0}$ are bounded. Obviously, in (3.25) the convergence holds for convergent subsequences. Let us make the elastic and residual stress problems more precise for a fixed $\varepsilon > 0$. For the sake of simplicity we assume $\boldsymbol{u}_0 = 0$, $\boldsymbol{s}_0 = 0$.

Elastic problem

$$s_{ij}^{e\varepsilon}(\boldsymbol{e}(\boldsymbol{u}^{e\varepsilon})) + f_i = 0 \ , \text{ in } \Omega \times [0, \infty) \tag{3.26}$$

$$s_{ij}^{e\varepsilon} = b_{ijkl} e_{kl}(\boldsymbol{u}^{e\varepsilon}) \ , \ \boldsymbol{b} = \boldsymbol{a}^{-1} \tag{3.27}$$

$$\boldsymbol{u}^{e\varepsilon} = 0 \ , \text{ on } \Gamma_0 \times [0, \infty) \tag{3.28}$$

$$s_{ij}^{e\varepsilon} N_j = F_i \ , \text{ on } \Gamma_1 \times [0, \infty) \tag{3.29}$$

$$\left.\begin{array}{r} s_N^{e\varepsilon} = p \\ \\ -\boldsymbol{u}_t^{e\varepsilon} = \varepsilon \boldsymbol{A} \boldsymbol{s}_t^{e\varepsilon} \end{array}\right\} \text{ on } \Gamma_2 \times [0, \infty) \ . \tag{3.30}$$

Residual stress problem

$$\rho_{ij,j}^\varepsilon = 0 \ , \text{ in } \Omega \times [0, \infty) \tag{3.31}$$

$$\rho_{ij,j}^\varepsilon = 0 \ , \text{ on } \Gamma_1 \times [0, \infty) \tag{3.32}$$

$$\boldsymbol{s}_t^{e\varepsilon} + \boldsymbol{\rho}_t^\varepsilon \in C(x) \ , \ x \in \Omega \ , \ t \geq 0 \tag{3.33}$$

$$\left.\begin{array}{r}\rho_N^\varepsilon = 0 \\ \boldsymbol{\rho}_t^\varepsilon + \boldsymbol{s}_t^{e\varepsilon} \in K(x,p)\end{array}\right\} \text{ on } \Gamma_2 \times [0,\infty) \ . \tag{3.34}$$

By introducing the bilinear form

$$a(\boldsymbol{u},\boldsymbol{v}) = \int_\Omega b_{ijkl}(x)e_{ij}(\boldsymbol{u})e_{kl}(\boldsymbol{v})dx \ , \tag{3.35}$$

the variational formulation of the elastic problem reads:

Find $\boldsymbol{u}^{e\varepsilon} \in H^1(\Omega)^3$, $\boldsymbol{u}^{e\varepsilon} = 0$ on Γ_0, such that

$$a(\boldsymbol{u}^{e\varepsilon},\boldsymbol{v}) + \frac{1}{\varepsilon}\int_{\Gamma_2}(\boldsymbol{u}_t^{e\varepsilon})^T\boldsymbol{B}\boldsymbol{v}d\Gamma = \mathsf{L}(\boldsymbol{v}) + \int_{\Gamma_2} pv_N d\Gamma \ , \tag{3.36}$$

for each $\boldsymbol{v} \in H^1(\Omega)^3$, $\boldsymbol{v} = 0$ on Γ_0, where $\boldsymbol{B} = \boldsymbol{A}^{-1}$ and L is given by (2.10). For each $t \geq 0$, $\boldsymbol{u}^{e\varepsilon}(\cdot,t)$ exists and is unique; consequently $\boldsymbol{s}^{e\varepsilon}(\cdot,t)$ is uniquely determined.

Setting $\boldsymbol{v} = \boldsymbol{u}^{e\varepsilon}$ in (3.26) we conclude

$$\|\boldsymbol{u}^{e\varepsilon}\|_{L^2(\Gamma_2)} \leq c\sqrt{\varepsilon} \ ,$$

where $c > 0$ is a constant independent of ε . Hence $\boldsymbol{u}^{e\varepsilon}(\cdot,t) \to 0$ a.e. $x \in \Gamma_2$ when $\varepsilon \to 0$.

The shakedown criterion (3.15) has now to be understood as being the limit, when $\varepsilon \to 0$ i.e.:

$$\begin{aligned} W_p(t) + \int_0^t\int_{\Gamma_2} D_f(x,p(x,\tau)),-\dot{\boldsymbol{u}}_t\,(x,\tau)]d\Gamma d\tau = \\ = \lim_{\varepsilon\to 0}\Big[\int_0^t\int_\Omega D(x,\boldsymbol{d}^{p\varepsilon}(x,\tau))dxd\tau + \\ + \int_0^t\int_{\Gamma_2} D_f(x,p(x,\tau),-\dot{\boldsymbol{u}}_t^{p\varepsilon}(x,\tau))d\Gamma d\tau\Big] < +\infty \ . \end{aligned} \tag{3.37}$$

If, for $\varepsilon > 0$ sufficiently small, $\boldsymbol{\rho}^* + m\boldsymbol{s}^{e\varepsilon} \notin C(x)$ and /or $\boldsymbol{\rho}_t^* + m\boldsymbol{s}_t^{e\varepsilon} \notin K(x,p)$, we construct a sequence $\boldsymbol{\rho}^{*\varepsilon} \to \boldsymbol{\rho}^*$, such that $\rho_N^{*\varepsilon}(x) = 0$ on Γ_2 and

$$\boldsymbol{s}^{*\varepsilon}(x,t) = \boldsymbol{\rho}^{*\varepsilon}(x) + m\boldsymbol{s}^{e\varepsilon}(x,t) \in C(x) \ , \ x \in \Omega \ , \ t \geq 0 \ ,$$

$$\boldsymbol{s}_t^{*\varepsilon}(x,t) \in K(x,p(x,t)) \ , \ x \in \Gamma_2 \ , \ t \geq 0 \ .$$

Since

$$\boldsymbol{s}^\varepsilon(x,t) = \boldsymbol{\rho}^\varepsilon(x,t) + \boldsymbol{s}^{e\varepsilon}(x,t) \ ,$$

hence

$$\boldsymbol{s}^\varepsilon(x,t) - \frac{1}{m}\boldsymbol{s}^{*\varepsilon}(x,t) = \boldsymbol{\rho}^\varepsilon(x,t) - \frac{1}{m}\boldsymbol{\rho}^{*\varepsilon}(x) \ . \tag{3.38}$$

Proceeding similarly as in the derivation of (2.36) we obtain

$$[s_{ij}^{\varepsilon}(x,t)-\frac{1}{m}s_{ij}^{*\varepsilon}(x,t)]d_{ij}^{p\varepsilon}(x,t)\geq\frac{m-1}{m}s_{ij}^{\varepsilon}(x,t)d_{ij}^{p\varepsilon}(x,t)\ , \tag{3.39}$$

$$[\boldsymbol{s}_t^{\varepsilon}(x,t)-\frac{1}{m}\boldsymbol{s}_t^{*\varepsilon}(x,t)]\cdot\dot{\boldsymbol{u}}_t{}^{p\varepsilon}(x,t)\geq\frac{m-1}{m}\boldsymbol{s}_t^{\varepsilon}(x,t)\cdot\dot{\boldsymbol{u}}_t{}^{p\varepsilon}(x,t)\ . \tag{3.40}$$

In (3.39) $x\in\Omega$, while in (3.40) $x\in\Gamma_2$. By adding (3.39) to (3.40), after integration and using the principle (3.19) for a fixed $\varepsilon>0$, we have

$$\begin{aligned}
&\frac{m-1}{m}\int_0^t\left[\int_\Omega D(x,\boldsymbol{d}^{p\varepsilon}(x,\tau))dx+\int_{\Gamma_2}D_f(x,p(x,\tau),-\dot{\boldsymbol{u}}_t^{p\varepsilon}(x,\tau))d\Gamma\right]d\tau\leq\\
&\leq\int_0^t\left\{\int_\Omega[\boldsymbol{s}^{\varepsilon}(x,\tau)-\frac{1}{m}\boldsymbol{s}^{*\varepsilon}(x,\tau)]:\boldsymbol{d}^{p\varepsilon}(x,\tau)dx+\right.\\
&\left.+\int_{\Gamma_2}[\boldsymbol{s}_t^{\varepsilon}(x,\tau)-\frac{1}{m}\boldsymbol{s}_t^{*\varepsilon}(x,\tau)]\cdot(-\dot{\boldsymbol{u}}^{p\varepsilon}(x,\tau))d\Gamma\right\}d\tau\leq\\
&\leq-\int_0^t\int_\Omega a_{ijkl}(x)(\rho_{ij}^{\varepsilon}(x,\tau)-\frac{1}{m}\rho_{ij}^{*\varepsilon}(x))\,\dot{\rho}_{kl}^{\varepsilon}(x,\tau)dxd\tau+\\
&+\int_0^t\int_{\Gamma_2}[\boldsymbol{\rho}_t^{\varepsilon}(x,\tau)-\frac{1}{m}\boldsymbol{\rho}_t^{*\varepsilon}(x)]\cdot\dot{\boldsymbol{u}}_t^{r\varepsilon}(x,\tau)d\Gamma d\tau+\\
&+\int_0^t\int_{\Gamma_2}[\boldsymbol{\rho}_t^{\varepsilon}(x,\tau)-\frac{1}{m}\boldsymbol{\rho}_t^{*\varepsilon}(x)]\cdot(-\dot{\boldsymbol{u}}_t^{p\varepsilon}(x,\tau))d\Gamma d\tau=\\
&=-\int_0^t\frac{1}{2}\frac{d}{d\tau}\|\boldsymbol{\rho}^{\varepsilon}(\tau)-\frac{1}{m}\boldsymbol{\rho}^{*\varepsilon}\|_a^2d\tau-\varepsilon\int_0^t\frac{1}{2}\frac{d}{d\tau}\|\boldsymbol{\rho}_t^{\varepsilon}(\tau)-\frac{1}{m}\boldsymbol{\rho}_t^{*\varepsilon}\|_A^2d\tau\ ,
\end{aligned}$$

since $\dot{\boldsymbol{u}}_t^{r\varepsilon}-\dot{\boldsymbol{u}}_t^{p\varepsilon}=-\varepsilon\boldsymbol{A}\boldsymbol{\rho}_t^{\varepsilon}$; here

$$\|\boldsymbol{\rho}_t^{\varepsilon}(\tau)-\frac{1}{m}\boldsymbol{\rho}_t^{*\varepsilon}\|_A^2=\int_{\Gamma_2}[\boldsymbol{\rho}_t^{\varepsilon}(x,\tau)-\frac{1}{m}\boldsymbol{\rho}_t^{*\varepsilon}(x)]^T\boldsymbol{A}[\boldsymbol{\rho}_t^{\varepsilon}(x,\tau)-\frac{1}{m}\boldsymbol{\rho}_t^{*\varepsilon}(x)]d\Gamma\ .$$

Thus we arrive at the estimation

$$\begin{aligned}
&\frac{m-1}{m}\int_0^t\left[\int_\Omega D(x,\boldsymbol{d}^{p\varepsilon}(x,\tau))dx+\int_{\Gamma_2}D_f(x,p(x,\tau),-\dot{\boldsymbol{u}}_t^{p\varepsilon}(x,\tau))d\Gamma\right]d\tau\leq\\
&\leq\frac{1}{2}\int_\Omega a_{ijkl}(x)[\rho_{ij}^{\varepsilon}(x,0)-\frac{1}{m}\rho_{ij}^{*\varepsilon}(x)][\rho_{kl}^{\varepsilon}(x,0)-\frac{1}{m}\rho_{kl}^{*\varepsilon}(x)]dx+\\
&+\frac{\varepsilon}{2}\int_{\Gamma_2}[\boldsymbol{\rho}_t^{\varepsilon}(x,0)-\frac{1}{m}\boldsymbol{\rho}_t^{*\varepsilon}(x)]^T\boldsymbol{A}[\boldsymbol{\rho}_t^{\varepsilon}(x,0)-\frac{1}{m}\boldsymbol{\rho}_t^{*\varepsilon}(x)]d\Gamma\ .
\end{aligned} \tag{3.41}$$

The desired result is obtained in the limit when $\varepsilon\to0$, provided that a subsequence exists, still denoted by $\boldsymbol{\rho}^{\varepsilon}$ such that $\boldsymbol{\rho}^{\varepsilon}\to\boldsymbol{\rho}$, in a suitable topology.

Remark 3.1

It is possible to weaken the condition (3.17) by assuming that independently of $\boldsymbol{\rho}^*$, there exist tangential contact residual stresses $\boldsymbol{\xi}(x)(x \in \Gamma_2)$ and $m_1 > 1$, such that

$$\boldsymbol{\xi} + m_1 \boldsymbol{s}_t^e(x,t) \in K(x, p(x,t)) \ , \ x \in \Gamma_2 \ , \ t \geq 0 \ . \tag{3.17a}$$

3.3. In the general case, when the contact conditions are given by (3.5) and (3.6), we propose the following approach, which should be treated rather as a conjecture. It seems natural to assume the shakedown criterion in the form

$$W_p(t) + \int_0^t \int_{\Gamma_2} D_f[x, s_N(x,\tau), -\dot{\boldsymbol{u}}_t(x,\tau)] d\Gamma d\tau < +\infty \ , \tag{3.42}$$

where $t \to \infty$. A plausible version of Melan's theorem which takes into account Signorini's conditions with friction, is formulated as follows:

"If there exist a field of residual stresses $\boldsymbol{\rho}^*(x)(x \in \Omega)$ and $m > 1$ such that

$$\boldsymbol{\rho}^*(x) + m\boldsymbol{s}^e(x,t) \in C(x) \ , \ x \in \Omega \ , \ t \geq 0 \tag{3.43}$$

$$\boldsymbol{\rho}_t^*(x) + m\boldsymbol{s}_t^e(x,t) \in K(x, s_N(x,t)) \ , \ x \in \Gamma_2 \ , \ t \geq 0 \tag{3.44}$$

$$\rho_N^*(x) \leq 0 \ , \ x \in \Gamma_2 \ , \tag{3.45}$$

then shakedown occurs".

The principle of virtual work rate has now the form

$$\begin{gathered}\int_\Omega a_{ijkl}(x)[\rho_{ij}(x,t) - \frac{1}{m}\rho_{ij}^*(x)] \, \dot{\rho}_{kl}(x,t)dx + \\ + \int_\Omega [s_{ij}(x,t) - \frac{1}{m}s_{ij}^*(x,t)] d_{ij}^p(x,t)dx = \\ = \int_{\Gamma_2} [\rho_N(x,t) - \frac{1}{m}\rho_N^*(x)] \, \dot{u}_N^r d\Gamma + \int_{\Gamma_2} [\boldsymbol{\rho}_t(x,t) - \frac{1}{m}\boldsymbol{\rho}_t^*(x)] \cdot \dot{\boldsymbol{u}}_t^r(x,t) d\Gamma \ .\end{gathered} \tag{3.46}$$

One can now envisage an iterative method for assessing shakedown:
(i) Step n : the normal stresses occuring in D_f are taken from the step $(n-1)$. By combining the developments of subsections 2.2 and 3.2 we conclude

$$\int_0^t \int_\Omega D(x, \boldsymbol{d}^{p(n)}(x,\tau)) dx d\tau + \int_0^t \int_{\Gamma_2} D_f(x, s_N^{(n-1)}(x,\tau), -\dot{\boldsymbol{u}}_t^{(n)}(x,\tau)) d\Gamma d\tau < +\infty \ ,$$

for $n = 1, 2, \ldots$.
The basic open problem now is to prove that a subsequence $s_N^{(n_k)}$ exists, such that $s_N^{(n_k)} \to s_N$ when $n_k \to \infty$ (in a suitable topology), where $s_N \leq 0$ stands for the actual normal stresses on Γ_2. If the answer is positive, then shakedown occurs. We think that even at this stage it would be interesting to implement the proposed iterative method.

Remark 3.2.
The condition (3.45) can be weakened by assuming that $\boldsymbol{\xi}(x)(x \in \Gamma_2)$ and $m_1 > 1$ exist such that

$$\boldsymbol{\xi} + m_1 \boldsymbol{s}_t^e(x,t) \in K(x, s_N(x,t)) \ , \ x \in \Gamma_2 \ , \ t \geq 0 \tag{3.45a}$$

Acknowledgement. The author was supported by the State Committee for Scientific Research, grant No. 309389101.

REFERENCES

1970 ROCKAFELLAR, R.T., Convex Analysis, Princeton University Press, Princeton.

1972 TELEGA, J.J., Perfectly plastic media acted upon by multi-parameter loads, Eng. Transactions, **20**, 241, in Polish.

1976 DÉBORDES, O., NAYROLES, B., Sur la théorie et le calcul à l'adaptation des structures élastoplastiques, J. de Méc., **15**, 1.

1977 MERCIER, B., Sur la théorie et l'analyse numérique des problèmes de plasticité, Thèse, Université Pierre et Marie Curie, Paris 6.

1980 WESFREID, E., Étude du comportement asymptotique pour quelques modèles de viscoplasticité, Thèse de 3-ème cycle, Université Pierre et Marie Curie, Paris 6.

1981 RYDHOLM, G., On inequalities and shakedown in contact problems, Linköping Studies in Science and Technology, Dissertations No 61, Linköping.

1985 JOHNSON, K.L., Contact Mechanics, CUP, Cambridge .

1985 TELEGA, J.J., Limit analysis theorems in the case of Signorini's boundary conditions and friction, Arch. Mech., **37**, 549.

1985 TEMAM, R., Mathematical Problems in Plasticity, Gauthier – Villars, Paris, Trans- -Inter–Scientia, Tonbridge.

1987 HALPHEN, B., SALENÇON, J., Élastoplasticité, Presses de l' École Nationale des Ponts et Chaussées, Paris .

1987 KÖNIG, J. A., Shakedown of Elastic-Plastic Structures, PWN-Elsevier, Warszawa- -Amsterdam.

1987 TELEGA, J.J., Variational methods in contact problems of mechanics, Uspekhi Mekhaniki (Adv. in Mech.), **10**, 3, in Russian.

1988 BOUCHITTÉ, G., VALADIER, M., Integral representation of convex functionals on a space of measures, J. Funct. Analysis, **80**, 398.

1988 FELDER, E., Étude expérimentale de l'anisotropie de frottement de tôles d'acier doux, J. de Méc. théorique et appl., **7**, 479.

1988 SUQUET, P.M., Discontinuities and plasticity in: Nonsmooth Mechanics and Applications, ed. by J.J. Moreau and P.D. Panagiotopoulos, Springer – Verlag, Wien – New York, p. 278.

1988 TELEGA, J.J., Topics on unilateral contact problems of elasticity and inelasticity, in: Nonsmooth Mechanics and Applications, ed. by J.J. Moreau and P.D. Panagiotopoulos, Springer – Verlag, Wien – New York, p. 341.

1989 ZMITROWICZ, A., On the thermodynamics of contact friction and wear, ZN Instytutu Maszyn Przepływowych PAN, Studia i Materiały, 287/12/89, Gdańsk.

1990 KADIJK, S.E., BROESSE VAN GROENOU, A., Wear anisotropy of MnZn ferrite. Part I: Recorder and sphere-on-tape experiments, Wear, **139**, 93; Part II: Sliding sphere experiments ibid., p. 115; Part III: A wear model, ibid., p. 133.

1990 REDDY, B.D., TOMARELLI, F., The obstacle problem for an elastoplastic body, Appl. Math. Optim., **21**, 89.

1990 TELEGA, J.J., Variational methods and convex analysis in contact problems and homogenization, IFTR Reports 38/90, Warszawa, in Polish.

1990 WEICHERT, D., Advances in the geometrically nonlinear shakedown theory, In: Inelastic Solids and Structures, ed. by M. Kleiber and J.A. König, Pineridge Press, Swansea, p. 489.

1991 ARIKAN, R, MURPHY, S., Anisotropic wear of planar-random metal matrix composites with zinc alloy matrix, Wear, **143**, 149.

1991 COMI, C., CORIGLIANO, A., Dynamic shakedown in elastoplastic structures with general internal variable constitutive laws, Int. J. Plasticity, **7**, , 679.

1991 NAYEB–HASHEMI, H., BLUCHER, J.T., MIRAGEAS, J., Friction and wear behavior of aluminium- -graphite composites as a function of interface and fiber direction, Wear, **150**, 21.

1992 JOHNSON, K.L., SHERCLIFF, H.R., Shakedown of 2-dimensional asperities in sliding contact, Int. J. Mech. Sci. **34**, 375.

1992 NORDGREN, R.P., Limit analysis of a stochastically inhomogenenous plastic medium with application to plane contact problems, J. Appl. Mech. **59**, 477.

1993 NAYROLES, B., WEICHERT, D., La notion de sanctuaire d' élasticité et d'adaptation des structures, C.R. Acad. Sci. Paris, Série II, **316**, 1493.

1993 POLIZZOTTO, C., On the conditions to prevent plastic shakedown of structures: Part I - Theory, J.Appl. Mech., **60**, 15; Part II - The plastic shakedown limit load, ibid., p. 20.

1994 HE, Q.-C., CURNIER, A., Anisotropic dry friction between two orthotropic surfaces undergoing large displacements, Eur. J. Mechanics, in print.

ON THEOREMS OF ADAPTATION OF ELASTIC–PLASTIC STRUCTURES

Jan SACZUK
Institute of Fluid–Flow Machinery
Polish Academy of Sciences
ul. J. Fiszera 14, PL–80–952 Gdańsk, Poland

Abstract: Based on a new consistent internal–variable theory of elasto–plasticity, the author's idea of adaptation (SACZUK [1992]) is generalized to non–linear problems for elastic–plastic structures. The underlying theory of elastic–plastic behaviour of materials in which, among others, no yield rule and intermediate configuration are assumed to exist, where the transition from micro- to macroscales is natural, and where the constitutive relations do not need the so–called loading criteria, is modelled by a metric generalization of the Riemannian geometry. It is used to reformulate the known statical approaches to path–dependent adaptation. The new adaptation theorems proposed, which have no counterparts in the available literature, are generalization of known versions to the finite shakedown theorems.

INTRODUCTION

One can argue that if we have a complete knowledge of a given physical system, its behaviour in space–time must always be predictable, perhaps, with the aid of sufficiently powerful computers. In practice this argument must be tempered due to the fact that the most probable behaviour of real processes is some average behaviour. For this and other reasons it is useful to have a mathematical model of the physical situation that leads to a reasonably accurate prediction of the long–term average behaviour of a given process. For that reason, the theory should produce good estimations of the average behaviour of real processes in an elastic–plastic structure.

When a given elastic–plastic structure is acted upon by loadings which vary with time in a non–proportional way, and if there are prescribed limits to the loads but the loading history is not given explicitly, then the classical limit analysis theory can sometimes give unsafe estimates of collapse loads since "incremental collapse" or low cycle fatigue can occur below the calculated load–carrying capacity of the structure

Z. Mróz et al. (eds.), Inelastic Behaviour of Structures under Variable Loads, 203–218.

(KOITER [1960]). Both phenomena of incremental collapse and alternating plasticity are collectively known as inadaptation.

In many practical applications, the loading and the initial state of a structure are not known with sufficient accuracy. In these cases, the knowledge of the whole evolution of the structure is of only limited interest. It may be interesting to know the asymptotic behaviour of the structure whatever its initial conditions. The answer to this question can be extracted to some degree from shakedown theory. The Shakedown theory provides a criterion of failure which may be considered as a more realistic basis for design than that of the limit analysis which assumes the failure to occur when a critical elastic stress is attained. This analysis is crucial for the assessment of structural behaviour under varying loads within the range of time–independent plasticity.

Among the earliest speculations on the foundations of shakedown theory, essentially based on the classic Melan and Koiter theorems, we find many attempts to use simple facts, mainly of yield rule properties, in the plasticity theory (KÖNIG [1982]), Although MELAN [1936] was the first to study the shakedown theorem, both his and his successors'contributions (HODGE [1959], KOITER [1960], KÖNIG [1982,1987], WEICHERT [1986], GROSS–WEEGE [1990], SACZUK and STUMPF [1990]) to the foundations of adaptation of structures are slightly different from the material presented here.

In the author's opinion, the present state of the shakedown formalism is too simple to grasp some salient features of elastic–plastic phenomena and to be helpful to a great extent in practice. Of course, agreement of the shakedown theory with experimental data has been found in certain cases (KÖNIG [1987]). On the other hand, the problem of adaptation is one of the more important in engineering activity. It is connected with the fact that the safety problem of elasto–plastic structures subjected to variable loads, traditionally identified with the boundedness of plastic work in time, is one of the major problems of structural design. While substantial progress has been made in the study of inelastic behaviour of structures under variable loads, their non–linear shakedown analysis is still at an early stage of development.

The present paper is intended to fill in this gap at least partially, and to offer the background for finding (approximate) solutions of shakedown problems, which is of great practical significance for engineering technology. We shall be primarily concerned with the mechanical side of the problem.

In the paper, we shall frequently use notation, conventions and terminology for differential theory of a Finsler space based on MATSUMOTO's [1986] and RUND's [1959] monographs. The continuum mechanics notation will be, in general, retained (TRUESDELL and NOLL [1965]). The main difference consists in interchanging the notation for the reference configuration (lower–case letters) and for a current configuration (upper–case letters).

A FINSLERIAN BACKGROUND OF THE ELASTO–PLASTICITY

The complex behaviour of a material is inevitably connected with the complexity of its microstructural rearrangement. Any advance in modelling should rely on a better understanding and schematization of the microstructural evolution. The best way towards a satisfactory compromise between realism, accuracy and manageability seems to be the use of models of material behaviour based on a Finlser–type elasto–plasticity (SACZUK [1993a,1993b]). A reader who wishes to get acquainted with Finsler geometry should consult RUND's [1959] and MATSUMOTO's [1986] monographs mentioned earlier.

The objective of this section is to present all the important concepts of a new internal theory of elasto–plasticity that need be known for a thorough understanding of our methodology of shakedown analysis.

Our version of elasto–plasticity is based on the following main premises:

1. The geometric nature of elasto–plastic phenomenon is non–Riemannian (locally non–Euclidean).
2. A Minkowskian geometry is categorical and consistent, just as Euclideanis.
3. The mathematical definition of the object under study should correspond as far as possible to its physical nature.
4. The underlying space for the elastic–plastic phenomenon has the nature of tangent bundle.

In the formulation presented below, material particles of the given body $\mathcal{B}$ are geometrically identified with the line–element of a Finsler space (SACZUK [1993a,1993b]). Due to the fact that the geometric nature of elasto–plasticity is generally non–Riemannian, the standard kinematic assumption, i.e. the classic multiplicative decomposition of the deformation gradient used in present–day elasto–plasticity, especially its geometro–physical aspects, is inconsistent with the nature of elasto- -plastic phenomenon (with the irreversible character of elastic–plastic deformations). Hence, the status of the multiplicative decomposition of the deformation gradient which is widely accepted in the literature is rather questionable (cf. its limitations cited in SACZUK [1993a]). It is also inadequate to grasp the salient features, like the hysteresis effect, of the finite elasto–plasticity. It also turns out that there is no simple superposition of the elastic and plastic constituents of finite elasto–plastic deformation.

In the classic continuum model for materials, a body is a set $\mathcal{B}$ an element x of which is called a particle or a material particle. In the sequel, we will identify the body $\mathcal{B}$ with an element of a set $\{\mathcal{B}_\chi : \chi \in \Upsilon\}$, where $\mathcal{B}$ is the (connected) oriented differential 3–dimensional manifold and $\chi \in \Upsilon$ is a diffeomorphism of $\mathcal{B}$ into R^3.

The family Υ of diffeomorphisms is called a family of configurations of the body $\mathcal{B}$. In practice, we usually identify each particle x with its position $\mathbf{x}$ in a fixed reference configuration $\kappa_0 : \mathcal{B} \to R^3$ and write

$$\mathbf{X} = \chi(\mathbf{x}) \tag{1}$$

where $\chi \equiv \kappa \circ \kappa_0$ denotes the deformation from $\kappa_0 \in \Upsilon$ to $\kappa \in \Upsilon$.

The geometrical relation (1) can be defined as follows within the Finsler formalism. Using the fact that an n–dimensional Finsler space may always be regarded as a (non–holonomic) subspace of $2n$–dimensional Riemannian space (RUND [1959], p. 251) by the diagonal operation $\Delta : \mathbf{X} \mapsto (\mathbf{X}, \mathbf{X})$, we can write the relation (1) as

$$\mathbf{X} = \hat{\chi}(\tilde{\mathbf{x}}). \tag{2}$$

Here $\tilde{\mathbf{x}} = (\mathbf{x}, \mathbf{u})$ = (a position vector, an internal variable vector) is the line–element of the Finsler space identified with $\mathcal{B}$, i.e. with the Finsler–type body, and the mapping $\hat{\chi}$ defines the deformation of $\mathcal{B}$.

To introduce the concept of a deformation gradient we start from the Finsler space equipped with the so–called Euclidean connection by means of Cartan's method, identified with the Cartan connection in the available literature (cf. RUND [1959], MATSUMOTO [1986]). The Cartan connection, in turn, enables us to define two operators of covariant derivatives ∇^h and ∇^v called h–derivative and v–derivative, respectively.

Under the identity (2), the direct sum of covariant derivatives, in the matrix notation, implies the following definition of the deformation gradient

$$\hat{\mathbf{F}} = \nabla^h\mathbf{X} + \nabla^v\mathbf{X} = (\mathbf{1}\;\;\mathbf{1})\begin{pmatrix} \nabla^h & \mathbf{0} \\ \mathbf{0} & \nabla^v \end{pmatrix}\begin{pmatrix} \mathbf{1} \\ \mathbf{1} \end{pmatrix}(\mathbf{X}) \tag{3}$$

where $\mathbf{1}$ is the identity matrix on $\mathcal{B}$. In continuum theory notation, $\hat{\mathbf{F}}$, termed the total deformation gradient, can be rewritten as

$$\hat{\mathbf{F}} = \mathbf{F}^h + \mathbf{F}^v \tag{4}$$

where

$$\mathbf{F}^h \equiv \nabla^h\mathbf{X} = X^i_{|h}\partial_i \otimes dx^h, \quad \mathbf{F}^v \equiv \nabla^v\mathbf{X} = X^i|_h\, \partial_i \otimes Dl^h \tag{5}$$

are horizontal and vertical parts of $\hat{\mathbf{F}}$, respectively; ∂_i is the unit vector in the current configuration κ of $\mathcal{B}$ and $\otimes$ denotes the tensor product. According to (5), the h–derivative and v–derivative of the position vector $\mathbf{X} = \mathbf{X}(\tilde{\mathbf{x}})$ are defined as follows

$$(\mathbf{F}^h)^i_h = \frac{\partial X^i}{\partial x^h} - \frac{\partial X^i}{\partial u^k}\frac{\partial G^k}{\partial u^h} + \Gamma^{\star i}_{kh} X^k, \tag{6}$$

$$(\mathbf{F}^v)^i_h = L\frac{\partial X^i}{\partial u^h} + A^i_{kh} X^k \tag{7}$$

and $Dl^i = dl^i + \Gamma^i_{kh} l^k dx^x$ is the covariant derivative of the unit vector $l^i = u^i(\tilde{\mathbf{x}})$; L is the fundamental function (an geometrical analogue of the classic yield function), $A^i_{kh}, \partial G^k/\partial u^h, \Gamma^{\star i}_{kh}$ are connection components of dislocated state defined by the fundamental function L, the torsion tensor $C_{ijk} = 1/2(\partial g_{ij}/\partial u^k)$, the metric tensor of the Finsler space $g_{ij} = 1/2(\partial^2 L^2/\partial u^i \partial u^j)$ and its Christoffel symbols of the first kind.

The decomposition (4), contrary to the one used in the classic plasticity, does not demand any artificial assumptions. The components $\mathbf{F}^h$ and $\mathbf{F}^v$ are not strictly elastic and plastic. In other words, the elastic–plastic behaviour of a material cannot be treated as a simple superposition of elastic and plastic constituents.

The fundamental equations of the static (mechanical) problem of elasto–plasticity can be summarized as follows (SACZUK [1993a,1993b]):

1. The balance of (static) linear momentum:

$$Div^h\mathbf{T} + \mathbf{f}^h = \mathbf{0}, \quad Div^v\mathbf{T} + \mathbf{f}^v = \mathbf{0} \tag{8}$$

where

$$\begin{aligned} (Div^h\mathbf{T})_k &= D_i({}_hT^i_k) - {}_hT^i_j\Gamma^{\star j}_{ki}, \\ (Div^v\mathbf{T})_k &= L\dot{D}_i({}_vT^i_k) - \dot{\partial}_i G^j \dot{D}_j({}_hT^i_k) - {}_vT^i_j A^j_{ki}; \end{aligned} \tag{9}$$

$$D_i(\cdot) = \partial_i(\cdot) + \frac{\partial(\cdot)}{\partial X^n}X^n_i, \quad \dot{D}_i(\cdot) = \dot{\partial}_i(\cdot) + \frac{\partial(\cdot)}{\partial X^n}Y^n_i, \quad \partial_i(\cdot) = \frac{\partial(\cdot)}{\partial x^i}, \quad \dot{\partial}_i(\cdot) = \frac{\partial(\cdot)}{\partial u^i} \tag{10}$$

and $\mathbf{T}$ is the first Piola–Kirchhoff stress tensor, $\mathbf{f}^h$ is the external body force and $\mathbf{f}^v$ can be identified with the internal source of the exchange of momentum between dislocated states.

2. The balance of (static) angular momentum:

$$\mathbf{T}^h(\mathbf{F}^h)^T = \mathbf{F}^h(\mathbf{T}^h)^T, \quad \mathbf{T}^v(\mathbf{F}^v)^T = \mathbf{F}^v(\mathbf{T}^v)^T \tag{11}$$

where the superscript ^{T} denotes the transpose.

3. The Clausius–Duhem inequalities:

$$\mathbf{T}^h \cdot (\bar{\mathbf{F}}^h - \mathbf{F}^h) \geq 0, \quad \mathbf{T}^v \cdot (\bar{\mathbf{F}}^v - \mathbf{F}^v) \geq 0 \tag{12}$$

and $\hat{\bar{\mathbf{F}}} \equiv (\bar{\mathbf{F}}^h, \bar{\mathbf{F}}^v)$ differ from $\hat{\mathbf{F}} \equiv (\mathbf{F}^h, \mathbf{F}^v)$ by any positive–definite tensor $\hat{\mathbf{W}} \equiv (\mathbf{W}^h, \mathbf{W}^v)$ of order 6.

4. Constitutive equations:

(a) The time–independent version:

$$\hat{\mathbf{T}}_S \equiv (\mathbf{T}^h_S, \mathbf{T}^v_S) = \hat{\mathcal{H}}_S(\mathbf{C}^h, \mathbf{C}^v) \tag{13}$$

where $\hat{\mathbf{T}}_S = \hat{\mathbf{F}}^{-1}\hat{\mathbf{T}}$ is the second Piola–Kirchhoff stress tensor and $\hat{\mathbf{C}} \equiv (\mathbf{C}^h, \mathbf{C}^v) = \hat{\mathbf{F}}^T\hat{\mathbf{F}}$ is the right Cauchy–Green strain tensor;

(b) The time–dependent version:

$$\hat{\mathbf{T}}^{\star}_K = \hat{\mathcal{G}}_K(\hat{\mathbf{T}}_K, \hat{\mathbf{F}}) \tag{14}$$

where $\hat{\mathbf{T}}_K = \hat{\mathbf{T}}\hat{\mathbf{F}}^T$ is the Kirchhoff stress tensor and the superscript * may represent one of the objective rates (cf. ATLURI [1984]).

The functionals $\hat{\mathcal{H}}_S$ and $\hat{\mathcal{G}}_K$ are assumed here to satisfy: (a) the principle of positiveness of entropy production, (b) the equipresence principle, (c) the material objectivity principle, and (d) a material symmetry. The detailed exposition of the restrictions may be found in TRUESDELL and NOLL [1965].

The arguments that support the presented internal variable theory of elasto--plasticity are the following:

1. There is no need to postulate a yield rule; its analogue is the indicatrix of the Finsler space.
2. There is no need to introduce an intermediate (stress–free) configuration in kinematic considerations.
3. There is no need to introduce restrictions as to the nature of a material, its loads, and its deformation process.
4. The internal variables appear in a natural manner in the presented version of elasto–plasticity.
5. The transition from micro- to macroscale is natural.
6. The elastic limit of the mathematical or the physical type appears in a natural manner.
7. The constitutive equations for elasto–plasticity have a consistent form and do not need the so–called loading criteria.
8. The Clausius–Duhem inequality is not required to be postulated, as it appears as the sufficient criterion of Weierstrass for an action integral under consideration.

STATICAL SHAKEDOWN THEOREMS

It is evident that, if a structure has been subjected to variable loads varying between prescribed limits, it is not easy to discover whether a state of the structure can be reached such that all subsequent load applications will produce only elastic changes of stress. If the structure does reach such a state then the structure is said to shake down under the given variable loads. In other words, the structure will shake down if a residual stress field exists such that all further changes of the load components, between their respective extreme values, will produce only elastic changes of stresses and deformation.

The classic shakedown criterion as the necessary condition of structural safety in the case of variable repeated loads is formulated as follows (KÖNIG [1982]):

> A certain domain of load variations is given and the question arises whether a given structure will shake down in an arbitrary sequence of the loads contained within this domain?

In practice, it is assumed that the regions of application of all loads are known and that the loads may vary completely arbitrarily between the prescribed limits.

The above problem can be solved by analyzing possible residual stress fields in the static approach (MELAN [1936], KOITER [1960]), or by considering possible mechanisms of plastic deformations in the kinematic methods (KOITER [1960]). The extensions of classic shakedown technique are discussed by WEICHERT [1986], GROSS–WEEGE [1990], SACZUK and STUMPF [1990] and SACZUK [1992]. Until now, no general (geometrically or/and physically) non–linear shakedown theory, exept for some special cases, has been constructed for which the shakedown limits of loads are path–dependent. In developing the shakedown theory, our strategy will be to begin with the fundamental admissible deformation process, considered within the Finsler–type elasto–plasticity.

The classic case

First, we restrict ourselves to a structure which is described by the continuous mathematical model of the Finlser–type space (a generalization of the configuration space) and where the classic methodology will be used (SACZUK [1993a,1993b]). (For simplicity the superscript ˆ will be omitted.) In this case, the kinematics of the problem is described by the total deformation gradient which is decomposed in such a way that

$$\mathbf{F} = \mathbf{F}^e\mathbf{F}^p \tag{15}$$

where $\mathbf{F}^e$ and $\mathbf{F}^p$ denote the so–called elastic and plastic deformation gradients, respectively. The above decomposition results from the concept of the internal configuration which is obtained by removing the stresses without further plastic deformation.

We adopt the principle of maximum plastic dissipation in the form

$$(\mathbf{T} - \mathbf{T}^\star) \cdot \dot{\mathbf{E}}^p \geq 0. \tag{16}$$

In this inequality, $\dot{\mathbf{E}}^p$ denotes the plastic strain–rate tensor, $\mathbf{T}$ the actual stress tensor, and $\mathbf{T}^\star$ any stress tensor lying on or inside the current yield locus.

In order to properly designate the influence of geometrical and physical non--linearities of the deformation process on the shakedown of the structure, we introduce the following assumptions:

1. The deformation process of the structure is non–singular and quasi–static.
2. There is a potential W for the velocity gradient of displacements.

3. The minimum velocity principle is satisfied along any kinematically admissible deformation path of the elastic–plastic structure.

Let $\mathcal{A}$ denote an optimization problem given as follows:

Find a pair $(\mathbf{T}, \mathbf{v})$ of time–dependent fields of the first Piola–Kirchhoff stress tensor and displacement as a solution to the problem:

maximize

$$\mathbf{T}(\tau) = \mathbf{T}(\tau_o) + \int_{\tau_o}^{\tau} (\mathcal{L} \cdot \nabla \mathbf{w})(\mathbf{x}, \tau) d\tau,$$

$$\mathbf{v}(\tau) = \mathbf{v}(\tau_o) + \int_{\tau_o}^{\tau} \mathbf{w}(\mathbf{x}, \tau) d\tau \tag{17}$$

over all $\mathbf{w} \in \mathcal{W}_v$ subjected (cf. PETRYK [1987]) to:

$$Div(\mathcal{L} \cdot \nabla \mathbf{w}) + \dot{\mathbf{b}} = \mathbf{0} \quad \text{on V},$$

$$(\mathcal{L} \cdot \nabla \mathbf{w})\mathbf{n} = \mathbf{k}[\mathbf{w}] \quad \text{on } S_t,$$

$$\int_V W(\mathbf{w}) dV - \int_{S_t} \mathbf{k}[\mathbf{w}] \cdot \mathbf{w} dS \geq 0, \quad \mathbf{w} \in \mathcal{W}_v, \tag{18}$$

$$\int_V \mathcal{L}[\nabla \mathbf{w}] \cdot \nabla \mathbf{w} dV - \int_{S_t} \mathbf{k}[\mathbf{w}] \cdot \mathbf{w} dS \geq 0, \quad \mathbf{w} \in \mathcal{W}_v$$

Here $\mathcal{L}$ is the tensor of tangent elastic–plastic moduli; $\mathbf{k}$ is a function of displacement velocity, its gradient and position; $\dot{\mathbf{b}}$ is the body force rate vector; $\mathcal{W}_v$ is a class of kinematically admissible velocity fields with zero boundary conditions.

By the class of kinematically admissible velocity fields $\mathcal{W}_v$ over $V \times [\tau_o, \tau)$ is understood a class of vector fields $\mathbf{v}$ satisfying the internal constraints imposed by continuity of the velocity field in V and the zero kinematic boundary conditions on S_v.

The given optimization problem defines suitable conditions to exclude, during a deformation process, such dynamical phenomena as jumps and geometric instabilities, i.e. buckling, necking, distortion, etc.

Let us take $\mathcal{B}$ to denote a class of pairs $(\bar{\mathbf{S}}, \bar{\mathbf{v}})$ of time–independent fields of the second Piola–Kirchhoff stress tensor and displacement such that

$$\bar{\mathbf{T}} = \bar{\mathbf{H}}\mathbf{T} + \bar{\mathbf{F}}\mathbf{F}\bar{\mathbf{S}}, \quad Div(\bar{\mathbf{T}}) = \mathbf{0} \quad \text{in V},$$

$$\bar{\mathbf{T}}\mathbf{n} = \mathbf{0} \quad \text{on } S_t,$$

$$\bar{\mathbf{F}} = \mathbf{1} + \bar{\mathbf{H}}, \quad \bar{\mathbf{H}} = \nabla\bar{\mathbf{v}} \quad \text{in V}, \tag{19}$$

$$\bar{\mathbf{v}} = \mathbf{0} \quad \text{on } S_u$$

for any pair $(\mathbf{T}, \mathbf{v}) \in \mathcal{A}$.

The tensor $\bar{\mathbf{T}}$ is the time–independent residual first Piola–Kirchhoff stress distribution in the structure referred to the reference configuration (the configuration ${}^R C$ in SACZUK and STUMPF [1990]). The residual stresses are defined here as the permanent stresses remaining in the body after unloading. In other words, the residual stresses are caused by the fact that the plastic strains are incompatible.

Now we come to a main result of this subsection.

THEOREM A. If at any point of a deformation path of the structure and for any pair $(\mathbf{T}, \mathbf{v}) \in \mathcal{A}$ there exists a pair $(\bar{\mathbf{S}}, \bar{\mathbf{v}}) \in \mathcal{B}$ such that

$$f\left(\mathbf{S}(\tau) + \bar{\mathbf{S}}, \mathbf{E}(\tau) + \bar{\mathbf{E}}\right) < 0 \tag{20}$$

where f is a non–negative convex scalar function, then the structure will shake down over any programme of loading.

Proof. In order to prove this theorem we first define the following time–dependent quadratic form Π

$$\Pi(\tau) = \frac{1}{2}\int_V \left(\mathbf{D}^{-1}[\tilde{\mathbf{S}} - \bar{\mathbf{S}}] \cdot (\tilde{\mathbf{S}} - \bar{\mathbf{S}}) + [(\tilde{\mathbf{F}} - \bar{\mathbf{H}})\mathbf{F}]^T(\tilde{\mathbf{F}} - \bar{\mathbf{H}})\mathbf{F} \cdot \mathbf{S}\right) dV \tag{21}$$

where the superscripts $\tilde{}$ and $\bar{}$ denote the actual residual state and the shakedown state of the structure, respectively. The form Π is constant when there is no plastic flow and decreases when plastic flow occurs. To show this, one may use standard procedures, see for instance SACZUK and STUMPF [1990]. Here superscripts $\tilde{}$ and $\bar{}$ have different status those in SACZUK and STUMPF [1990].

In the second step, one has to show that the total energy dissipated as required by the shakedown definition is bounded. In practice, it is assumed that the safety factor $\mu = constant > 1$, and

$$f\left[\mu(\mathbf{S}+\bar{\mathbf{S}}), \mathbf{E}+\bar{\mathbf{E}}\right] \leq 0. \tag{22}$$

There is no basic reason for limiting the safety factor to $\mu = constant$. In reality, for any deformation process χ, the inequality (22) implies

$$f\left[\mu(\chi)(\mathbf{S}+\bar{\mathbf{S}}), \mathbf{E}+\bar{\mathbf{E}}\right] \leq 0 \tag{23}$$

for $\mu(\chi) > 1$, i.e. the safety factor is path–dependent. The convexity of the yield surface and the validity of the normality rule (16) implies

$$(\mathbf{S} - \mu(\chi)\mathbf{S}^s) \cdot \dot{\tilde{\mathbf{E}}}^p \geq 0, \tag{24}$$

where $\mathbf{S}^s = \mathbf{S} + \bar{\mathbf{S}}$ and $\mathbf{S}^s$ is on or inside of the yield surface, and $\tilde{\mathbf{E}}^p$ is the actual residual strain calculated from $\tilde{\mathbf{E}}^p = \frac{1}{2}\left((\tilde{\mathbf{F}}^p)^T\tilde{\mathbf{F}}^p - \mathbf{1}\right)$. Then the integral (24), under the definition (21), implies the boundedness of $\int_0^T \int_V \mathbf{S} \cdot \dot{\tilde{\mathbf{E}}}^p dV d\tau$. This completes the proof of Theorem A. □

The following points about the shakedown theorem should be noted.

1. The postulated residual stress field need not necessarily be the same as the residual stress field which would actually exist in the structure after it had shaken down. It is sufficient to be able to postulate any set of stress fields satisfying the homogeneous equilibrium conditions.

2. The elastic stress field may include change of stress induced by variations of temperature.

3. No statement is made in the theory that a condition of shakedown can in fact be reached in a finite number of load variations (as is observed in experiments) — only that a state of shakedown is approached.

4. The order in which loads are applied has no effect on whether a structure can shake down.

5. It is possible to consider the influence of initial deformations or imperfections on the adaptation of the structure at hand.

The theorem presented here enables us to determine the shakedown of the structure made of an elastic–plastic material within finite deformations. It generalizes the recent static shakedown approaches (WEICHERT [1986], GROSS–WEEGE [1990], SACZUK and STUMPF [1990]) and the classical kinematical one (KOITER [1960]). For instance, if $\mathcal{A} = \emptyset$ then after some reinterpretations one can deduce its static version presented in SACZUK and STUMPF [1990]. Observe that the stress state $\mathbf{S}$ in (20) has been obtained under conditions which exclude such instabilities of the structure as buckling, necking, etc., and jump phenomena, which take place during finite deformations. The shakedown limits of loads resulting from the theorem are path–dependent in this case. This fact can easily be deduced from

$$\max_{\beta_s} f\left[\mu(\tau)\left(\mathbf{S}(\tau)+\bar{\mathbf{S}}\right), \mathbf{E}(\tau)+\bar{\mathbf{E}}\right] \leq 0,$$

as a simple consequence of (20), where $\mu(\tau) > 1$ for every τ is a safety function, and β_s denotes a load path. It should be emphasized that the variable τ can here mean either the time or a monotonically–increasing loading parameter, or even some other monotonically–increasing quantity like a characteristic dimension of the plastic zone.

A non–classical case

The main ideas of this subsection will be connected with the additive decomposition (4) of the total deformation gradient of the elastic–plastic phenomenon. Before going further, we need some modification to the optimization problem $\mathcal{A}$. For this purpose we denote by $\hat{\mathcal{A}}$ the optimization problem given as follows:

Find a pair $(\hat{\mathbf{T}}, \hat{\mathbf{v}}) \equiv ((\mathbf{T}^h, \mathbf{T}^v), (\mathbf{v}^h, \mathbf{v}^v))$ of time–dependent fields of the first Piola–Kirchhoff stress tensor and displacement as a solution to the problem:

maximize

$$\mathbf{T}^h(\tau) = \mathbf{T}^h(\tau_o) + \int_{\tau_o}^{\tau} (\mathcal{L}^h \cdot \nabla^h \mathbf{w})(\hat{\mathbf{x}}, \tau)d\tau,$$

$$\mathbf{T}^v(\tau) = \mathbf{T}^v(\tau_o) + \int_{\tau_o}^{\tau} (\mathcal{L}^v \cdot \nabla^v \mathbf{w})(\hat{\mathbf{x}}, \tau)d\tau,$$

(25)

$$\mathbf{v}^h(\tau) = \mathbf{v}^h(\tau_o) + \int_{\tau_o}^{\tau} \mathbf{w}^h(\hat{\mathbf{x}}, \tau) d\tau,$$

$$\mathbf{v}^v(\tau) = \mathbf{v}^v(\tau_o) + \int_{\tau_o}^{\tau} \mathbf{w}^v(\hat{\mathbf{x}}, \tau) d\tau$$

over all $\hat{\mathbf{w}} \equiv (\mathbf{w}^h, \mathbf{w}^v) \in \hat{\mathcal{W}}_v$ subjected to

$$\begin{aligned} &Div^h(\mathcal{L}^h \cdot \nabla^h \mathbf{w}) + \dot{\mathbf{b}}^h = \mathbf{0} \quad \text{on V}, \\ &Div^v(\mathcal{L}^v \cdot \nabla^v \mathbf{w}) + \dot{\mathbf{b}}^v = \mathbf{0} \quad \text{on V}, \\ &(\mathcal{L}^h \cdot \nabla^h \mathbf{w})\mathbf{n} = \mathbf{k}^h[\mathbf{w}^h] \quad \text{on } S_t, \\ &(\mathcal{L}^v \cdot \nabla^v \mathbf{w})\mathbf{n} = \mathbf{k}^v[\mathbf{w}^v] \quad \text{on } S_t, \end{aligned} \tag{26}$$

$$\begin{aligned} &\int_V \hat{W}(\hat{\mathbf{w}}) dV - \int_{S_t} \hat{\mathbf{k}}[\hat{\mathbf{w}}] \cdot \hat{\mathbf{w}} dS \geq 0, \quad \hat{\mathbf{w}} \in \hat{\mathcal{W}}_v, \\ &\int_V \hat{\mathcal{L}}[\nabla \hat{\mathbf{w}}] \cdot \nabla \hat{\mathbf{w}} dV - \int_{S_t} \hat{\mathbf{k}}[\hat{\mathbf{w}}] \cdot \hat{\mathbf{w}} dS \geq 0, \quad \hat{\mathbf{w}} \in \hat{\mathcal{W}}_v \end{aligned} \tag{27}$$

Here $\hat{\mathcal{L}} = (\mathcal{L}^h, \mathcal{L}^v)$ is the tensor of tangent elastic–plastic moduli; $\hat{\mathbf{k}} = (\mathbf{k}^h, \mathbf{k}^v)$ is a function of displacement velocity its gradient and position; $\dot{\hat{\mathbf{b}}} = (\dot{\mathbf{b}}^h, \dot{\mathbf{b}}^v)$ is the body force rate vector; $\hat{\mathcal{W}}_v$ is a class of kinematically admissible velocity fields with zero boundary conditions; and ∇^h, ∇^v are covariant derivatives, named the h–derivative and v–derivative (MATSUMOTO [1986]), respectively.

After this preparation, we come to the following theorem.

THEOREM B. If at any point of a deformation path of the structure and for any pair $(\hat{\mathbf{T}}, \hat{\mathbf{v}}) \in \hat{\mathcal{A}}$ there exists a pair $(\bar{\mathbf{S}}, \bar{\mathbf{v}})$ such that

$$\begin{aligned} &Div^h \bar{\mathbf{T}} = \mathbf{0} \text{ in } V, \quad \bar{\mathbf{T}}^h \mathbf{n} = \mathbf{0} \text{ on } S_t, \\ &Div^v \bar{\mathbf{T}} = \mathbf{0} \text{ in } V, \quad \bar{\mathbf{T}}^v \mathbf{n} = \mathbf{0} \text{ on } S_t, \end{aligned} \tag{28}$$

$$\bar{\mathbf{F}}^h = \mathbf{1} + \nabla^h \bar{\mathbf{v}}, \quad \bar{\mathbf{F}}^v = \mathbf{1} + \nabla^v \bar{\mathbf{v}} \text{ in } V, \quad \bar{\mathbf{v}}^h = \bar{\mathbf{v}}^v = \mathbf{0} \text{ on } S_u;$$

$$f\left(\hat{\mathbf{S}}(\tau) + \bar{\mathbf{S}}, \hat{\mathbf{E}}(\tau) + \bar{\mathbf{E}}\right) < 0 \tag{29}$$

where $\hat{\mathbf{S}} = \hat{\mathbf{F}}^{-1}\hat{\mathbf{T}}$ and f is a non–negative convex scalar function, then the structure will shake down over any programme of loading.

The proof is omitted, since, except for some technical details, the proof in SACZUK and STUMPF [1990] just reiterates what has to be said here. It is of interest to note that the v–deformation $\mathbf{F}^v$ is responsible for the irreversible part of the deformation $\hat{\mathbf{F}}$. From here, we do not need to impose any restriction on the deformation process, for instance, to define $\mathbf{E}^p$ identified with $\mathbf{E}^v$.

There are a few observations which we would like to make with regard to the presented technique. In the first place, the classical shakedown technique was adapted in a natural manner to the internal variable theory of elasto–plasticity. Secondly, the idea of adaptation within the Finsler type elasto–plasticity can be treated as an example, and its extensions to more complicated, say, thermo–mechanical cases are also possible. Finally the fundamental conditions in Theorems A and B depend on a particular problem under consideration, like the properties of the plastic energy, or the topological structure of the loads domain.

CONCLUSIONS

Shakedown theorems are of theoretical interest since they show that if an elastic-plastic structure shakes down under some particular programme of loading, it will shake down under any loading programme.

From our presentation it should be evident that the adaptation of a structure made of an elastic–plastic material can be best understood by considering and properly analyzing the underlying fundamental admissible deformation processes responsible for it. Our adaptation criteria combine geometric information of the deformation process with the properties of the structure material.

As final conclusions, we may also state:

1. Our adaptation theorems are generalizations of of classical adaptation criteria.

2. They give more realistic adaptation criteria than the classical ones.

3. They are based on internal variable theory of elasto–plasticity, and are consistent with the nature of elasto–plastic behaviour of materials observed in experiments.

4. Is possible to define an adaptation criterion in which a definite bound, established by calculations or experiments, of the plastic work is specified. In effect, it enables one to estimate a number of load cycles, and lower and upper limits of the plastic work, in the given time interval.

One is tempted to speak of thermal influences on the adaptation criteria, although almost nothing has been done as yet to solve this problem properly. It is more difficult because of the coupling between the qualitatively different thermal and mechanical phenomena. Hence, the thermo–mechanical adaptation of structures is, in general, still an open problem. Low–cycle thermal inadaptation seems to be one of the dominant failure modes in the high–temperature structural components. This type of problem is intensified in gas turbine blades.

REFERENCES

1984 ATLURI, S. N., "On constitutive relations at finite strains: hypo–elasticity and elastoplasticity with isotropic or kinematic hardening", Compt. Meth. Appl. Mech. Engng, 43, 137.

1990 GROSS–WEEGE, J., "A unified formulation of statical shakedown criteria for geometrically nonlinear problems", Int. J. Plasticity, 6, 433.

1959 HODGE, P. G., Plastic Analysis of Structures, Mc Graw–Hill Book Company, New York, Toronto, London.

1982 KÖNIG, J. A., "Shakedown analysis in structural design", in: MAHRENHOLTZ, O. and SAWCZUK, A. (eds), Mechanics of Inelastic Media and Structures, pp. 133–142, PWN — Polish Scientific Publishers, Warszawa.

1987 KÖNIG, J. A., Shakedown of Elastic–Plastic Structures, PWN — Polish Scientific Publishers, Warszawa.

1960 KOITER, W. T., "General theorems for elastic–plastic solids", in: SNEDDON, I. N. and HILL, R. (eds), Progress in Solid Mechanics, Vol. 1, pp. 165–221, North–Holland, Amsterdam.

1986 MATSUMOTO, M., Foundations of Finsler Geometry and Special Finsler Spaces, Kaiseisha Press.

1936 MELAN, E., "Theorie Statisch unbestimmter Systeme aus ideal–plastischem Baustoff", Sitzungsbericht der Akad. D. Wiss. (Wien), Akt. IIa, 145, 195.

1987 PETRYK, H., "Non–uniqueness and instability of plastic deformation processes", DSc Thesis, IFTR Report of PASci, Warszawa (in Polish).

1959 RUND, H., The Differential Geometry of Finsler Spaces, Springer, Berlin,

Göttingen, Heidelberg.
1990 SACZUK, J. and STUMPF, H., "On statical shakedown theorems for non-linear problems", Mitt. Inst. für Mech., Ruhr–Univ. Bochum, Germany, Vol. 74.
1993a SACZUK, J., "A contribution to the theory of elastic–plastic materials. I. Ideas of a new theory of elasto–plasticity", Int. J. Engng Sci., (submitted).
1993b SACZUK, J., "A contribution to the theory of elastic–plastic materials. II. The balance laws and constitutive equations", Int. J. Engng Sci., (submitted).
1992 SACZUK, J., "A version of path–dependent adaptation of elastic–plastic structures", EUROMECH 298, Warszawa, September 14–18.
1965 TRUESDELL, C. and NOLL, W., "The non–linear field theories of mechanics", in: FLÜGGE, S. (ed), Encyclopedia of Physics, Vol. III/3, Springer, Berlin, Heidelberg, New York.
1986 WEICHERT, D., " On the influence of geometrical nonlinearities on the shakedown of elastic–plastic structures", Int. J. Plasticity, 2, 135.

EXTREMUM PROBLEMS IN SHAKEDOWN THEORY

Jacov KAMENJARZH
School of Mechanics and Mathematics
Moscow State University
Moscow 119899, Russia

ABSTRACT

The problem of shakedown safety factor calculation is considered in the framework of convex analysis, which leads to the formulation of an upper bound kinematic method making use of time-independent velocity fields. An explicit formula for the upper bound is derived for the shakedown problem with a polyhedron set of variable loads. Conditions are established under which the infimum of upper bounds over a set of regular velocity fields equals the safety factor. Convergence of finite-element approximations to the safety factor is proved.

INTRODUCTION

Definitions of the safety factor in the shakedown theory and in limit analysis are similar. So are the static (lower bound) methods and the static extremum problems for calculating the safety factor in both theories. However, the kinematic method of limit analysis and the kinematic method originally proposed by KOITER [1960] for shakedown problems differ significantly. The former makes use of time-independent kinematic fields while the latter uses plastic deformation paths. Several approaches have been developed to eliminate the time variable from the formulation of the kinematic method in shakedown theory. A kinematic method similar to that of limit analysis was proposed by CHIRAS [1969] and CORRADI and ZAVELANI [1974]: discrete elastic-plastic systems were considered in the framework of the linear programming approach which is applicable when the yield surface is piece-wise linear. GOKHFELD and CHERNIAVSKI [1980] considered shakedown problems with the safety factor defined as the supremum of the multiplier determining (i) the value of the constant load or (ii) the limits of variation of the variable load. In case (i), the authors reduced the kinematic method of shakedown theory to that of limit analysis for a fictitious body and

Z. Mróz et al. (eds.), Inelastic Behaviour of Structures under Variable Loads, 219–236.

the time variable was eliminated. In case (ii), the time variable was also eliminated, however the method leads to a nonstandard problem inverse to the one of limit analysis. POLIZZOTTO et al. [1991] replaced the integration over the time interval by integration over the set of basic loads, and formulated a kinematic method similar to that of limit analysis. KAMENJARZH and WEICHERT [1992] and KAMENJARZH and MERZLJAKOV [1992] used the convex analysis approach and established a kinematic method for the safety factor calculation that makes use of time-independent kinematic fields. The latter approach is briefly reviewed here. The definition of the safety factor is reduced to the standard form of the convex optimization problem, and a dual extremum problem is considered. The dual problem is a kinematic one making use of time-independent kinematic fields. The problem is modified to obtain an explicit formula for the kinematic functional for the case of a shakedown problem with a polyhedron set of variable loads. We arrive at the formula proposed by POLIZZOTTO et al. [1991]. Every value of the kinematic functional calculated by the formula is an upper bound for the safety factor. In this connection there arises a kinematic extremum problem of obtaining the best estimation of the safety factor, i.e. the infimum of the functional. Most important in this respect is the question whether the infimum equals the safety factor or whether there is a gap between them. Conditions are given under which the infimum of the kinematic functional with respect to regular nonvanishing fields equals the safety factor. It may so happen that the infimum is unattainable, which, however, is no obstacle to calculating the safety factor by the method. A finite-element discretization of the kinematic extremum problem is considered. Convergence of minimums of the discretized problems to the safety factor is proved.

STATIC EXTREMUM PROBLEM OF SHAKEDOWN THEORY

Elastic-plastic problem. Consider the quasistatic problem for an elastic-plastic body $\boldsymbol{B}$ occupying a domain Ω. The body is subjected to body forces with a volume density $\boldsymbol{f}$ given in Ω and surface tractions with a density $\boldsymbol{q}$ given at a part S_q of the body boundary. The remaining part S_v of the boundary is fixed. We assume displacements and strains to be small, hence the geometrically linear theory is applicable.

We adopt usual constitutive relations for plastic strain rate $\boldsymbol{e}^p$ and stresses $\boldsymbol{\sigma}$

$$\boldsymbol{e}^p \cdot (\boldsymbol{\sigma} - \boldsymbol{\sigma}_*) \geq 0 \text{ for every } \boldsymbol{\sigma}_* \text{ satisfying } F_x(\boldsymbol{\sigma}_*) \leq 0.$$

Here, $\boldsymbol{F}_x$ is a yield function at point x of the body ($\boldsymbol{F}_x(\boldsymbol{s}) = \boldsymbol{0}$ is the equation for the corresponding yield surface); stresses $\boldsymbol{s}$ are admissible at point $\boldsymbol{x}$ if and only if they satisfy inequality $\boldsymbol{F}_x(\boldsymbol{s}) \leq \boldsymbol{0}$. We denote $\boldsymbol{C}_x$ the set of all admissible stresses at point $\boldsymbol{x}$, the boundary of $\boldsymbol{C}_x$ being the yield surface. We write the constitutive relations in the following form

$$e = e^e + e^p, \quad e_{ij} = \frac{1}{2}\left(\frac{\partial v_i}{\partial x_j} + \frac{\partial v_j}{\partial x_i}\right), \quad e_{ij}^e = A_{ijkl}\dot{\sigma}, \tag{1}$$

$$\sigma \in C_x, \quad e^p \cdot (\sigma - \sigma_*) \geq 0 \quad \text{for every} \quad \sigma_* \in C_x$$

where A_{ijkl} are elastic modules, e and e^e standing for total and elastic strain rate and v for velocity. Dependence of C_x upon x means that inhomogeneous bodies are considered as well as homogeneous ones. We assume that the mapping $x \to C_x$ is measurable which is always valid in applications of plasticity. We assume also that the set C_x is convex closed and includes zero stresses.

Adding to relations (1) the equilibrium equations, initial and boundary conditions we arrive at the usual formulation of the elastic-plastic problem.

Shakedown problem. Consider a load given by fields of body and surface forces $l = (f,q)$ that may vary with time. The shakedown problem is that of evaluating elastic-plastic body B behavior with respect to all loading histories within the given set L of loads. Consider the plastic work

$$\int\int_{\Omega} d_x(e^p(x,t))\, dV dt$$

Here, Ω is a domain occupied by the body B and d_x is the plastic dissipation $d_x(e^p) = e^p \cdot \sigma(e^p)$ that can also be expressed in the form

$$d_x(e^p) = \sup\{e^p \cdot \sigma_*: \ \sigma_* \in C_x\}$$

(we use the notation $a \cdot b = a_{ij} b_{ij}$). The question is whether the plastic work is bounded for all loading histories within L. If it is, the shakedown occurs with respect to the set L by definition. If it is not, the shakedown with respect to L does not take place.

Safety factor. MELAN [1938] gave the following condition for the shakedown. Consider a purely elastic reference body B^0. It occupies the same domain Ω as the elastic-plastic body B, its elastic modules being equal to those of B. For a given load l in L, there is a solution of the elasticity problem for B^0 with the elastic stress field σ^0. Let Δ be a set of such fields corresponding to all possible loads in L. According to Melan's theorem, the shakedown occurs with respect to L if there exists a time-independent field of self-equilibrated stresses ρ, such that the sum $\rho + \sigma^0$ is safe for every σ^0 in Δ (and it does not occur if for any self-equilibrated field ρ a field σ^0 can be found in Δ, such that the sum $\rho+\sigma^0$ is inadmissible).

The given set of possible loads may not satisfy the above- mentioned safety condition. However, a homotetic set λL, $\lambda>0$, with a sufficiently small multiplier λ satisfies it. The supremum of such multipliers is the safety factor s_Δ, i.e. it is determined by the extremum problem (DEBORDES and NAYROLES [1976])

$$s_\Delta = \sup\{\lambda \geq 0:\ \exists \rho \in \Sigma,\ \lambda(\rho+\Delta) \subset C\}. \tag{2}$$

Here, C is the set of admissible stress fields, i.e. those stress fields that take values inside the yield surface or on it for almost every point of the body

$$C = \{s \in S:\ s(x) \in C_x \ \text{ for a.e. } x \in \Omega\},$$

S is a space of stress fields, Σ is its subspace of self-equilibrated stress fields. According to Melan's theorem, shakedown occurs with respect to the set λL if $\lambda < s_\Delta$ (in particular, if $1 < s_\Delta$, shakedown occurs with respect to L). If $\lambda > s_\Delta$, then shakedown with respect to λL does not take place (in particular, if $1 > s_\Delta$, shakedown does not take place with respect to L).

Standard formulation of the static extremum problem. The problem (2) allows finding the safety factor and its lower bounds. Our aim is to formulate a dual kinematic problem for determining the safety factor and its upper bounds, the time variable will not be involved in its formulation. The first step in this direction is reducing the problem (2) to a standard form of a convex analysis problem. The reformulation was made by KAMENJARZH and WEICHERT [1992] for the case of spherical yield surface. In the case of general yield surface, the reformulation is obtained using the Minkowski function (KAMENJARZH and MERZLJAKOV [1992]).

The Minkowski function of the set C of admissible stress fields is defined on the space of stress fields in the following way. Consider a stress field s from S (it either belongs to C or not). Suppose first that there exists such a number $mu>0$ that s/mu belongs to C. Then we consider the infimum of all numbers satisfying this condition; the value of the Minkowski function is defined as this infimum

$$M_C(s) = \inf\{mu>0:\ \frac{1}{mu}s \in C\}, \quad s \in C.$$

If no such number $mu>0$ exists for which s/mu belongs to C, then the value of Minkowski function is defined as $M_C(s) = +\infty$. Note that, according to convention, the infimum of the empty set of numbers is $+\infty$. Hence, the above formula for M_C remains valid in the latter case.

Example 1. Consider a body with von Mises condition of stress admissibility $|s^d| \leq 1$, s

being dimensionless stress. The superscript d denotes the deviator component of a tensor, $|s|^2 = s \cdot s$. The set C is a set of all essentially bounded stress fields on Ω satisfying the condition $|s^d(x)| \leq 1$ for a.e. x in Ω. Then e.g. for a stress field s which is continuous on the closure $\bar{\Omega}$ of the domain Ω we have

$$M_C(x) = \max\{ |s^d(x)| : x \in \Omega \}.$$

Using the Minkowski function we write the definition (2) of the safety factor in the form

$$\frac{1}{s_\Delta} = \inf\{ mu > 0 : \exists \rho \in \Sigma, \frac{1}{mu}(\rho + \tau) \in C \text{ for every } \tau \in \Delta \}. \tag{3}$$

Under certain nonrestrictive hypotheses, the condition $\sigma / mu \in C$ is equivalent to the inequality $M_C(\sigma) \leq mu$. Then, introducing the function

$$F(\sigma) = \sup\{ M_C(\sigma + \tau) : \tau \in \Delta \}, \tag{4}$$

we reformulate the constraint in (3) as the inequality $F(\rho) \leq mu$ and present the static extremum problem (3) in the following form

$$\frac{1}{s_\Delta} = \inf\{ F(\rho) : \rho \in \Sigma \}. \tag{5}$$

Here, Σ is a subspace in S and F is a convex function. Therefore, problem (5) is a standard problem of convex optimization.

Example 2. A convex polyhedron $\mathbf{conv}\{ l_1, ..., l_m \}$ i.e. the convex hull of loads $l_1, ..., l_m$ is often considered as the set L of possible loads in a shakedown problem. The set Δ is, then, a polyhedron too

$$\Delta = \mathbf{conv}\{ \tau_1, ..., \tau_m \},$$

where τ_i is the elastic stress field in the reference body B^0 corresponding to the load l_i. Then expression (4) is reduced to a simpler one. Namely, $F(\sigma)$ is a supremum of the function $f(\tau) = M_C(\sigma + \tau)$ on Δ, σ being fixed. The function f is convex, therefore its supremum is attainable at one of the vertices of Δ. Hence,

$$F(\sigma) = \max\{ M_C(\sigma + \tau_1), ..., M_C(\sigma + \tau_m) \}$$

which means that one has to calculate m numbers $M_C(\sigma + \tau_i)$ and choose the largest one for $F(\boldsymbol{\sigma})$.

CONSTRUCTING THE DUAL EXTREMUM PROBLEM

Since (5) is a problem of convex optimization, the dual problem can be formulated by applying the standard approach of convex analysis (see GOLSTIEN [1971], EKELAND and TEMAM [1976]). Some of its applications to solid mechanics were given by NAYROLES [1971]. To construct the dual problem, one needs a bilinear form $\langle \boldsymbol{e},\boldsymbol{\sigma}\rangle$ where stresses $\boldsymbol{\sigma}$ belong to a space $\boldsymbol{S}$ and strain rates $\boldsymbol{e}$ to another space $\boldsymbol{E}$. The appropriate form for mechanics problems is $e_1\sigma_1 + \cdots + e_n\sigma_n$ in the case of discrete system (when stresses and strain rates are vectors) and

$$\langle \boldsymbol{e},\boldsymbol{\sigma}\rangle = \int_{\Omega} \boldsymbol{e}\cdot\boldsymbol{\sigma}\, dV, \quad \boldsymbol{e}\in \boldsymbol{E}, \boldsymbol{\sigma}\in \boldsymbol{S}$$

in the case of 3D problems (when stresses and strain rates are tensor fields on the domain Ω occupied by a body). The form can be defined similarly in the case of plates, shells, etc.

The bilinear form allows the introduction of the Fenchel transformation f^* of a function f defined on $\boldsymbol{S}$

$$f^*(\boldsymbol{e}) = \sup\{\langle \boldsymbol{e},\boldsymbol{\sigma}\rangle - f(\boldsymbol{\sigma}):\ \boldsymbol{\sigma}\in \boldsymbol{S}\}, \quad \boldsymbol{e}\in \boldsymbol{E}$$

and the polar set Π^0 for a subspace $\Pi \subset \boldsymbol{E}$

$$\Pi^0 = \{\ \boldsymbol{\sigma}\in \boldsymbol{S}:\ \langle \boldsymbol{e},\boldsymbol{\sigma}\rangle = 0 \quad \text{for every } \boldsymbol{e}\in \boldsymbol{E}\}.$$

In particular, consider the Fenchel transformation $G = F^*$

$$G(\boldsymbol{e}) = \sup\{\langle \boldsymbol{e},\boldsymbol{\sigma}\rangle - F(\boldsymbol{\sigma}):\ \boldsymbol{\sigma}\in \boldsymbol{S}\}.$$

By definition, the following inequality holds

$$F(\boldsymbol{\sigma}) \geq \langle \boldsymbol{e},\boldsymbol{\sigma}\rangle - G(\boldsymbol{e}) \quad \text{for every } \boldsymbol{\sigma}\in \boldsymbol{S} \text{ and every } \boldsymbol{e}\in \boldsymbol{E}. \tag{6}$$

We consider only self-equilibrated stress fields $\boldsymbol{\sigma}$ and only kinematically admissible strain rate fields $\boldsymbol{e}$ in (6). Recall that according to virtual work principle, the set Σ of self-equilibrated stress fields is the polar set for the subspace

Here, $\boldsymbol{V}$ is a space of virtual velocity fields and **Def** is the strain rate operator

$$\mathrm{Def}\, V = \{\, \boldsymbol{e} \in E\colon\ \boldsymbol{e} = \mathrm{Def}\, \boldsymbol{v},\ \ \boldsymbol{v} \in V \},$$

$$(\mathrm{Def}\, \boldsymbol{v})_{ij} = \frac{1}{2}\left(\frac{\partial v_i}{\partial x_j} + \frac{\partial v_j}{\partial x_i} \right).$$

It is sufficient to consider a set of regular velocity fields (e.g. those from Sobolev space satisfying the condition $\boldsymbol{v}|_{S_v} = \mathbf{0}$) for V. Thus, we have

$$\langle \boldsymbol{e}, \boldsymbol{\sigma} \rangle = 0 \text{ for every } \boldsymbol{\sigma} \in \Sigma \text{ and every } \boldsymbol{e} \in \mathrm{Def}\, V$$

and inequality (6) yields

$$\inf \{\, F(\boldsymbol{\rho})\colon\ \boldsymbol{\rho} \in \Sigma \,\} \geq -G(\boldsymbol{e}) \text{ for every } \boldsymbol{e} \in \mathrm{Def}\, V.$$

According to formula (5), we arrive at a lower bound for $1/s_\Delta$

$$\frac{1}{s_\Delta} \geq -G(\boldsymbol{e}) \text{ for every } \boldsymbol{e} \in \mathrm{Def}\, V.$$

The best estimation for $\frac{1}{s_\Delta}$ of this type is

$$\frac{1}{s_\Delta} \geq \sup \{ -G(\boldsymbol{e})\colon\ \boldsymbol{e} \in \mathrm{Def}\, V \} \qquad (7)$$

which gives a formulation of a kinematic method for estimating $1/s_\Delta$ from below (or the safety factor s_Δ from above) The extremum problem in the right hand side of (7) is referred to as a dual to the original problem (5). It is the kinematic extremum problem of the shakedown theory.

Note 1. Formulation (7) makes use of time-independent kinematic fields, unlike the formulation arising from Koiter's theorem.

Note 2. If L contains a single load, then (7) reduces to the standard formulation of the kinematic method of limit analysis. In that case, conditions are well known under which equality holds in (7) as well as explicit formula for calculating a value of the kinematic functional.

Two important questions arise from the formulation of the kinematic method (7). First, is the supremum (7) equal to $1/s_\Delta$ or is there is a gap between them? If the equality holds, then the kinematic method gives a precise value of the safety factor and not only its estimation. Some simple conditions for the equality

$$\frac{1}{s_\Delta} = \sup\{\, -G(\boldsymbol{e}):\ \boldsymbol{e} \in \mathrm{Def}\, V\} \tag{8}$$

are given below. Second, how are values of the kinematic functional $G(\boldsymbol{e})$ to be calculated? Explicit formulas for $G(\boldsymbol{e})$ were derived for some simple cases of discrete systems only (see KAMENJARZH and WEICHERT [1992] and KAMENJARZH and MERZLJAKOV [1994]). In a more general case, a value of the functional can be calculated just on the basis of its definition, which is not a very simple task. The aim of the next Section is to give an explicit formula for the kinematic functional. It is established for the case of a shakedown problem with a polyhedron set of variable loads. The price of obtaining an explicit formula is a larger number of arguments for the kinematic functional.

THE DUAL EXTREMUM PROBLEM IN THE CASE OF POLYHEDRON SET OF VARIABLE LOADS

Consider a shakedown problem with the polyhedron set of variable loads $L = \mathrm{conv}\{\boldsymbol{l}_1,\dots,\boldsymbol{l}_m\}$, the corresponding set

$$\Delta = \mathrm{conv}\{\boldsymbol{\tau}_1,\dots,\boldsymbol{\tau}_m\}$$

of stress fields in purely elastic reference body $\boldsymbol{B}^0$, and the function

$$F(\boldsymbol{\sigma}) = \max\{\, M_C(\boldsymbol{\sigma}+\boldsymbol{\tau}_1),\dots,M_C(\boldsymbol{\sigma}+\boldsymbol{\tau}_m)\} \tag{9}$$

in static extremum problem (see Example 2). We show that a kinematic method for estimating the safety factor can be formulated as follows

$$s_\Delta \le \int_\Omega \big(d_x(\boldsymbol{e}_1) + \cdots + d_x(\boldsymbol{e}_m)\big)\, dV \tag{10}$$

for every $(\boldsymbol{e}_1,\dots,\boldsymbol{e}_m)$, $\boldsymbol{e}_1,\dots,\boldsymbol{e}_m \in E$ satisfying the conditions

$$\boldsymbol{e}_1 + \dots + \boldsymbol{e}_m \in \mathrm{Def}\, V, \quad \int_\Omega (\boldsymbol{e}_1 + \dots + \boldsymbol{e}_m)\, dV = 1.$$

Moreover, we will see in the next section that the equality

$$s_\Delta = \inf\{ \int_\Omega \big(d_x(e_1) + \cdots + d_x(e_m)\big) dV: \quad e_1,\ldots,e_m \in E, \\ e_1 + \ldots + e_m \in \mathrm{Def}\, V, \quad \int_\Omega (e_1 \cdot \tau_1 + \cdots + e_m \cdot \tau_m) dV = 1 \} \tag{11}$$

holds under certain conditions. Hence, we arrive at a kinematic method for calculating the safety factor with an explicit formula for the upper bound functional and the time variable not involved in the formulation. The method is similar to that of limit analysis and is reduced to the latter when the set L of loads contains a single load.
To obtain formulation (10) we start again with the static extremum problem

$$\frac{1}{s_\Delta} = \inf\{ F(\rho): \ \rho \in \Sigma \}, \quad \Sigma = (\mathrm{Def}\, V)^0 \tag{12}$$

where the function F is given by the formula (9) in the case under consideration. We use the standard argumentation of the previous section. However, this time, we will consider the Fenchel transformation in the space $S^m = S \times \cdots \times S$. Namely, we introduce a variable $s = (s_1,\ldots,s_m) \in S^m$, the Minkowski function of the set $C^m = C \times \cdots \times C$ in S^m

$$M_{C^m}(s) = \max\{ M_C(s_1),\ldots,M_C(s_m)\},$$

and the function

$$F_{(m)}(s) = M_{C^m}(s + \tau), \quad s = (s_1,\ldots,s_m) \in S^m, \ \ \tau = (\tau_1,\ldots,\tau_m) \in S^m.$$

Function (9) is related to it by the following formula

$$F(\sigma) = F_{(m)}(\bar{\sigma} + \tau), \quad \text{where } \bar{\sigma} = (\sigma, \sigma,\ldots,\sigma), \ \sigma \in S, \quad \tau = (\tau_1,\ldots,\tau_m) \in S^m.$$

Due to that, we reformulate the static extremum problem (12) in the following form

$$\frac{1}{s_\Delta} = \inf\{ F_{(m)}(\bar{\rho}): \ \bar{\rho} \in \Sigma_{(m)} \} \tag{13}$$

where $\Sigma_{(m)}$ stands for the subspace

$$\Sigma_{(m)} = \{ \bar{r} \in S^m: \ \bar{r} = (r,r,\ldots,r), \ r \in \Sigma \}$$

To find the problem dual to (13), we introduce the space $E^m = E \times \cdots \times E$ and the bilinear form

$$\langle \boldsymbol{e},\boldsymbol{\sigma}\rangle_{(m)} = \langle \boldsymbol{e}_1,\boldsymbol{\sigma}_1\rangle + \cdots + \langle \boldsymbol{e}_m,\boldsymbol{\sigma}_m\rangle,$$

$$\boldsymbol{e} = (\boldsymbol{e}_1,\ldots,\boldsymbol{e}_m) \in \boldsymbol{E}^m, \quad \boldsymbol{\sigma} = (\boldsymbol{\sigma}_1,\ldots,\boldsymbol{\sigma}_m) \in \boldsymbol{S}^m.$$

The form allows consideration of the Fenchel transformation $G_{(m)} = F^*_{(m)}$

$$G_{(m)}(\boldsymbol{e}) = \sup\{\langle \boldsymbol{e},\boldsymbol{\sigma}\rangle_{(m)} - F_{(m)}(\boldsymbol{\sigma}):\ \boldsymbol{\sigma} \in \boldsymbol{S}^m\},$$

and the inequality

$$F_{(m)}(\boldsymbol{\sigma}) \geq \langle \boldsymbol{e},\boldsymbol{\sigma}\rangle_{(m)} - G_{(m)}(\boldsymbol{e})$$

holds for every $\boldsymbol{e}$ in $\boldsymbol{E}^m$ and every $\boldsymbol{\sigma}$ in $\boldsymbol{S}^m$. Consider a subspace

$$(\mathrm{Def}\,V)_{(m)} = \{(\boldsymbol{e}_1,\ldots,\boldsymbol{e}_m) \in \boldsymbol{E}^m:\ \boldsymbol{e}_1 + \ldots + \boldsymbol{e}_m \in \mathrm{Def}\,V\}.$$

It is easy to see that the equality $\langle \boldsymbol{e},\bar{\boldsymbol{r}}\rangle = 0$ holds for every $\bar{\boldsymbol{r}} = (\boldsymbol{r},\ldots,\boldsymbol{r})$ in $\Sigma_{(m)}$ and for every $\boldsymbol{e} = (\boldsymbol{e}_1,\ldots,\boldsymbol{e}_m)$ in $(\mathrm{Def}\,V)_{(m)}$. Then, we obtain the following estimation for the safety factor

$$\frac{1}{s_\Delta} \geq -G_{(m)}(\boldsymbol{e}) \quad \text{for every } \boldsymbol{e} \in (\mathrm{Def}\,\mathrm{V})_{(m)}$$

and the modified kinematic extremum problem for the best estimation of this type

$$\frac{1}{s_\Delta} \geq \sup\{-G_{(m)}(\boldsymbol{e}): \boldsymbol{e} \in (\mathrm{Def}\,\mathrm{V})_{(m)}\}. \tag{14}$$

There is an explicit formula for calculating values of the kinematic functional $G_{(m)}(\boldsymbol{e})$ for $\boldsymbol{e}$ in $\boldsymbol{E}^m$. In fact, the Fenchel transformation $G_{(m)}$ of $F_{(m)}$ is easy to calculate through the Fenchel transformation of the Minkowski function M_{C^m}, and we find

$$\begin{aligned} G_{(m)}(\boldsymbol{e}) &= -\langle \boldsymbol{e},\boldsymbol{\tau}\rangle_{(m)} && \text{if } D(\boldsymbol{e}_1) + \cdots + D(\boldsymbol{e}_m) \leq 1, \\ G_{(m)}(\boldsymbol{e}) &= +\infty && \text{if } D(\boldsymbol{e}_1) + \cdots + D(\boldsymbol{e}_m) > 1. \end{aligned}$$

Here, D is the dissipation

$$D(\boldsymbol{\epsilon}) = \sup\{\langle\boldsymbol{\epsilon},\boldsymbol{s}\rangle\colon\ \boldsymbol{s}\in C\},\quad \boldsymbol{\epsilon}\in E.$$

Then the kinematic extremum problem (14) is reduced to

$$s_\Delta \le \inf\Big\{\frac{D(\boldsymbol{e}_1)+\cdots+D(\boldsymbol{e}_m)}{\langle\boldsymbol{e}_1,\boldsymbol{\tau}_1\rangle+\cdots+\langle\boldsymbol{e}_m,\boldsymbol{\tau}_m\rangle}\colon\quad \boldsymbol{e}_1,\dots,\boldsymbol{e}_m\in E,$$

$$\boldsymbol{e}_1+\cdots+\boldsymbol{e}_m\in(\mathrm{Def}\,V)_{(m)},\quad \langle\boldsymbol{e}_1,\boldsymbol{\tau}_1\rangle+\cdots+\langle\boldsymbol{e}_m,\boldsymbol{\tau}_m\rangle>0\Big\}.$$

A simple formula is valid for calculating the dissipation $D(\boldsymbol{\epsilon})$ when $\boldsymbol{\epsilon}$ is an integrable field (ROCKAFELLAR [1968])

$$D(\boldsymbol{\epsilon}) = \int_\Omega d_x(\boldsymbol{\epsilon}(x))\,dV,\quad \boldsymbol{\epsilon}\in E$$

and obviously

$$\langle\boldsymbol{e}_i,\boldsymbol{\tau}_i\rangle = \int_\Omega \boldsymbol{e}_i(x)\cdot\boldsymbol{\tau}_i(x)\,dV.$$

Finally, we take into account degree 1 positive homogeneity of the functional

$$\int_\Omega d_x(\boldsymbol{\epsilon}(x))\,dV$$

and arrive at the estimation (10) provided that the fields $\boldsymbol{e}_1,\dots,\boldsymbol{e}_m$ are regular enough. A method making use of this formula for a kinematic functional was proposed by POLIZZOTTO et al. [1991]. However, the question of the functional's infimum over a set of regular fields being equal to the safety factor remains open. It acquires considerable importance since we have a rather simple kinematic method for calculating the safety factor and not only its upper bounds if the equality (11) is valid. We consider this question in the following section.

CONDITIONS FOR EQUALITY OF STATIC AND KINEMATIC EXTREMUMS

The safety factor is defined by static extremum problem (13) in the case of a polyhedron set of variable loads. Thus, we discuss conditions for the equality (11) in this Section. The equality is proved under assumption that the extremums in both kinematic and static problems are attainable on sufficiently smooth fields $\boldsymbol{e}=(\boldsymbol{e}_1,\dots,\boldsymbol{e}_m)\neq\boldsymbol{0}$ and $\boldsymbol{s}$ resp. (see

POLIZZOTTO et al. [1991]). However, the infimum in kinematic problem may be unattainable if only regular nonvanishing fields are considered in this problem. An example of such unattainability in a limit analysis problem with smooth inputs (yield surface, body boundary, and load) is given in KAMENJARZH [1980]. The example is relevant to the shakedown problem as well, since the former is a particular case of the latter. Therefore, we look for conditions under which the infimum of the kinematic functional over a set of regular fields equals the safety factor no matter whether the infimum is attained or not. When the equality holds, the kinematic method is applicable for calculating the safety factor despite eventual unattainability.

We establish the required conditions for (i) 3D bodies with bounded yield surfaces, (ii) 3D bodies with cylindrical yield surfaces, (iii) discrete elastic-plastic systems. The yield surface is bounded if it lies inside a certain ball in 6D stress space. The axis of cylindrical yield surface is assumed to have the direction of the unity tensor in 6D stress space. Hence, the axis corresponds to hydrostatic stresses, and the cylindrical yield surfaces are similar to von Mises or Tresca ones.

Spaces $\boldsymbol{E}$ and $\boldsymbol{V}$ of strain rate fields and kinematically admissible velocity fields can be chosen in different ways. We choose

$$\boldsymbol{V} = \{\boldsymbol{v} \in \boldsymbol{H}^1(\Omega):\ \boldsymbol{v}|_{S_v} = 0\}$$

where $\boldsymbol{H}^1(\Omega)$ is Sobolev space with the norm

$$\|\boldsymbol{v}\|^2 = \int_\Omega \left(|\boldsymbol{v}|^2 + \sum_{i,j} \left(\frac{\partial v_i}{\partial x_j} \right)^2 \right) dV$$

and $\boldsymbol{E}$ is the space of square integrable functions with the norm

$$\|\boldsymbol{e}\|^2 = \int_\Omega |\boldsymbol{e}|^2 dV.$$

This choice seems to be the most convenient for considering numerical methods for the safety factor calculation.

In the case of bodies with bounded yield surfaces, the following proposition results from standard argumentation of convex analysis.

Theorem 1. If (i) the sets $\boldsymbol{C}_x$ for all the points $\boldsymbol{x}$ of the body are contained in a certain ball of 6D space of stresses (the same for all the points $\boldsymbol{x}$) and (ii) the set of variable loads in a shakedown problem is presented by the polyhedron of elastic stress fields $\Delta = \mathbf{conv}\{\boldsymbol{\tau}_1,...,\boldsymbol{\tau}_m\}$, then equality (11) holds. It remains valid if $\boldsymbol{V}$ and $\boldsymbol{E}$ are replaced by their subspaces $\tilde{\boldsymbol{V}}$, $\tilde{\boldsymbol{E}}$ provided $\mathbf{Def}\tilde{\boldsymbol{V}}$ lies in $\tilde{\boldsymbol{E}}$ and $\mathbf{Def}\tilde{\boldsymbol{V}}$, $\tilde{\boldsymbol{E}}$ are dense in $\mathbf{Def}\boldsymbol{V}, \boldsymbol{E}$ resp.

The latter proposition allows calculation of the safety factor using only smooth velocity fields.
Additional consideration should be given to bodies with cylindrical yield surfaces. In this case, the hydrostatic components of stress fields can be excluded from the formulation of the original static problem (5) determining the safety factor. Namely, we consider the set

$$V_d = \{ \boldsymbol{v} \in V: \ \mathrm{div}\,\boldsymbol{v} = 0 \}$$

instead of the set V of virtual velocity fields, the deviator subspaces

$$S^d = \{ \boldsymbol{\sigma} \in S: \ \boldsymbol{\sigma} = \boldsymbol{\sigma}^d \}, \quad E^d = \{ \boldsymbol{e} \in E: \ \boldsymbol{e} = \boldsymbol{e}^d \},$$

and replace the set $\Sigma = (\mathrm{Def}\, V)^0$ by the set $\Sigma_d = (\mathrm{Def}\, V_d)^0$. We arrive at the static deviatoric extremum problem similar to (5). The problem determines the number $s_{\Delta d}$

$$\frac{1}{s_{\Delta d}} = \inf \{ F_d(\boldsymbol{\rho}): \ \boldsymbol{\rho} \in \Sigma_d \}. \tag{15}$$

Here, we use the function

$$F_d(\boldsymbol{\rho}) = \sup \{ M_{C^d}(\boldsymbol{\rho} + \boldsymbol{\tau}): \ \boldsymbol{\tau} \in \Delta^d \}, \quad \boldsymbol{\rho} \in C^d, \quad \Delta^d = \{ \boldsymbol{\tau}^d: \ \boldsymbol{\tau} \in \Delta \},$$

and the Minkowski function M_{C^d} of the set $C^d = \{ \boldsymbol{s} \in S: \ \boldsymbol{s} = \boldsymbol{s}^d \}$ is defined in the space S^d. Since the deviator $\boldsymbol{\sigma}^d$ of any $\boldsymbol{\sigma} \in \Sigma$ belongs to Σ, then $s_{\Delta d} \geq s_{\Delta}$.
Problem (15) can be reformulated so as to become similar to (13). Then the same reasoning as in the previous section leads to the dual kinematic problem for the best estimation for $s_{\Delta d}$

$$s_{\Delta d} \leq \inf \{ \int_{\Omega} \big(d_x(\boldsymbol{e}_1) + \cdots + d_x(\boldsymbol{e}_m)\big) dV: \ \boldsymbol{e}_1, \ldots, \boldsymbol{e}_m \in E^d,$$
$$\boldsymbol{e}_1 + \ldots + \boldsymbol{e}_m \in \mathrm{Def}\, V_d, \ \int_{\Omega} (\boldsymbol{e}_1 \cdot \boldsymbol{\tau}_1^d + \cdots + \boldsymbol{e}_m \cdot \boldsymbol{\tau}_m^d) dV = 1 \}.$$

Furthermore, the equality of the infimum and $s_{\Delta d}$ can be proved similarly to Theorem 1. Making use of results in KAMENJARZH [1981}, it is also possible to prove the equality $s_{\Delta d} = s_{\Delta}$. This leads to the following proposition for bodies with cylindrical yield surfaces.
Theorem 2. If (i) the sets C_x for all the points x of a body with cylindrical yield surfaces are contained in a certain ball of 5D space of stress deviators (the same for all the points x), (ii) the set of variable loads in a shakedown problem is presented by the

polyhedron of elastic stress fields $\Delta = \mathbf{conv}\{\boldsymbol{\tau}_1,...,\boldsymbol{\tau}_m\}$, (iii) the boundary $\partial\Omega$ satisfies certain regularity conditions and S_v is its sufficiently regular part, then the equality

$$s_\Delta = \inf\{\int_\Omega (d_x(\boldsymbol{e}_1) + \cdots + d_x(\boldsymbol{e}_m))dV: \ \boldsymbol{e}_1,...,\boldsymbol{e}_m \in \boldsymbol{E}^d, \quad (16)$$
$$\boldsymbol{e}_1 + ... + \boldsymbol{e}_m \in \mathbf{Def}\, \boldsymbol{V}_d, \ \int_\Omega (\boldsymbol{e}_1 \cdot \boldsymbol{\tau}_1^d + \cdots + \boldsymbol{e}_m \cdot \boldsymbol{\tau}_m^d)dV = 1\}$$

holds. It remains valid if $\boldsymbol{V}_d$ and $\boldsymbol{E}^d$ are replaced by their subspaces $\tilde{\boldsymbol{V}}_d$, $\tilde{\boldsymbol{E}}^d$ provided $\mathbf{Def}\,\tilde{\boldsymbol{V}}_d$ lies in $\tilde{\boldsymbol{E}}^d$, and $\mathbf{Def}\,\tilde{\boldsymbol{V}}_d$, $\tilde{\boldsymbol{E}}^d$ are dense in $\mathbf{Def}\,\boldsymbol{V}_d$, $\boldsymbol{E}^d$, respectively.

Note 3. The regularity conditions for $\partial\Omega$ and S_v in Theorem 2 are used to establish the equality $s_{\Delta d} = s_\Delta$. These conditions are of a technical nature and can be chosen in different ways. For example, the surface $\partial\Omega$ was assumed continuously differentiable in KAMENJARZH [1981]. However, minor changes in the argumentation also allows consideration of $\partial\Omega$ consisting of a finite number of continuously differentiable surfaces with nonzero angles of intersection, S_v being its part bounded by a finite number of continuously differentiable curves with non-zero angles of intersection.

Note 4. The variables $\boldsymbol{e}_1,...,\boldsymbol{e}_m$ in extremum problems (11) and (16) can be replaced by $\boldsymbol{v},\boldsymbol{e}_1,...,\boldsymbol{e}_{m-1}$ where

$$\boldsymbol{v} \in V \ \ (\boldsymbol{v} \in V_d), \quad \boldsymbol{e}_1,...,\boldsymbol{e}_{m-1} \in \boldsymbol{E} \ \ (\boldsymbol{e}_1,...,\boldsymbol{e}_{m-1} \in \boldsymbol{E}^d),$$

$$\boldsymbol{e}_m = \mathbf{Def}\,\boldsymbol{v} - \boldsymbol{e}_1 - \cdots - \boldsymbol{e}_{m-1}.$$

Thus, an upper bound in the kinematic method is determined by a kinematically admissible velocity field and $m-1$ fields of rank 2 symmetric tensors. A single linear constraint

$$\langle \boldsymbol{e}_1,\boldsymbol{\tau}_1\rangle + \cdots + \langle \boldsymbol{e}_m,\boldsymbol{\tau}_m\rangle + \langle \mathbf{Def}\,\boldsymbol{v} - \boldsymbol{e}_1 - \cdots - \boldsymbol{e}_{m-1},\boldsymbol{\tau}_m\rangle = 1$$

is involved in the formulation of the extremum problems.

Note 5. It may turn out to be useful to apply piece-wise smooth discontinuous velocity fields in the kinematic method. A corresponding explicit formula for the kinematic functional can be derived easily.

Note 6. The propositions similar to Theorems 1 and 2 are valid for kinematic problem (8). They are also valid for discrete systems with corresponding simplification.

FINITE ELEMENT METHOD IN SHAKEDOWN PROBLEM

Under the hypotheses of Theorem 1 or 2 the safety factor can be calculated from the extremum problems (11) or (16). We use internal approximations of the spaces V, E, V_d, E^d to descretize the problems. For example in the case of the space V, that means that for parameter h with its values in a neighborhood of zero, finite dimensional subspaces V_h are considered and there exists a subset $\tilde{V}$ dense in V such that for every $v \in V$ a sequence $\{v_h\}$, $v_h \in V_h$ can be found converging to v in V, $v_h \to v$ when $h \to 0$. We consider similarly internal approximations E_h, V_{dh}, E_h^d of the spaces E, V_d, E^d and adopt the following additional assumption: $\mathbf{Def}\, V_h \subset E_h$ $(\mathbf{Def}\, V_h^d \subset E_h^d)$. Finite element approximations of Sobolev space used as V and the space L_2 used as E are well known. Construction of approximations for the spaces V_d, E^d in the same sense poses no problem either.

Example 3. Consider the approximations in the case of a body with bounded yield surfaces. Suppose the body occupies a polyhedral domain Ω with fixed boundary, $v|_{\partial\Omega} = 0$. We consider a standard approximation for the Sobolev space

$$V = H_0^1(\Omega) = \{ v \in H^1(\Omega) \colon \; v|_{\partial\Omega} = 0 \},$$

by the spaces V_h of piece-wise linear vector-functions with tetrahedral finite elements (see e.g. CIARLET [1978]) and $\tilde{V}$
consisting of smooth vector fields vanishing on S_v (or even in a neighborhood of S_v). As far as an approximation of the space E is concerned, the condition $\mathbf{Def}\, V_h \subset E_h$ should be satisfied. We note that the set $\mathbf{Def}\, V_h$ consists of tensor-valued functions constant on every element of the considered triangulation T_h. That enables us to introduce the following approximation for E. For every element T of the triangulation we define the function $f_{(T)}$ that takes the value 1 on T and vanishes in its exterior. Then we consider the tensor-valued functions

$$f_{(T)ij}(x) = f_{(T)}(x) e_i \otimes e_j$$

where e_i is a basis in R^3. We introduce the space E_h of tensor-valued functions with the basis consisting of the functions $f_{(T)ij}$ for all the elements $T \in T_h$. We have obtained an approximation E_h of E with $\tilde{E}$ consisting of smooth tensor fields. The condition $\mathbf{Def}\, V_h \subset E_h$ is obviously satisfied.

Example 4. Consider approximations for the space V_d of solenoidal velocity fields from V

and for the space E^d of rank 2 symetric tensors' deviators. The spaces are involved in the formulation of the kinematic method (16) for bodies with cylindrical yield surfaces. Consider for example a body under the plane strain conditions occupying a polygonal domain Ω in R^2 with fixed boundary, $v|_{\partial\Omega}=0$. The internal approximation for the subspace

$$V_d = \{ v \in H_0^1(\Omega):\ \operatorname{div} v = 0 \}$$

is formed by finite dimensional spaces V_{dh} of continuous vector-functions that are solenoidal polynoms of degree 2 on every element T of a triangulation T_h, $\tilde{V}_d$ being the space of smooth solenoidal vector fields vanishing on $\partial\Omega$ or in its neighborhood (see MERCIER [1979]). To construct an approximation for E^d satisfying the condition $\operatorname{Def} V_{dh} \subset E_h^d$, we note that the fields $e = \operatorname{Def} v_h$ for $v_h \in V_{dh}$ are piece-wise linear tensor-valued functions. Besides, they satisfy Saint-Venant compatibility conditions as well as the relation $e = e^d$ arising from the incompressibility condition $\operatorname{div} v_h = 0$. All these conditions are certain constraints imposed on coefficients of the considered piece-wise linear functions. Leaving out the constraints arising from compatibility conditions and retaining those arising from the relation $e = e^d$, we arrive at an internal approximation E_h^d for E^d with $\tilde{E}^d$ consisting of smooth tensor fields. The condition $\operatorname{Def} V_h^d \subset E_h^d$ is obviously satisfied.

Replacing the spaces V, E in (11) with their approximations V_h, E_h we formulate the finite dimensional extremum problem

$$s_{\Delta h} = \inf \{ \int_\Omega \left(d_x(e_1) + \cdots + d_x(e_m)\right) dV:\ \ e_1,\dots,e_m \in E_h,$$
$$e_1 + \dots + e_m \in \operatorname{Def} V_h,\ \ \int_\Omega (e_1 \cdot \tau_1 + \cdots + e_m \cdot \tau_m)\, dV = 1 \}$$

Similarly, we obtain the discretization of the problem (16) using spaces V_d, E^d, V_{dh}, E_h^d instead of V, E, V_h, E_h and deviators $\tau_1^d,\dots,\tau_m^d$ instead of elastic stress fields $\tau_1,\dots,\tau_m$. We consider internal approximations, therefore $e = (e_1,\dots,e_m)$ satisfying the constraints of the discretized problems also satisfies those of original problems (11) or (16). Hence, the infimum $s_{\Delta h}$ gives an upper bound for the safety factor, $s_\Delta \le s_{\Delta h}$. Moreover, the following proposition states the convergence of finite element approximations.

Theorem 3. If the hypotheses of Theorem 1 or Theorem 2 are satisfied and V_h, E_h (V_{dh}, E_h^d) are internal approximations for V, E (V_d, E^d) with

$\mathrm{Def}\, V_h \subset E_h$ $(\mathrm{Def}\, V_{dh} \subset E_h^d)$, then

$$s_\Delta \leq S_{\Delta h} \quad \text{and} \quad s_{\Delta h} \to s_\Delta \quad \text{when} \quad h \to 0.$$

A proof of this proposition as well as proofs of Theorems 1, 2 and a more detailed discussion of the questions under consideration can be found in KAMENJARZH and MERZLJAKOV [1994].

ACKNOWLEDGMENTS
The research described in this paper was made possible in part by Grant No.ND6000 from the International Science Foundation.

REFERENCES

1938 MELAN, E., "Theorie Statisch Unbestimmter Tragewerke aus Idealplastischem Baustoff", Sitzungsbericht der Academie der Wissenschaften (Wien) Abt IIA, **145**, 195.

1960 KOITER, W.T., "General Theorems for elastic-plastic bodies", in: Progress in Solid Mechanics, Sneddon, I. and Hill, R.(eds.), Vol.1, North Holland, Amsterdam.

1968 ROCKAFELLAR, R.T., "Integrals which are Convex Functionals", Pacific J.Math., **24**, 525.

1969 CHIRAS, A.A., "Linear Programming Methods in Analysis of Elastic-Plastic Systems", Stroyizdat, Leningrad (in Russian).

1971 GOLSTEIN, E.G., Duality Theory in Mathematic Programming and its Applications, Nauka, Moscow (in Russian).

1971 NAYROLES, B. "Quelques Applications Variationelles de la Theorie des Fonctiones Duales a la Mecanique des Solids", J.Mecanique, **10**, 2, 263.

1973 MAIER, G., "A Shakedown Matrix Theory Allowing for Workhardening and Second-Order Geometrical Effects", in: SAWCZUK, A. (ed.), Foundations of Plasticity, p. 413, Noordhoff Int. Publ., Groningen.

1974 CORRADI, L. and ZAVELANI A., "A Linear Programming Approach to Shakedown Analysis of Structures", Comput. Meths. Appl. Mech. Engng., **3**, 37.

1975 MARTIN, J.B., "Plasticity, Fundamentals and General Results", The MIT Press, Cambridge, Ma.

1976 DEBORDES, O. and NAYROLES, B., "Sur la Theorie et le Calcul a l'Adaptation des Structures Elasto-plastiques", J.Mecanique **15**, 1, 1.

1976 EKELAND, I. and TEMAM, R.,"Convex Analysis and Variational Problems", North Holland, Amsterdam.

1978 CIARLET, P.," The Finite Element Method for Elliptic Problems", North Holland, Amsterdam.

1979 MERCIER, B., "Numerical Methods for Viscous-plastic Problems", in: Lions J.-L. and Marchuk, G.I. (eds.), Numerical Methods in Applied Mathematics, Nauka, Novosibirsk (in Russian).

1980 GOKHFELD, D.A. and CHERNIAVSKY, O.F., "Limit Analysis of Structures at Thermal Cycling", Sijthoff and Nordhoff, Alphen au den Rejn, The Netherlands.

1980 KAMENJARZH, . J., "On Dual Problems in the Theory of Limit Loads for Perfectly Plastic Bodies", Sov. Phys. Dokl., **24**(30),177.

1981 KAMENJARZH, J., "Stresses in Incompressible media. Equivalent Formulations of Perfect Plasticity Problem", Sov. Phys. Dokl., **26**(11), 1051.

1991 POLIZZOTTO, C., BORINO G.,CADDEMI, S., and FUSCHI, P., "Shakedown Problems for Material Models with Internal Variables", Eur. J. Mech., A: Solids, **10**, 6, 621.

1992 KAMENJARZH, J. and WEICHERT, D., "On Kinematic Upper Bounds for the Safety Factor in Shakedown Theory", Intern. J. Plasticity, **8,** 827.

1992 KAMENJARZH, J., and MERZLJAKOV, A., "On Dual Extremum Problems in Shakedown Theory", Sov.Phys.Dokl.,**325**, 1.

1994 KAMENJARZH, J. and MERZLJAKOV, A., "On Kinematic Methods in Shakedown Theory. Part I. Duality of Extremum Problems. Part II. Modified Kinematic Method." Intern.J. Plasticity. (to appear).

INFLUENCE OF GEOMETRICAL CHANGES ON CYCLIC RESPONSE OF STRUCTURES

GEOMETRIC EFFECTS ON SHAKEDOWN AND RATCHETTING OF AXISYMMETRIC CYLINDRICAL SHELLS SUBJECTED TO VARIABLE THERMAL LOADING

G. MAIER, L.G. PAN* and U. PEREGO
Department of Structural Engineering, Politecnico di Milano
Piazza Leonardo da Vinci
32-20133 Milano, Italy

INTRODUCTION

This paper presents results obtained from numerical simulations of the responses of an elastic-plastic thin cylindrical shell to fluctuating axisymmetric temperature in the presence of uniform axial stresses. The engineering situation considered has practical importance in nuclear reactors and has been the subject of a number of earlier studies. The main purpose is to assess quantitatively the influence of geometry changes, primarily due to plastic yielding, on shakedown and ratchetting (incremental collapse) phenomena. In particular, these phenomena are investigated with respect to both the stabilizing effects of tensile primary stresses on them, and their strong interference with elastoplastic buckling. The systematic evolutive analyses presented herein are also intended to critically assess the validity of earlier results (mainly condensed in the socalled Brussels diagrams) which have been established by simplified methods of shakedown based on the small deformation (no geometric effects) hypothesis.

Keywords : shakedown, ratchetting, buckling, thermal loading, shells, plasticity, nuclear reactors.

Engineering structures are often subjected to "variable repeated" loading (in particular, cyclic loading). This means that the external agents acting on the structure (body forces, tractions, thermal strains and/or imposed displacements) fluctuate significantly in time. Often the following hypotheses are reasonably realistic.
(1) The loading histories are slow enough to make inertia forces negligible and can be defined by a "load domain", i.e. by a given region in the load parameter space such that each point in that region represents a load condition attainable any number of times along an unknown time sequence.
(2) The materials are elastic-plastic, i.e. history-dependent ("nonholonomic") but temperature-independent and time-independent (or "inviscid").

* On leave of absence from Rsinghua University, Beijing, People's Republic of China

Z. Mróz et al. (eds.), Inelastic Behaviour of Structures under Variable Loads, 239–262.

Under variable repeated loading, after some time, plastic strains may cease to develop further and the accumulated dissipated energy in the whole structure remains bounded. This occurrence is traditionally referred to, after Prager, as (elastic) "shakedown" (SD), a term also traditionally used to label the whole area. Alternatively, plastic yielding can never stop developing (so that the accumulated dissipation is unbounded in time) by either one or the other of two distinct phenomena :

• Incremental collapse of ratchetting (R) : the deformed configuration is gradually amplified in the sense that meaningful displacements grow, in principle unboundedly, in time.

• Alternating plasticity (AP) or plastic shakedown (PSD) : plastic strains and, hence, displacements remain bounded in time.

A severe structural crisis, always to be avoided in design, is implied by R, which eventually leads to unserviceability, instability or local failure. A designer may or may not accept AP, depending on the material employed and on the number of cycles expected in the service life of the structure. A SD scenario does not necessarily imply that the structure is safe, because plastic yielding processes prior to SD might cause undesirable configuration changes and damage.

For a given elastoplastic structure under a given load domain, the SD theory and analysis methodology primarily aims at computing the critical load factor s (or safety factor) below which SD occurs and above which it does not occur. The analysis usually also provides, as by-products, a state of residual postshakedown stresses and/or a set of kinematic quantities associated to an R or AP mechanism.

Strictly understood, SD analysis avoids time-stepping evolutive solutions and represents, conceptually and computationally, an important generalization of classical (rigid-plastic) limit analysis.

Traditional SD theory preserves two basic assumptions of limit analysis : firstly, perfectly plastic (no hardening) constitutive models ; and secondly, "small displacements", i.e. equilibrium equations referred to the original, undeformed geometry. In a number of engineering situations, these assumptions turn out to be unrealistic and should be relaxed.

Hardening behaviour generally has a "stabilizing effect", in the sense that the safety factor s with respect to lack of SD (i.e. either AP or R) is generally increased over the one attainable with perfectly-plastic behaviour at equal original yield surfaces. For instance, linear kinematic and isotropic hardening models rule out the possibility of *R*. Sometimes materials (e.g. concrete) and structural components (e.g. compressed bars, reinforced concrete beams in bending) exhibit negative hardening, i.e. "softening", unstable behaviour. Such local instability generally has an unstabilizing effect on R thresholds. Stabilizing or unstabilizing effects in the above sense may be caused by the non-negligible influence of the deformations on the equilibrium relations.

An interplay may arise between constitutive post-yield behaviour (hardening or softening of materials or structural components) and the effects of configuration changes on equilibrium, when both these mechanical features are not negligible.

Many research contributions have been published with the purpose of extending the classical SD theory (centred on the Bleich-Melan theorem and on the Neal-Symonds-Koiter theorem) and the relevant computational methods, in order to allow for geometric effects, alone or in combination with hardening/softening.

Only some of these extensions are dealt with in available surveys of the SD area, such as those by KÖNIG [1987], and KÖNIG & MAIER [1981] or in books on plasticity, such as those by COHN et al [1979], ZYCZKOWSKI [1981] and KALISZKY [1989], where a broad conspectus of classical SD theory can be found. Therefore the SD literature explicitly intended to deal with geometric effects is concisely surveyed below.

To the writers' knowledge, so far the following general approaches have been considerably developed and have been applied to at least academic examples.

(1) A "second-order" theory proposed by MAIER and co-workers[1972, 1973, 1976, 1990], generalizes most results of the classical SD theory (static and kinematic theorems, bounds on post-shakedown quantities) to geometric and hardening effects, and preserves its computational simplicity by recourse to linearization of the geometric terms in the equilibrium equations and to piecewise-linearization of plastic constitutive laws.
(2) An "inadaptation" methodology, due to KÖNIG & SIEMASZKO [1984,1985, 1988, 1990], is centred on kinematic SD theorem which is applied iteratively and, hence, may become quite laborious. It permits a great generality as for the assumed constitutive laws, which are accounted for between steps consisting of conventional (small deformations) SD analysis. In the limit analysis context and interesting development with some conceptual similarity is due to TEXEIRA & SMITH [1989].
(3) A theory allowing for geometric nonlinearity in fairly general terms has been developed by WEICHERT and further extended by GROSS-WEEGE [1983, 1986, 1990]. Besides the customary constitutive hypotheses of convexity and normality, the additive decomposition of finite strains into an elastic and a plastic part was assumed. On this basis, the Beich-Melan theorem was generalized subject to a new condition which, in general, is hard to check numerically. In order to overcome this difficulty Weichert developed a less general theory, where the history of the external actions is conceived as consisting of two phases : a proportional increase of loads implying large displacement ; a subsequent variable repeated loading which cause "small" additional deformations and stresses. Small means here that all terms which are nonlinear in the time-dependent unknown additional quantities can be neglected. Geometric nonlinearity has rigorously been investigated with reference to shakedown of elastoplastic shells by STUMPF [to appear].

NGUYEN & GARY [1983] studied a one DOF model (Shanley model) of an elastic-plastic column under constant compressive axial load and cyclic thermal load acting on the central "Shanley cell". Despite the simplicity of the model, these authors pointed out a variety of possible alternative behaviours and, in particular, the possibility of interaction between ratchetting and buckling. In terms of theoretical general contributions, this study adopted and extended MAIER's second-order theory [1972].
PYCKO & KÖNIG [1991] extended to a geometrically non-linear context, a theorem concerning the convergence to steady plastic cycles of the response of structures subjected to cyclic loading when alternating plasticity inadaptation occurs.
Generalizations to geometric effects of various important contributions in the SD area might be possible, even if not explicitly considered so far. As representative of these contributions, we quote here those by POLIZZOTTO [1982], PONTER & KARADENIZ [1985] and STEIN et al [1990].
To the writers' knowledge, the only experimental research to investigate geometric effects on SD, was carried out by LEERS et al [1985], who tested thin tubes under radial loads and temperature gradient and slight axial compression. Unfortunately, no comments or information are given on the role of axial force and configuration changes in the published report [1985].
The purpose of the present paper is to elucidate and to quantitatively assess in a typical nuclear engineering structure, the influence of geometry changes on equilibrium in the presence of axial "primary" stresses and, hence, on shakedown or lack of shakedown under cyclic thermal loading. In a sense, the objectives pursued herein are similar to those of an experimental investigation. Numerical simulations are carried out systematically as parametric studies, by employing ABAQUS, a well established finite element code for step-by-step nonlinear analysis. The geometric effects are highlighted, while simplifying hypotheses are assumed on other mechanical features (perfect elastoplasticity, axial symmetry, no initial imperfections). Suitable comparisons are carried out with results achieved through time-stepping analyses based on the "small deformation" hypothesis and through Bree-type diagrams (so-called Brussels diagrams)

analyses based on an upper bound approach see KARADENIZ et al [1987]. A recent CEC report by SAVE et al [1991] contains a well organized comprehensive treatment of the whole nuclear-engineering-oriented shakedown analysis area in which the present study is inserted.

PROBLEM FORMULATION : AN AXISYMMETRIC SYSTEM WITH FLUCTUATING TEMPERATURE

Geometric, material and loading data.

The structure considered (*Figure 1 (a)*) is a thin-walled tube of circular section subjected to a constant axial mechanical load and to a cyclically varying temperature distribution along its length. Similar structures have also been analysed [1967, 1985, 1987, 1992] since they are typical of situations encountered in the design of liquid sodium cooled fast reactors.
The tube has a mean radius R = 350 mm, uniform thickness h = 0.875 mm, axial length L= 256.6 mm. These data correspond to a ratio of radius to thickness R/h = 400. The mechanical properties of the material, which is assumed to obey a time-independent elastic-perfectly plastic constitutive law, are as follows : Young's modulus $E = 1.84 \times 10^5$ MPa; Poisson's ratio $\nu = 0.3$; coefficient of thermal expansion $\alpha = 1.651 \times 10^{-5}$ °C^{-1} ; finally, a von Mises yield surface characterized by a uniaxial yield limit $\sigma_y = 1.42 \times 10^2$ MPa has been adopted. All material properties are assumed to be constant with temperature.
As mentioned above, two different kinds of load, mechanical and thermal, have been considered as acting upon the tube. The mechanical load is constant in time and acts in the axial direction. Its value will be characterized by the nondimensional variable $\sigma_p=\sigma_p/\sigma_y$, σ_p being the axial stress (primary stress), uniform throughout the tube, due to the applied axial load P. Tensile and compressive axial loads will be considered in the following sections.
The temperature distribution is assumed to be uniform and equal to a reference temperature θ_R = 150°C on the outer surface of the tube (*Figure 1 (b)*). Inside, due to the hypothetical presence of hot liquid, the temperature has a discontinuity across the midsection of the tube (while it remains uniform in the hoop direction) and therefore produces a discontinuity in the hoop stress (secondary stress). Denoting by Z the axial co-ordinate, and setting Z = 0 at the left end of the tube, the discontinuity of temperature is located at Z = 128.3 mm (midsection). The hot part of inner surface of the tube (the right half in *Figure 1(b)*) has a temperature which varies cyclically between θ_R and θ_{max} and can be viewed, for instance, as reproducing the thermal transients during start-up and shut-down conditions. The duration of cycle is immaterial since time only plays the role of an ordering variable and all time dependent effects are ruled out. The amplitude of the temperature jump $\Delta\theta = \theta_{max} - \theta_R$ is caracterized by the nondimensional variable $\sigma_t = E\alpha\Delta\theta\, \sigma_y^{-1}$; this represents the ratio between the maximum hoop stress (computed by assuming a perfectly elastic behaviour and located in correspondence to the temperature discontinuity) and the yield stress.
Owing to the difference between temperatures at the inner and outer surfaces of the tube, there will be an approximately linear variation of temperature across of thickness on the right half of the tube. The sharp temperature jump at a fixed location represents a rather crude idealization. However, it is suitable for focusing on geometric effects and will be smoothed by the finite element modelling (described later).

The left end of the tube is constrained in the axial direction while both ends are free in the radial direction. Fixed ends for radial displacements would represent a more severe condition but would add little to the present study.

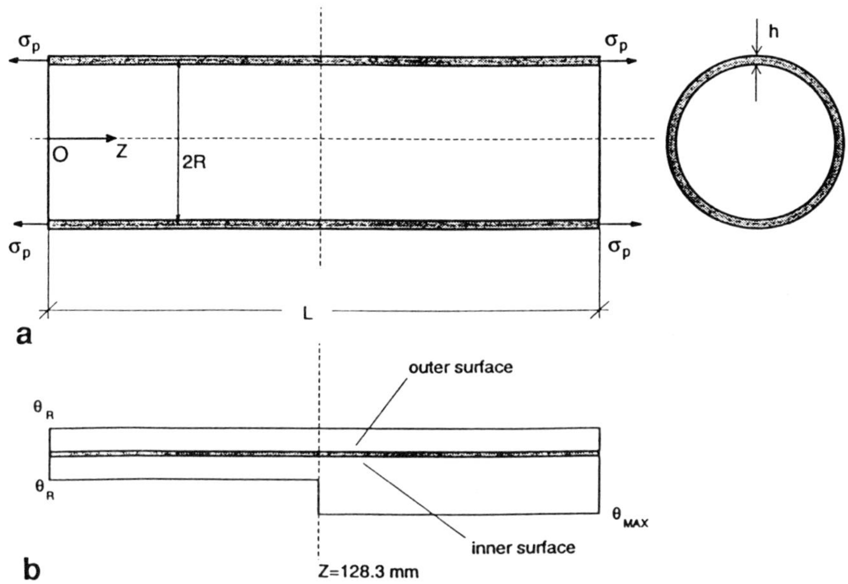

Figure 1 : (a) Geometry of cylindrical shell and "primary" axial load ; (b) temperature distribution on inner and outer surface of shell.

Bree (or Brussels) diagram by a "small-deformation" shakedown analysis.

The present analysis problem concerning the structure of *Figure 1 (a)* can be formulated as follows : for a given combination of external loads (mechanical, constant in time at the nondimensional value σ_p ; thermal, cyclically carying in time with nondimensional amplitude σ_t), determine whether, after an infinite number of cycles, plastic dissipation will cease (shakedown) or will grow unbounded either by plastic shakedown or by ratchetting.

The results of this type of analysis are conveniently represented in the plane of loading conditions σ_τ versus σ_p. The place is subdivided as in *Figure 2* into three regions : the region of load combinations such that the structure will elastically shakedown (marked by the initials ESD) ; the one such that a ratchetting mechanism will eventually develop (R), and the domain of load combinations such that a plastic shakedown mechanism will develop (PSD).

This type of diagram takes its name from BREE [1967]. Originally Bree devised this σ_t versus σ_p diagram with reference to a thin-walled tube subjected to internal pressure and/or axial load. The structure was also subjected to a cyclic temperature variation, uniform through the volume of the pipe wall. Owing to the effectiveness and the practical convenience of this representation of results, Bree-type diagrams were produced for structures other than Bree's and called "Brussels diagrams" after such extension (see Ponter et al [1985, 1987] and Save et al [1991]). Nowadays, in nuclear reactor structural design, Bree diagrams are particular Brussels diagrams concerning a specific kind of problem. *Figure 2* shows the Brussels diagram obtained by small displacement analysis relative to the problem described in the preceding section. The dashed curves define the boundaries between the three regions (elastic shakedown, plastic shakedown and ratchetting) for a material which obeys Tresca yield criterion.

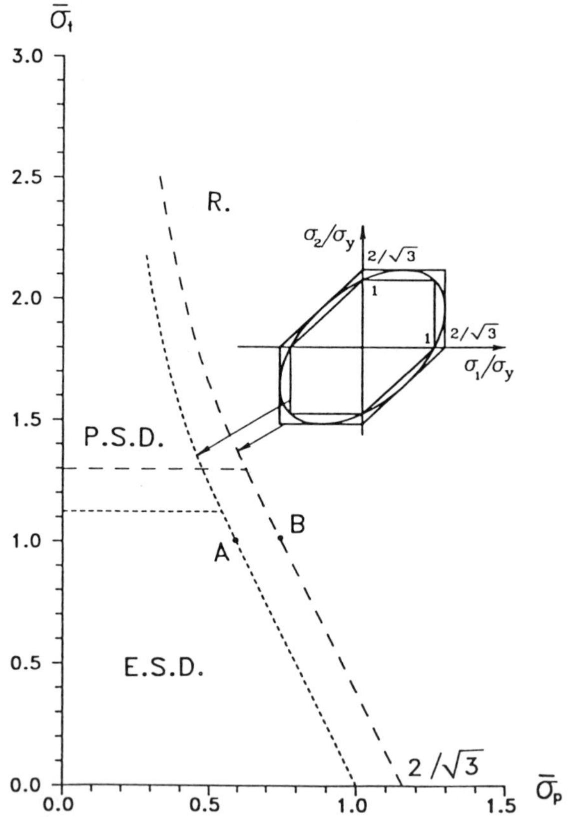

Figure 2 : Brussels diagram for a tube subjected to constant axial force and asymmetric stationary cycling thermal loading.

The curve with short dashes refers to a material which obeys a Tresca yield criterion with a yield limit equal to σy. This curve was made available by Carter and Ponter (University of Leicester, private communication), who computed it by their kinematic approach. The Tresca criterion is piecewise-linear in the space of principal stresses and since the principal directions can be regarded as fixed in the present case, the assumption of the Tresca criterion reduces the shakedown analysis to linear programming. Unfortunately, the Tresca criterion was not available in ABAQUS [1991] and all our computations were carried out using a von Mises yield condition. In order to make the comparison meaningful, the curve with longer dashes in *Figure 2* defines the shakedown boundaries obtained for a Tresca yield surface circumscribed on the Mises ellipse which intersects the principal axis at $\sigma_1/\sigma_y = \sigma_2/\sigma_y = 1$. The Mises boundaries lie between the inner and outer Tresca curves and, obviously, coincide with the former for $\sigma_t = 0$.

FINITE ELEMENT MODEL

The tube has been descretized in space by using two-dimensional, axisymmetric, solid finite elements (not shells). The mesh adopted is shown in *Figure 3.* The axial dimension has been subdivided into six fields of length 42.77 mm each. These fields have been descretized using and increasing number of elements as, starting from the left, the region of the temperature discontinuity is approached. Hence, the first field (A in *Figure 3*) has been descretized by 8 elements while the third field (D in *Figure 3*), which is close to the temperature discontinuity, has been descretized by 50 elements. All elements employed were either triangular or quadrilateral second-order elements (6 nodes, 3 Gauss points for triangles and 8 nodes, 9 Gauss points for quadrilateral elements). Overall, the mesh was composed of 589 nodes and 140 elements. Full integration was adopted over all elements, in order to have Gauss points as close as possible to the inner and outer surfaces of the tube.

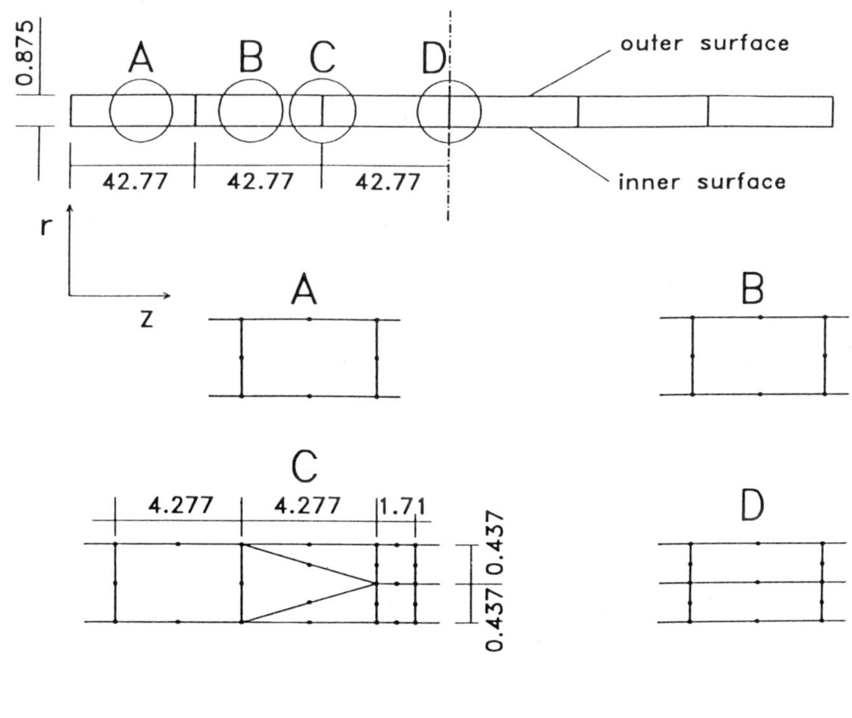

Figure 3 : Coarse finite element mesh. (not at real scale). Dimensions are in mm

In fact, owing to the temperature discontinuity along the length and to the presence of the temperature gradient across the thickness, the bending moment in the cylindrical wall plays an important role in the nonadaptation mechanism. The flexural behaviour is characterized by maximum (in modulus) stresses at the extreme fibres of the thickness. As usual in finite elements, the elastic-plastic constitutive law is enforced at Gauss points only. Trial tests with reduced integration have shown a small fictitious increase of strength due to the fact that the onset of plasticity at the edges of the wall thickness was missed by the Gauss points.

In order to investigate the sensitivity of the model to mesh refinement, a few analyses were also carried out using a finer mesh with 1561 nodes and 436 elements.

Figure 4 shows a comparison of results obtained by the coarse and fine mesh in the case of a tensile primary load. Small displacements are assumed here. The loading conditions are defined by an axial load $\sigma_p = 0.8$ and an amplitude of the thermal cycle $\sigma_t = 1$. The dashed line marked 2 represents the original configuration of the tube wall, before any thermal expansion at $\theta = \theta_R$. The dashed line marked 3 represents the configuration after the application of a tensile axial load which produces, as expected, a small inward radial displacement. Solid curves represent deformed configurations after 3 and 40 thermal cycles, respectively, obtained by using the fine mesh. As can be clearly seen, there is a R mechanism which develops as the thermal cycling progresses and which involves the flexural behaviour of the tube. It is worth observing that the plastic deformation due to the thermal cycling is almost completely confined to a very localized region of the tube, so that the maximum radial displacement takes place in the area located at about $Z = 150.5$ mm, at a distance of about 22 mm from the midsection. The dashed curve, which is almost superposed upon the solid curve, represents the deformed configuration after 40 cycles obtained by using the coarse mesh. The same curve for the coarse mesh after 3 cycles is not distinguishable from the corresponding solid one for the fine mesh.

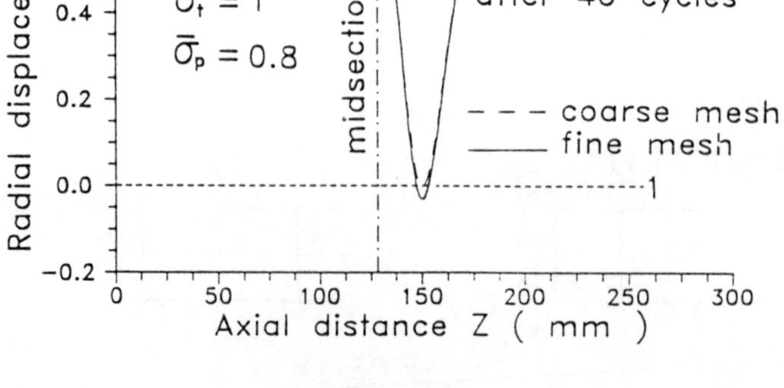

Figure 4 : Radial displacement of inner surface for coarse and fine meshes.

On the basis of these and other results, the coarse mesh of *Figure 3* has been judged accurate enough to assess the asymptotic structural behaviour under cyclic loading.
The finite element code ABAQUS [1991] employed herein assumes a continuous temperature distribution by interpolating among nodal values on the side of the elements. The lower side of the first 8-node element on the left of the temperature discontinuity, has a parabolic temperature which is equal to θ_R at the first two nodes from the left and is equal to θ_{max} at the third node.
As will be discussed in a later section for the case of compressive primary loads, buckling may occur before either a R or an AP mechanism can develop. Since the buckling load is imperfection-sensitive, its value is affected by the path followed in loading space (σ_t versus σ_p space). Different paths may lead to different buckling loads. The reference path which has been followed entails first the application of the axial load, after which the structure is brought to the uniform reference temperature θ_R (at this stage, the structure is still completely elastic) ; finally, the inner surface of the right half of the tube undergoes thermal cycles.

GEOMETRIC EFFECTS DUE TO TENSILE AXIAL STRESSES

The case of a tensile axial load is considered first. *Figures 5 (a)* and *5 (b)* show, for a temperature jump characterized by $\sigma_t = 1$, the development of the dissipated energy (dissipation) and its rate for several values of the axial load σ_p. In these pictures, solid and dashed lines refer to small and large displacement analyses, respectively. As it can be observed, for $\sigma_p = 0.7$, after an initial plastic dissipation in the first few thermal cycles, the system shakes down and the dissipation remains constant, while its rate plunges sharply to zero after an initial peak. The difference between small and large displacement analyses is in this case negligible. By contrast, for $\sigma_p = 0.8$ and $\sigma_p = 0.85$, the dissipation growth is apparently unbounded as the cycles progress. This is interpreted as a sign of lack of SD. As expected, when geometric effects are accounted for, the tensile load contributes to reduce significantly the stress peaks and, hence, the dissipated energy. Furthermore, the difference between the results of the two analyses increases with the number of thermal cycles.
As shown by MAIER [1972], nonadaptation due to R is ruled out from a theoretical standpoint when stabilizing second-order effects are present, but lack of SD due to AP is not ruled out. Other geometric effects, which are not allowed for by a second-order theory, are provided by the change (here reduction) of the wall section due to plastic and elastic lateral contraction, and could affect unfavourably the SD safety factor. However, in the structure and loading situation considered, it can be conjectured that, when a R mechanism is developing, the energy dissipation has eventually to stop because of prevailing stabilizing effects. Therefore, a nonvanishing trend of dissipation like the one in *Figure 5* means that either a PSD mechanism will eventually prevail (as is possibly hinted from *Figure 6*) or SD will occur, even though this may happen after a large number of cycles. From a practical point of view, a trend of dissipation, as in *Figure 5*, can be reasonably regarded as critical no matter whether, in reality, the dissipation eventually reaches an upper bound or tends to a finite asymptotic value.

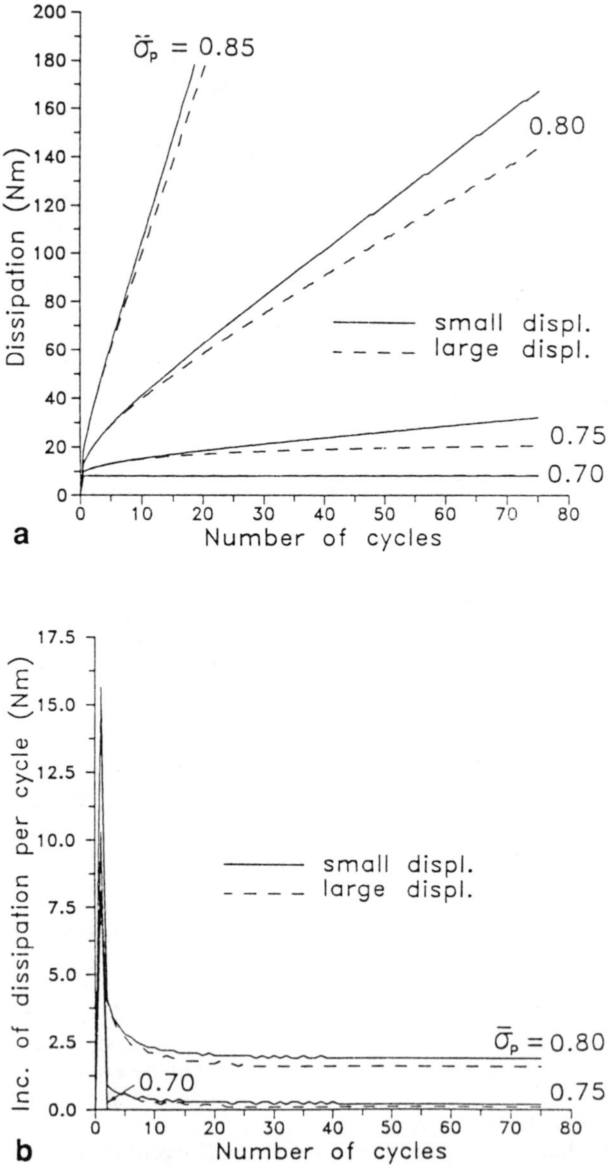

Figure 5 : Small and large displacement analyses for varying tensile axial load and thermal load $\sigma_t = 1$; (a) dissipation versus number of cycles; (b) increment of dissipation per cycle versus number of cycles.

The stiffening effect of the axial tensile load is apparent from the results illustrated in *Figure 6*. In *Figure 6 (a)*, for $\sigma_p = 0.8$ and $\sigma_t = 1$, the deflection of the point located at $Z = 150.5$ mm on the inner surface of the tube is plotted against the number of cycles. The structure exhibits a R flexural mechanism with inward displacements which increase unboundedly in the small displacement analysis. As anticipated, in the large displacement analysis, the axial load significantly reduces the lateral deflection. Results for a load combination, $\sigma_t = 1$, $\sigma_p = 0.2$, which produces ESD are also plotted as reference. In *Figure 6 (b)*, similar results are shown for the axial displacement. It can be observed that, in this case, the flexural mechanism is combined with an axial mechanism.

The case for $\sigma_p = 0.75$ is more intriguing (*Figure 5*). Here the effect of the axial load is such that, if geometric effects are taken into account (dashed curve), the dissipation remains "almost constant" (the dissipation rate almost goes to zero after about 25 cycles). On the other hand, if geometric effects are neglected (solid curve), the dissipation shows an almost constant nonzero rate of increase. In the first case (dashed curve), the structure seems to shake down. In the second case (solid curve) it does not.
A possible, empirical criterion to ascertain whether shakedown occurs is as follows. Choose a number of cycles which is sufficiently large (we have chosen 75 cycles in the present case) to allow for the structure to reach an almost steady state. Let ΔD_{75th} be the dissipation increment over the 75th cycle. Let ΔD_{max} be the maximum dissipation increment in any cycle from 0 to 75. If the ratio $\Delta D_{75th}/\Delta D_{max}$ is less than a certain reference value (say 0.05), the structure is shaking down. If $\Delta D_{75th}/\Delta D_{max} > 0.05$, the structure will fail due to either R or AP. According to this criterion, the shakedown boundary for $\sigma_t = 1$ is at $\sigma_p = 0.78$ for large displacement analysis. This value may be compared to the values $\sigma_t = 1$, $\sigma_p = 0.594$ (point A in *Figure 2*) and $\sigma_t = 1$, $\sigma_p = 0.75$ (point B in *Figure 2*) computed by Carter and Ponter (private communication) assuming small displacements, the kinematic approach and a Tresca material.

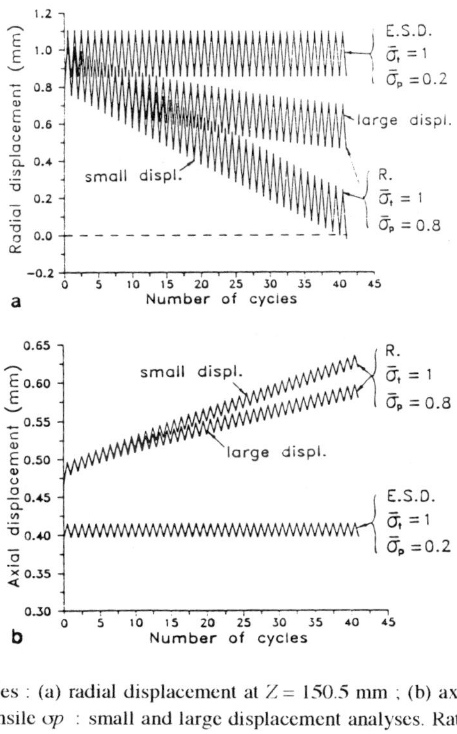

Figure 6 : Deformation histories : (a) radial displacement at $Z = 150.5$ mm ; (b) axial displacement at right end of tube for $\sigma t = 1$ and varying tensile σp : small and large displacement analyses. Ratchetting mechanism.

If the axial primary load is decreased to $\sigma_p = 0.2$ and the thermal load is raised to $\sigma_t = 2.5$, a different structural behaviour is observed. *Figure 7* shows that the energy dissipation increases unboundedly but also that the difference between small and large displacement analyses is almost negligible. The diagrams of *Figure 8 (a)* and *8 (b)* show that both the radial (*Figure 8 (a)*) and axial (*Figure 8 (b)*) displacements remain bounded and that accounting for the geometric effects results in a small, almost negligible, reduction of the maximum displacement. The radial displacement concerns a point on the inner surface of the tube at $Z = 150.5$ mm, while the axial displacement refers to a point on the right end of the tube ($Z = 256.6$ mm). The combination of unbounded dissipated energy and bounded cyclic displacement is interpreted as the development of an AP mechanism.

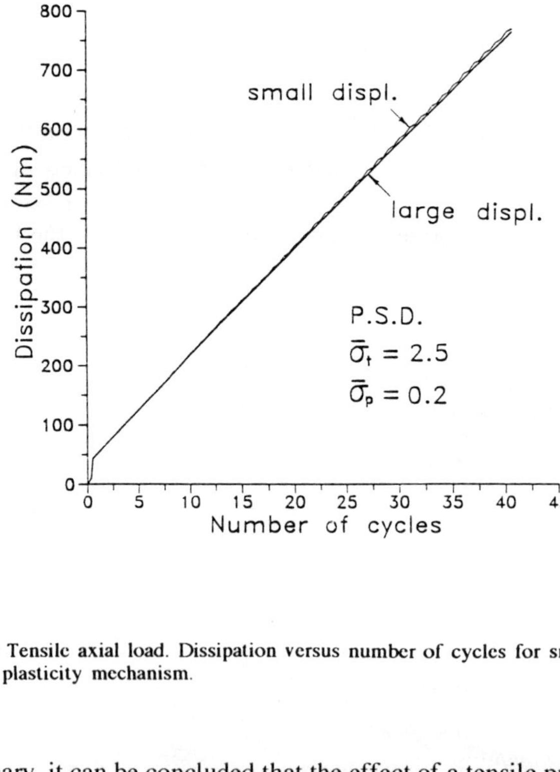

Figure 7 : Tensile axial load. Dissipation versus number of cycles for small and large displacement models : alternating plasticity mechanism.

In summary, it can be concluded that the effect of a tensile primary load is to enlarge the SD domain in load space when large displacement analyses are performed. However, this favourable effect is not very significant, not at least for the structural problem considered here. Therefore, neglecting geometric effects would give rise, in this case, to a slightly conservative design.

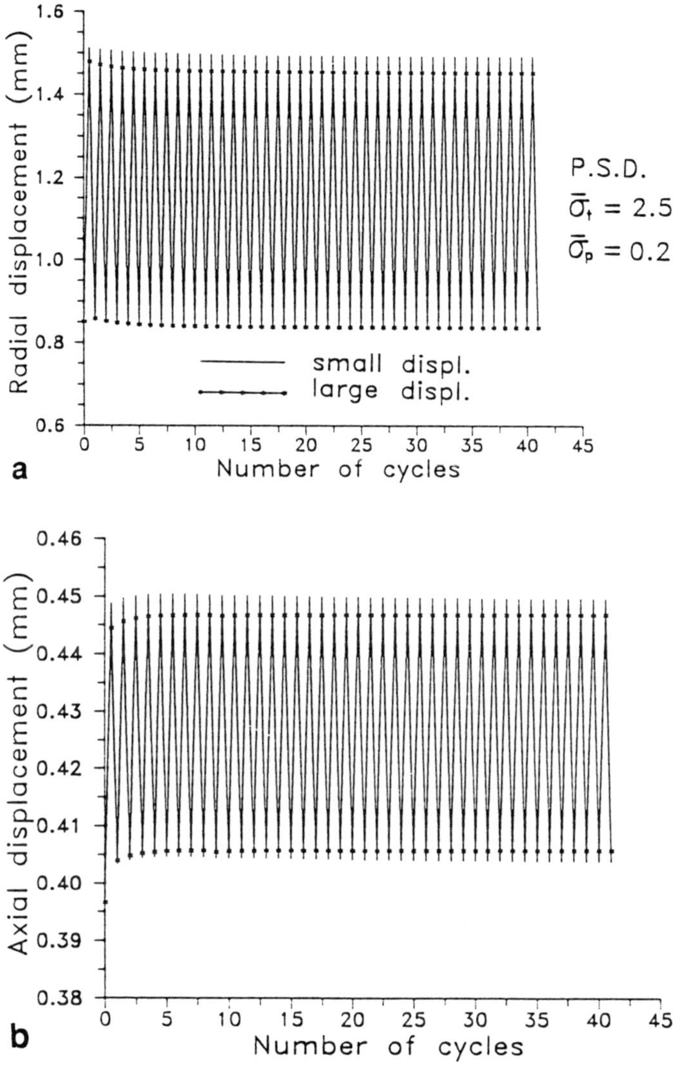

Figure 8 : Tensile axial load. Deformation histories : (a) radial displacement at Z = 150.5 mm ; (b) axial displacement at right end of tube from small and large displacement analyses. Alternating plasticity mechanism.

GEOMETRIC EFFECTS DUE TO COMPRESSIVE AXIAL STRESSES : RATCHETTING-INDUCED BUCKLING

Buckling load of cylindrical shells with imperfections.

When the axial load is compressive, the scenario changes completely as is apparent from *Figure 9* where the dissipation corresponding to σ_p = 0.6 and to several values of σ_t is

plotted against the number of cycles. This analysis, like all the others that will be discussed in this section, was carried out accounting for large displacement effects.
For lower values of ($\sigma t \leq 1.1$), after an initial plastic dissipation, the structure shakes down. For higher values (σt _ 1.125), after a certain critical number of cycles, the diagram shows a sudden increase of dissipation which terminates with a vertical slope. For instance, for $\sigma t = 1.125$, the structural response in the first few cycles is such that shakedown seems likely to occur. The trend changes dramatically after about 10 cycles and the structure finally collapses. The phenomenon is interpreted as a manifestation of the interaction between cyclic loading and buckling.

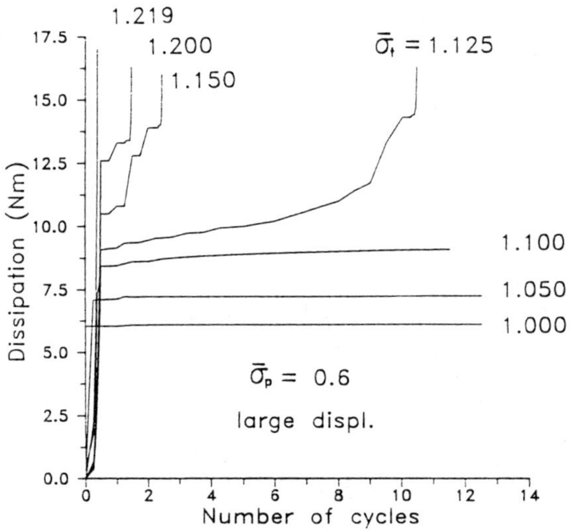

Figure 9 : Dissipation versus number of cycles for varying thermal load and compressive $\sigma_p = 0.6$: large displacement analysis.

It should be noted that the analysis has been carried out with reference to the finite element model of *Figure 3*, which enforces axisymmetry. However, the buckling of cylindrical shells is characterized by the interaction of different buckling modes (see BAZANT and CEDOLIN [1991], CALLADINE [1983] or PIGNATARO et al [1991] for a review of the existing literature regarding both theoretical and experimental studies). Typically, for a cylindrical shell subjected to axial load, the first small-strain bifurcation involves an axisymmetric extensible pattern with a positive slope in a load-displacement curve. For larger deformations, this pattern changes to a nonsymmetric diamond-shaped, inextensible pattern (the so-called "Yoshimura pattern") which provides the critical load for the shell. Therefore, the results presented in this section are qualitative only. The real buckling loads will, in general, be lower than those shown in the following figures. Alternatively, one can imagine having a cylinder reinforced by stiffening rings in the hoop direction spread along its length. The stiffening rings prevent nonaxisymmetric modes and raise the buckling load. However, this increase is likely to

be small in view of the multiplicity of buckling modes, including axisymmetric of quasi-axisymmetric modes, which correspond to the same load [1991].
From a computational standpoint, the buckling load combination (axial load and temperature discontinuity) corresponds to a lack of convergence of the iterative procedure employed for the step solution, owing to onset of negative eigenvalues of the tangent stiffness matrix.

Empirical formulae

The buckling load of a shell structure, such as the cylinder considered here, is heavily affected by geometric imperfections [1972, 1991]. The theoretical buckling load obtained by the so-called "linear theory" (which only accounts for second-order effects) is substantially reduced by the presence of imperfections whose effects are accounted for, in a "nonlinear theory" by higher-order terms.
The following buckling loads may be useful as a reference. Classical (Euler's) elastic stability theory applied to cylindrical shells without imperfections provides the popular formula $\sigma_{rc} = (3\ (1 - \nu^2))^{-1/2}\ EhR^{-1}$ which for the present situation gives $\sigma_{cr} = 1.96\sigma_y$ (i.e. $\sigma_{cr} = 1.96$). It is worth noting that this formula is exact only for a shell of length equal to integer multiples of the half wavelength of the buckling mode and that the excess of σ_{cr} over the yield stress σ_y points out that buckling will occur in the plastic range at a lower load.
The effect of imperfections on elastic buckling can be captured by empirical formulae (obtained as the lower bound envelope of experiments) such as the one reported in BAZANT Z.P. [1991] which makes use of a "knockdown factor" ϕ in order to reduce the classical critical load, namely $\sigma'_{cr} = \phi\ \sigma_{cr}$, $\phi\ \sigma_{cr,}$, ϕ being defined as

$$\phi = 1 - 0.9\ [1 - \exp(-\ 0.0625\ \sqrt{R/h}\ )]$$

for $0.5 \leq L/R \leq 5$
$100 \leq R/h \leq 3000$

The present case is included within the applicability ranges of this formula which yields $\phi = 0.358$ and hence, $\sigma'_{cr} = 0.701\ \sigma_y$.
The combined effects of imperfections and material plastic behaviour can again be expressed by empirical formulae such as those suggested by the European Convention of Constructional Steelwork (ECCS) [1983], namely (see also GALETLY & BLACHUT [1990])

$$\sigma''_{cr} = \sigma_y\ [1 - 0.4123\ (\sigma_y\ / \alpha\ \sigma_{cr})^{0.6}] \text{ for } \alpha\ \sigma_{cr} ,> \sigma_y\ /2$$

where α is a knockdown factor defined as $\alpha = 0.70(0.1 + 0.01\ R/h)^{-1/2}$. Since the present case is included in its validity range (except for the end conditions which do not permit radial restraints), the above formula provides an estimate of the actual buckling load at the value $\sigma''_{cr} = 0.479\ \sigma_y$ ($\alpha = 0.345$), (see GALLETLY & BLACHUT [1990]).

Finite element analysis

The effect of axisymmetric imperfections of various shapes on the elastic-plastic critical load of an axially loaded pressurized cylinder has been recently investigated by extensive numerical analysis in the elastic range [1992]. The results of a similar finite element investigation for the present tube thermally and axially loaded in the elastic-plastic range are reported in *Figure 10*, where the nondimensional buckling load, denoted by $\sigma_{cr,}$ is plotted against the size of the thermal imperfection. Here, the term "thermal imperfection" on the abscissa denotes the maximum deflection, due to the combined action of the axial load and of the thermal discontinuity, at a point located at Z = 150.0 mm. The plot refers to a monotonic (not cyclic) thermal loading (and is marked by the label $N = 0$, N being the number of thermal cycles). As observed in *Figure 6 (a)*, thermal cycles produce an increasing lateral deflection which acts as an imperfection for the evaluation of the buckling load. Diagrams of the type shown in *Figure 10*, for $N > 0$ would by slightly different, although they would reproduce the same behaviour qualitatively. From *Figure 10*, it turns out that the buckling load undergoes a dramatic reduction (with an almost infinite slope) even for very small lateral deflections ("imperfections"). These deflections may initially increase with thermal cycling even if the structure would finally shakedown in a small displacement analysis. As a consequence, the primary compressive load, which was originally safe, can become critical.

The ECCS formula [1983] used above is based on standard manufacturing imperfections, therefore it is expected that the asymptotic value of the buckling load resulting from the present analysis and affected by thermally induced imperfections will be lower than the ECCS value $\sigma''cr = 0.479\ \sigma y$, as is apparent in *Figure 10*.

The occurrence of buckling corresponds, in *Figure 9*, to an infinite slope of the dissipation curve. The higher σt, the sooner (i.e. the smaller the number of cycles) the radial deflection reaches a value which makes the compressive load critical. This has important design implications : in the presence of cyclic thermal loading, an estimate of the buckling load, on the bases of the undeformed structure with fabrication imperfections only, may lead to seriously unconservative designs.

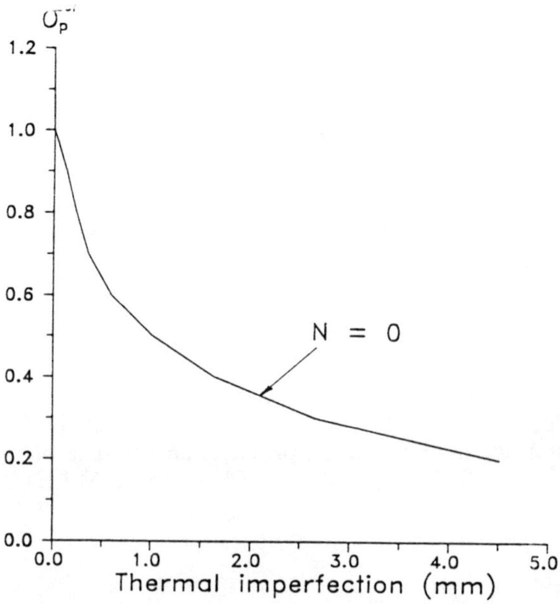

Figure 10 : Buckling load versus thermal imperfection for monotonically increasing temperature discontinuity (N = 0).

A summary of the structural response under compressive loading is illustrated in *Figure 11*. Here, the loading combinations which lead to buckling, are plotted in a plane σ_t versus σ_p (solid lines). The figure also shows the shakedown boundaries (dashed lines) already discussed with reference to *Figure 2*. Each of the solid curves corresponds to a different number of cycles. More specifically, the rightmost curve of this type corresponds a monotonic increase of the amplitude of the temperature discontinuity (noncyclic thermal load). We recall that the loading path is such that the primary load is applied first, together with the uniform thermal expansion. Then the temperature discontinuity is applied, its amplitude being characterized by σ_t. Points belonging to this rigthmost curve represent loading combinations such that the structure buckles before even one cycle can be performed. Likewise, the second curve from the right is the locus of loading combinations such that the structure buckles after 20 cycles, and so on. These curves tend to become closer and closer as the critical number of cycles increases so that the last curve, corresponding to 70 cycles, can be considered as a reasonable reference for design purposes.

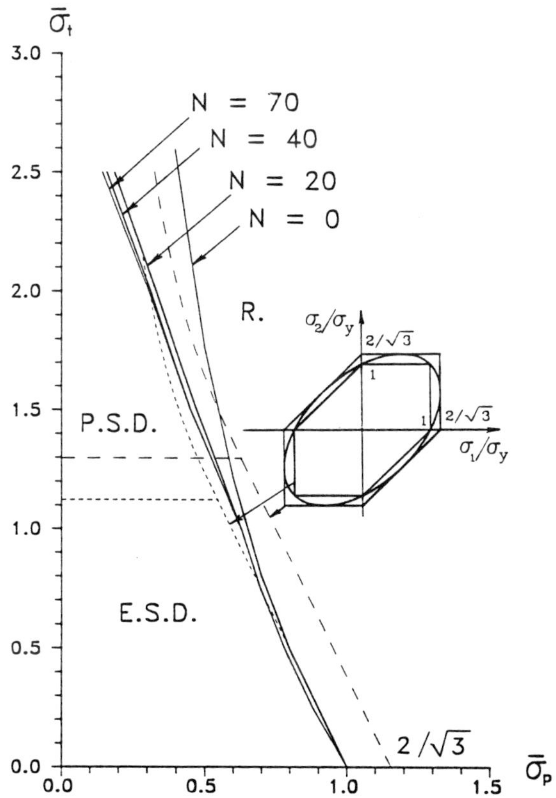

Figure 11 : Brussels diagram for compressive primary load : solid lines are loci of buckling load combinations.

An alternative view of the interaction between the size of the thermal load and the number of cycles is given in *Figure 12*. Again, it can be observed that, after about 70 cycles, the buckling load becomes almost insensitive to the number of cycles. On the other hand, the buckling load depends heavily on the amplitude of the temperature discontinuity : for $\sigma_t = 2.5$, the buckling load tends asymptotically to a value which is about one sixth of the corresponding asymptotic value for $\sigma_t = 0.5$. The structural response for $\sigma_p = 0.6$ and for σ_t increasing from $\sigma_t = 1.125$ to $\sigma_t = 1.15$ is shown in greater detail in *Figure 13 (a)* and *13 (b)*, respectively. Both figures show the outward radial displacement of a point at $Z = 150.5$ mm belonging to the inner surface. In *Figure 13 (b)*, it can be seen that, after the initial uniform expansion (point A), the radial deflection increases more than linearly due to the geometric effect in the first half-cycle until the temperature rate changes sign (point B in *Figure 13 (b)*) before the tube can buckle. Therefore, the structure undergoes further cycles until it reaches the buckling point, marked by F, during the first half of the third cycle.
The same pattern can also be recognized in *Figure 13 (a)*, the only difference being that buckling occurs after a larger number of cycles. The radial displacement computed by a small displacement analysis is also plotted as a reference in *Figure 13(a)* and *13 (b)*. In this case for both load combinations the outcome is elastic shakedown.
A different loading condition, $\sigma_p = 0.2$, $\sigma_t = 2.5$, is considered in *Figures 14-15*. *Figure 14* shows the increment of dissipation against the number of cycles in small (dashed line) and large (solid line) displacement analyses. The small displacement analysis predicts a final failure due to lack of SD and, of course, cannot predict the onset of buckling, which is clearly caught by the large displacement analysis. The radial displacement at $Z = 150.5$ mm and the axial displacement at $Z = 256.6$ mm (the right end of the tube) are plotted in *Figures 15 (a) and 15 (b)*, respectively. The small displacement analysis predicts a final state of PSD : this is shown by the unbounded dissipation of *Figure 14* and by the bounded radial and axial displacements of *Figures 15 (a)* and *15 (b)*. On the contrary, the large displacement analysis, captures the geometric interplay between the axial load and the cyclic radial deflection : this produces a gradual growth of deformation and leads to the final instability with unbounded displacements.

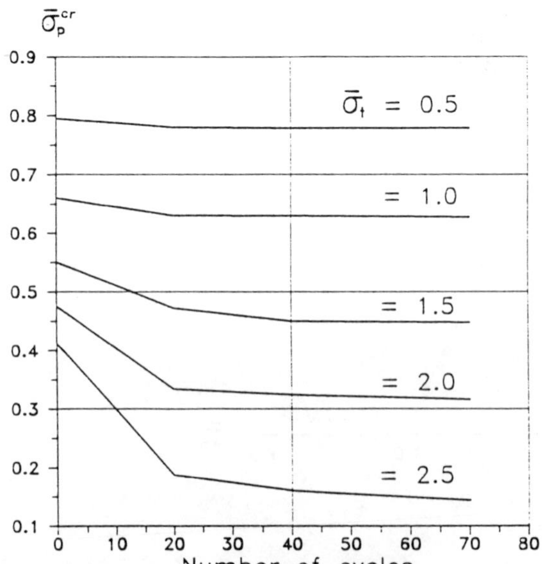

Figure 12 : Buckling load versus number of cycles for varying thermal load.

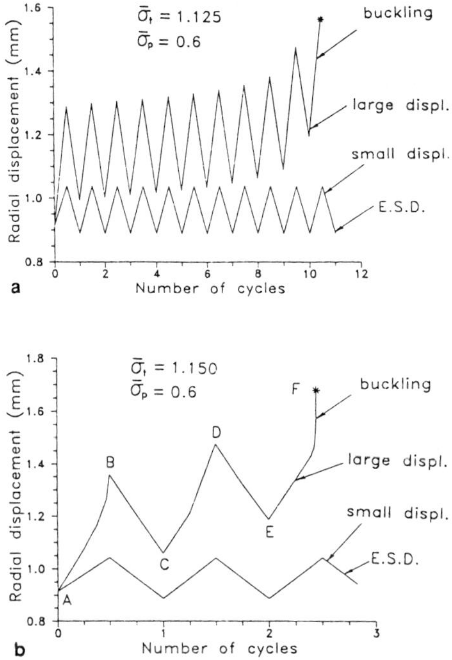

Figure 13. Compressive axial load. Deformation histories of radial displacement at $Z = 150.5$ mm for small and large displacement models: (a) $\sigma_t = 1.125$, $\sigma_p = 0.6$; (b) $\sigma_t = 1.150$, $\sigma_p = 0.6$.

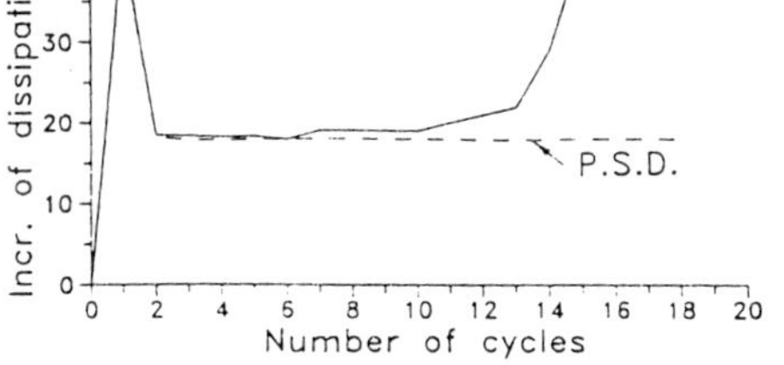

Figure 14. Increment of dissipation per cycles versus number of cycles for compressive axial load $\sigma_p = 0.2$ and for thermal load $\sigma_t = 2.5$: small and large displacement analyses.

In conclusion, unlike a tensile primary load, a compressive primary load has significant effects on structural safety. Evaluation of the SD boundaries bases on small displacement assumptions may be dramatically nonconservative owing to the important interaction between the radial deflections caused by cyclic loading and the geometric effects of the axial compressive load. Likewise, estimates of the buckling load which do not consider the cyclic nature of the thermal load may also be substantially nonconservative.

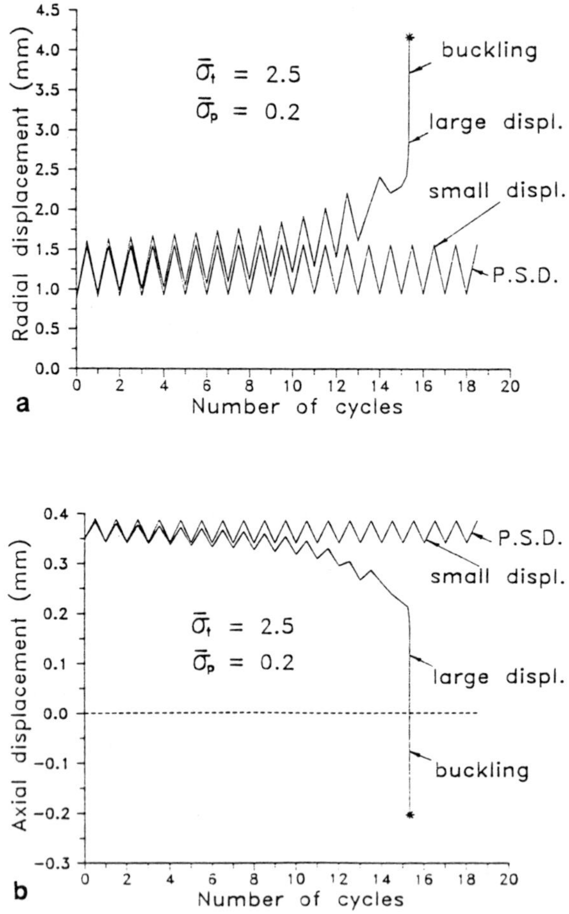

Figure 15 : Deformation histories : (a) radial displacement at Z = 150.5 mm ; (b) axial displacement at right end of tube for compressive axial load : small and large displacement analyses.

For the problem considered in our tests, it appears from *Figure 11*, that buckling and not R, is almost always responsible for the failure of the cylinder. On the contrary, there seems to be a limited region of load combinations in the σ_t versus σ_p plane such that PSD is the final outcome of the loading history. At this stage, for this kind of problem, in the absence of simplified approaches which correctly treat the interaction between the different loadings, the only available analysis tool seems to be a finite element, step-by-step evolutive analysis.

CONCLUSIONS

The study conducted herein by finite element, time stepping, elastic-perfectly plastic analyses of the engineering structure considered in the preceding sections, led to the following conclusions.
Firstly, tensile axial loads enlarge the elastic and plastic shakedown regions, i.e. give rise to the expected stabilizing effects. However, to ignore such effects by adopting the conventional small deformation hypothesis turns out to be legitimate in most engineering situations, because this hypothesis is slightly conservative throughout and its effect on the shakedown threshold does not exceed a few percent.
Secondly, compressive axial loads, even if small compared to the yield limit value, may have a very strong influence on the behaviour of the structure under variable-repeated thermal loading. Basically, plastic strains developing during the first loading cycles act like new imperfections generated in a strongly imperfection-sensitive structure. Buckling and ratchetting may combine in situations for which shakedown would be predicted by simplified or evolutive analyses resting on the usual hypothesis of small deformations. Therefore, in general, geometric effects must be allowed for in this case to avoid seriously unconservative designs. This conclusion would be strengthened by analyses carried out by removing the present assumption of axisymmetric response (which might be justified by the presence of circumferential stiffening ribs in the cylinder).
Thirdly, step-by-step elastic-plastic finite element analysis of engineering situations considered allowing for finite strains and onset of buckling, turns out to represent a practical proposition, at least as long as axisymmetry is assumed and computationally exploited. In fact, with commercially available nonlinear software and hardware, not only is feasibility guaranteed, but computing times and costs can also be regarded as acceptable in an industrial environment. More specifically, the following CPU times were required on a small workstation Silicon Graphics 4D25 to run problems which are representative of the category dealt with in the present paper. The small and large displacement analyses of *Figure 6*, for tensile axial load (75 cycles), ran in about 6600 s (less than 2 h) and about 12000 s (less than 3.5 h), respectively. In both cases the outcome was ratchetting. In compression, the analysis of *Figure 13 (a)* (11 cycles) required about 2900 s (less than 1 h) of CPU time to run. Nowadays, these CPU times appear to be quite affordable, especially when the performances of the new generation of reduced instruction set computer (RISC) workstations are considered.
Finally, "simplified methods" proposed in the literature as extensions of the classical shakedown theory and methodology to geometric effects, are hardly a practical proposition for the class of structural systems considered herein. In fact, the most critical compressive case and, in particular, ratchetting-buckling phenomena, are out of their applicability ranges. On the other hand, "Brussels diagrams", widely accepted in nuclear engineering design and resting on small deformation assumptions and on a heuristic approach, are basically confirmed for tensile axial loads, but invalidated for compressive primary stresses.

ACKNOWLEDGEMENTS

This paper is dedicated to the memory of Professor J.A. KÖNIG.
The results presented in this paper have been obtained in a CEC-sponsored research project conducted jointly by the Engineering Research Centre of GEC, Leicester, UK (Contractor) and the Department of Structural Engineering, Politecnico di Milano (Subcontractor). The writers wish to acknowledge with thanks : the fruitful and friendly

cooperation with Dr. P. WHITE in this project ; useful information provided by Dr. K.F. Carter and Professor A.R.S. PONTER, University of Leicester, and by Dr. A. SIEMASZKO, Polish Academy of Sciences ; the support from W.G. "Codes and Standards", AG2, through subcontract of Study Contract RA1-0148-UK, in the context of the CEC Fast Reactor Coordinating Committee ; finally, the permission of the above sponsoring agencies to publish the present paper. The second author gratefully acknowledges a fellowship from the Italian Ministry of Foreign Affairs as part of a joint project between Tsinghua University, People's Republic of China and Politecnico di Milano, Italy.

REFERENCES

1987 KÖNIG, J.A., Shakedown of elastic-plastic structures, Elsevier, Amsterdam.

1981 KÖNIG, J.A. and MAIER, G., "Shakedown analysis of elastoplastic structures : a review of recent developments", Nuclear Engng Design, 66, pp. 81-95.

1979 COHN, M.Z., and MAIER, G., Engineering plasticity by mathematical programming, Pergamon, New York.

1981 ZYCZKOWSKI, M., Combined loadings in the theory of plasticity, PWN, Pol. Sci. Publ., Warsaw.

1989 KALISZKY, S., Plasticity : theory and engineering applications, Elsevier, Amsterdam.

1972 MAIER, G., "A shakedown matrix theory allowing for workhardening and second-order geometric effects", in : Foundations of plasticity, Vol. I (A. Sawczuk, Ed.), Noordhoff, Leyden, pp. 417-433.

1973 MAIER, G., "Upper bounds on deformations of elastic work-hardening structures in the presence of dynamic and second order effects", J. Struct. Mech., 2 (4), pp. 265-280.

1976 KÖNIG, J.A. and MAIER, G., "Adaptation of rigid-work-hardening discrete structures subjected to load and temperature cycles and second-order geometric effects", Comput. Meth. Appl. Mech. Engng, 8, pp. 37-50.

1990 MAIER, G. and NOVATI, G., "A shakedown and bounding theory allowing for nonlinear hardening and second-order geometric effects with reference to discrete structural models", in : Inelastic solids and structures, A. Sawczuk Memorial Volume (M. Kleiber and J.A. König, Eds) Pineridge, Swansea, pp. 451-471.

1984 KÖNIG, J.A., "Stability of the incremental collapse" in : Inelastic structures under variable loads, (C. Polizotto and A. Swaczuk Eds), COGRAS, Palermo, pp. 329-334.

1988 KÖNIG, J.A. and SIEMASZKO, A., "Strain hardening effects in shakedown processes", Ing. Archiv., 58, pp. 58-66.

1985 SIEMASZKO, A. and KÖNIG, J.A., "An analysis of stability of incremental collapse of skeletal structures", J. Struct. Mech., 13, pp. 301-321.

1990 SIEMASZKO, A. and KÖNIG, J.A., "Geometric effects in shakedown of optimum structures", in : Inelastic solids and structures, A. Sawczuk Memorial Volume, (M. Kleiber and J.A. König, Eds) Pineridge, Swansea, pp. 503-515.

1989 TEXEIRA DE FREITAS, J.A. and SMITH, D.L. "The post-collapse behaviour of rigid-plastic structures", Eur. J. Mech. A/Mech. A/Solid, 1, pp. 35-52.

1983 WEICHERT, D., "Shakedown at finite displacement ; a note on Melan's theorem", Mech. Res. Comm., 11 pp. 121-127.

1986 WEICHERT, D., "On the influence of geometrical nonlinearities on the shakedown of elastic-plastic structures", Int. J. Plasticity, 2, 135-148.

1986 WEICHERT, D., "Elastic-plastic structures under variable loads at small strain and moderate rotations", in : Finite Rotations in Structural Mechanics, (W. Pietraszkiewicz, Ed.), Springer, Berlin, pp. 343-356.

1990 WEICHERT, D., "Advances in the geometrically nonlinear shakedown theory", in : Inelastic solid and structures, A. Swaczuk Memorial Volume, (M. Kleiber and J.A. König, Eds) Pineridge, Swansea, pp. 489-502.

1990 GROSS-WEEGE, J., "A unified formulation of statical shakedown criteria for geometrically nonlinear problems", Int. J. Plasticity, 6, pp. 433-477.

STUMPF, H., "Theoretical and computational aspects in the shakedown analysis of finite elastoplasticity", Int. J. Plasticity, (To appear).

1983 NGUYEN QUOC SON and GARY, G. "Flambage par deformations plastiques cumulées sous charge cyclique additionnelle", J. de Méc. Théorique Appl., 3, pp. 351-373.

1991 PYCKO, S. and KÖNIG, J.A., "Steady plastic cycles on reference configuration in the presence of second-order geometric effects", J. Mech. A/Solids, 10, pp. 563-574.

1982 POLIZZOTTO, C., "A unified treatment of shakedown theory and related boundary techniques", Solid Mech. Arch., 1, pp 19-75.

1985 PONTER A.R.S. and KARADENIZ, S., "An extended shakedown theory for structures that suffer cyclic thermal loading, part 1 : theory", J. Appl. Mech., 52, pp. 877-882.

1990 PONTER, A.R.S. and KARADENIZ, S., "An extended shakedown theory for structures that suffer cyclic thermal loading, part 2 : applications", J. Appl. Mech., 52, pp. 883-889.

1990 STEIN, E., ZHANG, G., MAHNKEN, R. and KÖNIG, J.A., "Micromechanical modelling and computation of shakedown with nonlinear kinematic hardening including examples for 2-D problems", Proc. of CSME, Mech. Eng. Forum., Toronto, pp. 425-430.

1985 LEERS, K., KLIE W., KÖNIG, J.A. and MAHRENHOLTZ, O. "Experimental investigations on shakedown of tubes", Plasticity Today, (A. Sawczuk and G. Bianchi, Eds), Elsevier, London, pp. 259-275.

1987 KARADENIZ, S. PONTER, A.R.S. and CARTER, K.F., "The plastic ratchetting of thin cylindrical shells subjected to axisymmetric thermal and mechanical loading", ASME J. Press. Vessel Tech., 109, pp. 387-393.

1991 SAVE, M., DE SAXCÉ, G. and BORKOWSKI, A. "Computation of shakedown loads : feasibility study", CEC Dir. Gen. Science, Res. and Dev., Nuclear Sci. and Techn., Report EUR13618EN., Brussels.

1967 BREE, J. "Elastoplastic behaviour of thin tubes subjected to internal pressure and intermittent high heat fluxes with application to fast nuclear reactor fuel elements", J. Strain Analysis, 2, pp. 226-238.

1985 COCKS, A.C.F. and PONTER, A.R.S., "The plastic behaviour of components subjected to constant primary stresses and cyclic secondary strains", J. Strain Analysis, 20 (1), pp. 7-14.

1991 HIBBITT, KARLSSON and SORENSEN, "ABAQUS, Version 4.9", Providence, RI.

1991 BAZANT, Z. P. and CEDOLIN, L., Stability of structures, Oxford University Press, Oxford.

1983 CALLADINE, C.R., Theory of shell structures, Cambridge University Press, London.

1991 PIGNATARO, M., RIZZI, N. and LUONGO, A., Stability, bifurcation and poscritical behaviour of elastic structures, Elsevier, Amsterdam.

1972 HUTCHINSON, J.W., "On the postbuckling behaviour of imperfection-sensitive structures in the plastic range", J. Appl. Mech., 39, pp. 155-162.

1991 SHEN, H.S. and CHEN, T.Y., "Buckling and postbuckling behaviour of cylindrical shells under combined external pressure and axial compression", Thin-walled Structures, 12, pp. 321-334.

1983 ECCS, "European Recommendations for Steel Construction : Buckling of Shells", CCCS Publ., N° 29, Brussels.

1990 GALLETLY, G.D. and BLACHUT, J., "Axially-compressed cylindrical shells : a comparison of experiment and theory" in : Inelastic Solid and Structures, A. Sawczuk Memorial Volume, (M. Kleiber and J.A. König, Eds) Pineridge, Swansea, pp. 257-276.

1992 TENG, J. and ROTTER, J.M. "Buckling of pressurized axisymmetrically imperfect cylinders under axial loads", J. eng. Mech. ASCE, 118, pp. 229-247.

SHAKEDOWN OF SHELLS UNDERGOING MODERATE ROTATIONS

Johannes GROSS-WEEGE, Dieter WEICHERT*
Freysestrasse 25, D-47802 Krefeld, Germany
* LML (EUDIL), CNRS URA 1441, Université de Lille 1,
Cité Scientifique, F-59655 Villeneuve d'Ascq Cedex, France

ABSTRACT

A shakedown theorem for elasto-plastic shells undergoing moderate rotations is presented, particularly adapted to situations where perturbations of a permanent mechanical load may induce progressive deterioration. The numerical examples illustrate the importance of the effects of change of geometry during the process.

INTRODUCTION

In recent years, safety concepts based on failure loads have been more and more integrated in the design codes for mechanical and engineering structures. This is motivated by the fact, that only these loads allow a realistic assessment of the safety margin with respect to failure. Concepts which are based on admissible stresses and/or strains are often simpler in practice but do not give precise information on the safety level. As a consequence, the classical methods of limit analysis have increasingly gained importance. These methods provide the limit loads with respect to failure due to global plastic flow. An essential restriction for the applicability of the theory of limit analysis is the assumption that the external loading is time-independent.
In practice, however, the external loading generally is variable in time. In this case, structures may fail under loads which are considerably lower than those obtained by limit analysis.

The extension of the methods of limit analysis to the case of variable repeated loading is the shakedown theory. Its foundations are given by the classical theorems of MELAN [1938] and KOITER [1960]. The most characteristic feature of the shakedown theory is the fact, that the loading history may be completely arbitrary, which makes it very attractive in practice. However, these theorems in their classical form are restricted to geometrically linear problems and elastic-perfectly plastic material behaviour.

Z. Mróz et al. (eds.), Inelastic Behaviour of Structures under Variable Loads, 263–277.

The influence of material hardening on the shakedown behaviour has been studied, among others, by MAIER [1973], PONTER [1975], MANDEL [1976] and, more recently, by KÖNIG & SIEMASZKO [1988], WEICHERT & GROSS-WEEGE [1988] and STEIN et al. [1990]. On the other hand, only very few papers are concerned with the influence of geometrical changes on shakedown, although in many practical problems failure due to non-shakedown or even shakedown may occur only when the displacements are considerable. This behaviour applies especially to thin-walled shell-like structures, which are of particular importance in mechanical engineering. The first to treated this problem was MAIER [1973], who extended Melan's theorem, taking into account so-called "second order geometrical effects". WEICHERT [1984, 1986] investigated in several papers the influence of geometrical nonlinearities on the shakedown of elastic-plastic structures within the framework of continuum mechanics. On the basis of an additive decomposition of Green's strain tensor he formulated a static theorem for arbitrary finite displacements. The practical application in the general form, however, is only possible if the complete loading history is known. Consequently, special cases were considered [1986, 1989]. A more practical theorem has been presented by GROSS-WEEGE [1990] for loading cases, where the state of deformation and the state of stress in the body is subjected to small variations in time. The loading history, however, is completely arbitrary.

In the present paper, the practical approach (cf. GROSS-WEEGE [1990]) will be applied to shell-like structures. Due to the additive decomposition of the strain tensor, the formulation is restricted to small total strains and moderate rotations. We use Donnell-Mushtari-Vlasov shell theory (cf. SAWCZUK [1982]), which gives sufficiently accurate results within the imposed range of applicability. Analogously to GROSS-WEEGE [1990] we presume loading cases, where the external loads are subjected to arbitrary oscillations with small amplitudes. We use a generalised material law expressed in terms of the stress and strain resultants. This enables us to derive a strictly two-dimensional formulation, which is advantageous from a computational point of view. Only for simplicity we restrict our considerations to elastic-perfectly plastic material behaviour. An extension of the formulation presented to hardening material behaviour in the sense of the work of GROSS-WEEGE & WEICHERT [1992] is immediately possible.

A simple example of a circular plate under a uniformly distributed surface load demonstrates the practicability of the proposed method.

BASIC SHELL RELATIONS

Let *P* be the region of the three-dimensional Euclidean space *E* occupied by the shell in its undeformed configuration.
In *P* we introduce a system of curvilinear co-ordinates $(\Theta^1, \Theta^2, \Theta^3)$ such that $-h/2 \leq \Theta^3 \leq +h/2$ is the distance from the middle surface *M* of *P*. The undeformed shell thickness h is assumed to be small as compared to the smallest radius of curvature R of *M*, i.e. $h/R \ll 1$. The surface *M* is described by the position vector $\mathbf{r}(\Theta^\alpha)$, $\alpha=1,2$. With each point $M \in M$ we associate covariant base vectors $\mathbf{a}_\alpha = \partial \mathbf{r} / \partial \Theta^\alpha = \mathbf{r},_\alpha$, the covariant components $a_{\alpha\beta} = \mathbf{a}_\alpha \cdot \mathbf{a}_\beta$ of the

The influence of material hardening on the shakedown behaviour has been studied, among others, by MAIER [1973], PONTER [1975], MANDEL [1976] and, more recently, by KÖNIG & SIEMASZKO [1988], WEICHERT & GROSS-WEEGE [1988] and STEIN et al. [1990]. On the other hand, only very few papers are concerned with the influence of geometrical changes on shakedown, although in many practical problems failure due to non-shakedown or even shakedown may occur only when the displacements are considerable. This behaviour applies especially to thin-walled shell-like structures, which are of particular importance in mechanical engineering. The first to treated this problem was MAIER [1973], who extended Melan's theorem, taking into account so-called "second order geometrical effects". WEICHERT [1983, 1986] investigated in several papers the influence of geometrical nonlinearities on the shakedown of elastic-plastic structures within the framework of continuum mechanics. On the basis of an additive decomposition of Green's strain tensor he formulated a static theorem for arbitrary finite displacements. The practical application in the general form, however, is only possible if the complete loading history is known. Consequently, special cases were considered [1986, 1989]. A more practical theorem has been presented by GROSS-WEEGE [1990] for loading cases, where the state of deformation and the state of stress in the body is subjected to small variations in time. The loading history, however, is completely arbitrary.

In the present paper, the practical approach (cf. GROSS-WEEGE [1990]) will be applied to shell-like structures. Due to the additive decomposition of the strain tensor, the formulation is restricted to small total strains and moderate rotations. We use Donnell-Mushtari-Vlasov shell theory (cf. SAWCZUK [1982]), which gives sufficiently accurate results within the imposed range of applicability. Analogously to GROSS-WEEGE [1990] we presume loading cases, where the external loads are subjected to arbitrary oscillations with small amplitudes. We use a generalised material law expressed in terms of the stress and strain resultants. This enables us to derive a strictly two-dimensional formulation, which is advantageous from a computational point of view. Only for simplicity we restrict our considerations to elastic-perfectly plastic material behaviour. An extension of the formulation presented to hardening material behaviour in the sense of the work of GROSS-WEEGE & WEICHERT [1992] is immediately possible.

A simple example of a circular plate under a uniformly distributed surface load demonstrates the practicability of the proposed method.

BASIC SHELL RELATIONS

Let P be the region of the three-dimensional Euclidean space E occupied by the shell in its undeformed configuration.
In P we introduce a system of curvilinear co-ordinates $(\Theta^1,\Theta^2,\Theta^3)$ such that $-h/2 \leq \Theta^3 \leq +h/2$ is the distance from the middle surface M of P. The undeformed shell thickness h is assumed to be small as compared to the smallest radius of curvature R of M, i.e. $h/R \ll 1$. The surface M is described by the position vector $\mathbf{r}(\Theta^\alpha)$, $\alpha=1,2$. With each point $M \in M$ we associate covariant base vectors $\mathbf{a}_\alpha = \partial\mathbf{r} / \partial\Theta^\alpha = \mathbf{r},_\alpha$, the covariant components $a_{\alpha\beta} = \mathbf{a}_\alpha \cdot \mathbf{a}_\beta$ of the

surface metric tensor **a** with the determinant $a = |a_{\alpha\beta}|$, the contravariant components $\varepsilon^{\alpha\beta}$ of the permutation tensor such that $\varepsilon^{12} = -\varepsilon^{21} = 1/\sqrt{\mathbf{a}}$, $\varepsilon^{11} = \varepsilon^{22} = 0$, the unit normal vector $\mathbf{a}_3 = \frac{1}{2}\varepsilon^{\alpha\beta}\mathbf{a}_\alpha \times \mathbf{a}_\beta$ and the covariant components $b_{\alpha\beta} = -\mathbf{a}_\alpha \cdot \mathbf{a}_{3,\beta}$ of the curvature tensor **b**.
Here, and in all following formulae, $(\ )_{,\alpha}$ denotes partial derivatives with respect to Θ^α. Greek and Latin indices take values 1,2 and 1,2,3, respectively. The symbols · and x between two vectors represent scalar product and cross product.
At line C, bounding the middle surface M, we introduce an orthonormal vector system **v**, **t**, $\mathbf{a}_3$, where $\mathbf{t} = \partial\mathbf{r}/\partial s$ is the unit tangent vector, s being the arc length along C, and $\mathbf{v} = \mathbf{t} \times \mathbf{a}_3$ denotes the outward unit normal vector.

The deformation of the shell middle surface from the undeformed configuration M into the deformed configuration $\overline{M}$ can be described by the displacement field $\mathbf{u} = \overline{\mathbf{r}} - \mathbf{r} = u^\alpha \mathbf{a}_\alpha + u^3 \mathbf{a}_3$.
Assuming the validity of the Kirchhoff-Love hypothesis, the shell deformation can further be defined by the middle surface strain tensor $\boldsymbol{\gamma}$ and the change of curvature tensor $\boldsymbol{\chi}$.
In this article the classical non-linear shell theory given by Donnell, Mushtari and Vlasov will be considered. Its range of validity is given by the following restrictions:

- moderate rotations about tangents to M
- small rotations about normals to M
- small tangential displacements
- shallow shells

In this case the field equations are given in the following form (see e.g. SAWCZUK [1982], BASAR & KRÄTZIG [1985]), where all quantities are referred to the initial, undeformed configuration M.

Kinematic equations

$$\gamma_{\alpha\beta} = \frac{1}{2}(u_\alpha|_\beta + u_\beta|_\alpha - 2b_{\alpha\beta}u_3 + u_{3,\alpha}u_{3,\beta}) \qquad \text{in } M$$

$$\chi_{\alpha\beta} = -u_3|_{\alpha\beta} \qquad \text{in } M \tag{1}$$

$$u^\alpha v_\alpha = \overset{*}{u}_v \qquad \text{on } C_u$$

$$u^\alpha t_\alpha = \overset{*}{u}_t \qquad \text{on } C_u$$

$$u_3 = \overset{*}{u}_3 \qquad \text{on } C_u \tag{2}$$

$$-u_{3,\alpha}v^\alpha = \overset{*}{\beta}_v \qquad \text{on } C_u$$

In eqns (2) C_u denotes the part of C, where geometrical boundary conditions are given. Only for simplicity, geometrical corner conditions are neglected. Prescribed quantities are indicated by an asterisk and β_v denotes the rotation about the tangent to the boundary line. A vertical stroke preceding a subscript means covariant differentiation with respect to the corresponding direction.

Static equations

$$N^{\alpha\beta}\big|_{\beta} + p^{\alpha *} = 0 \qquad \text{in } M$$

$$(M^{\alpha\beta}\big|_{\alpha} + u_{3,\alpha}N^{\alpha\beta})\big|_{\beta} + b_{\alpha\beta}N^{\alpha\beta} + p^{3*} = 0 \qquad \text{in } M \tag{3}$$

$$\nu_{\alpha}\nu_{\beta}N^{\alpha\beta} = N^{*}_{\nu\nu} \qquad \text{on } C_f$$

$$t_{\alpha}\nu_{\beta}N^{\alpha\beta} = N^{*}_{t\nu} \qquad \text{on } C_f$$

$$\nu_{\beta}(M^{\alpha\beta}\big|_{\alpha} + u_{3,\alpha}N^{\alpha\beta}) - M_{t\nu,\alpha}t^{\alpha} = N^{*}_{n\nu} - M^{*}_{t\nu,t} \qquad \text{on } C_f \tag{4}$$

$$\nu_{\alpha}\nu_{\beta}M^{\alpha\beta} = M^{*}_{\nu\nu} \qquad \text{on } C_f$$

In eqns (4) C_f denotes the part of C, where static boundary conditions are prescribed. Static corner conditions are neglected for simplicity. $p^{\alpha *}$ and p^{3*} are the components of the distributed surface loads per unit area of the undeformed middle surface M. **N** and **M** denote the symmetric stress resultant tensor and the symmetric stress couple tensor, respectively.

Constitutive relations

We assume that the strain measures $\boldsymbol{\gamma}$ and $\boldsymbol{\chi}$ can be additively decomposed into purely elastic parts $\boldsymbol{\gamma}^e$, $\boldsymbol{\chi}^e$ and purely plastic parts $\boldsymbol{\gamma}^p$, $\boldsymbol{\chi}^p$, such that

$$\gamma_{\alpha\beta} = \gamma^{e}_{\alpha\beta} + \gamma^{p}_{\alpha\beta}$$

$$\chi_{\alpha\beta} = \chi^{e}_{\alpha\beta} + \chi^{p}_{\alpha\beta} \tag{5}$$

For isotropic linear elastic material behaviour, the strain measures $\boldsymbol{\gamma}^e$ and $\boldsymbol{\chi}^e$ are given by the relations

$$\gamma^{e}_{\alpha\beta} = \frac{1}{D}G_{\alpha\beta\gamma\lambda}N^{\gamma\lambda}$$

$$\chi^{e}_{\alpha\beta} = \frac{1}{B}G_{\alpha\beta\gamma\lambda}M^{\gamma\lambda} \tag{6}$$

where $D=Eh/(1-\nu^2)$ and $B=Eh^3/(12(1-\nu^2))$ denote the extensional and bending stiffness, respectively. E denotes Young's modulus and ν is Poisson's ratio. The positive-definite and time-independent elasticity tensor **G**, defined by

$$G_{\alpha\beta\gamma\lambda} = \frac{1}{2(1-\nu)}(a_{\alpha\gamma}a_{\beta\lambda} + a_{\alpha\lambda}a_{\beta\gamma} - \frac{2\nu}{1+\nu}a_{\alpha\beta}a_{\gamma\lambda}) \tag{7}$$

obeys the usual symmetry conditions.

For the plastic part of the material law we assume the existence of a convex elastic domain C of purely elastic states, bounded by the yield function

$$F(\mathbf{N}, \mathbf{M}, k) = 0 \tag{8}$$

Furthermore, we assume the validity of the normality rule for plastic flow, such that

$$\dot{\gamma}^{p}_{\alpha\beta} = \dot{\lambda}_N \frac{\partial F}{\partial N^{\alpha\beta}}$$
$$\dot{\chi}^{p}_{\alpha\beta} = \dot{\lambda}_M \frac{\partial F}{\partial M^{\alpha\beta}}, \tag{9}$$

where λ_N, λ_M are non-negative proportionality factors; A superimposed dot denotes the rate of the considered quantity.

In the present paper we use a yield condition, first derived by BIENIEK & FUNARO [1976] and later modified by several authors. Recently, GROSS-WEEGE & WEICHERT [1992] have applied this condition to shakedown problems. Following the derivation in that work, we obtain a simplified two-surface yield condition, expressed exclusively in terms of the stress-resultants

$$F_I = I_{NN} + \frac{9}{4} I^*_{MM} \pm 3 I^*_{NM} - 1 = 0$$
$$F_L = I_{NN} + I_{MM} \pm \frac{1}{\sqrt{3}} I_{NM} - 1 = 0, \tag{10}$$

where F_I and F_L denote the initial yield surface and the limit yield surface, respectively. I_{NN}, I_{MM} and I_{NM} are defined as

$$I_{NN} = I_{\alpha\beta\lambda\mu} \frac{N^{\alpha\beta} N^{\lambda\mu}}{N_y^2}$$
$$I_{MM} = I_{\alpha\beta\lambda\mu} \frac{M^{\alpha\beta} M^{\lambda\mu}}{M_y^2} \tag{11}$$
$$I_{NM} = I_{\alpha\beta\lambda\mu} \frac{N^{\alpha\beta} M^{\lambda\mu}}{N_y M_y},$$

where

$$I_{\alpha\beta\lambda\mu} = \frac{3}{2} a_{\alpha\lambda} a_{\beta\mu} - \frac{1}{2} a_{\alpha\beta} a_{\lambda\mu}. \tag{12}$$

In eqns (11), $N_y = \sigma_y h$ and $M_y = \sigma_y h^2/4$ denote the uniaxial yield force and the uniaxial yield moment, respectively, and σ_y is the uniaxial yield stress. I^*_{NM} and I^*_{MM} are also given by eqns (11b, 11c) if we replace **M** by the differences **M-M***, where the evolution of **M*** is defined by

$$\dot{M}^{*\alpha\beta} = \begin{cases} C^{\alpha\beta\gamma\lambda} \dot{\chi}^{p}_{\gamma\lambda} \ ; & \text{for } F_L < 0,\ F_S = 0 \text{ and } \dot{F}_S > 0 \\ 0 \ ; & \text{for } F_L = 0 \text{ or } F_S < 0 \text{ or } F_S = 0 \text{ and } \dot{F}_S < 0 \end{cases} \tag{13}$$

The components of the tensor **C** are given by

$$C^{\alpha\beta\gamma\lambda} = a^{\alpha\gamma} a^{\beta\lambda} \frac{Eh^3}{12} \tag{14}$$

SPECIAL KIND OF LOADING

Let **a*** denote the external agencies acting on the shell *P*. We assume, that **a*** is of a special type: Up to an instant τ^R the shell *P* undergoes finite displacements $\mathbf{u}^R$ under time-independent loads $\mathbf{a}^R$. For times $\tau > \tau^R$ the shell is subjected to additional variable loads $\tilde{\mathbf{a}}$, such that $\mathbf{a}^*(\tau) = \mathbf{a}^R + \tilde{\mathbf{a}}(\tau)$. Analogously, all field quantities can be decomposed into a time-independent part $(\,)^R$, resulting from $\mathbf{a}^R$, and a variable part $(\tilde{\ })$ due to the additional loads $\tilde{\mathbf{a}}$. Assuming that the shell is in equilibrium under the loads $\mathbf{a}^R$, the additional field quantities $(\tilde{\ })$ have to satisfy the following set of equations:

$$\begin{aligned} \tilde{\gamma}_{\alpha\beta} &= \frac{1}{2}(\tilde{u}_\alpha|_\beta + \tilde{u}_\beta|_\alpha - 2b_{\alpha\beta}\tilde{u}_3 + \\ &\quad u^R_{3,\alpha}\tilde{u}_{3,\beta} + \tilde{u}_{3,\alpha}u^R_{3,\beta} + \underline{\tilde{u}_{3,\alpha}\tilde{u}_{3,\beta}}) && \text{in } M \\ \tilde{\chi}_{\alpha\beta} &= -\tilde{u}_3|_{\alpha\beta} && \text{in } M \end{aligned} \tag{15}$$

$$\begin{aligned} &\tilde{u}^\alpha \nu_\alpha = \tilde{u}^*_\nu, \quad \tilde{u}^\alpha t_\alpha = \tilde{u}^*_t, && \text{on } C_u \\ &\tilde{u}_3 = \tilde{u}^*_3, \quad -\tilde{u}_{3,\alpha}\nu^\alpha = \tilde{\beta}^*_\nu && \text{on } C_u \end{aligned} \tag{16}$$

$$\begin{aligned} &\tilde{N}^{\alpha\beta}|_\beta + \tilde{p}^{\alpha*} = 0 && \text{in } M \\ &(\tilde{M}^{\alpha\beta}|_\alpha + u^R_{3,\alpha}\tilde{N}^{\alpha\beta} + \tilde{u}_{3,\alpha}N^{R\alpha\beta} + \underline{\tilde{u}_{3,\alpha}\tilde{N}^{\alpha\beta}})|_\beta + \\ &\quad b_{\alpha\beta}\tilde{N}^{\alpha\beta} + \tilde{p}^{3*} = 0 && \text{in } M \end{aligned} \tag{17}$$

$$\nu_\alpha \nu_\beta \tilde{N}^{\alpha\beta} = \tilde{N}^*_{\nu\nu}, \qquad t_\alpha \nu_\beta \tilde{N}^{\alpha\beta} = \tilde{N}^*_{t\nu} \qquad \text{on } C_f$$

$$\begin{aligned} &\nu_\beta(\tilde{M}^{\alpha\beta}|_\alpha + u^R_{3,\alpha}\tilde{N}^{\alpha\beta} + \tilde{u}_{3,\alpha}N^{R\alpha\beta} + \underline{\tilde{u}_{3,\alpha}\tilde{N}^{\alpha\beta}}) \\ &\quad - \tilde{M}_{t\nu,\alpha}t^\alpha = \tilde{N}^*_{n\nu} - \tilde{M}^*_{t\nu,t} \end{aligned} \qquad \text{on } C_f \tag{18}$$

$$\nu_\alpha \nu_\beta \tilde{M}^{\alpha\beta} = \tilde{M}^*_{\nu\nu} \qquad \text{on } C_f$$

$$\tilde{\gamma}^e_{\alpha\beta} = \frac{1}{D} G_{\alpha\beta\gamma\lambda}\tilde{N}^{\gamma\lambda}, \qquad \tilde{\chi}^e_{\alpha\beta} = \frac{1}{B} G_{\alpha\beta\gamma\lambda}\tilde{M}^{\gamma\lambda} \tag{19}$$

$$\tilde{\gamma}^p_{\alpha\beta} = \tilde{\gamma}^p_{\alpha\beta}(\mathbf{a}^R, \tilde{\mathbf{a}}(\tau)), \qquad \tilde{\chi}^p_{\alpha\beta} = \tilde{\chi}^p_{\alpha\beta}(\mathbf{a}^R, \tilde{\mathbf{a}}(\tau)) \tag{20}$$

We assume that the solution to a reference problem differing from the original one only by the fact that for times $\tau > \tau^R$ the considered reference shell $\overset{\circ}{P}$ behaves in a purely elastic manner, is given. Henceforth, all quantities related to this reference problem are indicated by the superscript "o". In particular, we assume that $\overset{\circ}{P}$ occupies at the time $\tau = \tau^R$ the same configuration Ω^R as P and is subjected to the same state of stress resultants $(\mathbf{N}^R, \mathbf{M}^R)$ as P. Moreover, for times $\tau > \tau^R$, reference $\overset{\circ}{P}$ is subjected to the same additional loading $\tilde{\mathbf{a}}$. Obviously eqns (15-19) then also hold for the reference problem with the only exception that in reference $\overset{\circ}{P}$ no additional plastic strains $(\boldsymbol{\gamma}^p, \boldsymbol{\chi}^p)$ can occur. The differences between the states in P and $\overset{\circ}{P}$ are then described by the following difference fields

$$\begin{aligned} &\Delta u_i = \tilde{u}_i - \overset{o}{u}_i \\ &\Delta\gamma_{\alpha\beta} = \tilde{\gamma}_{\alpha\beta} - \overset{o}{\gamma}_{\alpha\beta}, \qquad \Delta\chi_{\alpha\beta} = \tilde{\chi}_{\alpha\beta} - \overset{o}{\chi}_{\alpha\beta} \\ &\Delta N^{\alpha\beta} = \tilde{N}^{\alpha\beta} - \overset{o}{N}{}^{\alpha\beta}, \qquad \Delta M^{\alpha\beta} = \tilde{M}^{\alpha\beta} - \overset{o}{M}{}^{\alpha\beta} \end{aligned} \tag{21}$$

In view of eqns (15-19) the difference fields have to satisfy the following set of equations:

$$\begin{aligned} \Delta\gamma_{\alpha\beta} &= \frac{1}{2}(\Delta u_\alpha|_\beta + \Delta u_\beta|_\alpha - 2b_{\alpha\beta}\Delta u_3 + u^R_{3,\alpha}\Delta u_{3,\beta} + \\ &\qquad \Delta u_{3,\alpha}u^R_{3,\beta} + \underline{\tilde{u}_{3,\alpha}\tilde{u}_{3,\beta} - \overset{o}{u}_{3,\alpha}\overset{o}{u}_{3,\beta}}) \qquad \text{in } M \\ \Delta\chi_{\alpha\beta} &= -\Delta u_3|_{\alpha\beta} \qquad \text{in } M \end{aligned} \tag{22}$$

$$\Delta u^{\alpha} \nu_{\alpha} = 0, \quad \Delta u^{\alpha} t_{\alpha} = 0, \quad \Delta u_3 = 0, \quad -\Delta u_{3,\alpha} \nu^{\alpha} = 0 \qquad \text{on } C_u \tag{23}$$

$$\Delta N^{\alpha\beta}\big|_{\beta} = 0 \qquad \text{in } M$$
$$(\Delta M^{\alpha\beta}\big|_{\alpha} + u^{R}_{3,\alpha} \Delta N^{\alpha\beta} + \Delta u_{3,\alpha} N^{R\alpha\beta} + \underline{\tilde{u}_{3,\alpha} \tilde{N}^{\alpha\beta} - \overset{o}{u}_{3,\alpha} \overset{o}{N}{}^{\alpha\beta}})\big|_{\beta} + b_{\alpha\beta} \Delta N^{\alpha\beta} = 0 \qquad \text{in } M \tag{24}$$

$$\nu_{\alpha} \nu_{\beta} \Delta N^{\alpha\beta} = 0, \qquad t_{\alpha} \nu_{\beta} \Delta N^{\alpha\beta} = 0 \qquad \text{on } C_f$$
$$\nu_{\beta} (\Delta M^{\alpha\beta}\big|_{\alpha} + u^{R}_{3,\alpha} \Delta N^{\alpha\beta} + \Delta u_{3,\alpha} N^{R\alpha\beta} + \underline{\tilde{u}_{3,\alpha} \tilde{N}^{\alpha\beta} - \overset{o}{u}_{3,\alpha} \overset{o}{N}{}^{\alpha\beta}}) - \Delta M_{t\nu,\alpha} t^{\alpha} = 0 \qquad \text{on } C_f \tag{25}$$
$$\nu_{\alpha} \nu_{\beta} \Delta M^{\alpha\beta} = 0 \qquad \text{on } C_f$$

$$\Delta \gamma^{e}_{\alpha\beta} = \frac{1}{D} G_{\alpha\beta\gamma\lambda} \Delta N^{\gamma\lambda}, \qquad \Delta \chi^{e}_{\alpha\beta} = \frac{1}{B} G_{\alpha\beta\gamma\lambda} \Delta M^{\gamma\lambda} \tag{26}$$

$$\Delta \gamma^{p}_{\alpha\beta} = \tilde{\gamma}^{p}_{\alpha\beta}, \qquad \Delta \chi^{p}_{\alpha\beta} = \tilde{\chi}^{p}_{\alpha\beta} \tag{27}$$

We restrict now our considerations to situations where the state of deformation and the state of stress in P are subjected to small variations in time. Consequently, we neglect in the governing relations (15-27) all terms which are non-linear in the time-dependent additional field quantities $((\tilde{\ }),(\overset{o}{.}))$. The corresponding terms are underlined by a solid line. Under this assumption we can formulate the following extended theorem.

AN EXTENSION OF MELAN'S THEOREM

The problem

An elastic-perfectly plastic shell P is in the configuration $\mathbf{\Omega^R}$ at the time $\tau = \tau^R$ in equilibrium with time-independent external loads $\mathbf{a^R}$. Will P shake down under the action of additional variable loads $\tilde{\mathbf{a}}$?

Proposition

The shell shakes down with respect to the given loading history if there are a time-independent field of parameters $\overline{\mathbf{M}}^*$ and time-independent fields of residual stress resultants $\Delta\overline{\mathbf{N}}, \Delta\overline{\mathbf{M}}$, such that for all times $\tau > 0$ the following conditions hold:

$$\Delta\overline{N}^{\alpha\beta}\Big|_{\beta} = 0 \qquad \text{in } M$$

$$(\Delta\overline{M}^{\alpha\beta}\big|_{\alpha} + u^{R}_{3.\alpha}\Delta\overline{N}^{\alpha\beta} + \Delta\overline{u}_{3.\alpha}N^{R\alpha\beta})\Big|_{\beta} + b_{\alpha\beta}\Delta\overline{N}^{\alpha\beta} = 0 \qquad \text{in } M \tag{28}$$

$$\nu_{\alpha}\nu_{\beta}\Delta\overline{N}^{\alpha\beta} = 0, \qquad t_{\alpha}\nu_{\beta}\Delta\overline{N}^{\alpha\beta} = 0 \qquad \text{on } C_f$$

$$\nu_{\beta}(\Delta\overline{M}^{\alpha\beta}\big|_{\alpha} + u^{R}_{3.\alpha}\Delta\overline{N}^{\alpha\beta} + \Delta\overline{u}_{3.\alpha}N^{R\alpha\beta}) - \Delta\overline{M}_{tv.\alpha}t^{\alpha} = 0 \qquad \text{on } C_f \tag{29}$$

$$\nu_{\alpha}\nu_{\beta}\Delta\overline{M}^{\alpha\beta} = 0 \qquad \text{on } C_f$$

$$\Delta\overline{u}_3 = 0, \qquad -\Delta\overline{u}_{3.\alpha}\nu^{\alpha} = 0 \qquad \text{on } C_u \tag{30}$$

$$\Delta\dot{\overline{u}}_3 = 0 \qquad \text{on } C_u \tag{31}$$

$$F_I(N^R + \overset{o}{N}(\tau) + \Delta\overline{N}, M^R - M^{R*} + \overset{o}{M}(t) + \Delta\overline{M} - \overline{M}^*, \sigma_y) < 0 \qquad \text{in } M$$

$$F_L(N^R + \overset{o}{N}(\tau) + \Delta\overline{N}, M^R + \overset{o}{M}(\tau) + \Delta\overline{M}, \sigma_y) < 0 \qquad \text{in } M \tag{32}$$

for all times $\tau > \tau^R$.

$$\int_M (\Delta N^{\alpha\beta}\frac{1}{D}G_{\alpha\beta\gamma\lambda}\Delta N^{\gamma\lambda} + \Delta M^{\alpha\beta}\frac{1}{B}G_{\alpha\beta\gamma\lambda}\Delta M^{\gamma\lambda} + M^{*\alpha\beta}(C^{-1})_{\alpha\beta\gamma\lambda}M^{*\gamma\lambda} + \Delta u_{3.\alpha}N^{R\alpha\beta}\Delta u_{3,\beta})\, dM > 0 \tag{33}$$

for arbitrary non-vanishing $\Delta\mathbf{N}$, $\Delta\mathbf{M}$, $\mathbf{M}^*$ and Δu_3 fulfilling eqns (28, 29,30).

We note, that the first three terms of the integrand in ineq (33) are always positive due to the fact that $\mathbf{G}$ and $\mathbf{C}^{-1}$ are positive definite and D and B are positive valued scalars. On the other hand, the fourth term is only positive for membrane stress states $\mathbf{N}^R$ having non negative principle stresses.

Proof
We introduce the quadratic form W, defined by

$$\begin{aligned} W = \frac{1}{2}\int_M (&(\Delta N^{\alpha\beta} - \Delta\overline{N}^{\alpha\beta})\frac{1}{D}G_{\alpha\beta\gamma\lambda}(\Delta N^{\gamma\lambda} - \Delta\overline{N}^{\gamma\lambda}) + \\ &(\Delta M^{\alpha\beta} - \Delta\overline{M}^{\alpha\beta})\frac{1}{B}G_{\alpha\beta\gamma\lambda}(\Delta M^{\gamma\lambda} - \Delta\overline{M}^{\gamma\lambda}) + \\ &(M^{*\alpha\beta} - M^{R*\alpha\beta} - \overline{M}^{*\alpha\beta})(C^{-1})_{\alpha\beta\gamma\lambda}(M^{*\gamma\lambda} - M^{R*\gamma\lambda} - \overline{M}^{*\gamma\lambda})\ )\, dM \\ + \frac{1}{2}\int_M &(\Delta u_{3,\alpha} - \Delta\overline{u}_{3,\alpha})N^{R\alpha\beta}(\Delta u_{3,\beta} - \Delta\overline{u}_{3,\beta})\ dM \end{aligned} \tag{34}$$

In view of condition (33) W is always non-negative. Taking into account that $\mathbf{G}$, $\mathbf{C}^{-1}$ and $\mathbf{N}^R$ are symmetric and time independent, the rate $\dot{W}$ of W is given by

$$\dot{W} = \frac{1}{2}\int_M \Big((\Delta N^{\alpha\beta} - \Delta\overline{N}^{\alpha\beta})\frac{1}{D}G_{\alpha\beta\gamma\lambda}\Delta\dot{N}^{\gamma\lambda} + (\Delta M^{\alpha\beta} - \Delta\overline{M}^{\alpha\beta})\frac{1}{B}G_{\alpha\beta\gamma\lambda}\Delta\dot{M}^{\gamma\lambda} + (M^{*\alpha\beta} - M^{R*\alpha\beta} - \overline{M}^{*\alpha\beta})(C^{-1})_{\alpha\beta\gamma\lambda}\dot{M}^{*\gamma\lambda} \Big)\, dM + \frac{1}{2}\int_M (\Delta u_{3,\alpha} - \Delta\overline{u}_{3,\alpha})N^{R\alpha\beta}\Delta\dot{u}_{3,\beta}\; dM \tag{35}$$

Substituting the time derivatives of eqns (26) and eqn (13) it follows that

$$\dot{W} = \int_M \Big((\Delta N^{\alpha\beta} - \Delta\overline{N}^{\alpha\beta})\Delta\dot{\gamma}^{e}_{\alpha\beta} + (\Delta M^{\alpha\beta} - \Delta\overline{M}^{\alpha\beta})\Delta\dot{\chi}^{e}_{\alpha\beta} + (M^{*\alpha\beta} - M^{R*\alpha\beta} - \overline{M}^{*\alpha\beta})\,\dot{\chi}^{p}_{\alpha\beta} \Big)\, dM + \frac{1}{2}\int_M (\Delta u_{3,\alpha} - \Delta\overline{u}_{3,\alpha})N^{R\alpha\beta}\Delta\dot{u}_{3,\beta}\; dM \tag{36}$$

Taking into account eqns (5), W can be transformed

$$\dot{W} = \int_M \Big((\Delta N^{\alpha\beta} - \Delta\overline{N}^{\alpha\beta})\Delta\dot{\gamma}_{\alpha\beta} + (\Delta M^{\alpha\beta} - \Delta\overline{M}^{\alpha\beta})\Delta\dot{\chi}_{\alpha\beta} + (\Delta u_{3,\alpha} - \Delta\overline{u}_{3,\alpha})N^{R\alpha\beta}\Delta\dot{u}_{3,\beta} \Big)\, dM - \int_M \Big((\Delta N^{\alpha\beta} - \Delta\overline{N}^{\alpha\beta})\,\dot{\tilde{\gamma}}^{p}_{\alpha\beta} + (\Delta M^{\alpha\beta} - \Delta\overline{M}^{\alpha\beta} - \Delta\overline{M}^{R*\alpha\beta} + \overline{M}^{*\alpha\beta} - M^{*\alpha\beta})\,\Delta\dot{\tilde{\chi}}^{p}_{\alpha\beta} \Big)\, dM \tag{37}$$

After the substitution of the strain-displacement relations (22) into eqn (37) and the application of Gauss' theorem, it follows that the first integral on the right-hand side of (37) vanishes identically because $\Delta\mathbf{N}$, $\Delta\mathbf{M}$ as well as $\Delta\overline{\mathbf{N}}$, $\Delta\overline{\mathbf{M}}$ are self-equilibrated fields of residual stress resultants in the sense of equations (24), (25) and (28), (29). For proof see e.g. [16]. Thus, equation (37) reduces to

$$W = \int_M ((\Delta N^{\alpha\beta} - \Delta \overline{N}^{\alpha\beta}) \Delta \dot{\bar{\gamma}}_{\alpha\beta} + (\Delta M^{\alpha\beta} - \Delta \overline{M}^{\alpha\beta} + \Delta \overline{M}^{R*\alpha\beta} + \overline{M}^{*\alpha\beta} - M^{*\alpha\beta}) \Delta \dot{\bar{\chi}}^{p}_{\alpha\beta}) \, dM \tag{38}$$

Inserting equations (21d,e) into equation (38) we arrive at

$$W = \int_M ((\tilde{N}^{\alpha\beta} - \overset{o}{N}^{\alpha\beta} - \Delta \overline{N}^{\alpha\beta}) \Delta \dot{\bar{\gamma}}^{p}{}_{\alpha\beta} + (\tilde{M}^{\alpha\beta} - \overset{o}{M}^{\alpha\beta} - \Delta \overline{M}^{\alpha\beta} + M^{R*\alpha\beta} + \overline{M}^{*\alpha\beta} - M^{*\alpha\beta}) \, \dot{\bar{\chi}}^{p}_{\alpha\beta}) \, dM \tag{39}$$

Conditions (32) assure, that the stress-resultant states $\mathbf{N}^{\mathbf{R}} + \overset{o}{\mathbf{N}} + \Delta\overline{\mathbf{N}}$ as well as $\mathbf{M}^{R} - \mathbf{M}^{*} + \overset{o}{\mathbf{M}} + \Delta\overline{\mathbf{M}} - \overline{\mathbf{M}}^{*}$ are inside the elastic domain, i. e. safe states $\mathbf{N}^{\mathrm{S}}$ and $\mathbf{M}^{\mathrm{S}}$, respectively. Thus, eqn (39) can be transformed to

$$W = -\int_M ((N^{R\alpha\beta} + \tilde{N}^{\alpha\beta} - N^{s\alpha\beta}) \Delta \dot{\bar{\gamma}}^{p}_{\alpha\beta} + (M^{R\alpha\beta} + \tilde{M}^{\alpha\beta} - M^{*\alpha\beta} - M^{s\alpha\beta}) \, \dot{\bar{\chi}}^{p}_{\alpha\beta}) \, dM \tag{40}$$

With $\mathbf{N}^{\mathbf{R}} + \tilde{\mathbf{N}}$ as well as $\mathbf{M}^{\mathbf{R}} + \tilde{\mathbf{M}}$ as actual states of stress resultants in the shell it follows from the convexity of the yield surface and the validity of the normality rule that the integrand of eqn (40) is always non-negative. Therefore, it follows that $\dot{W} \leq 0$ and, furthermore, $\dot{W}$ is equal to zero only if $\dot{\bar{\gamma}}^{p}$ and $\dot{\bar{\chi}}^{p}$ are equal to zero. Thus, since W is always non-negative, $\dot{W}$ and $\dot{\bar{\gamma}}^{p}$ and $\dot{\bar{\chi}}^{p}$ tend to zero when τ tends to infinity.

NUMERICAL EXAMPLE

We consider a circular plate (Fig. 1) with radius R and wall-thickness h, where h/R=5.5/150. Young's modulus $E = 2.06\ 10^5$ N/mm² and Poisson's ratio $\nu = 0.3$ are given. The plate is hinged at the outer edge and subjected to a uniformly distributed surface load p

$$p = \left[1 + \mu(\tau) p^{R} \right],$$

where p^R is a time-independent reference load and $\mu(\tau)$ varies between fixed bounds μ^- and μ^+, so that $\mu^- \leq \mu(\tau) \leq \mu^+$. For convenience, we assume that $-\mu^- = \mu^+ = \overline{\mu} \leq 0{,}2$.

This kind of loading is of great practical significance because, in many technical situations high pressure loads may be subjected to arbitrary oscillations with small amplitudes. For simplicity, we use von Mises sandwich yield condition (see GROSS-WEEGE & WEICHERT [1992]). Details of the numerical procedure can be found in the work of GROSS-WEEGE [1988].

The shakedown load domains for different values of the uniaxial yield stress σ_y are shown in Fig. 1. For the sake of comparison, the geometrically linear solution is also represented.
Using this diagram the shakedown limit loads p^R_{sn} can be determined for different values of $\bar{\mu}$. The results are shown in Fig. 2, where a remarkable difference between the non-linear solutions and the linear solution can be observed. The difference increases for increasing values of σ_y and decreasing values of m. For $\bar{\mu}$ <2,5% the deformations exceed the range of applicability of the Donnell-Mushtari-Vlasov-theory. Moreover, it has to be kept in mind that $\bar{\mu}$ should be small enough in order to justify the proposed linearization of the governing equations.

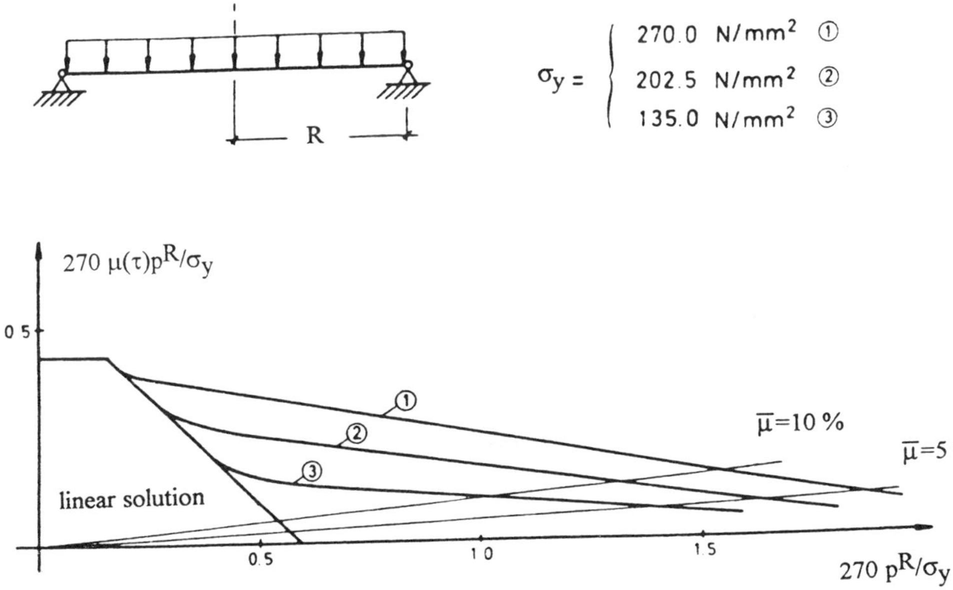

Fig. 1: Circular plate under distributed surface load - shakedown load domain for various values of σ_y.

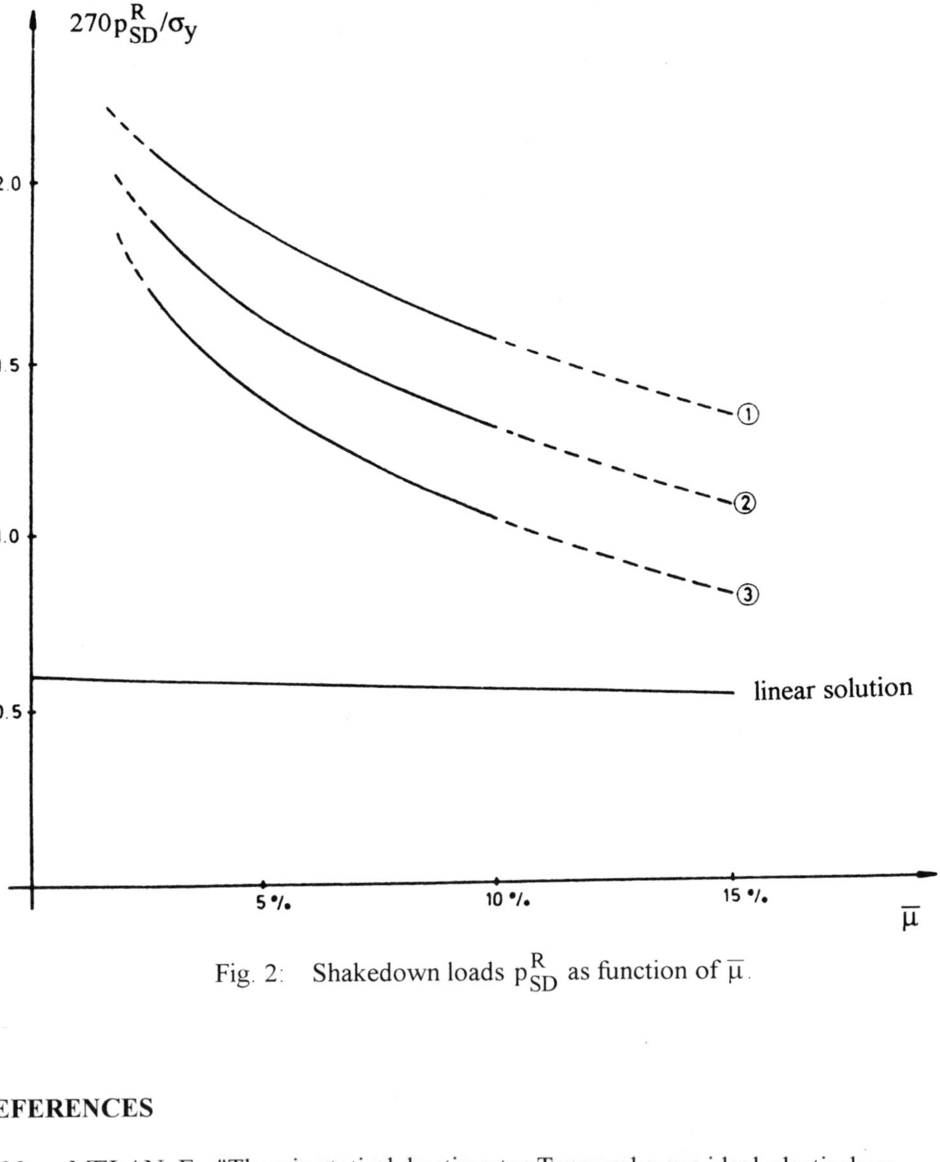

Fig. 2: Shakedown loads p_{SD}^{R} as function of $\bar{\mu}$.

REFERENCES

1938 MELAN, E., "Theorie statisch bestimmter Tragwerke aus ideal-plastischem Baustoff", Sitzungsbericht der Akademie der Wissenschaften (Wien) Abt. IIA, 145,195.

1960 KOITER, W. T., "General theorems for elastic-plastic solids", In: Progress in Solid Mechanics (Eds.: I. N. SNEDDON and R. HILL), pp. 165-221, North-Holland, Amsterdam.

1973 MAIER, G., "A Shakedown matrix theory allowing for workhardening and second-order geometric effects", In: Foundations of Plasticity (Eds.: A. SAWCZUK), pp. 417-433. North-Holland, Amsterdam.

1975 POINTER, A.R.S., "A general shakedown theorem for elastic/plastic bodies with work hardening", Proc. SMiRT-3, London, Paper L5/2.

1976 MANDEL, J., "Adaptation d'une structure plastique écrouissable et approximation", Mech. Res. Commun. 3, 483.

1976 BIENIEK, M.P., FUNARO, J.R., "Elastic-plastic theory of plates and shells", Report NO. DNA 3954 T, Weidlinger Associate, New York.

1982 SAWCZUK, A., "On plastic shell theories at large strains and displacements, Int. J. Mech. Sci., 24, 231.

1984 WEICHERT, D., "Shakedown at finite displacements; a note on Melan's theorem", Mech. Res. Comm. 11, 121.

1985 BASAR, Y., KRÄTZIG, W.B., "Mechanik der Flächentragwerke", Vieweg, Braunschweig.

1986 WEICHERT, D., "On the Influence of geometrical nonlinearities on the shakedown of elastic-plastic structures", Int. J. Plasticity, 2, 135.

1988 GROSS-WEEGE, J. "Zum Einspielverhalten von Flächentragwerken", Report 58, Institute of Mechanics, Ruhr-University Bochum.

1988 KÖNIG, J.A., SIEMASZKO, A., "Strainhardening effects in shakedown processes", Ing. Arch. 58, 58.

1988 WEICHERT, D., GROSS-WEEGE, J., "The numerical assessment of elastic-plastic sheets under variable mechanical and thermal loads using a simplified two-surface yield-condition", Int. J. Mech. Sci. 30, 757.

1989 WEICHERT, D., "Shakedown of shell-like structures allowing for certain geometrical nonlinearities", Arch. Mech. 41 (1), 61.

1990 STEIN, E., ZHANG, G., KÖNIG, J.A., "Micromechanical modelling and computation of shakedown with nonlinear kinematic hardening including examples for 2D-problems", In: Recent Developments of Micromechanics (Eds.: D. R. AXELRAD and W. MUSCHIK). Springer, Berlin.

1990 GROSS-WEEGE, J., "A unified formulation of statical shakedown criteria for geometrically nonlinear problems". Int. J. Plasticity 6, 433.

1992 GROSS-WEEGE, J., WEICHERT, D., "Elastic-plastic shells under variable mechanical and thermal loads", Int. J. Mech. Sci., 34, 863.

LIMIT, SHAKEDOWN, POST-YIELD, AND INADAPTATION ANALYSES OF DISCRETE PLASTIC STRUCTURES

Andrzej SIEMASZKO
Polish Academy of Sciences
Institute of Fundamental Techological Research
ul. Świętokrzyska 21, 00-049 Warsaw Poland

Abstract: This paper proposes a uniform formulation of limit, shakedown, post-yield, and inadaptation analyses of discrete plastic structures. Classical limit and shakedown problems are presented as linear programs, whereas the post-yield and inadaptation problems are sequences of linear programs. The formulation is based on the kinematic approach. Nonlinear geometric effects and nonlinear material hardening-softening are accounted for.

INTRODUCTION

The majority of engineering structures of practical importance are made of materials manifesting distinct plastic properties. Two of the most important global methods which allow us to asses the behaviour and safety of such structures are limit analysis and shakedown (adaptation) analysis, cf. GOKHFELD and CHERNIAVSKY [1980], KÖNIG and MAIER [1981], KÖNIG [1987]. The former considers rigid-perfectly plastic structures subjected to monotonic and proportional loading, the latter considers elastic-perfectly plastic ones subjected to variable repeated loading. The classical limit and shakedown theories are based upon an assumption of geometrical linearity.

Extensions of limit analysis to nonlinear geometrical effects have been studied by DUSZEK and SAWCZUK [1976], ŁODYGOWSKI [1980], ONAT and HAYTHORNTHWAITE [1956], and JANAS [1978]. The most common approach is called the post-yield behaviour analysis; it considers the post-yield equilibrium path of rigid-plastic structures under proportional loading. By subjecting the structure to loads corresponding to the limit load multiplier, an unlimited plastic flow should be initiated. However, nonlinear geometrical effects and, also, material hardening (softening) will stabilize or destabilize the process. In the first case we have to increase the load multiplier to

Z. Mróz et al. (eds.), Inelastic Behaviour of Structures under Variable Loads, 279–291.

continue plastic flow in the structure. In the second case, the plastic flow is possible at a decreasing load multiplier. The behaviour of structure can be described by a post-yield curve showing the dependence of the limit load multiplier on changes of structural geometry.

Extensions of shakedown theory to nonlinear geometrical effects and hardening have been presented. MAIER [1973], modifying the linear static approach, proposed an "a priori" method in which the initial stiffness matrix is modified to account for the foreseen geometric effects. WEICHERT [1986], GROSS-WEEGE [1990], and SACZUK and STUMPF [1990] presented static "a posteriori" approaches where the shakedown condition is checked after a known large deformation history.

A different approach, based upon the kinematic formulation, was proposed by KÖNIG and SIEMASZKO [1980,1985,1988,1993]. This new method, which is a generalization of the post-yield behaviour method to the case of variable repeated loading, is called inadaptation analysis. An elastic strain hardening-softening material model is considered. As for post-yield analysis, the dependence of shakedown multiplier on changes of structural geometry is determined. If the nonlinear geometrical effects and material hardening stabilize the adaptation process, the growing inadaptation curve will be obtained and the structure will be able to adapt itself (shake down) at higher multipliers than the initial shakedown multiplier. For the destabilizing effects, a decreasing inadaptation curve is obtained, and the structure will not adapt itself at the initial value of shakedown multiplier. In this case, the inadaptation process (alternating plasticity or incremental collapse) should manifest itself, which will lead to structural collapse.

These four types of global methods of inelastic analysis can be presented in the following scheme, see Table 1. At the time of the creation, the inadaptation analysis was the missing link in this scheme. It can be considered as a generalization of post-yield analysis to variable repeated loading and, also, as a generalization of classical shakedown analysis to nonlinear geometrical effects, nonlinear hardening and damage.

	monotonic proportional loading	variable repeated loading
geometricaly linear	*rigid-perfectly plastic model* **LIMIT ANALYSIS**	*elastic-perfectly plastic model* **SHAKEDOWN ANALYSIS**
geometricaly nonlinear	*rigid-plastic model, nonlinear hardening damage* **POST-YIELD ANALYSIS**	*elastic-plastic model, nonlinear hardening damage* **INADAPTATION ANALYSIS**

Table 1. Limit and shakedown analyses of discrete plastic structures

The classical limit and shakedown (adaptation) analyses of discrete structures can be formulated as dual linear programs, cf. MAIER [1970]. The post-yield and inadaptation analyses, accounting for nonlinear geometric effects and nonlinear hardening and softening, are in the present paper formulated as a sequence of linear programming tasks. For the description of nonlinear material properties, three internal state variables, namely α, ϵ, and π are introduced which account for the kinematic hardening, isotropic hardening and progressive damage of the material, respectively. The present paper proposes a unified treatment of the geometrically linear limit and shakedown and geometrically nonlinear post-yield and inadaptation analyses for discrete structures. An example of as elastic-plastic hardening-softening frame structure subjected to monotonic or variable repeated loads is studied in detail.

SHAKEDOWN ANALYSIS

Let us consider an elastic-perfectly plastic discrete structure subjected to variable repeated loads. For such structures, shakedown is a widely recognized necessary condition of safe structural behaviour. The shakedown (adaptation) means that after some initial period of plastic deformations a certain residual stress field will be produced, which when imposed on the actual stresses, will allow the structure to respond further exclusively in a purely elastic way.

It is assumed that loads $\mathbf{p}(t)$ may vary on arbitrary paths contained in a prescribed load domain Δ . The shakedown multiplier η is a maximal multiplier of the load domain which ensures shakedown. Stress, strain and displacement fields within the structure are assumed to be fully described by finite-dimensional vectors $\mathbf{s}, \mathbf{q}$ and $\mathbf{u}$, respectively.

Within the framework of geometrical linearity, the fundamental shakedown theorems can be expressed as dual linear programs, cf. MAIER, [1970]:

$$\eta = \max\{\xi \mid \xi d + N^T \mathbf{s} \leq \mathbf{k} \ , \quad C^T \mathbf{s} = 0\} \tag{1}$$

$$\eta = \min\{\mathbf{k}^T \dot{\lambda} \mid \mathbf{d}^T \dot{\lambda} = 1 \ , \quad \mathbf{C}\dot{\mathbf{u}} = \mathbf{N}\dot{\lambda} \ , \ \dot{\lambda} \geq 0\} \tag{2}$$

where $\mathbf{C}$ and $\mathbf{N}$ are rectangular matrices of compatibility and gradients of yield polyhedrans (the model assumes a piece-wise linear yield condition), $\mathbf{k}, \mathbf{d}$ and λ are the vectors of yield moduli, elastic stress envelope and plastic multipliers, ξ and η are the load multiplier and shakedown multiplier, respectively.

For the given load domain Δ, the vector $\mathbf{d}$ of the elastic stress envelope can be found as a solution of the following problem, cf. MAIER [1970]:

$$\mathbf{d} = \max\{\mathbf{N}^T \mathbf{E} \mathbf{C} (\mathbf{C}^T \mathbf{E} \mathbf{C})^{-1} \mathbf{p}(t) \mid \mathbf{p}(t) \in \Delta\} \tag{3}$$

where $\mathbf{E}$ is the matrix of elasticity.

The problem (1) consists in the determination of the maximal load multiplier ξ (strictly speaking the load domain multiplier) such that loads $\xi \mathbf{p}(t)$ still allow for shakedown, i.e. for the stabilization of plastic strains. This maximal multiplier is called the shakedown (adaptation) multiplier, η .

The problem (2) implies that the inadaptation will take place for a given load program $(\xi = 1)$ if we find any kinematically admissible vector of plastic multipliers $\dot{\lambda}$ for which the dissipation $\mathbf{k}^T\dot{\lambda}$ is lower than the rate of internal work $\mathbf{d}^T\dot{\lambda}$. On the other hand, to ensure the adaptation (shakedown), we should analyze all possible mechanisms and find the one with the lowest value of the ratio $\mathbf{k}^T\dot{\lambda}/\mathbf{d}^T\dot{\lambda}$.

LIMIT ANALYSIS

In many practical cases, variations of loads in time are negligible and it is sufficient to analyze a rigid-perfectly plastic structure under proportionally increasing loading $\xi\hat{\mathbf{p}}$, where $\hat{\mathbf{p}}$ is the reference load. The dual linear programs which allow us to determine the limit load multiplier ζ are special cases of the shakedown problems (1) and (2). They are formulated as follows, cf. MAIER [1970]

$$\zeta = \max\{\xi \mid \mathbf{C}^T\mathbf{s} = \xi\hat{\mathbf{p}} \ , \ \mathbf{N}^T\mathbf{s} \le \mathbf{k}\} \tag{4}$$

$$\zeta = \min\{\mathbf{k}^T\dot{\lambda} \mid \hat{\mathbf{p}}^T\dot{\mathbf{u}} = 1 \ , \ \mathbf{C}\dot{\mathbf{u}} = \mathbf{N}\dot{\lambda} \ , \ \dot{\lambda} \ge \mathbf{0}\} \tag{5}$$

In the problem (4), the maximal load multiplier ξ is sought, provided that the stress field $\mathbf{s}$ is statically admissible. The problem (5) consists in the determination of the kinematically admissible vector of plastic multipliers $\dot{\lambda}$ which gives the minimal value of the ratio $\mathbf{k}^T\dot{\lambda}/\mathbf{p}^T\dot{\mathbf{u}}$.

INADAPTATION ANALYSIS

The majority of engineering structures suffer from destabilization of post-yield or adaptation (shakedown) processes due to geometrically nonlinear effects (geometrical softening), cf. SIEMASZKO [1988]. On the other hand, the majority of structural materials, as for example steel, manifest substantial hardening properties. If hardening effects prevail, we can obtain stable structural behaviour. The influence of both these factors on structural behaviour can be determined by the inadaptation analysis proposed by SIEMASZKO and KÖNIG [1985,1988]. It is a step-by-step procedure based on a kinematic approach, which allows us to determine the most stringent plastic deformation path of elastic-strain hardening structures. The procedure can be applied under the following assumptions:

i) small displacements and strains due to a single load cycle;

ii) kinematically in admissible strains are small in comparison with kinematically admissible ones.

These assumptions seem to be realistic for the majority of sufficiently stiff engineering structures. The second assumption means that structural plastic deformation can be described by plastic flow mechanisms with straight structural elements taken into consideration. The first assumption allows us to use the "linear" formulation of the shakedown problem on each step of the method.

For the kinematic shakedown problem (2) for the structure in an initial undeformed configuration $\mathbf{x}^0$, the corresponding initial shakedown multiplier η^0 and the most stringent plastic deformation mechanism described by a vector of displacement rates $\dot{\mathbf{u}}^0$ and plastic multipliers $\dot{\lambda}^0$ can be found. This mechanism allows us to determine neighbouring deformed configurations. The next neighbour structural configuration $\mathbf{x}^1$ is defined by displacement increments $\mu\dot{\mathbf{u}}^0$ imposed on the initial structural configuration $\mathbf{x}^0$. The transformation to the configuration $\mathbf{x}^1$ is associated with the increment of plastic strains described by the increment of plastic multipliers $\mu\dot{\lambda}^0$. Hence, plastic moduli may be updated to account for effects of hardening and damage. For the actual deformed configuration $\mathbf{x}^1$, the linear shakedown problem (2) may be solved again to find if changes of geometry and hardening have stabilized or destabilized the shakedown process.

Repeating this procedure sequentially, we can obtaina so-called inadaptation curve which shows the dependence of the shakedown multiplier on structural deformation. An increasing inadaptation curve indicates a stabilizing influence of nonlinear geometrical and material hardening effects, a decreasing one a destabilizing influence.

For the description of elastic-plastic hardening-softening material we use three internal state variables: backstress α, equivalent plastic strain ϵ and damage parameter π. This material function is discussed in the Appendix 1.

The inadaptation analysis accounting for nonlinear geometrical effects, nonlinear hardening and damage may be formulated as the following sequence of linear programming tasks, cf. SIEMASZKO [1993]:

$$\left\{\begin{array}{ll} \eta(\mathbf{x}^0) = \min\{\mathbf{k}^T(\alpha^0, \epsilon^0, \pi^0 \mid \mathbf{d}^T(\mathbf{x}^0)\dot{\lambda}^0 = 1 \ , \ \mathbf{N}\dot{\lambda} = \mathbf{C}(\mathbf{x}^0)\mathbf{u}^0, \dot{\lambda}^0 \geq 0\} & \\ ------------------------------- & \\ \mathbf{x}^\delta = \mathbf{x}^{\delta-1} + \mu^{\delta-1}\dot{\mathbf{u}}^{\delta-1} & \delta = 1, 2, \ldots \\ \alpha^\delta = c_K \sum_{i=0}^{\delta-1} \mu^i N^T N \dot{\lambda}^i & \\ \epsilon^\delta = \sum_{i=0}^{\delta-1} \mu^i \mid N^T N \mid \dot{\lambda}^i & \\ \pi^\delta = \pi(\epsilon^\delta) & \\ \eta(\mathbf{x}^\delta) = \min\{\mathbf{k}^T(\alpha^\delta, \epsilon^\delta, \pi^\delta \mid \mathbf{d}^T(\mathbf{x}^\delta)\dot{\lambda}^\delta = 1 \ , \ \mathbf{N}\dot{\lambda} = \mathbf{C}(\mathbf{x}^\delta)\mathbf{u}^\delta, \dot{\lambda}^\delta \geq \delta\} & \end{array}\right. \tag{6}$$

Here $\mathbf{x}^0$ is a vector of nodal coordinates for the initial undeformed structural configuration; $\mathbf{x}^\delta$ are vectors for the subsequent deformed configurations; μ is a small control parameter; $\alpha^\delta, \epsilon^\delta$ and π^δ are vectors of internal state variables corresponding to generalized backstresses, equivalent plastic strains and damage parameters, respectively; and c_K is a kinematical hardening modulus.

The inadaptation analysis determinesthe most stringent plastic deformation path; thus the plastic strain history is known, and the arbitrary strain-hardening and arbitrary strain-dependent evolution equation of the damage parameter can be adopted. Appendix 1 provides basic kinematical and isotropic hardening rules for updating the vector of plastic moduli **k** for the case of skeletal structures.

POST-YIELD ANALYSIS

When effects of variation of loading can be neglected, problem (6) of inadaptation analysis can be reduced to an analogous problem of the post-yield analysis. Similarly, solving "linear" limit analysis problem (5) for the undeformed structural configuration $\mathbf{x}^0$, we obtain the most stringent plastic deformation mechanism, defined by the vector $\dot{\mathbf{u}}^0$. Scaling it by the factor μ^0,we can define the neighbouring deformed configuration $\mathbf{x}^1$. At the same time, the vector of plastic moduli **k** must be updated to account for hardening and damage caused by increments of strain defined by the vector $\mu^0\lambda^0$. By repeating computations, the post-yield dependence of limit multiplier ζ on changes of geometry can be determined from the following sequence of linear programming tasks:

$$\left\{\begin{array}{ll}
\zeta(\mathbf{x}^0) = \min\{\mathbf{k}^T(\alpha^0, \epsilon^0, \pi^0 \mid \mathbf{d}^T(\mathbf{x}^0)\dot{\lambda}^0 = 1 \ , \ \mathbf{N}\dot{\lambda} = \mathbf{C}(\mathbf{x}^0)\mathbf{u}^0, \dot{\lambda}^0 \geq 0\} & \\
------------------------------ & \\
\mathbf{x}^\delta = \mathbf{x}^{\delta-1} + \mu^{\delta-1}\dot{\mathbf{u}}^{\delta-1} & \delta = 1, 2, \ldots \\
\alpha^\delta = c_K \sum_{i=0}^{\delta-1} \mu^i N^T N \dot{\lambda}^i & \\
\epsilon^\delta = \sum_{i=0}^{\delta-1} \mu^i \mid N^T N \mid \dot{\lambda}^i & \\
\pi^\delta = \pi(\epsilon^\delta) & \\
\zeta(\mathbf{x}^\delta) = \min\{\mathbf{k}^T(\alpha^\delta, \epsilon^\delta, \pi^\delta \mid \mathbf{d}^T(\mathbf{x}^\delta)\dot{\lambda}^\delta = 1 \ , \ \mathbf{N}\dot{\lambda} = \mathbf{C}(\mathbf{x}^\delta)\mathbf{u}^\delta, \dot{\lambda}^\delta \geq \delta\} &
\end{array}\right. \tag{7}$$

where $\alpha^\delta, \epsilon^\delta$ and π^δ are vectors of internal state variables corresponding to generalized backstresses, equivalent plastic strains and damage parameters.

The post-yield analysis is possible under the assumption of small strains and displacements on each step of computations. Since the most stringent post-yield deformation path is determined, the plastic strain history is known and an arbitrary strain-hardening and arbitrary strain-depended evolution equation of the damage parameter can be adopted. To account for nonlinear hardening and damage, the vector of plastic moduli, **k** , should be updated according to the formulae presented in Appendix 1.

EXAMPLE

Let us perform the limit, shakedown, post-yield and inadaptation analyses for the five story frame, Fig. 1, subjected to six independent variable repeated load systems P_i, $i = 1, 2, \ldots, 6$. Limits of variation of each load are shown in Fig. 1. We assume four sandwich cross-sections of area F_i, $i = 1, \ldots, 4$, with the rhombic yield condition defined in the bending moment-axial force space, Fig. 1.

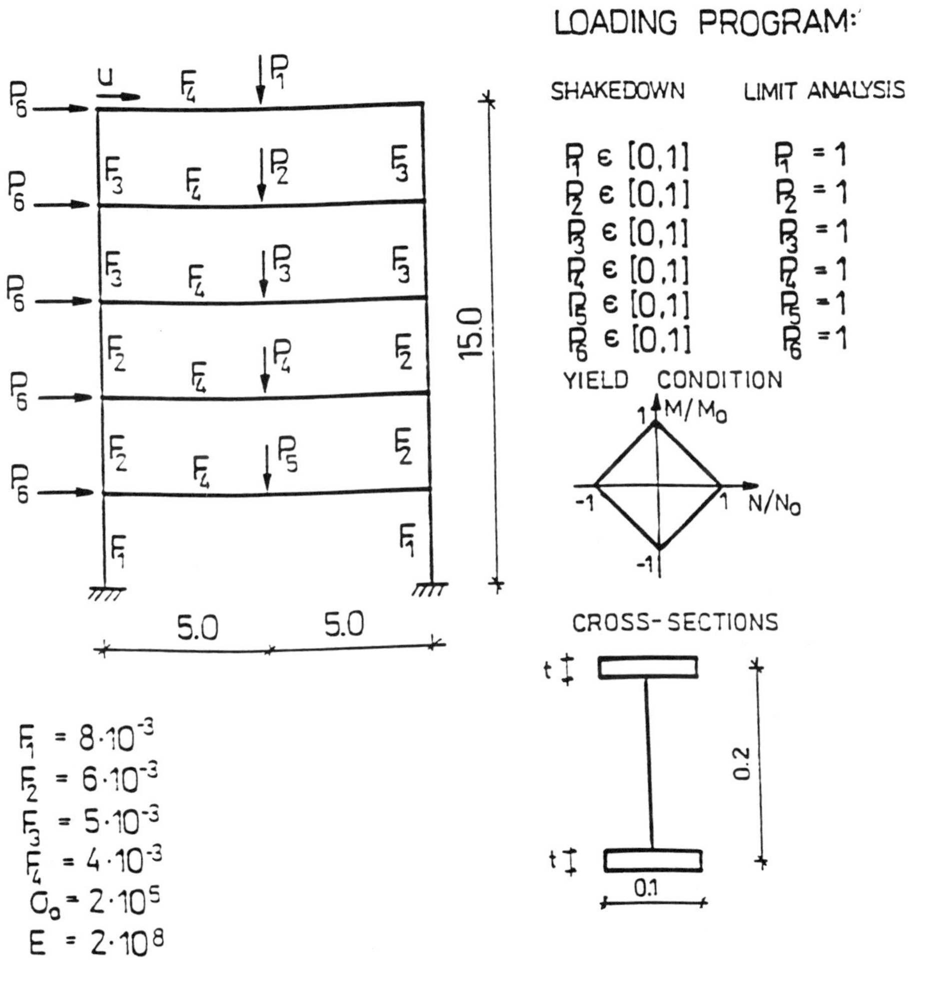

Fig. 1. Five story frame example

Let us consider the elastic-linear isotropic and kinematic hardening-damaging material with hardening moduli $c_I = c_K = 10^6$, cf. line "a", Fig. 2, and the piece-wise linear damage function $(\pi/\pi_F)^{1/2}$, cf. curve "b", Fig. 2.

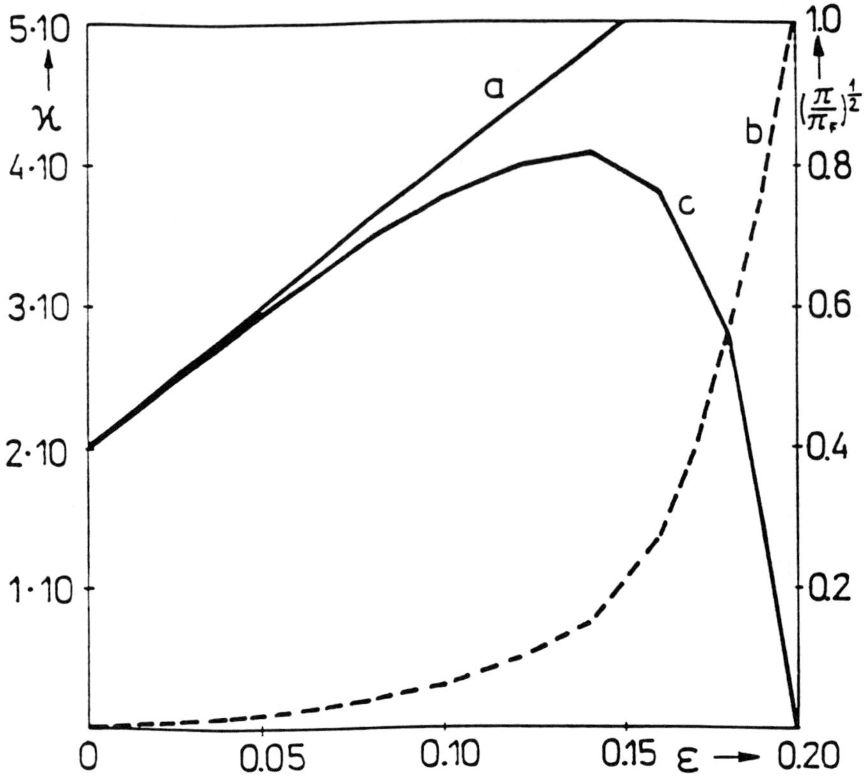

Fig. 2. Material model. Linear kinematic and isotropic hardening "a", damage function "b", resulting material function "c".

The final hardening-softening material function is described by the curve "c", Fig. 2. The ideal plastic hinge theory is generalized by the assumption that plastic strains are distributed uniformly in the zone of prescribed length next to the hinge.

Let us perform the four types of analysies. The classical limit and shakedown analyses give us of the limit load multiplier $\zeta^0 = 23.3$ and the shakedown multiplier $\eta^0 = 20.7$, cf. Fig. 3. Of course, the shakedown multiplier is lower than the limit one, since the introduction of loading variations always reduces the load multiplier. Unfortunately, the safety of the structure is not fully determined. The results obtained give no information on the influence of changes of geometry, post-critical structural behaviour, and structural sensitivity to imperfections.

The above questions can be answered by the post-yield or inadaptation analyses. Fig. 3 shows the results obtained. In both cases the structural plastic deformation proceeds according to the same side-sway mechanism and very similar post-yield "a" and inadaptation "b" curves are obtained. It is easy to see that material hardening combined with geometrical effects stabilize the processes considered (strictly speaking, the effects of material hardening prevail over the effects of geometrical softening, producing overall hardening). Fig. 3 shows the maximal limit and shakedown multipliers

($\zeta_{\max} = 36.2, \eta_{\max} = 33.4$) which are significantly greater than the "linear" multipliers. It can be concluded that, in this case, the results of the classical limit and shakedown analyses were safe and recommendable for the design purposes.

By a detailed analysis of the process of plastic deformation we can observe the phenomenon of changes of instantaneous side-sway mechanisms. At the maximal values of load multiplier, the phenomenon of plastic strain localization at the support nodes occurs. The localization is observed immediately after the material of these nodes enters the softening range corresponding to the decreasing part of the material function "c", Fig. 2. Despite significant reserves of strength at the other nodes, only these two softening nodes determine the behaviour of the structure. The resulting destabilization leads quickly to the structural collapse.

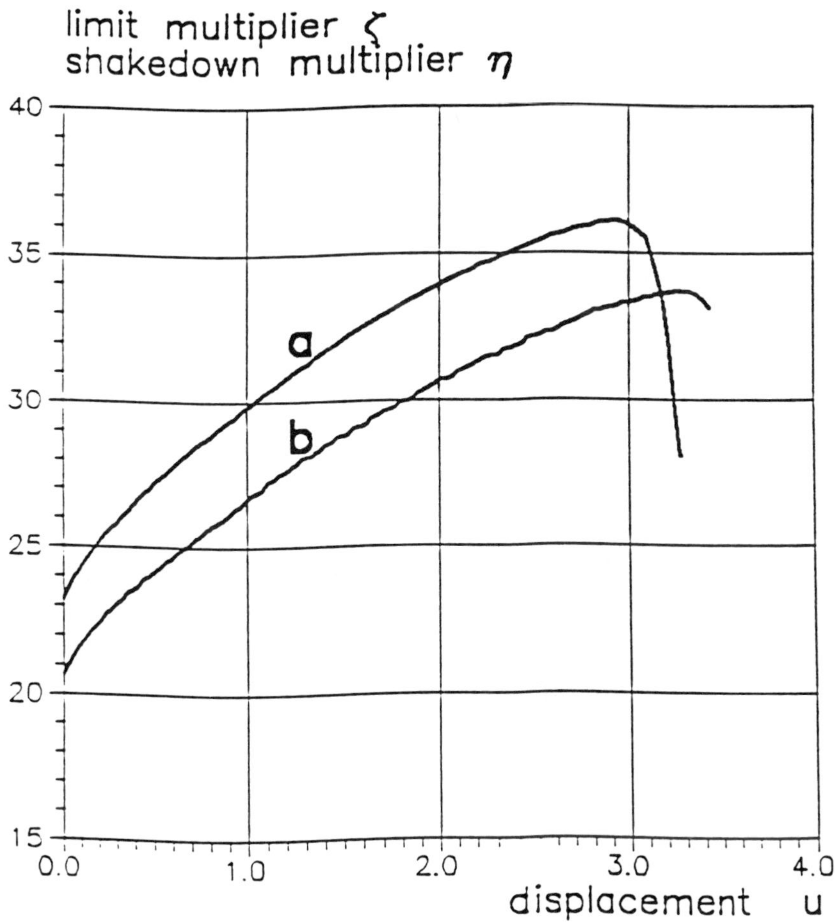

Fig. 3. Post-yield "a" and inadaptation "b" curves

FINAL REMARKS

This paper presents aunified treatment of the shakedown, limit, post-yield, and inadaptation analyses of plastic discrete structures. The classical, geometrically linear shakedown and limit analyses are formulated as linear programs. Using the kinematical approach, we formulate the geometrically nonlinear post-yield and inadaptation analyses as sequences of linear programs. For the description of nonlinear hardening and material damage, three internal state variables are introduced, namely α, ϵ, and π, describing kinematic hardening, isotropic hardening, and progressive damage, respectively.

It is shown that the post-yield and inadaptation analyses provide much more information necessary for safe structural design than do the classical limit and shakedown ones. They define the most stringent plastic deformation path, describe the most dangerous mechanisms, changes of instantaneous mechanisms and the way structural collapse occurs. For the stable structural behaviour it is possible to determine the maximal load multiplier (safety factor) which accounts for the nonlinear geometrical effects, hardening and softening of material. For the unstable behaviour, the sensitivity of load multiplier to changes of geometry can be defined. In this case, the "classical" load multiplier should be reduced to avoid an overestimation of structural safety. For this purpose a suitable method should be worked out.

ACKNOWLEDGEMENT

The work was supported by the National Research Council (KBN) under the Reserch Project 309789101.

REFERENCES

1956 ONAT, E.T., HAYTHORNTHWAITE, R.M.,"The load-carrying capacity of circular plates at large deflection", J.Appl.Mech., 23.

1984 PERZYNA, P.,"Constitutive modelling of dissipative solids for postcritical behaviour and fracture", ASME, J.Eng.Mat.Techn., 106, 410.

1970 MAIER, G.,"A matrix structural theory of piecewise-linear elastoplasticity with interacting yield planes", Meccanica, 5, 54.

1973 MAIER, G.,"A shakedown matrix theory allowing for work-hardening and second-order geometric effects", Foundations of Plasticity, ed. A.Sawczuk, Noordhoff, Leyden, 417.

1973 MRÓZ, Z.,"A description of workhardening of metals with application to variable loading", Found. of Plasticity, ed.A.Sawczuk, Noordhoff, Leyden, 551.

1976 DUSZEK, M., SAWCZUK, A.,"Stable and unstable states of rigid-plastic frames at the yield point load", J. Struct. Mech. <u>4</u>, 33.

1978 JANAS, M.,"Changes of plastic mechanism in the post-yield process of bar structures", IFTR Reports, 48.

1980 FISHER, J.R., "Void nucleation in spheroidized steels during tensile deformations", Ph.D. thesis, Brown University.

1980 GOKHFELD, D.A., CHERNIAVSKY, O.F.,"Limit analysis of structures at thermal cycling", Sijthoff and Nordhoff, Leyden.

1980 KÖNIG, J.A.,"On stability of the incremental collapse process", Arch. Inz.Lad., <u>26</u>, 219.

1981 KÖNIG, J.A., MAIER, G., "Shakedown analysis of elastoplastic structures: a review of recent developments", Nuclear Engng. Des., <u>66</u>, 81.

1981 LE ROY, G., EMBURY, J.D., EDWARD, G., ASHBY, M.F.,"A model of ductile fracture based on the nucleation and growth of voids", Acta Metall., <u>29</u>, 1509.

1982 ŁODYGOWSKI, T.,"Geometrically nonlinear analysis of rigid-plastic plane beams and frames", IFTR Reports, 9.

1984 PERZYNA, P., DRABIK, A.,"Analysis of the evolution equation describing the postcritical behaviour of dissipative solids", Arch.Mech., <u>36</u>, 5, 733.

1985 SIEMASZKO, A., KÖNIG, J.A.,"Analysis of stability of incremental collapse of skeletal structures", J.Struct.Mech., <u>13</u>, 301.

1986 CHABOCHE, J.L.,"Time-independent constitutive theories for cyclic plasticity", Int.J.Plasticity, <u>2</u>, 149.

1986 WEICHERT, D.,"On the influence of geometrical nonlinearities on the shakedown of elastic-plastic structures", Int.J.Plast., <u>2</u>, 135.

1987 KÖNIG, J.A.,"Shakedown of elastic-plastic structures", PWN - Elsevier, Warszawa - Amsterdam.

1988 KÖNIG, J.A., SIEMASZKO, A.,"Strainhardening effects in shakedown process", Ing.-Archiv, <u>58</u>, 58.

1988 SIEMASZKO, A.,"Stability analysis of the shakedown processes of the skeletal structures", IFTR Reports, <u>12</u>, 1 (in Polish).

1990 GROSS-WEEGE, J.,"A unified formulation of statical shakedown criteria for geometrically nonlinear problem", Int.J.Plast. <u>6</u>, 433.

1990 SACZUK, J., STUMPF, H.,"On statical shakedown theorems for nonlinear problems", Inst.Mech., Ruhr-Univ. Bochum, 47, 1.

1993 SIEMASZKO, A.,"Inadaptation analysis with hardening and damage", Submitted to Eur.J.Mech. 12, 2.

Appendix 1

Let us consider elastic-plastic hardening and damaging material. The material function for a one-dimensional case is postulated in the form, cf. PERZYNA [1984]:

$$\kappa(\alpha, \epsilon, \pi) = [\kappa_0 + \alpha + \bar{\kappa}(\epsilon)\epsilon] \left[1 - \left(\frac{\pi(\epsilon)}{\pi_F}\right)^{1/2}\right] \tag{8}$$

where κ_0 is an initial yield point, α is the backstress parameter defining kinematical hardening, $\bar{\kappa}(\epsilon)$ is the nonlinear isotropic hardening function depending on equivalent plastic strain ϵ, $\pi(\epsilon)$ is the porosity parameter function describing material damage and π_F is the critical porosity (fracture point). From (8) it follows that loads tend to zero at the fracture point π_F .

For linear kinematical hardening, it can be assumed that the rate of backstress parameter $\dot{\alpha}$ is proportional to the rate of strain $\dot{\mathbf{q}}$

$$\dot{\alpha} = c_K \dot{\mathbf{q}} \tag{9}$$

where c_K is the kinematic hardening modulus. We can propose the following rule of updating the vector $\mathbf{k}$, cf. KÖNIG and SIEMASZKO [1988]:

$$\begin{aligned} \mathbf{k}(\alpha^0) &= \mathbf{k}^0 \\ \mathbf{k}(\alpha^\delta) &= \mathbf{k}^0 + \alpha^\delta \quad \delta = 1, 2, \ldots \end{aligned} \tag{10}$$

where $\mathbf{k}^0$ is a vector of initial plastic moduli, and α^δ is a vector of generalized backstresses. It is also possible to incorporate here the more general nonlinear kinematic model by MRÓZ [1973].

In the case of nonlinear isotropic hardening, we will have the following formula:

$$\begin{aligned} \mathbf{k}(\epsilon^0) &= \mathbf{k}^0 \\ \mathbf{k}(\epsilon^\delta) &= \bar{\kappa}(\epsilon^\delta) \quad \delta = 1, 2, \ldots \end{aligned} \tag{11}$$

where $\bar{\kappa}$ is a given nonlinear isotropic hardening function whereas ϵ^δ is a vector of equivalent plastic strains. In the case of linear isotropic hardening, we have

$$\mathbf{k}(\epsilon^\delta) = \mathbf{k}^0 + c_I \epsilon^\delta \quad \delta = 1, 2, \ldots \tag{12}$$

where c_I is an isotropic hardening modulus.

The damage of material can be described by the porosity parameter π interpreted as a void volume fraction. The admissible domain of the porosity parameter π is the interval [0,1]. The general evolution equation of this parameter describes the changes of the internal imperfections caused by nucleation, growth, linkage and also compaction of voids. The final stage of the process corresponds to the fracture of material. The evolution equation can be postulated in the form, cf. PERZYNA [1984]:

$$\dot{\pi} = (\dot{\pi})_{diffusion} + (\dot{\pi})_{nucleation} + (\dot{\pi})_{growth} \tag{13}$$

In general, by integrating the evolution equation (13), the dependence of the porosity parameter ζ on the equivalent plastic strain ϵ is obtained. Following the results obtained by PERZYNA and DRABIK [1984], based on experimental results of FISHER [1980] and LE ROY et al.[1981], we can also use the approximation function providing the explicit relation between π and equivalent plastic strain ϵ :

$$\pi = \pi(\epsilon) \tag{14}$$

More detailed discussion of hardening rules can be found in papers by KÖNIG and SIEMASZKO [1988], CHABOCHE [1986] and MRÓZ [1973]. For the evolution equations of the porosity parameter $\pi = \pi(\epsilon)$ we refer to papers of PERZYNA [1984], PERZYNA, DRABIK [1984], FISHER [1980] and LE ROY et al.[1981].

CONSTITUTIVE MODEL AND INCREMENTAL SHAKEDOWN ANALYSIS IN FINITE ELASTOPLASTICITY

HELMUT STUMPF

Lehrstuhl für Allgemeine Mechanik, Ruhr-Universität Bochum
44780 Bochum, Germany

ABSTRACT

For the deformation analysis of structures undergoing finite elastic-plastic strains, an appropriate constitutive model is presented for isotropic elastic and plastic material behavior with isotropic and kinematic hardening. The model is based on the unique decomposition of the deformation gradient into a plastic stretch, a back-rotated elastic stretch and a rotation, which can be considered as macroscopic substructure rotation. Elastic and plastic spin are determined exactly as functions of the elastic and plastic strain rate, respectively, which enables the calculation of the macroscopic substructure spin. This spin is used to define the objective corotational rate adequate to formulate the constitutive equations of finite elastoplasticity with an appropriate evolution law for kinematic hardening.

Next, we present constitutive and equilibrium equations of an elastic-plastic body at finite strain taking into account the microstructure of the material. All equations are given with reference to the undeformed configuration.

Finally, we discuss some new possibilities for a generalization of Melan's linear shakedown theorem for finite strains and we propose a theorem for an incremental shakedown analysis.

1. INTRODUCTION

In linear elastoplasticity, efficient solution algorithms are known for the analysis of structures subject to small elastic-plastic strains and rotations. For loading histories with variable repeated load cycles, the static shakedown theorem of Melan (1936) provides an elegant method to determine lower bounds to the shakedown load factor assuring adaptation of the structure. An upper bound can be obtained by the kinematic theorem of Koiter (1960) defining non-shakedown. Both methods are dual to each other (Debordes (1977)), which enables the construction of converging error bounds for the shakedown load factor (e. g. Stumpf and Le (1991)).

Z. Mróz et al. (eds.), Inelastic Behaviour of Structures under Variable Loads, 293–307.

In the analysis of thin structures as beams, plates and shells we have to distinguish between geometrical nonlinearities caused by moderate or finite rotations of material elements, and those due to finite strain deformation. For small strains and moderate or finite rotations, many appropriate theories with efficient finite element codes are known in the literature for elastic shells (see the review article by Pietraszkiewicz (1989)), while only a few papers deal with elastic-plastic shells (see e. g. Schmidt and Weichert (1989), Simo and Kennedy (1989), Basar and Weichert (1991)). For elastic-plastic structures subject to small strains and moderate rotations, generalizations of Melan's linear shakedown theorem were given by Maier (1973), Weichert (1986), Groß-Weege (1990) and Saczuk and Stumpf (1992). Two models were presented in the literature for the analysis of shells undergoing finite elastic strains and rotations, a 6-parameter (3 displacemental and 3 rotational) model by Makowski and Stumpf (1990) with a corresponding general finite element code by Chroscielewski et al. (1992), and a 3-parameter displacemental model by Schieck et al. (1992) with finite element application as well. The latter theory was generalized by Stumpf and Schieck (1994) for shells subject to finite elastic-plastic strains and rotations. Theoretical and numerical aspects for an incremental shakedown analysis in the frame of finite elastoplasticity are investigated in Stumpf (1993).

Many discussions and controversial assumptions can be found in the literature concerning finite elastoplasticity. Lee and Liu (1967) introduced the multiplicative decomposition of the deformation gradient $\mathbf{F}$ into an elastic, $\mathbf{F}^e$, and a plastic $\mathbf{F}^p$, part, $\mathbf{F} = \mathbf{F}^e\mathbf{F}^p$, as a basic kinematical assumption. As is well-known, this decomposition is unique only up to an arbitrary rotation. Therefore, various additional assumptions were introduced in the literature concerning the elastic and plastic rotation, respectively, or the so-called plastic spin $\mathbf{w}^p$. Many authors (e. g. Dafalias (1983), van der Giessen (1991), Paulun and Pecherski (1992)) formulated additional constitutive equations for the plastic spin, while others (e. g. Simo (1988), Etorovic and Bathe (1990)) assumed the plastic spin to be zero. Contrary to these papers, it was stated in Stumpf and Badur (1990) and Nemat-Nasser (1990) that the plastic spin for Cauchy continua is a dependent function and that therefore assumptions of the type described above cannot be made.

A Lagrangean type constitutive model for the analysis of structures subject to finite elastic - finite plastic strains is presented in Schieck and Stumpf (1993) using logarithmic elastic and logarithmic plastic strain measures for superposed deformations in an incremental solution algorithm. It is shown that assumptions concerning the elastic or plastic rotation are superfluous. Dashner (1993), intending to show that our exact kinematical results can be obtained also in a simpler way, derived formulae which are valid only for deformations subject to moderate elastic and moderate plastic strains. This can be seen from Dashner (1993), eqn $(8)_3$ defining an "elastic" deformation gradient $\mathbf{F}_e$, which depends also on the plastic stretch $\overset{+}{\mathbf{U}}_p$. This dependency on the plastic stretch $\overset{+}{\mathbf{U}}_p$ exists also in all his further definitions of the elastic stretch tensor $\overset{+}{\mathbf{V}}_e$, logarithmic strain tensor $\overset{+}{\mathbf{H}}_e$ and also in the strain energy density ψ , while in our results all elastic measures are purely elastic quantities. The influence of the plastic stretch $\overset{+}{\mathbf{U}}_p$ in Dashner's formulae is canceled, if the deformation is restricted to moderate elastic and moderate plastic strains, a special case of our general result. This special case was first considered by Etorovic and Bathe (1990).

In Schieck and Stumpf (1993) the plastic spin and its influence on an appropriate formulation of the evolution law for the backstress tensor is not considered.

In this paper we present kinematics and constitutive equations for an incremental analysis of elastic-plastic structures at finite strain. The model is based on the unique decomposition of the deformation gradient $\mathbf{F}$ into a plastic stretch $\mathbf{U}^p$, *a* back-rotated elastic stretch $\mathring{\mathbf{U}}^e$ and a rotation tensor $\mathbf{Q}$, $\mathbf{F} = \mathbf{Q}\mathring{\mathbf{U}}^e\mathbf{U}^p$, where $\mathbf{Q}$ can be considered as a macroscopic substructure rotation. Using this decomposition, we determine exactly the elastic spin $\mathbf{w}^e$ as a function of the elastic strain rate $\mathbf{d}^e$, and the plastic spin $\mathbf{w}^p$ as a function of the plastic strain rate $\mathbf{d}^p$. With known spins $\mathbf{w}^e, \mathbf{w}^p$, we are able to calculate the macroscopic substructure spin $\boldsymbol{\Omega} = \dot{\mathbf{Q}}\mathbf{Q}^{-1}$ also exactly. Then we use the spin $\boldsymbol{\Omega}$ to construct the objective corotational rate $(.\overset{\nabla}{.}.) = (.\dot{.}.) - \boldsymbol{\Omega}(\ldots) + (\ldots)\boldsymbol{\Omega}$ appropriate for formulating the elastic and plastic constitutive equations and the evolution law for kinematic hardening.

Next, we present constitutive and equilibrium equations in a Lagrangean description for elastic-plastic bodies with microstructure. It is shown that, in addition to the macroscopic variables of Cauchy continua, in our model of oriented media we have to take into account, as microscopic variables, the plastic torsion (1,2)-tensor (contravariant of order 1, covariant of order 2), the microstress (2,1)-tensor, the microforce (1,1)-tensor and the microcouple (2,1)-tensor.

Finally, we discuss some new possibilities for a generalization of Melan's linear shakedown theorem for finite strains using the results of this paper. For an incremental deformation analysis, a shakedown theorem is proposed generalizing results of Borkowski and Kleiber (1980) for the linear theory.

2. EXACT KINEMATICS

As a basic assumption of finite elastoplasticity, Lee and Liu (1967) introduced a current, relaxed intermediate configuration with an associated multiplicative decomposition of the total deformation gradient into an elastic, $\mathbf{F}^e$, and a plastic, $\mathbf{F}^p$, part,

$$\mathbf{F} = \mathbf{F}^e\mathbf{F}^p. \tag{2.1}$$

It is well known that decomposition (2.1) is not unique, and that the intermediate configuration for Cauchy continua can be determined only up to an arbitrary rotation. Contrary to (2.1) we proposed in Schieck and Stumpf (1993) a unique decomposition of the total deformation gradient $\mathbf{F}$ derived as follows. First we apply the polar decomposition theorem to $\mathbf{F}^e$ and $\mathbf{F}^p$,

$$\mathbf{F}^e = \mathbf{R}^e\mathbf{U}^e, \qquad \mathbf{F}^p = \mathbf{R}^p\mathbf{U}^p, \tag{2.2}$$

where $\mathbf{U}^e, \mathbf{U}^p$ are the Lagrangean-type elastic and plastic stretches, and $\mathbf{R}^e, \mathbf{R}^p$ the elastic and plastic rotations. Defining now a back-rotated elastic stretch tensor $\mathring{\mathbf{U}}^e$ and a rotation tensor $\mathbf{Q}$,

$$\mathring{\mathbf{U}}^e := \mathbf{R}^{pT}\mathbf{U}^e\mathbf{R}^p, \qquad \mathbf{Q} := \mathbf{R}^e\mathbf{R}^p, \tag{2.3}$$

then we obtain by introducing (2.2), (2.3) into (2.1) the unique decomposition,

$$\mathbf{F} = \mathbf{Q}\mathring{\mathbf{U}}^e\mathbf{U}^p. \tag{2.4}$$

The rotation tensor $\mathbf{Q}$ can be considered as a macroscopic substructure rotation. Representation (2.4) of the total deformation gradient $\mathbf{F}$ suggests that we consider a deformation $\mathring{\mathbf{F}}$ composed of a successive application of the plastic stretch $\mathbf{U}^p$ and the elastic stretch $\mathring{\mathbf{U}}^e$,

$$\mathring{\mathbf{F}} := \mathring{\mathbf{U}}^e\mathbf{U}^p. \tag{2.5}$$

We have to point out that in the general case of non-coaxial deformations composition, (2.5) is not symmetric.

For the configuration without macroscopic substructure rotation $\mathbf{Q}$ defined by (2.5), we determine deformation rate, strain rate and spin, and decompose them into elastic and plastic parts:

$$\mathring{\mathbf{l}} := \dot{\mathring{\mathbf{F}}}\mathring{\mathbf{F}}^{-1} = \mathring{\mathbf{l}}^e + \mathring{\mathbf{l}}^p \tag{2.6}$$

with

$$\mathring{\mathbf{l}}^e = \dot{\mathring{\mathbf{U}}}^e\mathring{\mathbf{U}}^{e-1}, \qquad \mathring{\mathbf{l}}^p = \mathring{\mathbf{U}}^e\dot{\mathbf{U}}^p\mathbf{U}^{p-1}\mathring{\mathbf{U}}^{e-1} \tag{2.7}$$

The symmetric and antisymmetric parts of (2.6) and (2.7) lead to the strain rate $\mathring{\mathbf{d}}$ and to the spin $\mathring{\mathbf{w}}$, with their additive decompositions into elastic and plastic parts:

$$\mathring{\mathbf{l}} = \mathring{\mathbf{d}} + \mathring{\mathbf{w}}, \qquad \mathring{\mathbf{d}} = \mathring{\mathbf{d}}^e + \mathring{\mathbf{d}}^p, \qquad \mathring{\mathbf{w}} = \mathring{\mathbf{w}}^e + \mathring{\mathbf{w}}^p, \tag{2.8}$$

where the elastic and plastic strain rates are obtained as

$$\mathring{\mathbf{d}}^e = \frac{1}{2}\left(\dot{\mathring{\mathbf{U}}}^e\mathring{\mathbf{U}}^{e-1} + \mathring{\mathbf{U}}^{e-1}\dot{\mathring{\mathbf{U}}}^e\right)$$

$$\tag{2.9}$$

$$\mathring{\mathbf{d}}^p = \frac{1}{2}\left(\mathring{\mathbf{U}}^e\dot{\mathbf{U}}^p\mathbf{U}^{p-1}\mathring{\mathbf{U}}^{e-1} + \mathring{\mathbf{U}}^{e-1}\mathring{\mathbf{U}}^{p-1}\dot{\mathbf{U}}^p\mathring{\mathbf{U}}^e\right)$$

and the elastic and plastic spins as

$$\mathring{\mathbf{w}}^e = \tfrac{1}{2}\,(\dot{\mathring{\mathbf{U}}}^e\mathring{\mathbf{U}}^{e-1} - \mathring{\mathbf{U}}^{e-1}\dot{\mathring{\mathbf{U}}}^e)$$

$$\mathring{\mathbf{w}}^p = \tfrac{1}{2}\,(\mathring{\mathbf{U}}^e\dot{\mathbf{U}}^p\mathbf{U}^{p-1}\mathring{\mathbf{U}}^{e-1} - \mathring{\mathbf{U}}^{e-1}\mathbf{U}^{p-1}\dot{\mathbf{U}}^p\mathring{\mathbf{U}}^e). \tag{2.10}$$

Introducing the fourth order tensors $\mathring{\mathbb{A}}^e$ and $\mathring{\mathbb{A}}^p$ we can write eqns (2.9) as (for details see Schieck and Stumpf (1994)),

$$\mathring{\mathbf{d}}^e = \mathring{\mathbb{A}}^e \cdot \dot{\mathring{\mathbf{U}}}^e, \qquad \mathring{\mathbf{d}}^p = \mathring{\mathbb{A}}^p \cdot \dot{\mathbf{U}}^p \tag{2.11}$$

Using convected coordinates with covariant and contravariant base vectors $\mathring{\mathbf{g}}_i, \mathring{\mathbf{g}}^i$, we obtain the fourth order tensor $\mathring{\mathbb{A}}$ in the form,

$$\mathring{\mathbb{A}} = \mathring{A}_{ij}^{kl}\mathbf{g}^i \otimes \mathbf{g}^j \otimes \mathbf{g}_k \otimes \mathbf{g}_l, \tag{2.12}$$

where the usual summation convention is applied. We use a dot to denote the tensor product with contraction of two indices.

Inverting eqns (2.11) and introducing them into (2.10) we derive the elastic spin $\mathring{\mathbf{w}}^e$ and the plastic spin $\mathring{\mathbf{w}}^p$ in operator form as functions of the strain rates,

$$\mathring{\mathbf{w}}^e = \tfrac{1}{2}\,[(\mathring{\mathbf{d}}^e \cdot \mathring{\mathbb{A}}^{e-T})\mathring{\mathbf{U}}^{e-1} - \mathring{\mathbf{U}}^{e-1}(\mathring{\mathbb{A}}^{e-1} \cdot \mathring{\mathbf{d}}^e)],$$

$$\mathring{\mathbf{w}}^p = \tfrac{1}{2}\,[\mathring{\mathbf{U}}^e(\mathring{\mathbf{d}}^p \cdot \mathring{\mathbb{A}}^{p-T})\mathbf{U}^{p-1}\mathring{\mathbf{U}}^{e-1} - \mathring{\mathbf{U}}^{e-1}\mathbf{U}^{p-1}(\mathring{\mathbb{A}}^{p-1} \cdot \mathring{\mathbf{d}}^p)\mathring{\mathbf{U}}^e]. \tag{2.13}$$

It is seen that the elastic spin $\mathring{\mathbf{w}}^e$ depends on the elastic strain rate $\mathring{\mathbf{d}}^e$ and the elastic stretch $\mathring{\mathbf{U}}^e$, and that the plastic spin $\mathring{\mathbf{w}}^p$ depends on the plastic strain rate $\mathring{\mathbf{d}}^p$ and on the elastic and plastic stretches $\mathring{\mathbf{U}}^e, \mathbf{U}^p$.

The Eulerian form of the elastic and plastic strain rates and spins can be obtained by push-forward of eqns (2.9), (2.10) with the rotation tensor $\mathbf{Q}$. First we define the Eulerian elastic stretch $\mathbf{V}^e$ and the push-forward $\mathbf{V}^p$ of the plastic stretch $\mathbf{U}^p$

$$\mathbf{V}^e := \mathbf{Q}\mathring{\mathbf{U}}^e\mathbf{Q}^T, \qquad \mathbf{V}^p := \mathbf{Q}\mathbf{U}^p\mathbf{Q}^T, \tag{2.14}$$

and we introduce the corotational rate

$$(\overset{\nabla}{.}) := \mathbf{Q}(\dot{\ldots})\mathbf{Q}^T = (\dot{\ldots}) - \boldsymbol{\Omega}(\ldots) + (\ldots)\boldsymbol{\Omega}, \tag{2.15}$$

where we call $\boldsymbol{\Omega}$ the macroscopic substructure spin,

$$\boldsymbol{\Omega} := \dot{\mathbf{Q}}\mathbf{Q}^{-1}. \tag{2.16}$$

Then the push-forward of the strain rates (2.9) with $\mathbf{Q}$ leads to

$$\mathbf{d}^e := \mathbf{Q}\mathring{\mathbf{d}}^e\mathbf{Q}^T = \frac{1}{2}(\overset{\nabla}{\mathbf{V}}{}^e\,\mathbf{V}^{e-1} + \mathbf{V}^{e-1}\,\overset{\nabla}{\mathbf{V}}{}^e),$$

$$\mathbf{d}^p := \mathbf{Q}\mathring{\mathbf{d}}^p\mathbf{Q}^T = \frac{1}{2}(\mathbf{V}^e \overset{\nabla}{\mathbf{V}}{}^p\, \mathbf{V}^{p-1}\mathbf{V}^{e-1} + \mathbf{V}^{e-1}\mathbf{V}^{p-1}\, \overset{\nabla}{\mathbf{V}}{}^p\, \mathbf{V}^e), \tag{2.17}$$

and the push-forward of the spins (2.10) to

$$\mathbf{w}^e := \mathbf{Q}\mathring{\mathbf{w}}^e\mathbf{Q}^T = \frac{1}{2}(\overset{\nabla}{\mathbf{V}}{}^e\, \mathbf{V}^{e-1} - \mathbf{V}^{e-1}\, \overset{\nabla}{\mathbf{V}}{}^e)$$

$$\mathbf{w}^p := \mathbf{Q}\mathring{\mathbf{w}}^p\mathbf{Q}^T = \frac{1}{2}(\mathbf{V}^e \overset{\nabla}{\mathbf{V}}{}^p\, \mathbf{V}^{p-1}\mathbf{V}^{e-1} - \mathbf{V}^{e-1}\mathbf{V}^{p-1}\, \overset{\nabla}{\mathbf{V}}{}^p\, \mathbf{V}^e). \tag{2.18}$$

Using the same procedure as described above, we introduce fourth order tensors $\mathbb{A}^e$ and $\mathbb{A}^p$ and rewrite eqns (2.17) in operator form,

$$\mathbf{d}^e = \mathbb{A}^e \cdot \overset{\nabla}{\mathbf{V}}{}^e, \qquad \mathbf{d}^p = \mathbb{A}^p \cdot \overset{\nabla}{\mathbf{V}}{}^p\,. \tag{2.19}$$

Inverting eqns (2.19) and introducing them into (2.18) yields

$$\mathbf{w}^e = \frac{1}{2}[(\mathbf{d}^e \cdot \mathbb{A}^{e-T})\mathbf{V}^{e-1} - \mathbf{V}^{e-1}(\mathbb{A}^{e-1} \cdot \mathbf{d}^e)]$$

$$\mathbf{w}^p = \frac{1}{2}[\mathbf{V}^e(\mathbf{d}^p \cdot \mathbb{A}^{p-T})\mathbf{V}^{p-1}\mathbf{V}^{e-1} - \mathbf{V}^{e-1}\mathbf{V}^{p-1}(\mathbb{A}^{p-1} \cdot \mathbf{d}^p)\mathbf{V}^e]. \tag{2.20}$$

It can be seen that the elastic spin is a function of the elastic strain rate $\mathbf{d}^e$ and of the elastic stretch $\mathbf{V}^e$, and that the plastic spin $\mathbf{w}^p$ is a function of the plastic strain rate $\mathbf{d}^p$ and of the elastic and plastic stretches $\mathbf{V}^e, \mathbf{V}^p$. It should be pointed out that the eqns (2.20) are exact.

Now let us consider the total deformation $\mathbf{F}$ with the unique decomposition (2.4). We can determine the total deformation rate $\mathbf{l}$,

$$\mathbf{l} = \dot{\mathbf{F}}\mathbf{F}^{-1} = \boldsymbol{\Omega} + \mathbf{Q}\mathring{\mathbf{l}}\mathbf{Q}^T$$

$$= \boldsymbol{\Omega} + \mathbf{Q}\mathring{\mathbf{l}}^e\mathbf{Q}^T + \mathbf{Q}\mathring{\mathbf{l}}^p\mathbf{Q}^T, \tag{2.21}$$

where $\mathring{\mathbf{l}}$ is the deformation rate referred to $\mathring{\mathbf{F}}$ with its elastic-plastic decomposition according to (2.6).

Introducing the push-forward quantities

$$\mathbf{l}^e := \mathbf{Q}\mathring{\mathbf{l}}^e\mathbf{Q}^T, \qquad \mathbf{l}^p := \mathbf{Q}\mathring{\mathbf{l}}^p\mathbf{Q}^T \tag{2.22}$$

into (2.21) we obtain

$$\mathbf{l} = \boldsymbol{\Omega} + \mathbf{l}^e + \mathbf{l}^p = \mathbf{d} + \mathbf{w} = \boldsymbol{\Omega} + (\mathbf{d}^e + \mathbf{w}^e) + (\mathbf{d}^p + \mathbf{w}^p) \tag{2.23}$$

with its symmetric and antisymmetric parts

$$\mathbf{d} = \mathbf{d}^e + \mathbf{d}^p, \tag{2.24}$$

$$\mathbf{w} = \boldsymbol{\Omega} + \mathbf{w}^e + \mathbf{w}^p, \tag{2.25}$$

where $\mathbf{d}$ is the total strain rate, $\mathbf{w}$ the total spin and $\mathbf{d}^e, \mathbf{d}^p, \mathbf{w}^e, \mathbf{w}^p$ are given by (2.17), (2.18).

From eqn (2.25) we can calculate the macroscopic substructure spin $\boldsymbol{\Omega}$,

$$\boldsymbol{\Omega} = \mathbf{w} - \mathbf{w}^e - \mathbf{w}^p. \tag{2.26}$$

As a result we can use equations (2.20) to show that the elastic spin $\mathbf{w}^e$ can be determined exactly as a function of the elastic strain rate $\mathbf{d}^e$, and that the plastic spin $\mathbf{w}^p$ can be determined also exactly as a function of the plastic strain rate $\mathbf{d}^p$. The spin $\boldsymbol{\Omega} = \dot{\mathbf{Q}}\mathbf{Q}^{-1}$, needed to formulate the appropriate objective corotational rate (2.15), is then given by (2.26). Therefore, constitutive assumptions with unknown coefficients introduced in the literature to determine the plastic spin (see e. g. Dafalias (1983), Van der Giessen (1991), Paulun and Pecherski (1992)) are superfluous and in contradiction to the exact kinematic dependency.

All the results presented above are exact. To analyse structures undergoing finite elastic and finite plastic strains, we have to construct an efficient incremental solution algorithm with an associated update procedure. For simplicity, we restrict our further considerations in this section and in the following one to moderate elastic, but finite plastic deformations covering a large class of problems of engineering interest.

For moderately large elastic strains, we can estimate as follows,

$$\mathring{\mathbf{U}}^e = \mathbf{1} + \boldsymbol{\epsilon}^e, \qquad \boldsymbol{\epsilon}^e = \mathbf{0}(\theta), \qquad \theta^2 \ll 1, \tag{2.27}$$

where $\mathbf{0}(\theta)$ is a tensor with eigenvalues of order θ, with θ a small number.

With (2.27) and neglecting higher order terms we obtain an additive decompositon of the total strain rate $\mathbf{d}$ into a purely elastic, $\mathbf{d}^e$, and a purely plastic part, $\mathbf{d}^p$, according to (2.24), with $\mathbf{d}^e$ given by $(2.17)_1$, and $\mathbf{d}^p$ approximated by

$$\mathbf{d}^p =\tfrac{1}{2} (\overset{\nabla}{\mathbf{V}}{}^p\, \mathbf{V}^{p-1} + \mathbf{V}^{p-1}\, \overset{\nabla}{\mathbf{V}}{}^p). \tag{2.28}$$

Similarly, we obtain a decomposition of the total spin $\mathbf{w}$, (2.25), into the macroscopic substructure spin $\boldsymbol{\Omega}$ according to (2.16), the elastic spin $\mathbf{w}^e$ given by $(2.18)_1$ and the plastic spin $\mathbf{w}^p$ approximated by

$$\mathbf{w}^p =\tfrac{1}{2} (\overset{\nabla}{\mathbf{V}}{}^p\, \mathbf{V}^{p-1} - \mathbf{V}^{p-1}\, \overset{\nabla}{\mathbf{V}}{}^p), \tag{2.29}$$

depending now on the plastic strain rate $\overset{\nabla}{\mathbf{V}}{}^p$ and the plastic stretch $\mathbf{V}^p$ only. The additive decomposition of $\mathbf{d}$ and $\mathbf{w}$ into purely elastic and purely plastic parts simplifies the formulation of the appropriate constitutive equations for moderate elastic - finite plastic deformations including metal elastoplasticity.

3. ELASTIC-PLASTIC CONSTITUTIVE EQUATIONS FOR CAUCHY CONTINUA

In this section we want to formulate the constitutive equations for moderately large elastic strains and finite plastic strains with isotropic and kinematic hardening, including the important special case of metal plasticity. We assume that the elastic material behavior is isotropic and remains so during the deformation. For simplicity we ignore the influence of texturing on the elastic constitutive equations.

As was shown by Anand (1979), an appropriate form of the elastic constitutive equation for moderate elastic strains can be given by using the constitutive equation of linear theory, if the stress tensor there is replaced by the Kirchhoff stress tensor, and the linear strain tensor by the logarithmic strain tensor. Let $\boldsymbol{\tau}$ denote the Kirchhoff stress tensor referred to the deformed configuration, and $\mathring{\boldsymbol{\tau}}$ the back-rotated Kirchhoff stress tensor referred to the configuration defined by (2.5),

$$\mathring{\boldsymbol{\tau}} = \mathbf{Q}^T \boldsymbol{\tau} \mathbf{Q}, \tag{3.1}$$

then for moderate elastic strains the elastic potential can be given in the form

$$W = \tfrac{1}{2} \ln \mathring{\mathbf{U}}^e \cdot \mathbb{C}^e \cdot \ln \mathring{\mathbf{U}}^e \tag{3.2}$$

leading to the hyper-elastic constitutive equation in the form,

$$\mathring{\boldsymbol{\tau}} = \frac{\partial W}{\partial (\ln \mathring{\mathbf{U}}^e)} = \mathbb{C}^e \cdot \ln \mathring{\mathbf{U}}^e, \tag{3.3}$$

where $\mathbb{C}^e$ is the constant elasticity tensor of the linear elasticity theory.
Push-forward of (3.3) with the macroscopic substructure rotation $\mathbf{Q}$ leads to the elastic constitutive equation referred to the deformed configuration

$$\boldsymbol{\tau} = \mathbf{Q} \mathring{\boldsymbol{\tau}} \mathbf{Q}^T = \mathbb{C}^e \cdot \ln \mathbf{V}^e. \tag{3.4}$$

The rate form of eqn (3.3) is obtained as

$$\dot{\mathring{\boldsymbol{\tau}}} = \mathbb{C}^e \cdot \frac{\partial (\ln \mathring{\mathbf{U}}^e)}{\partial \mathring{\mathbf{U}}^e} \cdot \frac{\partial \dot{\mathring{\mathbf{U}}}^e}{\partial \mathring{\mathbf{d}}^e} \cdot \mathring{\mathbf{d}}^e \tag{3.5}$$

with $\mathring{\mathbf{d}}^e$ given by (2.9). Formulae for the calculation of the rate of the logarithm of tensors and a proof of the symmetry of the differentiation operator in (3.5) can be found in Schieck and Stumpf (1994).

Push-forward of (3.5) with $\mathbf{Q}$ yields the corotational rate

$$\overset{\nabla}{\boldsymbol{\tau}} = \mathbf{Q}\dot{\overset{\circ}{\boldsymbol{\tau}}}\mathbf{Q}^T = \dot{\boldsymbol{\tau}} - \boldsymbol{\Omega}\boldsymbol{\tau} + \boldsymbol{\tau}\boldsymbol{\Omega}$$

$$= \mathbb{C}^e \cdot \frac{\partial(\ln \mathbf{V}^e)}{\partial \mathbf{V}^e} \cdot \mathbb{A}^{e-1} \cdot \mathbf{d}^e \tag{3.6}$$

with

$$\mathbb{A}^{e-1} = \frac{\partial \overset{\nabla}{\mathbf{V}}{}^e}{\partial \mathbf{d}^e} = \mathbf{Q}\frac{\partial \dot{\overset{\circ}{\mathbf{U}}}{}^e}{\partial \overset{\circ}{\mathbf{d}}{}^e}\mathbf{Q}^T. \tag{3.7}$$

Using some estimations we can show that the elastic constitutive equation (3.6) can be simplified for moderate elastic strains to

$$\overset{\nabla}{\boldsymbol{\tau}} = \mathbb{C}^e \cdot \mathbf{d}^e. \tag{3.8}$$

To formulate appropriate macroscopic plastic constitutive equations, we may take into account also the microscopic plastic material behavior, as will be done in the following section. From the results presented there, we conclude that for the description of the hardening phenomena by two macroscopic quantities, the isotropic hardening parameter τ_y and the backstress tensor $\boldsymbol{\alpha}$ for kinematic hardening, we may assume the existence of a potential ψ for the backstress tensor $\boldsymbol{\alpha}$, and may also assume that in the framework of Cauchy continua the potential ψ depends only on the symmetric part of $\mathbf{F}^p$, leading to the proposition

$$\psi = \psi(\ln \mathbf{U}^p). \tag{3.9}$$

The simplest functional form of the backstress tensor is then

$$\overset{\circ}{\boldsymbol{\alpha}} = \zeta \ln \mathbf{U}^p, \tag{3.10}$$

where ζ is the kinematic hardening modulus. If we assume furthermore that the backstress does not depend on the elastic deformation, then referred to the deformed configuration, we obtain $\boldsymbol{\alpha}$ by push-forward with $\mathbf{Q}$,

$$\boldsymbol{\alpha} = \mathbf{Q}\overset{\circ}{\boldsymbol{\alpha}}\mathbf{Q}^T. \tag{3.11}$$

The rate form of eqn (3.10) is

$$\dot{\overset{\circ}{\boldsymbol{\alpha}}} = \zeta\frac{\partial(\ln \mathbf{U}^p)}{\partial \mathbf{U}^p} \cdot \frac{\partial \dot{\mathbf{U}}^p}{\partial \overset{\circ}{\mathbf{d}}{}^p} \cdot \overset{\circ}{\mathbf{d}}{}^p, \tag{3.12}$$

leading to the corotational rate of the backstress tensor

$$\overset{\nabla}{\boldsymbol{\alpha}} = \mathbf{Q}\dot{\overset{\circ}{\boldsymbol{\alpha}}}\mathbf{Q}^T = \dot{\boldsymbol{\alpha}} - \boldsymbol{\Omega}\boldsymbol{\alpha} + \boldsymbol{\alpha}\boldsymbol{\Omega}$$

$$= \zeta\frac{\partial(\ln \mathbf{V}^p)}{\partial \mathbf{V}^p} \cdot \mathbb{A}^{p-1} \cdot \mathbf{d}^p \tag{3.13}$$

with the fourth order tensor $\mathbb{A}^{p-1}$ following from $(2.19)_2$.

For moderately large plastic strains eqn (3.13) can be estimated and simplified to

$$\overset{\nabla}{\boldsymbol{\alpha}} = \zeta \mathbf{d}^p, \tag{3.14}$$

where higher order terms are neglected.

With the Kirchhoff stress tensor $\boldsymbol{\tau}$, the backstress tensor $\boldsymbol{\alpha}$, the average yield stress τ_y, and assuming for simplicity that the influence of texturing on the yield condition can be neglected we can formulate the yield condition in the following form,

$$F = F(\boldsymbol{\tau}, \boldsymbol{\alpha}, \tau_y) \begin{cases} < 0 & : \text{ no yielding} \\ = 0 & : \text{ yielding} \end{cases} \tag{3.15}$$

The elastic and the plastic constitutitve equations so derived lead to the elastic-plastic constitutive equation

$$\overset{\nabla}{\boldsymbol{\tau}} = \mathbb{C}^{ep} \cdot \mathbf{d} \tag{3.16}$$

with the elastic-plastic tangential operator

$$\mathbb{C}^{ep} = \mathbb{C}^e - \frac{(\mathbb{C}^e \cdot F,_{\boldsymbol{\tau}}) \otimes (F,_{\boldsymbol{\tau}} \cdot \mathbb{C}^e)}{F,_{\boldsymbol{\tau}} \cdot \mathbb{C}^e \cdot F,_{\boldsymbol{\tau}} - \left(F,_{\tau_y} \frac{\partial \dot{\tau}_y}{\partial (\ln \dot{\mathbf{V}}^p)} + \zeta F,_{\boldsymbol{\alpha}} \right) \cdot \frac{\partial (\ln \mathbf{V}^p)}{\partial \mathbf{V}^p} \cdot \mathbb{A}^{p-1} \cdot F,_{\boldsymbol{\tau}}} \tag{3.17}$$

Further details of the considerations , some approximations, numerical applications and a comparison of the numerical results with those of the literature can be found in Schieck and Stumpf (1994).

4. CONSTITUTIVE AND EQUILIBRIUM EQUATIONS OF AN ELASTIC-PLASTIC BODY WITH MICROSTRUCTURE

In section 3 we considered finite elastoplasticity within the framework of Cauchy continua; in this section we present constitutive and equilibrium equations of an elastic-plastic body within the theory of oriented media. Details can be found in Le and Stumpf (1994,1995).

Denoting the derivative with respect to the undeformed configuration by $\mathcal{D}$, we assume the following functional form of the strain energy density W ,

$$W = W(\mathbf{F}, \mathbf{F}^p, \mathcal{D}\mathbf{F}^p). \tag{4.1}$$

Ignoring the inertia of the dislocation motion we can derive the following equilibrium equations,

$$\text{DIV}\ \mathbf{P} + \rho_0 \mathbf{B} = \mathbf{0},$$

$$(4.2)$$

$$\text{DIV}\ \mathbf{P}^d - \mathbf{J} = \mathbf{0}.$$

In (4.2) $\mathbf{P}$ is the (first Piola-Kirchhoff) macrostress tensor, $\mathbf{B}$ the body macroforce, $\mathbf{P}^d$ the (2,1)-microstress tensor (contravariant of order 2 and covariant of order 1),

$$\mathbf{P} = \mathbf{g}^{-1}\frac{\partial W}{\partial \mathbf{F}}, \qquad \mathbf{P}^d = \frac{\partial W}{\partial \mathcal{D}\mathbf{F}_p}, \tag{4.3}$$

and $\mathbf{J}$ is the internal microforce,

$$\mathbf{J} = \frac{\partial W}{\partial \mathbf{F}^p} \tag{4.4}$$

with $\mathbf{g}$ the metric tensor.
Let us introduce the right Cauchy-Green deformation tensor $\mathbf{C}$, the symmetric (second Piola-Kirchhoff) macrostress tensor $\mathbf{S}$ and the (2,1)-microcouple tensor $\mathbf{S}^d$

$$\mathbf{C} = \mathbf{F}^T\mathbf{g}\mathbf{F}, \qquad \mathbf{P} = \mathbf{F}\mathbf{S}, \qquad \mathbf{P}^d = \mathbf{F}^{p-T}\mathbf{S}^d, \tag{4.5}$$

where $\mathbf{S}^d$ is skew-symmetric with respect to its contravariant components. With some transformations of (4.1) and with (4.6), eqns (4.3) can be obtained in the alternative form

$$\mathbf{S} = 2\frac{\partial W}{\partial \mathbf{C}}, \qquad \mathbf{S}^d = 2\frac{\partial W}{\partial \mathbf{T}^p}, \tag{4.6}$$

where $\mathbf{T}^p$ is the plastic torsion referred to the undeformed configuration. The (1,2)-tensor $\mathbf{T}^p$ is skew-symmetric with respect to the covariant indices.

In Le and Stumpf (1994) it is shown that the microforce $\mathbf{J}$ can be determined as follows

$$\mathbf{J} = -\mathbf{F}^{p-T}[(\mathbf{C}\mathbf{S} - W\mathbf{1}) + \tfrac{1}{2}\,\mathbf{S}^d : \mathbf{T}^p], \tag{4.7}$$

where $(\mathbf{CS} - W\mathbf{1})$ is the Eshelby tensor (the reader may also refer to Eshelby (1951), Epstein and Maugin (1990), Maugin (1992)).

Applying appropriate invariance requirements to the strain energy density (4.1) we can derive the balance equations of moment of macromomentum and micromomentum

$$\mathbf{P}\mathbf{F}^T = \mathbf{F}\mathbf{P}^T,$$

$$(4.8)$$

$$\bar{\mathbf{k}} = (\det \mathbf{F}^p)^{-1}[\mathbf{g}^{-1}\mathbf{J}\mathbf{F}^{pT} + \mathbf{g}^{-1}(\mathbf{P}^d : \mathcal{D}\mathbf{F}^p)] = \bar{\mathbf{k}}^T.$$

With (4.1) - (4.8) we present a set of static equations for a material body with continuously distributed dislocations at finite strain. They can be used to investigate the basic balance equations and the Clausius-Duhem inequality relative to the current, initial and crystal reference frames and to determine uniquely the intermediate configuration $\mathbf{F}^p$ (see Le and Stumpf (1995)).

5. SHAKEDOWN ANALYSIS IN FINITE ELASTOPLASTICITY

One crucial point in the large deformation analysis of structures is the existence of critical equilibrium states, where bifurcation or snap-through of the structure may occur. Therefore, a nonlinear shakedown analysis for cyclic loading has to take into account the stability boundaries of the structure. In the generalizations of Melan's linear shakedown theorem for moderate rotations by Weichert (1986), Groß-Weege (1990) and Saczuk and Stumpf (1990) an additional integral restriction was introduced assuring uniqueness of the solution. This restriction corresponds to a positive second Gâteaux derivative of the strain energy, which is satisfied if the stress tensor and the membrane stress tensor are positive throughout the structure. (see Stumpf (1978), (1979)). This is a very strong restriction, because it excludes all deformations of shells and shell-like structures loaded from "outside". A limitation of this type can be avoided by an incremental shakedown analysis using finite element codes with the usual critical point calculation. Then, as a result, a shakedown domain is obtained, which is partly restricted by the stability boundaries of the structure.

In Stumpf (1993) it was stated that efforts to derive a generalization of Melan's linear shakedown theorem for finite elastoplasticity are not very promising. This statement was made because it was shown that the total strain tensor referred to the deformed or undeformed configuration and also the Lie derivative of the strain tensor referred to the deformed, intermediate (according to the Lee decomposition) and undeformed configuration cannot be decomposed into purely elastic and purely plastic parts. So, using the Lee decomposition of the deformation gradient and the Lie derivatives as objective rates, first introduced by Simo (1988), there is no additivity of strains or strain rates and no linear stress-strain relation. With the Lagrangean-type constitutive model of Schieck and Stumpf (1993) and the Eulerian-type model of this paper we have a better basis for the formulation of a shakedown theorem or for an incremental shakedown analysis for two reasons:

1) Introducing our decomposition $\mathbf{F} = \mathbf{Q}\overset{\circ}{\mathbf{U}}{}^e\mathbf{U}^p$ and the logarithmic strain tensor, we obtain additivity of elastic and plastic logarithmic strains and their rates for superposed moderately large strains and finite rotations,

2) Using the logarithmic strain tensor we obtain a linear stress-strain relation for superposed moderately large elastic strains and finite rotations.

With the additivity of elastic and plastic strain measures and the linear stress-strain relation we arrive at two characteristics of a linear theory, which should enable us to generalize Melan's linear theorem to finite elastoplasticity.

Finally, we can formulate the following theorem for an incremental shakedown analysis.

Proposition - If for the extremal load cycles enclosing the load domain and for their inverse cycles the same residual state is obtained, the structure will shake down for arbitrary loading paths within the load domain.

Numerical applications of this incremental shakedown theorem can be found in Stumpf (1993).

Acknowledgement - Support of this work provided by Deutsche Forschungsgemeinschaft (DFG) under contract number $St - 135/1 - 2$ is gratefully acknowledged.

REFERENCES

1951 ESHELBY, J. D.: The force on an elastic singularity. Phil. Trans. Royal Soc. (London) **A 244,** 87-112.

1960 KOITER, W. T.: General theorems for elastic-plastic solids. In: Sneddon, I. N. and Hill, R. (eds). Progress in Solid Mechanics, pp. 165-221. North-Holland, Amsterdam.

1967 LEE, E. H. and LIU, D. T.: Finite strain elastic-plastic theory with application to plane-wave analysis. J. Appl. Phys. **38,** 19-27.

1973 MAIER, G.: A shakedown matrix theory allowing for work-hardening and second-order geometric effects. In: Sawczuk, A. (ed.). Foundations of Plasticity, pp. 417-433. North-Holland, Amsterdam.

1977 DEBORDES, O.: Duality: some results in asymptotical elastoplasticity. In: Auslender, A. (ed.). Convex Analysis and Its Application, pp.100-114. Springer, Berlin.

1978 STUMPF, H.: Dual extremum principles and error bounds in nonlinear elasticity theory. J. Elasticity **8/4,** 425-438.

1979 ANAND, L.: On the Hencky's approximate strain energy function for moderate deformations. Trans. ASME, J. Appl. Mech. **76,** 78-82.

1979 STUMPF, H.: The derivation of dual extremum and complementary stationary principles in geometrical nonlinear shell theory. Ing.-Archiv **48,** 221-237.

1980 BORKOWSKI, A. and KLEIBER, M.: On a numerical approach to shakedown analysis of structures. Comp. Meth. Appl. Mech. Engng. **22,** 101-119.

1983 DAFALIAS, Y. F.: Corotational rates for kinematic hardening at large plastic deformations. J. Appl. Mech. **50,** 561-565.

1986 WEICHERT, D.: On the influence of geometrical nonlinearities on the shakedown of elasto-plastic structures. Int. J. Plasticity **2,** 135-148.

1988 SIMO, J. C.: A framework for finite strain elastoplasticity based on the maximum plastic dissipation and multiplicative decomposition: Part I: Continuum formulation. Comp. Meth. Appl. Mech. Engng. **66,** 199-219.

1989 SCHMIDT, R. and WEICHERT, D.: A refined theory of elastic-plastic shells at moderate rotations. ZAMM **69,** 11-21.

1989 SIMO, J. C. and KENNEDY, J. G.: Finite strain elastoplasticity in stress resultant geometrically exact shells. In: Owen, D. R., Hinton, E., and Onate, E. (eds). Computational Plasticity, pp. 651-673. Pineridge Press Ltd., Swansea, U. K.

1990 EPSTEIN, M. and MAUGIN, G. A.: The energy-momentum tensor and material uniformity in finite elasticity. Acta Mechanica **83,** 127-133.

1990 ETOROVIC, A. L. and BATHE, K.-J.: A hyperelastic-based large strain elastoplastic constitutive formulation with combined isotropic-kinematic hardening using the logarithmic stress and strain measures. Int. J. Num. Meth. Engng. **30,** 1099-1114.

1990 GROSS-WEEGE, J.: A unified formulation of statical shakedown criteria for geometrically nonlinear problems. Int. J. Plasticity **6,** 433-447.

1990 NEMAT-NASSER, S.: Certain basic issues in finite-deformation continuum plasticity. Meccanica **25/4,** 223-229.

1990 SACZUK, J. and STUMPF, H.: On statical shakedown theorems for nonlinear problems. Mitt. Inst. Mechanik **74,** Ruhr-Univ. Bochum.

1990 STUMPF, H. and BADUR, J.: On missing links of rate-independent elastoplasticity at finite strain. Mech. Res. Comm. **17,** 353-364.

1991 BASAR, Y. and WEICHERT, D.: A finite rotation theory for elastic- plastic shells under consideration of shear deformation. ZAMM **71,** 379-389.

1991 STUMPF, H. and LE, K. Ch.: On shakedown of elastoplastic shells. Quart. Appl. Math. XLIX **4,** 781-793.

1991 VAN DER GIESSEN, E.: Micromechanical and thermodynamic aspects of the plastic spin. Int. J. Plasticity **7,** 365-386.

1992 CHROSCIELEWSKI, J., MAKOWSKI, J. and STUMPF, H.: Genuinely resultant shell finite elements accounting for geometric and material non-linearity. Int. J. Num. Meth. Engng. **35,** 63-94.

1992 MAUGIN, G. A.: Thermomechanics of Plasticity and Fracture. Cambridge University Press.

1992 PAULUN, J. E. and PECHERSKI, R. B.: On the relation of the plastic spin. Arch. Appl. Mech. **62,** 376-385.

1992 SCHIECK, B., PIETRASZKIEWICZ, W. and STUMPF, H.: Theory and numerical analysis of shells undergoing large elastic strains. Int. J. Solids Structures **29,** 689-709.

1993 DASHNER, P. A.: An objective kinematical formalism for the modeling of elastic-plastic materials subject to large deformation. Int. J. Solids Structures **30/19,** 2661-2672.

1993 LE, K. Ch. and STUMPF, H.: Constitutive equations for elastoplastic bodies at finite strain. Thermodynamic implementation. Acta Mechancia **100,** 155-170.

1993 SCHIECK, B. and STUMPF, H.: Deformation analysis for finite elastic-plastic strains in a Lagrangean-type description. Int. J. Solids Structures **30/19,** 2639-2660.

1993 STUMPF, H.: Theoretical and computational aspects in the shakedown analysis of finite elastoplasticity. Int. J. Plasticity **9,** 583-602.

1994 LE, K. Ch. and STUMPF, H.: Nonlinear continuum theory of dislocations. Proc. Roy. Soc. (London) (to appear).

1994 SCHIECK, B. and STUMPF, H.: Exact formulae of the plastic and elastic spin and constitutive model for finite elastoplasticity. Int. J. Solids Structures (to appear).

1994 STUMPF, H. and SCHIECK, B.: Theory and analysis of shells undergoing finite elastic-plastic strains and rotations. Acta Mechanica **106,** 1-21.

1994 LE, K.Ch.and STUMPF, H.: Finite elasto-plasticity with microstructure. Mitt. Inst. Mechanik Nr. **92**, Ruhr-Universität Bochum.

1995 LE, K.Ch.and STUMPF, H.: On the determination of the crystal reference in nonlinear continuum theory of dislocations (submitted).

CASE STUDIES ON THE INFLUENCE OF GEOMETRIC EFFECTS ON THE SHAKEDOWN OF STRUCTURES

Jean-Bernard TRITSCH and Dieter WEICHERT
L.M.L. (EUDIL)-CNRS URA 1441,
Université des Sciences et Technologies de Lille
F-59655 Villeneuve d'Ascq, France

INTRODUCTION

The classical theory of shakedown as formulated by MELAN [1938] and KOITER [1960] is based on the assumptions of elastic-perfectly plastic or unlimited linear hardening material behaviour within the framework of geometrically linear theory. Out of the different extensions of these theorems (for review see, e.g. COHN & MAIER [1979], KÖNIG & MAIER [1981], MAIER & LLOYD SMITH [1986], KÖNIG [1987], GROSS-WEEGE [1990]), here the problem of the influence of geometrical changes of the considered structures during the loading procedure is addressed. Indeed, geometrically linear theory is invalid already for thinwalled plates and shells when the displacements of the midsurface are of the same order as the wall thickness. This, however, is a very common situation if thinwalled structures are loaded beyond their elastic limit.
Geometrical effects on the shakedown behaviour of structures were first investigated by MAIER [1973] using a matrix formulation. A similar approach was followed by KÖNIG [1980, 1984], KÖNIG & SIEMASZKO [1988] to study the problem of instabilities.The tensorial approach to geometrically non-linear problems was initiated by WEICHERT [1984, 1986] and then further developed by GROSS-WEEGE [1988], SACZUK & STUMPF [1990], TRITSCH [1993] and others. These formulations have similar theoretical structures; however, they differ by the considered types of geometrical effects taken into account. A comparative study is found in TRITSCH [1993] and TRITSCH & WEICHERT [1993].
The present paper investigates a particular kind of geometrical effects in many mechanical structures applied loads are composed of main constant service loads superposed by small additional fluctuating loads. In this case, the problem of shakedown can be studied according to propositions given by WEICHERT [1986], where the total process is split into a control evolution of the structure, represented in the sequel by a constant preloaded reference state and a fluctuation of the structure in the vicinity of this state.
Here, a hybrid numerical method for axisymetric problems based on a finite element discretisation is proposed combining an incremental approach for the determination of

Z. Mróz et al. (eds.), Inelastic Behaviour of Structures under Variable Loads, 309–320.

deformed reference states corresponding to service load states and a non-linear optimization procedure for the determination of the shakedown loading factor under these circumstances. The theoretical background of the proposed method is not repeated in this article; readers may refer to the abovementioned literature and the paper by GROSS-WEEGE & WEICHERT in the present volume.

DESCRIPTION OF THE METHOD

Determination of reference states

To determine displacements and stress state of a structure in a preloaded reference configuration C_R, a modified version of the finite element program NONSAP (see BATHE & al. [1974]) with degenerated finite elements for axisymmetric thinwalled shells was used. Due to the symmetry of the problem, the rotational motion of the cross sections of the shell must not be considered and the discretisation is to be performed only in the longitudinal direction of the shell (fig. 1).

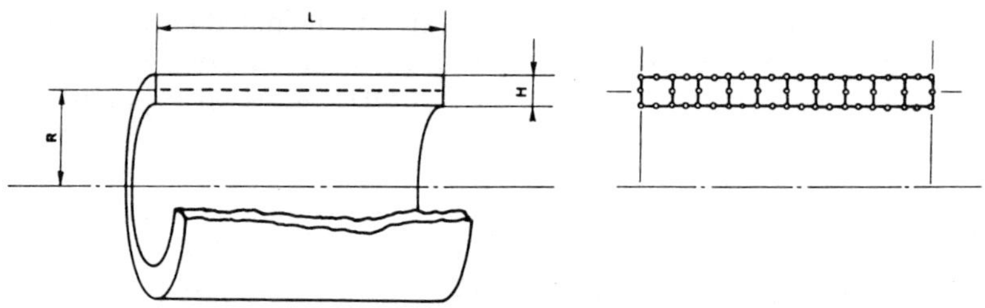

Figure 1: A shell segment and finite element representation

As usual, pressure loads are taken into account by equivalent nodal loads. The configuration C_R is computed incrementally by applying the RIKS-WEMPNER path following procedure and the Total Lagrangian approach.

Determination of shakedown loading factors

To study the shakedown behaviour of the structure in the vicinity of C_R, a finite element method based on the principle of complementary energy and developed by GROSS-WEEGE [1988] has been used. This program deals with axisymmetrical thinwalled shells and takes into account the state of deformation and stresses of a preloaded configuration. C_R.

The shakedown factor α_{SD} is obtained by an optimization procedure using the augmented Lagrangian formulation proposed by PIERRE & LOWE [1975]. The von Mises yield criterion for shell-like structures following VLASOV [1964] leads then to a non-linear optimizing problem. The two relevant failure mechanisms, Incremental Collapse and Alternating Plasticity are considered, giving two different values for the shakedown factor the least of which is to be retained as global shakedown factor $\alpha_{SD} = \min(\alpha_{IC}, \alpha_{AP})$.

Dimension of the optimization problem

The structures studied in this paper are discretized by 10 elements with 8 nodes each; the yield criterion is checked at 3 points per element. In the space of generalized loads, four loading states, corresponding to the corners of a rectangle, have to be controlled. Because two conditions are requested for the incremental collapse criterion, the corresponding optimizing problem counts 240 subsidary conditions.

STUDY OF A SHORT CYLINDRICAL SHELL

The first investigated structure is a short cylindrical shell with the following dimensions: H/R=0,1, L/R=0,3, where R, L and H are respectively radius, length and wall-thickness of the shell. A similar problem has been studied by GROSS-WEEGE [1988]. Both ends of the shell are free.The material is assumed to be elastic perfectly plastic with Young's modulus, $E = 2{,}1.10^7$ N/cm^2, Poisson's ratio $\nu = 0$ and yield stress $\sigma y = 36000$ N/cm^2. The loading consists of two variable loads which may vary independently. The first load is an internal pressure P, and the second load is equivalent to a ring load Q on one edge. The maximal value of each load which the structure may support separately are given within the framework of geometrically linear theory by limit analysis:

$$P_o = \sigma_y \frac{H}{R} \tag{1}$$

$$Q_o = \sigma_y\left(\sqrt{2}-1\right)\frac{HL}{R} \tag{2}$$

The structure is subjected to a reference loading state (P^R,Q^R), (first step), which may exhibit independent fluctuations of load $P^r(t)$ and $Q^r(t)$, (second step).

Solution strategy

The loads P and Q are considered to vary independently. The optimization procedure consists of finding the biggest rectangle in the space of generalized loads defined by (0, 0), (0, Qmaxi), (Pmaxi, Qmaxi), (Pmaxi, 0), according to a prescribed ratio Pmaxi/Qmaxi. The optimization problem may then be formulated as follows:

Determine α_{max} for which it is possible to find a time-independent residual stress field $\overline{\rho}$ such that $(\alpha\sigma^e + \overline{\rho})$ is a safe state of stress in all points of the structure, where σ^e is the purely elastic stress field corresponding to the geometrically linear response of a purely elastic comparison structure in the configuration C_R and loaded by the unit loads (P*,Q*). Because of proportionality between loads and stresses in this case we have:

$$\text{Pmaxi} = \alpha_{max} P^*$$

$$\text{Qmaxi} = \alpha_{max} Q^*$$

Time independence of the residual stress field is assured by the fact that the field of residual stresses so determined is common to all corners of the biggest rectangle.
By repeating this procedure for several ratio P/Q (scanning), we get several loading points which are on the limit of the shakedown domain. According to MORELLE [1980], we deduce that the shakedown domain may be determined by the polygonal line joining all obtained points (figure 2).

Description of the hybrid numerical method
The aim of this method is to exploit, as far as possible, the advantages of a two step loading approach for a geometrically non-linear shakedown analysis. This, however, makes it necessary to link the two rather different numerical procedures corresponding to the two steps of loading and to implement data converting interfaces. Here, the following strategy has been used (figure 3):

1. Assessment of the shakedown domain by use of the classical geometrically linear approach. (KZS: calculation of reference purely elastic solution for each independent load. KZSPLAS: assessement of α_{SD} through optimization)
2. Assessment of displacements and stress state by NONSAP for several configurations corresponding to scanning directions such that the load state is close to the previous shakedown limit (calulation of preloads by PREC).
3. New assessment of the shakedown limit accounting for the displacements and stress state (data EPSOL) given by NONSAP. The flow chart of the procedure is shown in figure 3.

Because of the differences between the two finite element formulations, conversion interfaces had to be developed in order to allow for the communication of data between the two software parts.

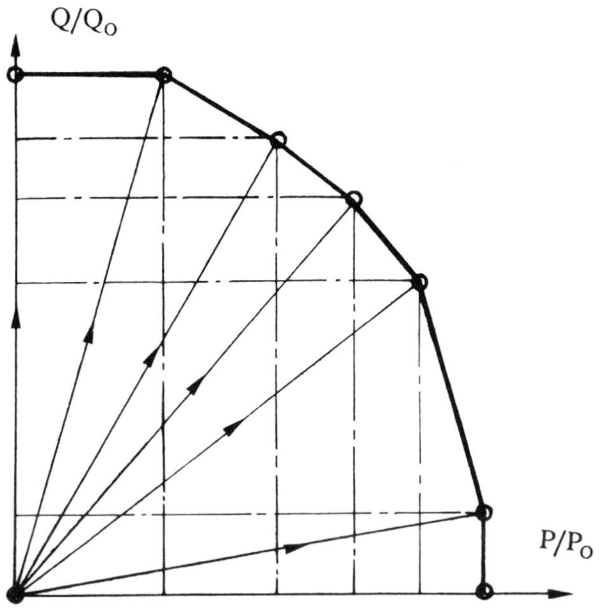

Figure 2: Determination of the shakedown domain

The first interface (CONV-KN) converts the data of the geometrically linear formulation (DGLF) into data for NONSAP. It calculates the nodal coordinates for NONSAP from the nodal characteristics of the simplified shell model. Each element of this model corresponds to one element of NONSAP. Then, the equivalent nodal load coefficients may be deduced from the description of load conditions of DGLF.
The second interface (NK-CONV) converts the nodal displacements calculated by NONSAP into displacements and rotations of the middle surface for the simplified shell model and translates the stress state given by NONSAP into the resultant forces and moments of the considered cross section of the shell model.
A third software part (PREC) determines the magnitude of the nodal load coefficients of NONSAP from the magnitude of the shakedown limit given by the simplified model.

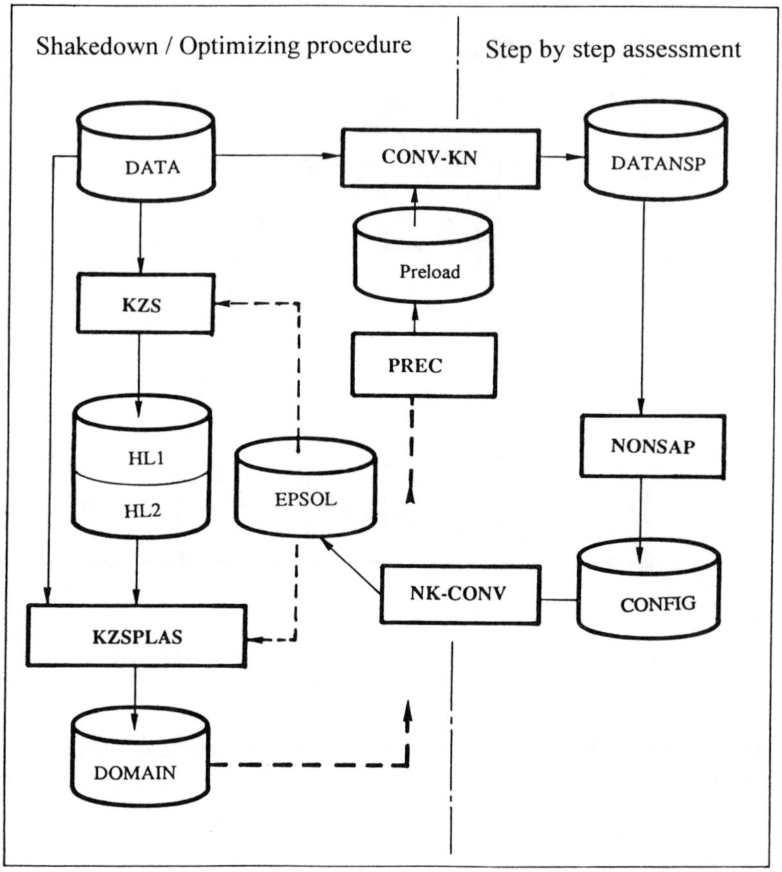

Figure 3: Flow chart of the program

Conversion of data from the simplified shell model into data for NONSAP:

	Simplified Model	NONSAP Model
Mesh Generation	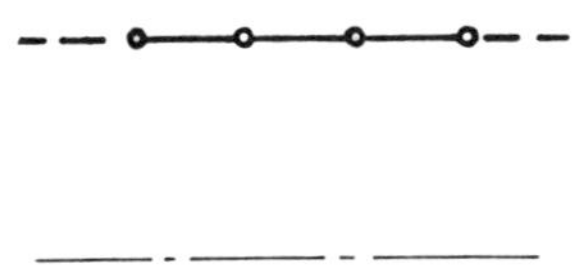	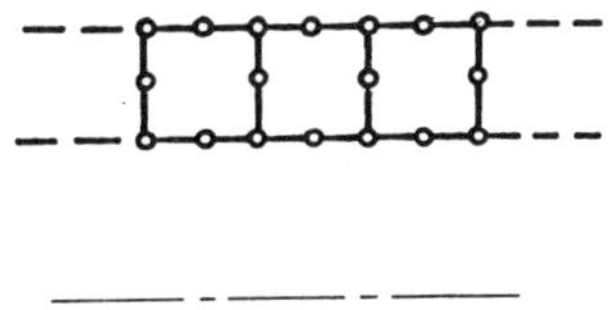
Loading Conditions	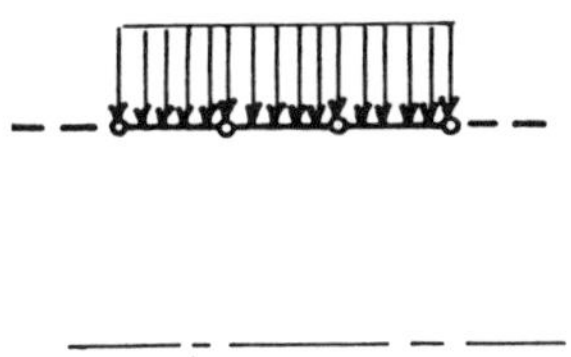	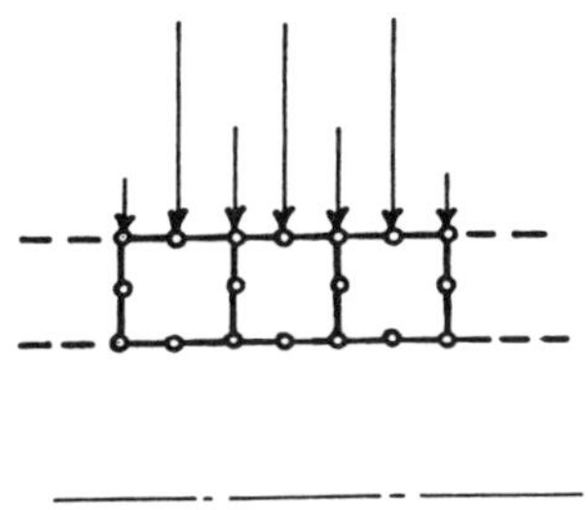
Kinematic Boundary conditions	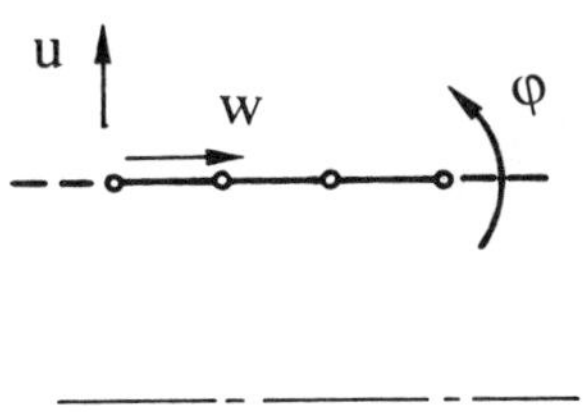	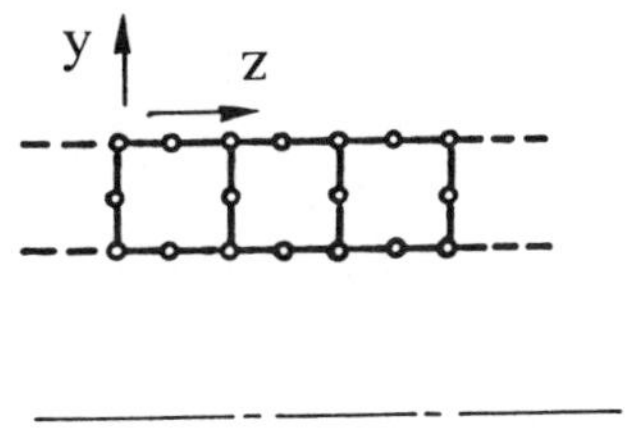

Table 1

Numerical result for the short shell

The described method has been applied to the problem of a short tube as defined above. The preloaded configurations C_R were obtained by applying a load monotonously increasing up to 90% of the shakedown limit as determined by a geometrically linear analysis. Also, the effect of an eventual axial tosion which may exist in real tubes after assembly has been studied. Torsional buckling was not considered. Then, the limit value for the torsional moment is given by:

$$C_o = \frac{1}{\sqrt{3}}\left(2\pi R^2 H\right)\sigma_y$$

The results of the calculations on the Incremental Collapse mechanism are shown in figure 4.

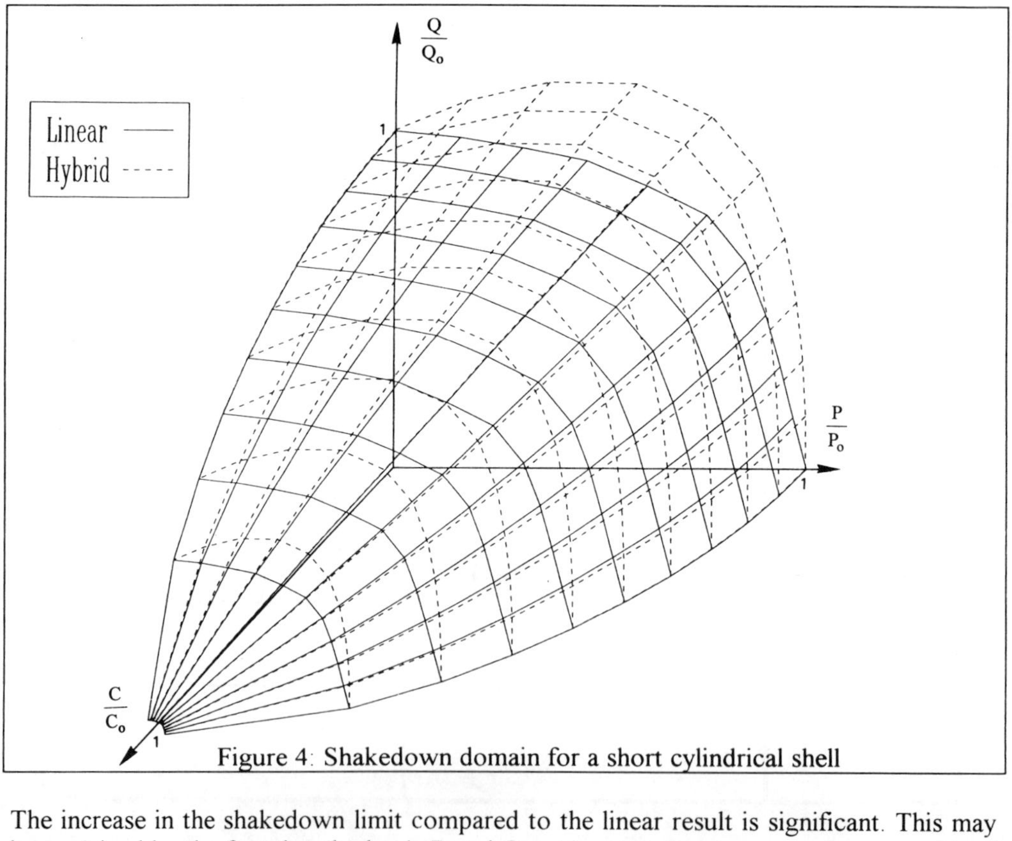

Figure 4: Shakedown domain for a short cylindrical shell

The increase in the shakedown limit compared to the linear result is significant. This may be explained by the fact that the loads P and Q act in opposite senses, so that a combined loading state may be safer than a single load state, remembering that transverse forces had been neglected.

Calculation time

Each surface representing in figure 4 the two different limits for the shakedown domain is defined by 121 calculated points. For the hybrid assessment, 121 preloaded configurations are then requested. Knowing that each point needs between 200 to 600 iterations, each iteration taking approximately 1,5 seconds of CPU time, a total of nearly 40 hours of CPU time is needed (IBM RS/6000). The 121 assessments of the configuration by NONSAP must be added (about 1 minute for each step-by-step assessment). All calculations were automatically executed by batch procedures.

STUDY OF A FAMILY OF CONICAL SHELLS

The second example is the problem of conical shells for which not only the effects of geometrical changes, but also the influence of the opening angle of the cone are studied. The characteristic dimensions are:

Ri/Re=0,75, H/Re=0,035 where Ri is the smaller radius; Re is the external radius .This kind of mechanical structure is found in valve systems. Cone angles between 30 and 60 degrees are investigated, where 0 degrees corresponds to a circular plate. The characteristics of the 45-degree shell will be used as reference to normalize the axis of the loading domain. The cone is simply supported on the little radius edge. At the large radius edge, the cone is in contact with the interior of a cylinder so that the radial degree of freedom is suppressed. The material properties of steel are assumed with the following values for Young's modulus $E = 2{,}1.10^7$ N/cm^2, Poisson's ratio $\nu = 0{,}3$ and yield stress, σ_y=36000 N/cm^2.

Pressure loads are applied on the concave side of the shell. As the cone performs a translation inside a cylinder, friction forces and thus axial forces on the external edge are induced. The friction problem is not addressed here and the shell is supposed to stay in contact with the interior of the cylinder. Even if, in reality, friction forces depend on pressure, the present analysis considers them as fully independent. The yield value for the pressure P and the axial ring load Q obtained from a geometrically linear formulation for the 45-degree shell are:

$$P_o = 1236 \text{ N/cm}^2$$
$$Q_o = 737{,}7 \text{ N/cm}$$

The same solution strategy and the same discretisation by finite elements as in the case of the short shell are used.

It can be seen that the shakedown limit does not increase due to geometric effects when the axial force Q acts in the same axial direction as pressure P. However, differences occur when these forces act in the opposite direction. The presented results are normalised by the uniaxial shakedown load of the shell with a cone angle of 45°.

CONCLUDING REMARKS

A systematic hybrid numerical procedure for the investigation of the influence of geometrical changes during the deformation procedure on the shakedown limit of axisymetric structures has been presented. Technically relevant examples illustrate the practical interest of the proposed method. However, quite a number of interesting questions remain open for further research:

In the present work, elastic perfectly-plastic material behaviour has been used, however hardening effects may easily be accounted for through the use of the generalised standard material model proposed by HALPHEN & NGUYEN QUOC SON [1975] in a similar way as proposed by WEICHERT & GROSS-WEEGE [1988]. The influence of the changes in geometry on the stability of the structure are ignored by the present method in its actual form. Therefore, it should only be used in the particular case when the problem of loss of stability can be disregarded from an engineering point of view. Also, damage and crack propagation phenomena should be integrated in the shakedown analysis in order to increase its utility for practical engineering purposes.

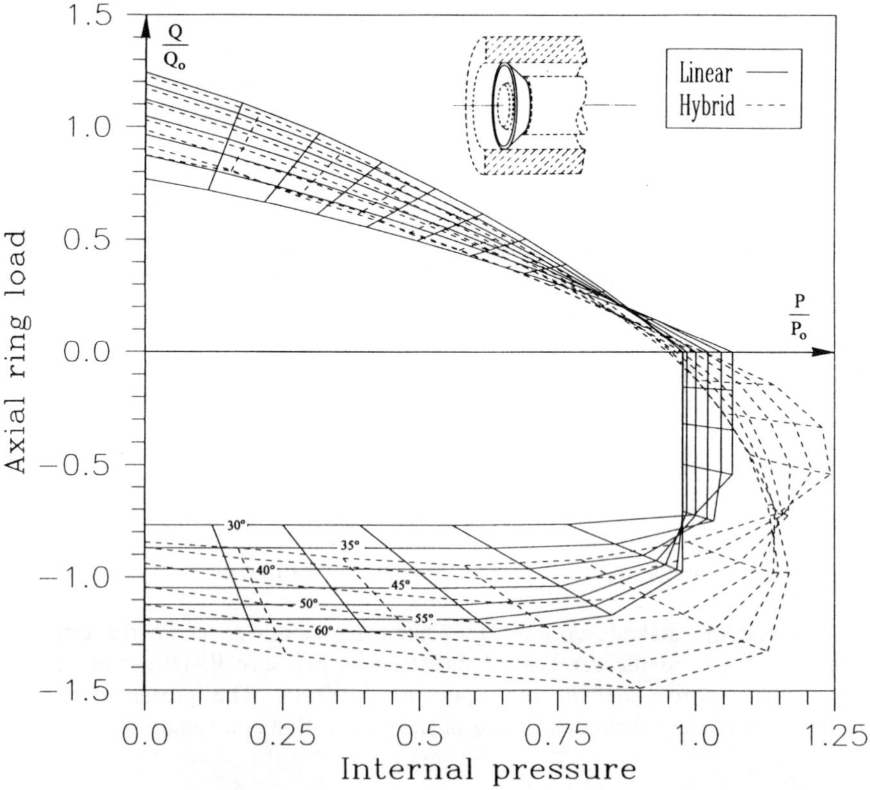

Figure 5: Shakedown domain for conical shells

REFERENCES

1938 MELAN, E., "Zur Plastizität des räumlichen Kontinuums", Ing.Arch.9, pp 116-126.

1960 KOITER, W.T., "General theorems for elastic-plastic solids",In Progress in solid mechanics, (Edited by I.N. SNEDDON and R. HILL), 165-221, North Holland, Amsterdam.

1964 VLASOV, V.Z., "General theory of shells and its applications in engineering", NASA Technical Translation TTF-99, Washington.

1973 MAIER, G., "A shakedown matrix theory allowing for work-hardening and second-order geometrical effects", in Foundation of Plasticity, 417-433.

1975 HALPHEN, B. & NGUYEN QUOC SON, "Sur les matériaux standards généralisés", Journal de mécanique, Vol 14, n°1, 39-63.

1975 PIERRE, D.A., LOWE M.J., "Mathematical programming via augmented Lagrangians", London: Addison-Wesley.

1979 COHN, M.Z., MAIER, G., "Engineering plasticity by Mathematical Programming", Pergamon Press, New York.

1980 KÖNIG J. A., "On the stability of the incremental collapse process", Arch. Inz. Lad., 27, 1, 219-229.

1980 MORELLE, P., "Etude expérimentale et numérique (par éléments finis) de l'adaptation plastique (shakedown) des structures minces axisymétriques", travail de fin d'étude, Laboratoire de Mécanique des matériaux et stabilité des constructions, Liège.

1981 KÖNIG, J.A. & MAIER, G., "Shakedown analysis of elastoplastic structures: A review of recent developments", Nuclear Engineering and design, 66, 81-95.

1984 KÖNIG, J.A., "Stability of the incremental collapse", In: Inelastic Structures under Variable Loads (Eds. C. Polizzotto, A. Sawczuk), 329-344, COGRAS, Palerme, (1984).

1984 WEICHERT, D., "Shakedown at finite displacements; a note on Melan's theorem", Mech. Res. Comm. 11, 121-127.

1986 MAIER, G., LLOYD SMITH, D., "Update to Mathematical Programming Applications to Engineering Plastic Analysis", In STEELE, C.R. & SPRINGER, G.S. (eds.), Applied Mechanics Update, ASME, New York, 377-383.

1986 WEICHERT, D.: "On the influence of geometrical nonlinearities on the shakedown of elastic-plastic structures", Int.J.Plasticity, Vol.2, n°2, 135-148.

1987 KÖNIG, J.A., "Shakedown of Elastic-Plastic Structures", Elsevier, Amsterdam.

1988 GROSS-WEEGE, J.: "Zum Einspielverhalten von Flächentragwerken", Mitteilungen aus dem Institut für Mechanik n°58, Ruhr-Universität Bochum.

1988 KÖNIG J. A. & SIEMASZKO A., "Strainhardening effects in shakedown processes", Ing. Archiv, 58, 58-66.

1988 WEICHERT, D. & GROSS-WEEGE, J.: "The numerical assessement of elastic-plastic sheets under variable mechanical and thermal loads using a simplified two-surface yield condition", Int.J.Mech.Sci, Vol.30, n°10, 757-767.

1990 GROSS-WEEGE, J., "A unified formulation of statical shakedown criteria for geometrically nonlinear problems", Int.J.Plasticity, Vol 6, n°4, 433-447.

1990 SACZUK, J. & STUMPF, H.: "On statical shakedown theorems for non-linear problems", Mitteilungen aus dem Institut für Mechanik, Nr.47, Ruhr-Universität Bochum.

1993 TRITSCH, J.-B. & WEICHERT, D., "Shakedown of Elastic-Plastic Structures at Finite Deformations - a Comparative Study of Static Shakedown Theorems", ZAMM, Z. angew. Math. Mech., 73, 4-5, T309-T312.

1993 TRITSCH, J.-B., "Analyse d'adaptation des structures élasto-plastiques avec prise en compte des effets géométrique / Shakedown analysis of elastic-plastic structures accounting for geomtrical effects", Ph.D-Thesis, Université des Sciences et Technologies de Lille, Lille.

DISCUSSION OF THE CHAOTIC BEHAVIOUR OF AN ELASTIC-PLASTIC STRUCTURE

Franco BONTEMPI[1], Fabio CASCIATI[2]
[1]Department of Structural Engineering, Politecnico di Milano,
P.zza L. da Vinci, 32, 20133 Milan,Italy.
[2]Department of Structural Mechanics, Universita' di Pavia,
Via Abbiategrasso, 211, 27100 Pavia,Italy.

ABSTRACT. The dynamic behaviour of an elastic-plastic structure undergoing large displacements is considered. Emphasis is put on its possible chaotic behaviour. This has important consequences for the shakedown of the structure, on the predictability of its response and on the estimation of its lifetime.

INTRODUCTION

In the last few years, a new kind of dynamic behaviour, called *chaos* has been recognized (c.f. RASBAND [1990], BAKER & GOLLUB [1990]). Essentially, a dynamic system is said to be chaotic when the behaviour of the system itself is, in some sense, unpredictable. More specifically, it means that the motion of the system appears quasi-random after some time interval. Even if one has knowledge of the initial status of the system, the information on the current system status is lost as the experiment proceeds. The randomness shown by a chaotic system is not homogeneous, as in a fully random walk, but reveals some inner structure. There is a hidden order that distinguishes a chaotic system from a perfectly stochastic system (for the analogies and the differences between a chaotic and a stochastic system c.f. WEDIG [1988], BONTEMPI et al. [1991]).

Z. Mróz et al. (eds.), Inelastic Behaviour of Structures under Variable Loads, 321–338.

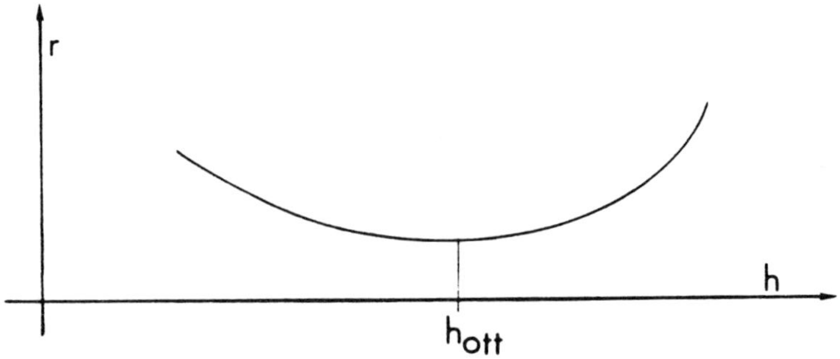

Figure 1: The effect of the discretization on the solution for a non chaotic system.

A condition necessary to have chaos is the presence of nonlinearities in the system. Since nonlinear dynamical problems can be handled only in a numerical way, chaos investigations started with the development of computers. The fact that one only encounters chaos within the numerical solution of a problem, rather than in the problem itself, is astonishing. A reaction could be to believe that chaos is only an effect of the inevitable discretization process needed to reduce the original problem into a computer form. Actually, there are some remarkable examples in which a numerical error (typically a too large integration step) has led to a quasi chaotic behaviour (c.f. REINHALL et al. [1989]). Nevertheless, due to the basic difference between chaotic and stochastic systems, one must conclude that chaos is a real phenomenon: numerical errors of a random nature make the solution behaviour like that of a system with a lot of noise, but no structure is shown.

The diagram of Fig.1 concerns a generic non chaotic problem. It shows a relationship between the error r (exact solution versus numerical result) and the parameter h by which the quality of the numerical discretization is measured: for instance h can represent the number of integration steps considered in a given time interval. For a small value of h, i.e. for coarse discretization, one has a large error decreasing as h increases. For a large value of h the error grows again, due to the finite number of digits with which the computer works. This gives rise to an optimal point $h \approx h_{ott}$: the minimum is generally smooth enough to provide a small error in a broad interval around h_{ott}.

Assume now the system is chaotic: the solutions corresponding to two values h_1 and h_2 of the parameter h, are always different, at least locally. This means that it is no longer possible to recognize the convergence of the solution to the problem by the empirical technique (widely used in engineering) of repeating the analysis with a finer resolution. The chaotic nature of the motion manifests on a macroscopic scale the small numerical differences due to the different discretization, and/or to small perturbations in the initial conditions or in the parameters of the system. The sensitivity of a chaotic system to perturbations makes a single dynamic analysis meaningless: it would require assigning the actual values of the system parameters and the actual initial conditions, which cannot be measured. Such a situation leads to a probabilistic description over several realizations of the system behaviour (c.f. BONTEMPI & CASCIATI [1993]). The methodology of approaching a nonlinear dynamic system which may show a chaotic behaviour (like a structure with geometrical and material non linearities), consists of two steps (c.f. BONTEMPI [1993]):

- identification of the *regions of chaoticity* in the space of the parameters which define the structure: these zones of irregular behaviour are not necessarily connected and, hence, the understanding of the behaviour of the system can be jeopardized;
- probabilistic description of the structural response in these regions of chaoticity (c.f. AUGUSTI et al. [1984], CASCIATI & FARAVELLI [1991]).

This methodology is applied in this paper to the study of an elastic-plastic Euler arch in large displacements.

REMARKS ON THE ELASTIC-PLASTIC MATERIAL LAW

The general equation of motion of a structure, with a finite number, say n, of degrees of freedom, can be put in the form

$$\underline{\underline{M}} \cdot \underline{\ddot{q}} + \underline{R}(\underline{q}, \underline{\dot{q}}) = \underline{F}(t) \tag{1}$$

where $\underline{q}$, $\underline{\dot{q}}$ and $\underline{\ddot{q}}$ are respectively the displacement, the velocity and the acceleration vectors, $\underline{\underline{M}}$ is the mass matrix, $\underline{R}(\underline{q}, \underline{\dot{q}})$ is the restoring force vector and $\underline{F}(t)$ is the external forcing vector. The restoring force vector $\underline{R}$ is an explicit function of both displacements and velocity. The presence of other dissipative factors besides that due to plastic deformations, is not considered here. The following incremental constitutive law between increments of the stress σ and the strain ϵ holds for elastic-plastic materials (see Fig.2):

$$d\sigma = \begin{cases} E \cdot d\epsilon & \text{for elastic behavior} \\ H \cdot d\epsilon & \text{for plastic behavior} \end{cases} \tag{2}$$

In eq.(2), E is the initial elastic modulus of the material and H is the hardening modulus. The stress $\sigma(t)$ in a generic point of the structure at time t, results from the expression:

$$\sigma(t) = \int_{t_0}^{t} d\sigma \tag{3}$$

and, at a structure level, the restoring force is

$$R(t) = \int_{t_0}^{t} dR(\underline{q}, \underline{\dot{q}}) \tag{4}$$

Depending on the value of parameter H, one distinguishes:

- $H > 0$: *hardening* behaviour;

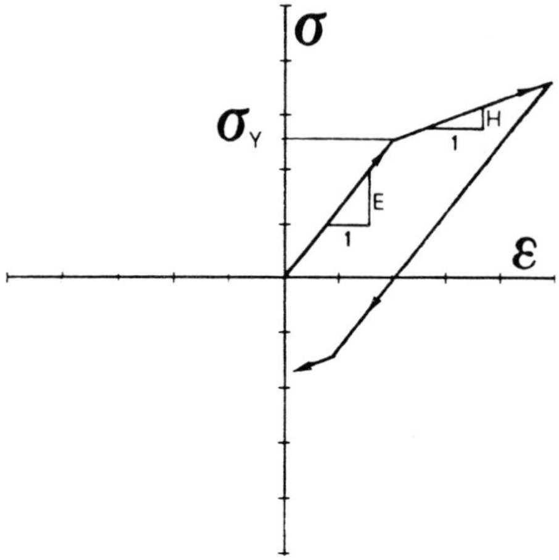

Figure 2: The elastic-plastic constitutive relation.

- $H = 0$: *perfectly plastic* behaviour;
- $H < 0$: *softening* behaviour;

The softening case is not investigated because it always leads to an unstable motion. The *dissipative work* cumulated during the motion by the structure is locally defined in terms of strain and deformation as

$$L_P = \int_{t_0}^{t} dL_P = \int_{t_0}^{t} \sigma \cdot d\epsilon_P = \int_{t_0}^{t} \sigma \cdot (d\epsilon - d\epsilon_E) \tag{5}$$

where the *plastic deformation* is given by $\epsilon_P = \epsilon - \epsilon_E = \epsilon - \sigma/E$. Depending on the evolution of the plastic work, *elastic shakedown* occurs when, after some time, the structure no longer develops anymore plastic deformations. Otherwise, the system can reach collapse by *alternating plasticity* (i.e plastic fatigue) or by *incremental plasticity* (i.e. accumulation of plastic deformations) (c.f. KÖNIG [1987]).
The goal of this paper is to show, with a relatively simple example, that chaotic behaviour is possible for elastic-plastic structures (even with hardening behaviour, i.e. $H > 0$) that undergo large displacements. This implies:

- sensitivity to initial conditions, i.e. unpredictability of the dynamic behaviour of the structure;
- unpredictability of the lifetime of the structure.

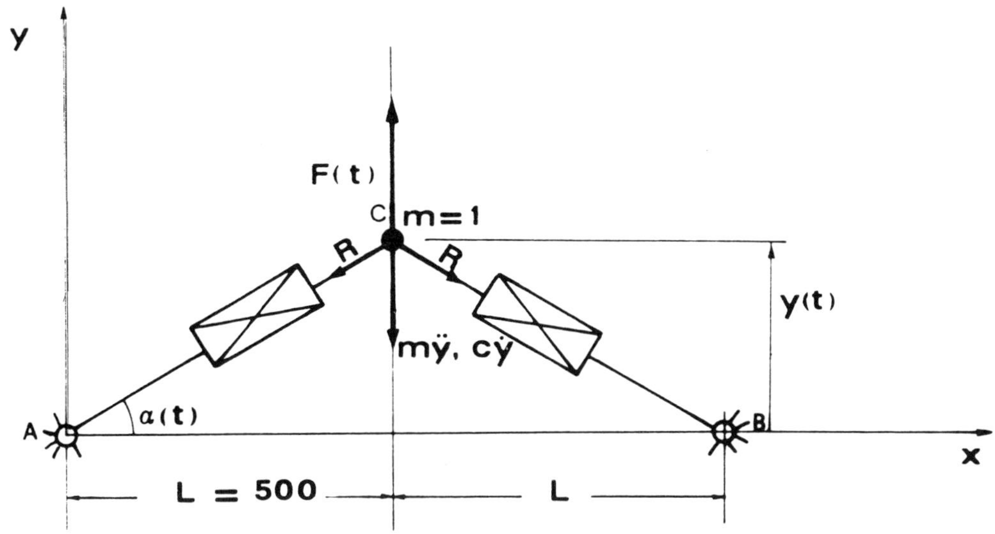

Figure 3: The Euler arch.

THE EULER ARCH

The structure under study is presented in Fig.3. There is a mass m in C and two bars: the first bar connects hinge A to hinge C, and the second hinge B to hinge C. The structure is loaded by a harmonic vertical force $F(t)$

$$F(t) = A \cdot \sin(\omega_F \cdot t) \tag{6}$$

with period $T_F = 2 \cdot \pi / \omega_F$. We consider the vertical displacement $y(t)$ of the hinge C to be the only degree of freedom. The equation of motion of the structure is written

$$m \cdot \ddot{y} + 2R \sin(\alpha(t)) = F(t) \tag{7}$$

where $y(t) = L \cdot \tan(\alpha(t))$. The following numerical values are considered for the parameters: mass $m = 1$; area of the bars $a = 1$; distance between the hinges A and B, $2 \cdot L = 1000$; elastic modulus $E = 510$ and yield stress $\sigma_Y = 10$. The initial position of the hinge C is $y(0) = 10$: in this initial configuration the bars are unloaded and the material does not present deformations. The natural pulsation for small elastic vibrations of the structure is $\omega_0 = 0.02$ and the pulsation of the harmonic force is $\omega_F = 0.6$, with amplitude $A = 9.0$.
The structure is studied by integrating eq.(7) with the initial conditions $y(t = 0) = y(0); \dot{y}(t = 0) = \dot{y}(0)$. The Newmark method of numerical integration (c.f. ARGYRIS [1991], BATHE [1982]) is adopted with parameters $\beta = 0.25$ and $\gamma = 0.50$; the time step ranges from $T_F/100$ to $T_F/400$. For validation, checks have been performed by a general purpose computer code: it makes use of the updated Lagrangian formulation in order to take into account large displacements.

BIFURCATION DIAGRAMS

The behaviour of the dynamical system is first investigated by the bifurcation diagram with the parameter h as abscissa. One sample H with $K = 500$ uniformly distributed values $H_k, (k = 1, ..K)$ in the range between the value $H = 0.0$ (elastic-perfectly plastic structure) and $H = 510.0 = E$ (perfectly elastic case). For each of these values H_k, one

- integrates the equation of motion of the Euler arch for a time period $T_I = 480 \cdot T_F$, to bypass the initial transient;
- continues to integrate the motion along an additional period $T_P = 320 \cdot T_F$, marking in the bifurcation diagram (Fig.4.a) the value of y at the end of each period T_F for any value H_k of the abscissa.

The plastic work dissipated after the first $80 \cdot T_F$ is drawn in Fig.4.b. A further bifurcation diagram is provided in Fig.4.c: the same structure, with the same values of the parameters, is studied, but the material constitutive law is supposed to be holonomous (i.e. the loading and unloading paths coincide). The points of this last diagram are distributed in a homogenous way, due to the non dissipative nature of the system. As a consequence, the system explores a significant part of the *phase space* $\{y(t), \dot{y}(t)\}$, around the two points of stable equilibrium (upper and lower position of hinge C).
An overview of Figs.4.a and .b, leads one to distinguish three different situations:

- two zones ($50 < H < 150$ and $306 < H < 510$) where an initial plasticization is followed by a mainly elastic behaviour. The system shows shakedown and its bifurcation diagram, even if less complex, is like the one in Fig.4.c;
- two zones (one for $35 < H < 50$ and the other around $H = 255$), in which plastic deformations progress, and, therefore, high values of the plastic work are achieved. A collapse for alternate plasticity characterizes these regions;
- a remaining region, for $135 < H < 255$, in which the behaviour looks irregular, with a disordered arrangement of the points in the bifurcation diagram.

Moreover, along all the bifurcation diagram situations of periodic behaviour show themselves.

DYNAMIC BEHAVIOUR

The dynamic behaviour of the elastic-plastic system has been simulated for a period of $30000 \cdot T_F$: this is generally enough to check the presence or absence of shakedown. The results are shown in Fig.5 to Fig.12: each of the rows shows the results for a fixed value of H. The following three plots are provided:

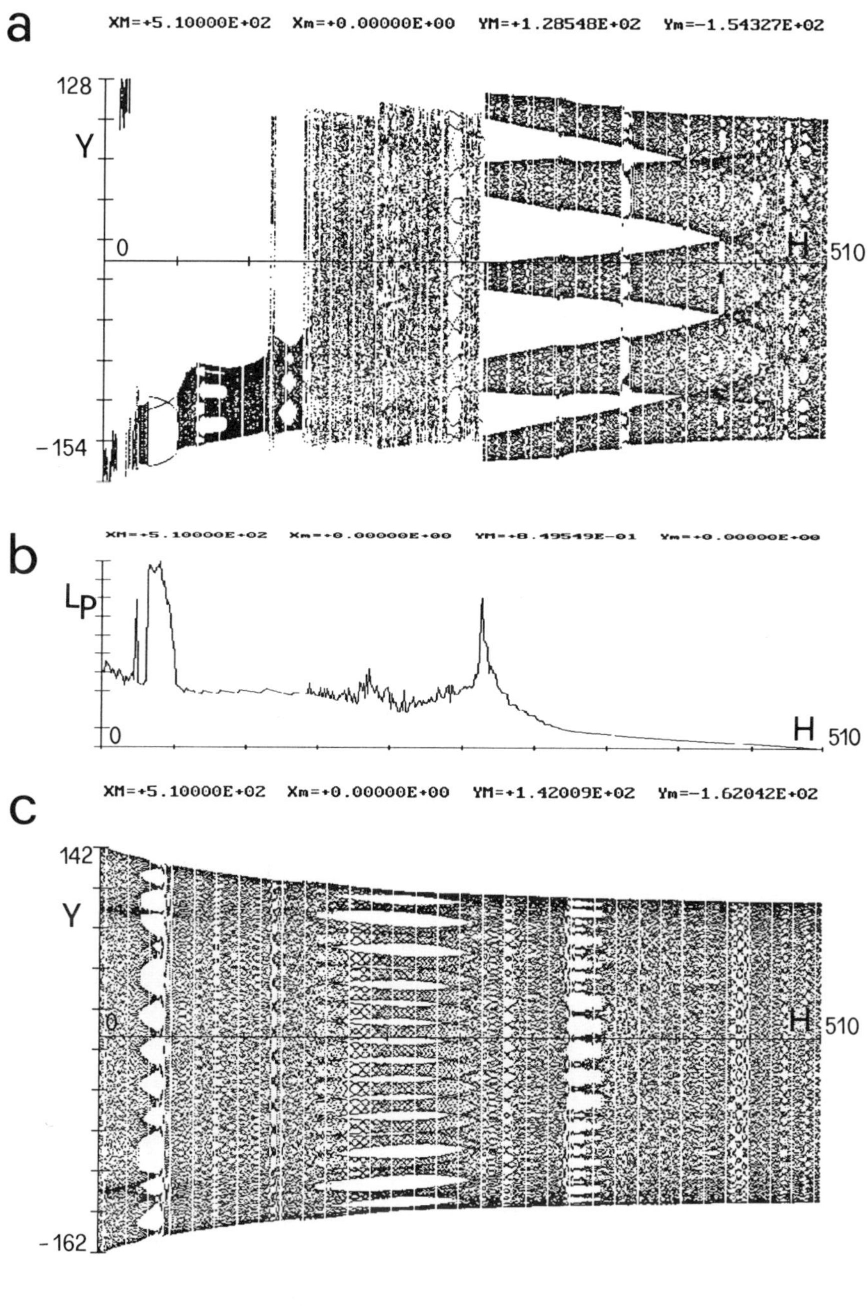

Figure 4: For $H = 0.0$ to $H = 510.0$, a) the bifurcation diagram for the elastic-plastic Euler Arch; b) the plastic work L_P developed after $80 \cdot T_F$; c) the bifurcation diagram for the perfectly elastic Euler arch.

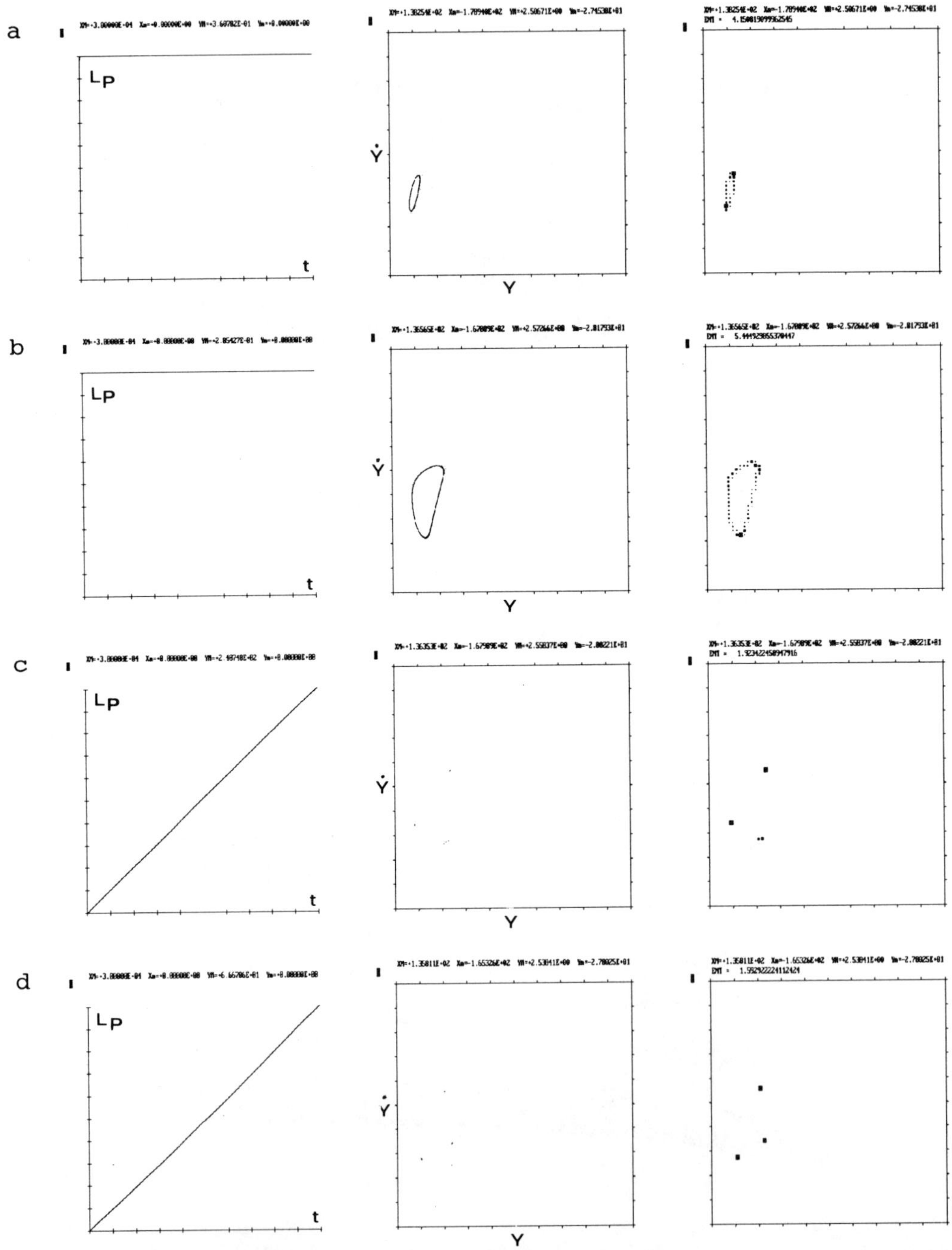

Figure 5: Numerical simulations: a) $H = 0.0000$; b) $H = 25.500$; c) $H = 35.700$; d) $H = 51.000$.

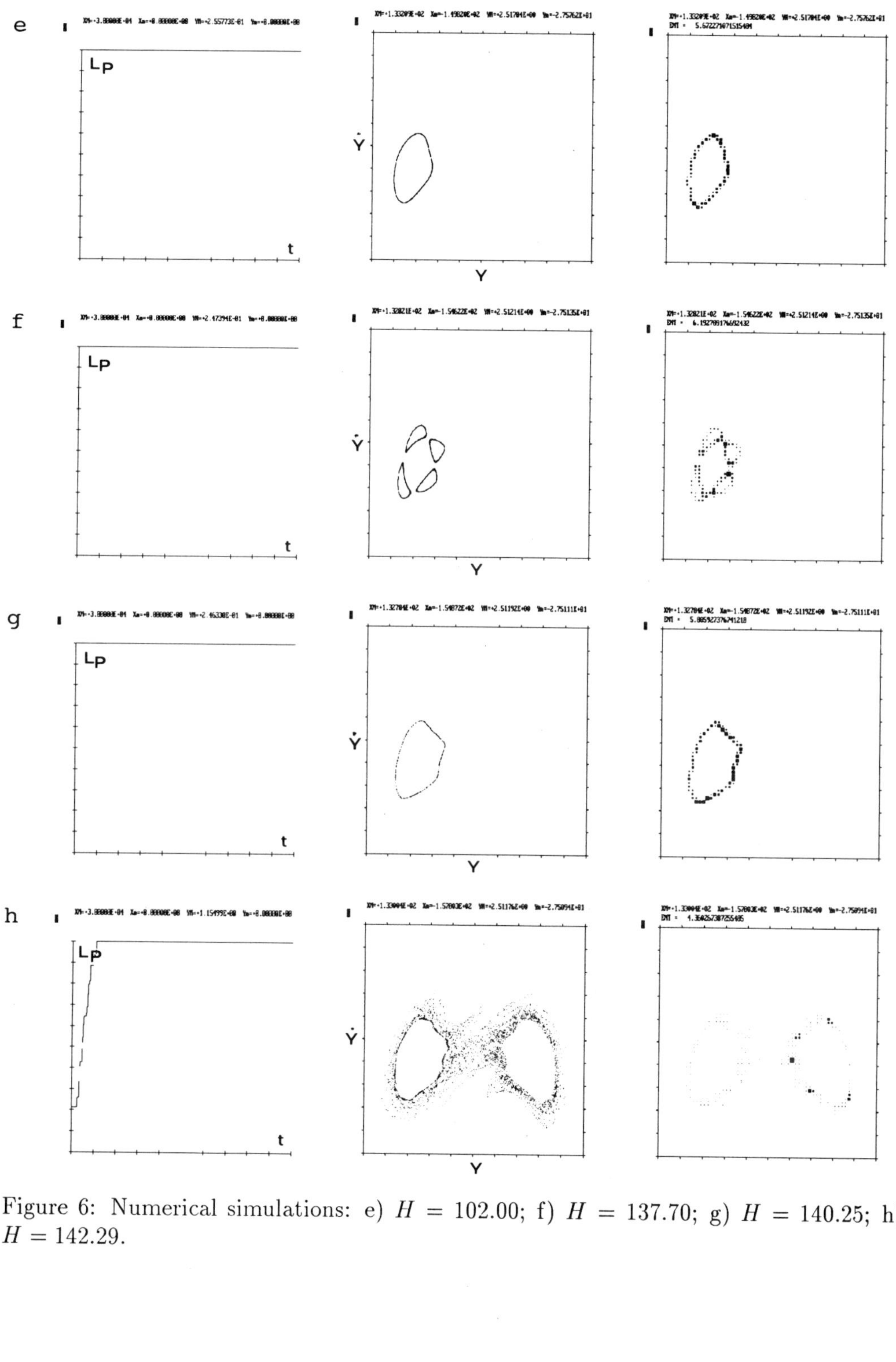

Figure 6: Numerical simulations: e) $H = 102.00$; f) $H = 137.70$; g) $H = 140.25$; h) $H = 142.29$.

- on the left, one has the progress in time of the plastic work L_P; from this diagram it is possible to argue whether the structure shows shakedown or not;
- in the middle, one has the *Poincare' section* in the phase space $\{y(t), \dot{y}(t)\}$, from which it is possible to classify the behaviour of the system: *periodic*, if in the section only a finite number of points appears, *quasi-periodic*, if the points are arranged along one or more closed lines; *chaotic*, if the points are disposed on a strange attractor;
- on the right, one has the *probability density function* of finding a point of the Poincare' section, in a partition of the phase space: 60×60 boxes were adopted for the discretization. In this way one measures the disorder in the dislocation of the points on the Poincare' section.

Many illustrations, are made necessary by the variety and the complexity of the dynamics of the system. They show in fact the presence of different types of motion as the parameter H increases.
The value $H = 0.0$ corresponds to the elastic-perfectly plastic case. The structure shows shakedown and the motion appears quasi-periodic, the Poincare' section being a closed line. As the hardening parameter increases, this closed line changes, until the value $H = 35.7$ is reached; here the line itself is broken, a periodic solution appears and shakedown is lost. This window of periodic behaviour, as the bifurcation diagram in Fig.4.a shows, holds until the value $H = 51.0$, where a quasi-periodic regime again appears.
A quasiperiodic motion can be represented on a torus, the closed line of the Poincare' section being just the intersection of the torus with the plane of the Poincare' section. The motion assumes over the torus the form of a spiral whose points can be defined, in a parametric way, by two angles: one measures the rotation of a plane around the velocity axis, the other the rotation in this plane. For a quasi-periodic motion, periodicity of the first angle may occur as shown for $H = 137.7$.
The region of quasi-periodic motion ends just after the value $H = 142.29$, which represents a case of transient chaos. After an initial transient, the motion appears periodic as is revealed by the intensity of the probability density function that shows only three peaks. From $H = 142.9$ until $H = 250.0$ the motion is chaotic: some periodic solution still occurs and in this case one has either shakedown, as for $H = 204.0$, or continuously increasing plastic deformations, as for $H = 219.3$. Note that the intensity of the plastic work accumulation is stronger in the periodic case than in the chaotic case. This is due to the fact that with the periodic motion one periodically (that means systematically) develops plastic deformations while in the chaotic case, only sporadic developments of plastic deformations occur.

The shape of the strange attractor changes a lot in this interval of chaotic behaviour, since the system presents dissipation only for its plastic behaviour and the development of the plastic deformations is not continuous. Between two periods of dissipation, the

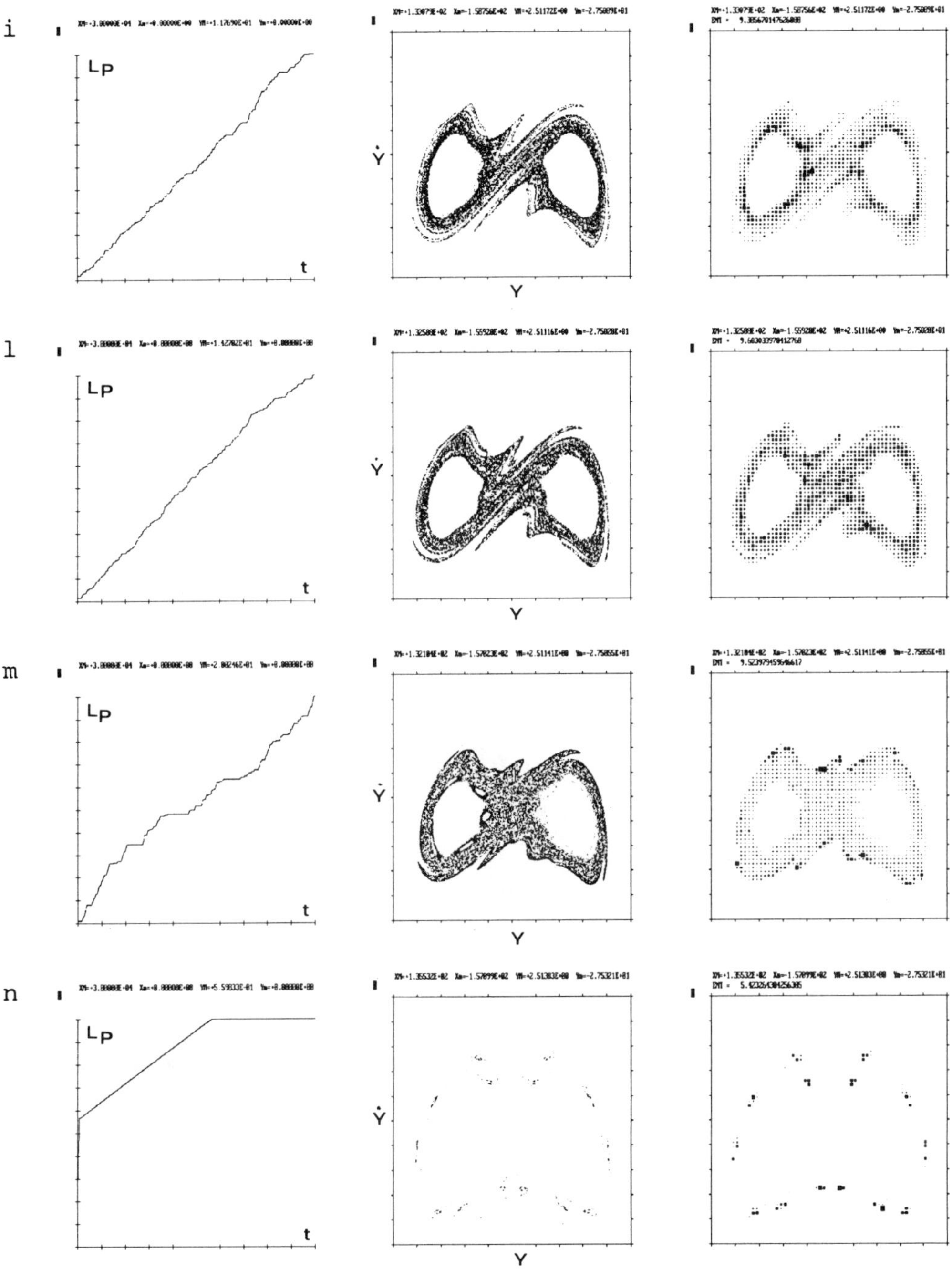

Figure 7: Numerical simulations: i) $H = 142.80$; l) $H = 153.00$; m) $H = 178.50$; n) $H = 204.00$.

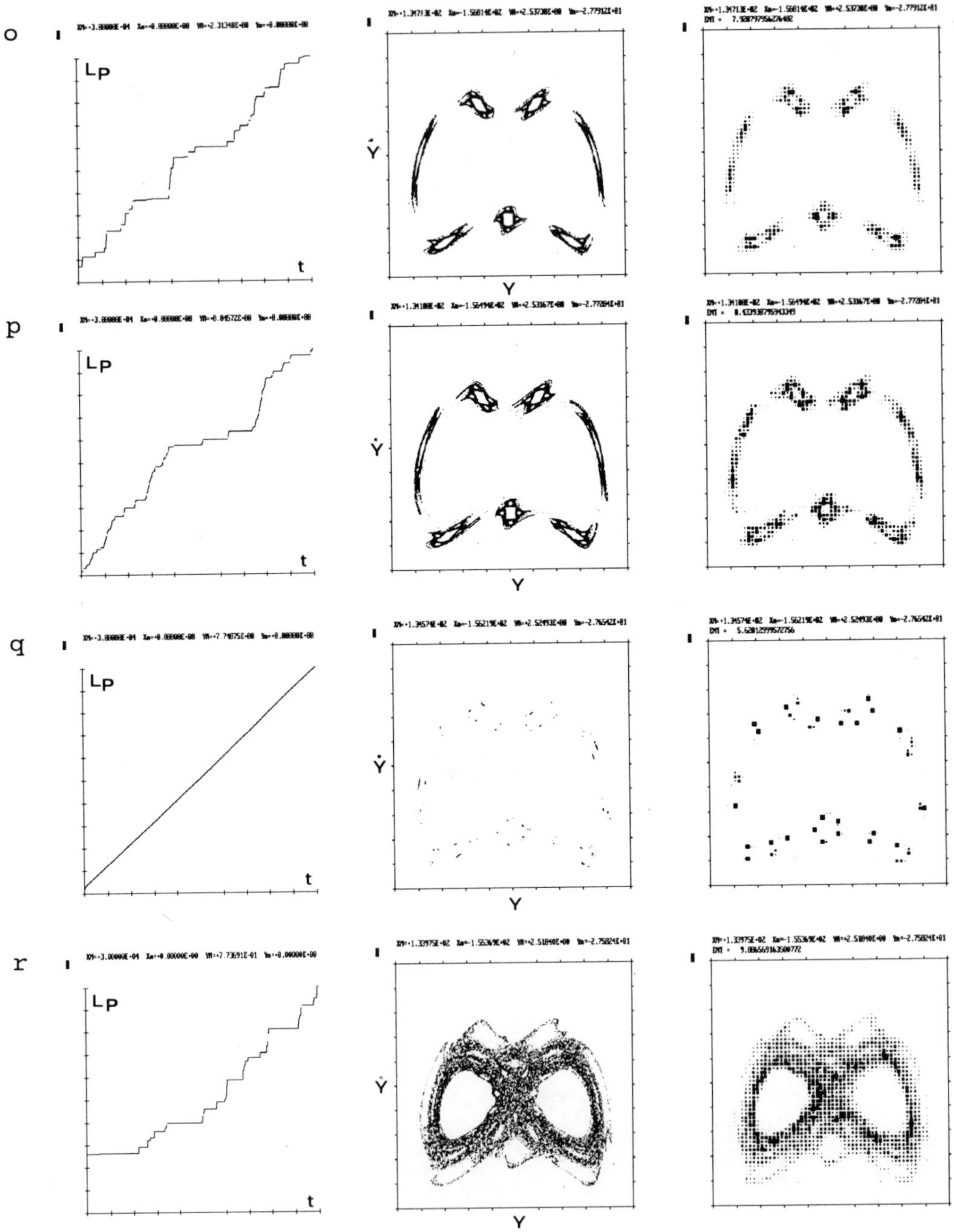

Figure 8: Numerical simulations: o) $H = 214.20$; p) $H = 216.75$; q) $H = 219.30$; r) $H = 229.50$.

structure behaves like a Hamiltonian system. Comparison of the three cases $H = 142.8$, $H = 153.0$, $H = 178.5$, shows that the rate of work dissipation regularly decreases from the first to the last of these three cases. The last analysis also shows long periods without plastic deformations, giving the erroneous impression that shakedown was achieved. The corresponding attractors appear to be less defined (that is less *fractal*) as the dissipation rate decreases. A further example is recorded for $H = 229.5$: the bursts of plastic dissipation are further apart and the strange attractors are wider.
The sequence $H = 204$, $H = 214.2$ and $H = 219.3$ appears as the strange attractor forms itself around the discrete points of the periodic motion: from this loss of periodicity the chaos is achieved for $H = 214.2$. If H increases to the value $H = 265.2$, the chaoticity disappears and the motion is first quasi-periodic and then periodic. From the value $H = 266.22$ to the value $H = 267.75$, the points which characterize the periodicity of the solution explode into closed lines, which implies of a quasi-periodic motion.
The stretching of these closed lines ends with their fracture around the value $H = 459$. After this point, the plastic phenomena become more and more negligible, and the Euler arch behaves like a Hamiltonian system. Attention is focused now on the different ways the plastic work is developed. In Fig.5.c, the periodic motion of the Euler arch for $H = 35.7$ is presented. The development of plastic deformations appears regular, and the function $L_P(t)$ is a straight line. The evolution of the plastic deformation $\epsilon_P(t)$ in Fig.13.a, shows a regular trend, i.e. the motion is periodic. In contrast, for $H = 153.0$, Fig.7.1 shows chaotic motion with a strange attractor and $L_P(t)$ appears as the Devil's staircase function. The trend of the plastic deformation is shown in this case in Fig.13.b, where one sees some bursts of plastic behaviour, with a totally irregular path.

CONCLUSIONS

Through the observation of the complex mechanical behaviour of an elastic-plastic structure, i.e. a Euler arch that undergoes large displacements, possible chaotic behaviour has been emphasized. The presence of chaos in a dynamic system involves the impossibility to foresee its behaviour, which implies, in the case of elastic-plastic structures, the impossibility of predicting the collapse due to the accumulation of plastic work.
An additional difficulty arises because the regions in which the irregular behaviour appear are not connected, and there is the possibility that they have a fractal aspect. Such a situation can jeopardize the overall structural response.

Acknowledgment.
This research has been supported by a grant from the Italian Ministry of the University and of the Scientific and Technological Research.

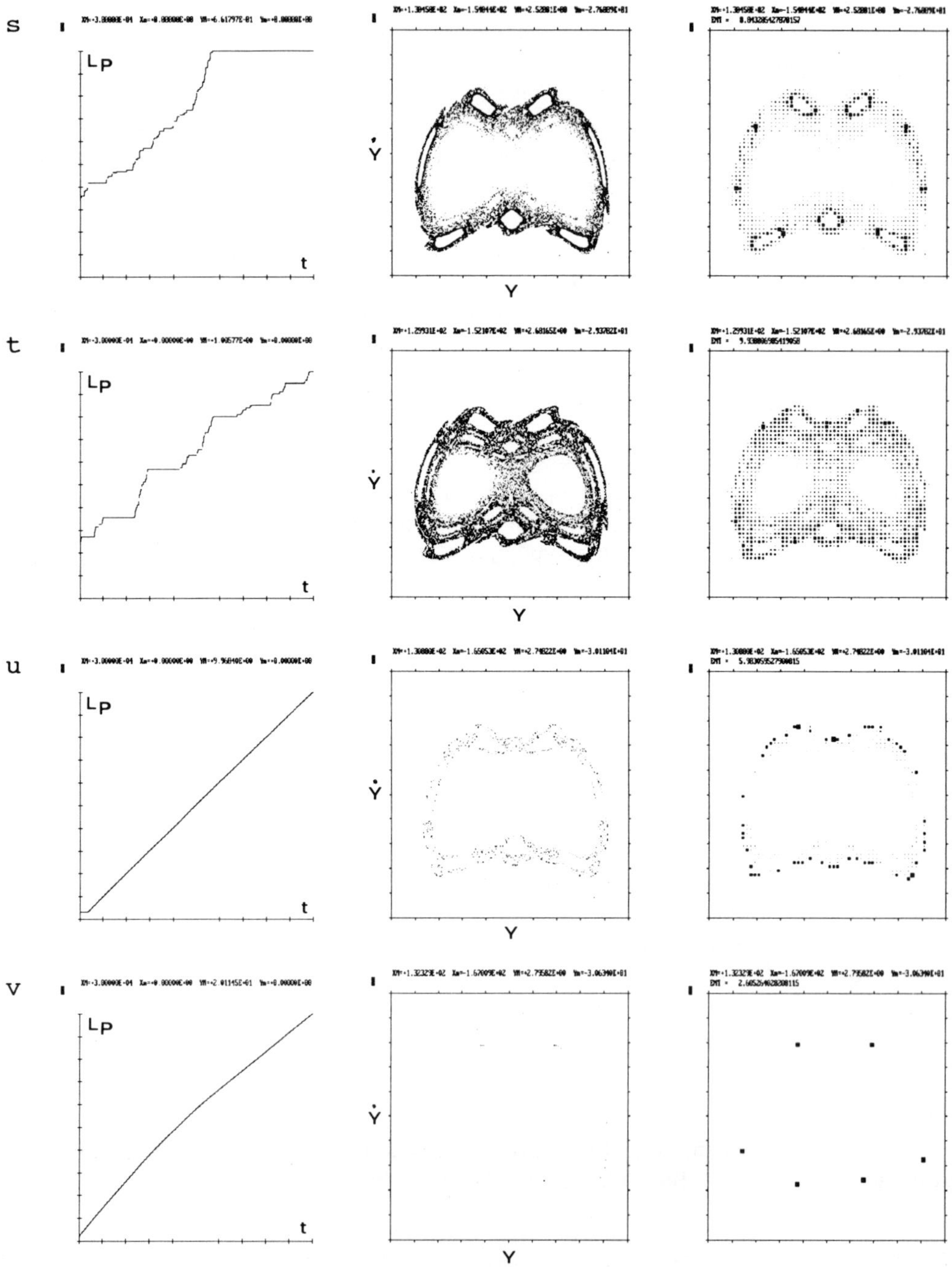

Figure 9: Numerical simulations: s) $H = 239.70$; t) $H = 255.00$; u) $H = 265.20$; v) $H = 265.71$.

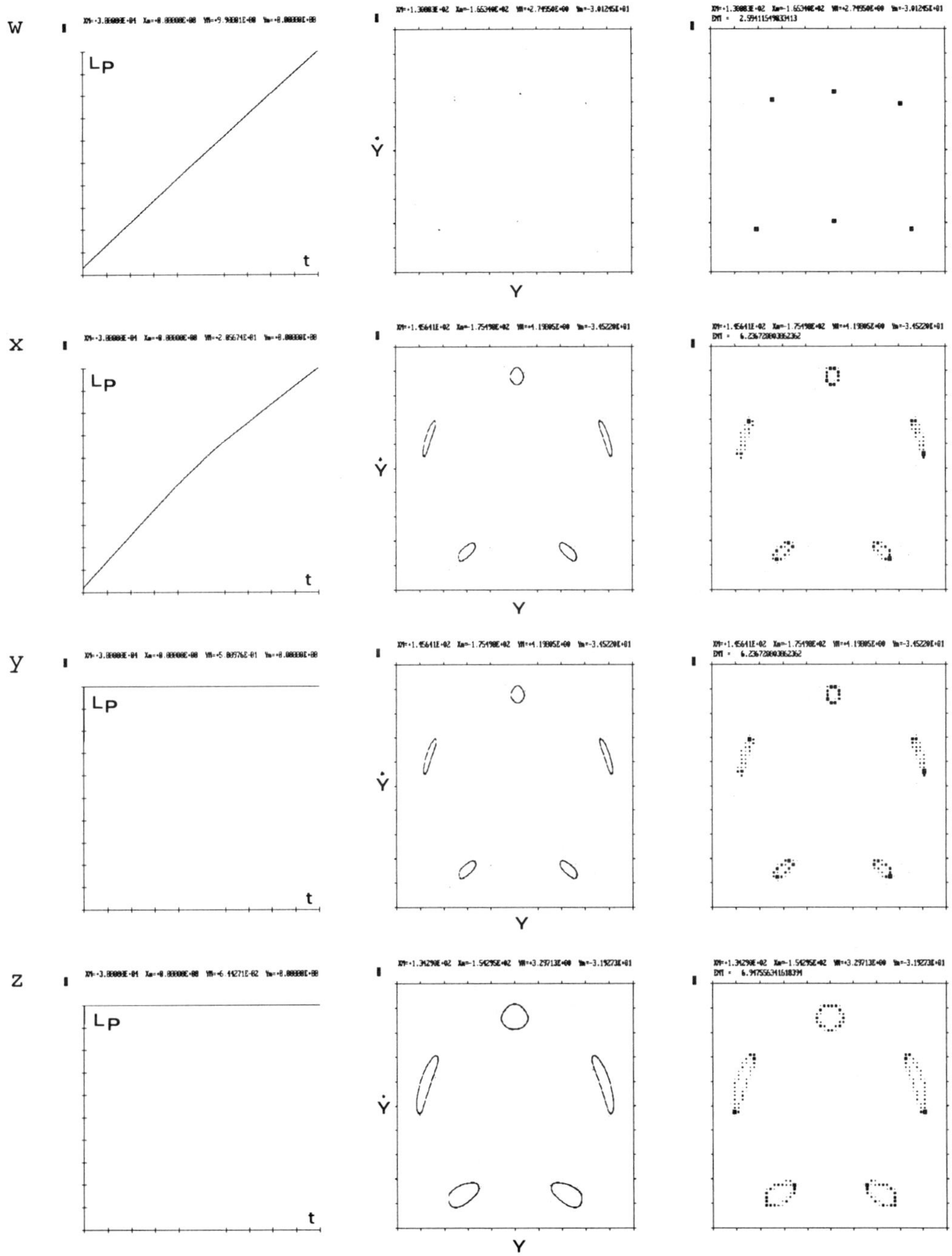

Figure 10: Numerical simulations: w) $H = 266.22$; x) $H = 267.75$; y) $H = 270.30$; z) $H = 357.00$.

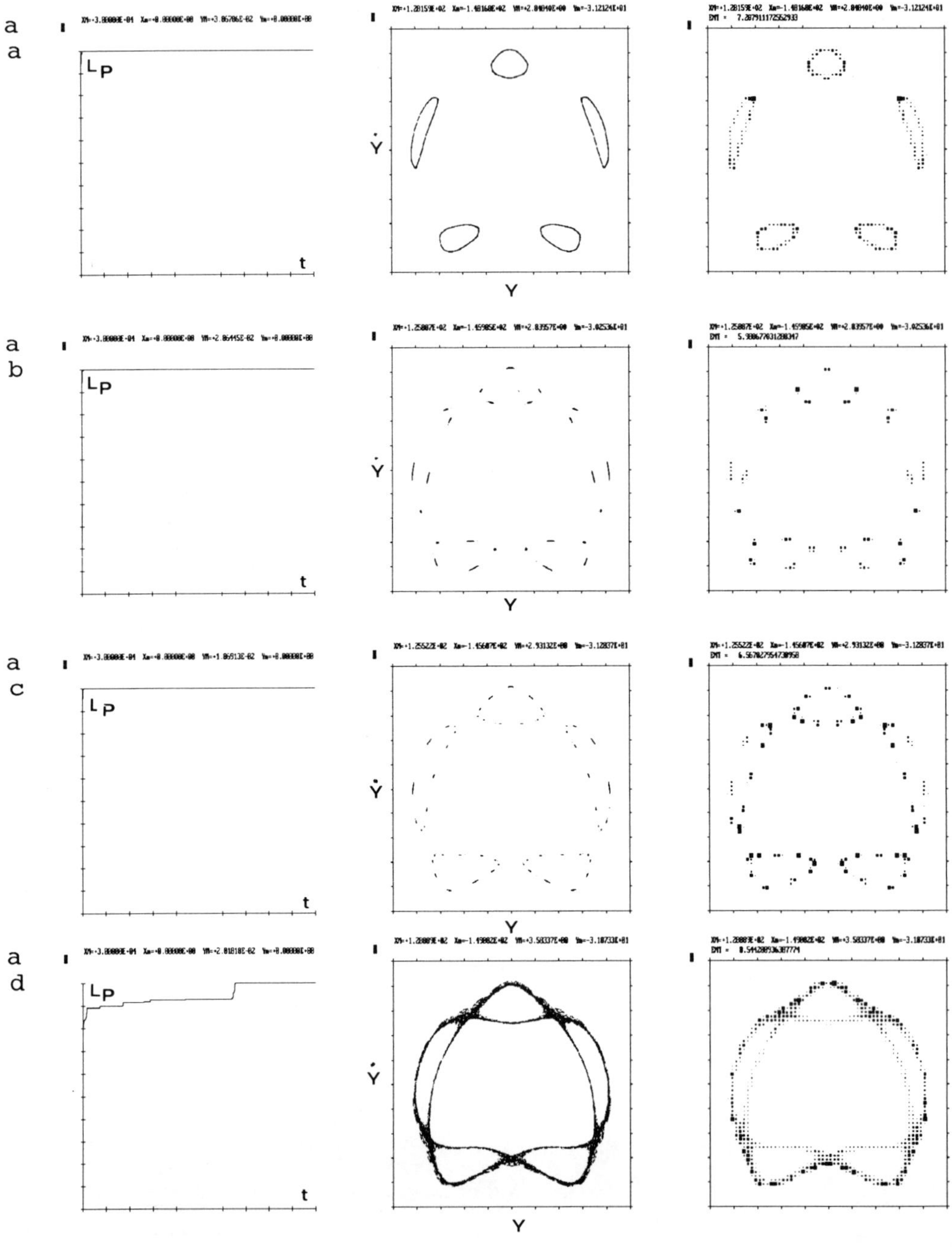

Figure 11: Numerical simulations: aa) $H = 408.00$; ab) $H = 459.00$; ac) $H = 464.10$; ad) $H = 469.20$.

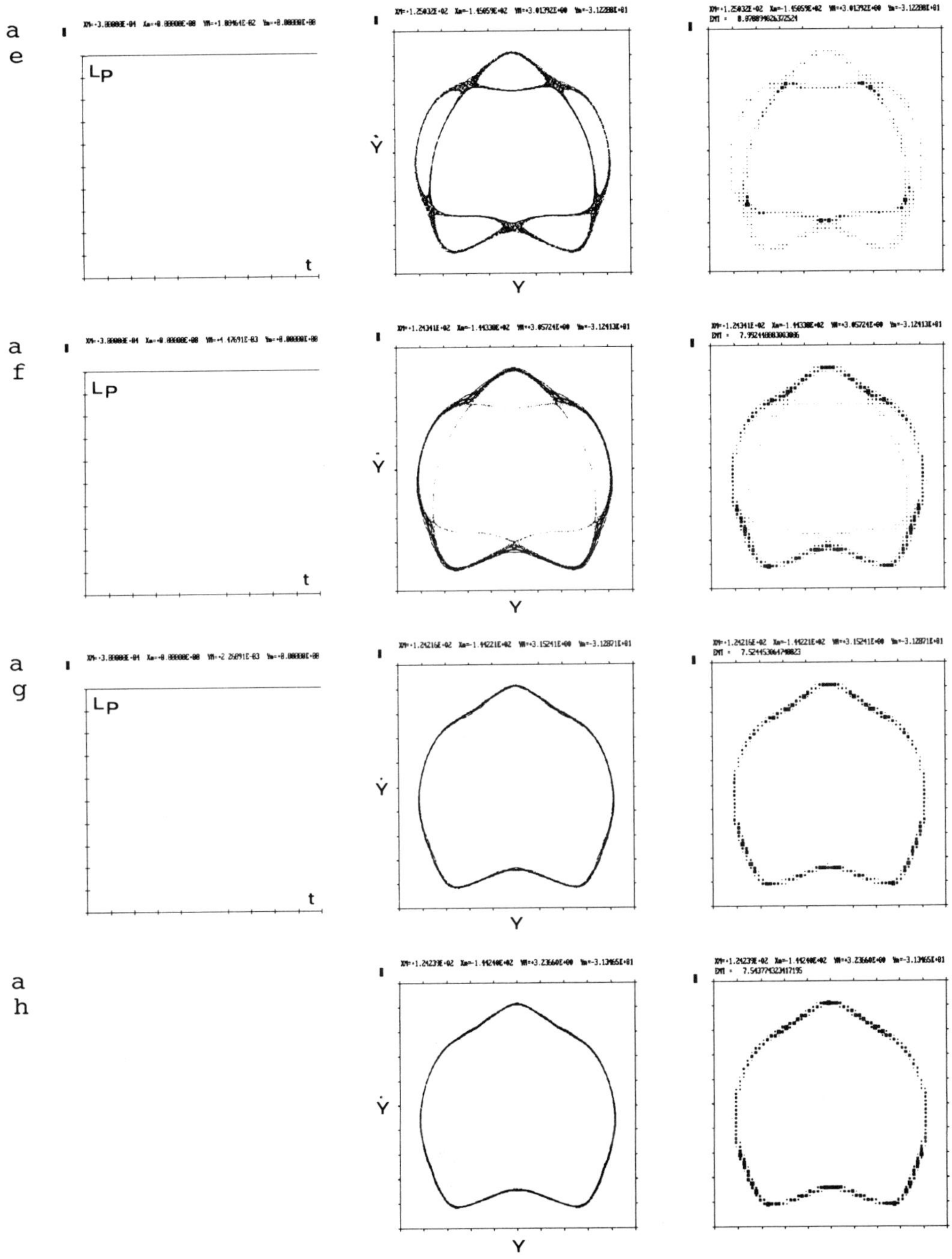

Figure 12: Numerical simulations: ae) $H = 484.50$; af) $H = 499.80$; ag) $H = 504.90$; ah) $H = 510.00$.

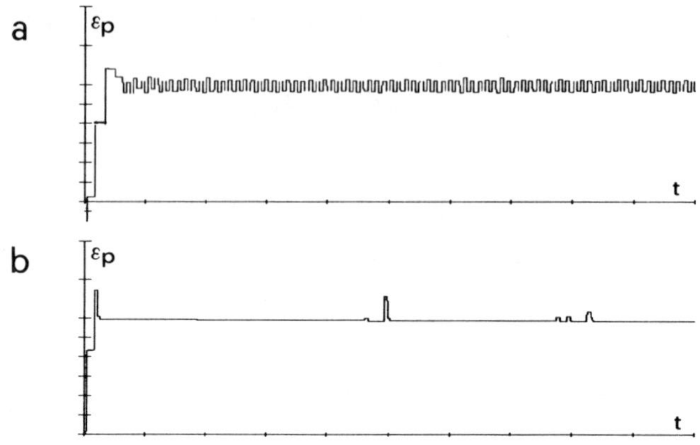

Figure 13: Plastic deformation as function of time: a) $H = 35.7$; b) $H = 153.0$.

REFERENCES.

1982 BATHE K.J., Finite Element Procedures in Engineering Analysis., Prentice-Hall.

1984 AUGUSTI G., BARATTA A., CASCIATI F., Probabilistic Methods in Structural Engineering., Chapman & Hall.

1987 KÖNIG J.A., Shakedown of Elastic-Plastic Structures. Elsevier, Amsterdam.

1988 WEDIG W., "Vom Chaos zur Ordnung.", GAMM-Mitteilungen,Heft 2, 3.

1989 REINHALL P.G., CAUGHEY T.K., STORTI D.W., "Order and Chaos in a Discrete Duffing Oscillator: Implications on Numerical Integration." Journal of Applied Mechanics,162/Vol. 56,March,162.

1990 RASBAND S.N., Chaotic Dynamics of Nonlinear Systems., John Wiley & Sons.

1990 BAKER G.L. & GOLLUB J.P., Chaotic Dynamics. An Introduction., Cambridge University Press.

1991 CASCIATI F., FARAVELLI L., Fragility Analysis of Complex Structural Systems., Research Studies Press, Tauton,.

1991 ARGYRIS J., MLEJNEK H.P., Dynamics of Structures., Text on Computational Mechanics Volume V, North-Holland.

1991 BONTEMPI F., CASCIATI F., FARAVELLI L., "Chaotic Motion versus Stochastic Excitation.", Chaos, Solitons and Fractals, Vol.1,no.6,491.

1993 BONTEMPI F., Transition to Chaos in Structural Dynamic Problems., Ph.D. dissertation, Politecnico di Milano (in Italian).

1993 BONTEMPI F., CASCIATI F., "Stochastic and Chaotic Dynamics.", in CHENG A.H.D. and YANG C.Y. (eds.), Computational Stochastic Mechanics, Computational Pubblications Ashurst Lodge, pp.215-238.

NUMERICAL TECHNIQUES APPLIED TO INDUSTRIAL PROBLEMS, DAMAGED STRUCTURES

INADAPTATION MECHANISMS IN BELLOWS SUBJECT TO SUSTAINED PRESSURE AND CYCLIC AXIAL LOADINGS IN TERMS OF FINITE DEFORMATIONS

Błażej SKOCZEŃ, Jacek SKRZYPEK
Cracow University of Technology
ul. Warszawska 24, 31-155 Cracow, Poland

ABSTRACT

Failure mechanisms in S- and U-type bellows subjected to sustained pressure and cyclic axial loadings are discussed. The analysis is based upon geometrically nonlinear III-rd order theory (large deflections but small strains) as well as on the IV-th order theory with large strains expressed by means of the logarithmic measure taken into account. Elastic- perfectly plastic and bilinear kinematic hardening material models were adopted, respectively. Progressive deformation in the absence of internal pressure due to strong anisosensitivity of S-type bellows is explained in the framework of geometrically nonlinear theory. Force, displacement and mixed controlled processes are analysed. Two different mechanisms of collapse: ultimate load mechanism and formation of an inadmissible discontinuity line are discussed. Effect of a stiff ring installed at the root of U-type bellows on the accumulation of large plastic strains is presented.

INTRODUCTION

Structural ratcheting has recently gained much attention, especially for shell structures. Progressive deformation in shells is caused by nonhomogenous distribution of stress that leads to the accumulation of plastic strains in some parts of the body. This kind of ratcheting, resulting from purely structural causes, is observed also in the absence of the material ratcheting effect.

Failure of thin shells subjected to cyclic loadings in the elastic-plastic regime has been analysed mainly for of cyclic thermal stresses. Let us mention here the classic paper by BREE [1967] on the behaviour of thin tubes subjected to cyclic temperature gradient across the wall in the presence of internal pressure. The criteria for the onset of ratcheting were derived, and some expressions for the plastic strains, incurred by each cycle, were obtained. This paper has initiated a strong wave of theoretical and experimental investigations in the field of shakedown and inadaptation problems.

The first step in the shakedown analysis of structures consists in calculating the interaction diagrams based on various deformation mechanisms under cyclic loadings.

Z. Mróz et al. (eds.), Inelastic Behaviour of Structures under Variable Loads, 341–361.

PONTER & CARTER [1989] have applied upper bound methods to compute interaction diagrams for axisymmetric shells under cyclic thermal loading in the presence of constant mechanical load. Recently JIANG & LECKIE [1992] presented a direct method for determination of the steady solutions in shakedown analysis. The kinematic hardening material model was used. As an example the ratcheting mode for a long cylindrical tube subjected to mechanical and thermal loads was analysed. Closed-form shakedown solutions were found. However, this efficient method is restricted to the shakedown analysis with small deformations.

For corrugated flexible shells (bellows expansion joints), subjected to sustained and cyclic mechanical loads, a large deformation approach is required. A finite difference technique for large deformation problems of U-shaped bellows was applied by HAMADA & TANAKA [1974]. The authors presented the analysis of the low-cycle fatigue failure of bellows under a deflection-controlled cyclic loading. The extended Manson-Coffin formula was used in order to estimate the number of cycles to failure. Creep ratcheting of bellows under sustained pressure and cyclic displacements was studied by POROWSKI & O'DONNELL [1978]. The ratcheting mechanism for bellows was assumed to be the same as for a pressurized cylinder. Results of the analysis were incorporated into the ASME Code. Another model was proposed by YAMASHITA et al. [1989], where the ratchetting mechanism in bellows subjected to pressure and cyclic displacement loads was considered. The authors found that the model adopted by Porowski and O'Donnell cannot properly describe the plastic ratcheting mechanism of bellows, and often leads to non-conservative results. The substitutive rectangular beam model subjected to distributed lateral force and cyclic lateral deflection was developed in order to replace bending moment of bellows due to pressure and axial displacement, respectively. The proper interaction diagrams based on this model were presented. These two papers tackle the problem of ratchetting of bellows in the framework of small deformation theory. The simplified approach based on the elastic-perfectly plastic material yields the effect of infinite ratcheting, due to constant plastic strain increment accumulated cycle-by-cycle.

Finite ratcheting of bellows was described by SKOCZEŃ et al. [1992a] for S-shaped bellows subjected to constant internal pressure and cyclic axial force. The behaviour of the shell was analysed on the basis of large deflection, but small strain, theory. The localization of intensive yielding was observed at the root and crest of the shell, leading to formation of an inadmissible discontinuity line. The ultimate load mechanism corresponding to the force controlled process was discussed. The mixed mechanism of failure, composed of nonsymmetric alternating plasticity and incremental collapse, was explained.

In the present paper the authors would like to compare failure mechanisms for S-shaped elastic-perfectly plastic bellows subjected to force or deflection controlled processes of cyclic loading in the framework of the large deflection, small strain, theory. Moreover, we discuss the influence of boundary conditions (the effect of a stiff ring at the root) on the progressive deformation of U-shaped bellows on the basis of the large rotation, large strain, theory for a kinematic hardening material model. The analysis presented in the paper is based either on the III-rd order theory of large displacements and rotations, but small strains, or on the IV-th order theory of large rotations and large strains. The III-rd order theory was previously applied by BIELSKI & SKRZYPEK [1989] to the analysis of finite elastic-plastic deformations of curved tubes. A particular case in which the change of the unit toroidal angle vanishes is considered in the present paper. In order to analyse thin corrugated shells within the large plastic strains regime the ANSYS 5.0 finite element code [1992] has been applied. Basic results for both approaches are presented in the sequal.

LARGE DEFLECTIONS, SMALL STRAINS ANALYSIS - PRINCIPAL ASSUMPTIONS

The third order theory of large displacements and rotations but small strains, based on the REISSNER [1981] nonlinear geometrical relations, is used to describe the large rotationally-symmetric deformation of the bellows segment. A more general approach, in which overall small bending is imposed upon large axisymmetric deformation of the bellows, was applied by the authors to the stability analysis of axially compressed bellows [1992b]. In the present paper nonsymmetric deformation of bellows due to transverse bending is not taken into account. Therefore, strains and stresses as well as displacements depend on the meridional coordinate only. The shell is subjected to normal internal pressure and cyclically varying axial force. Pressure is assumed as the follower, with normality retained, and its magnitude is kept constant during the cyclic loading process. Load or deflection control with respect to the axial force or the axial displacement parameters is applied.

In order to simplify the elastoplastic analysis, the substitutive sandwich section, considered as a discrete approximation of the uniform section of the wall, has been introduced. The tensile and bending rigidities of the sandwich wall must be the same as for the uniform wall. The loading is carried by the working layers only. The core is perfectly rigid in the direction normal to the middle surface and perfectly flexible for bending. The Love-Kirchhoff hypothesis is applied. The material of working layers is assumed to be perfectly elastoplastic and incompressible, as described by the associated flow rule and the Huber-Mises-Hencky yield condition.

Superscripts + and - refer to the outer and inner sandwich layers, respectively. Moreover, the following notation is used:

φ, ϑ	subscripts, for to meridional and circumferential directions
ϕ, φ	initial and current slope of normal to the middle surface
$(A)'=dA/d\phi$	derivative with respect to meridional angular coordinate
$R_{\varphi,\vartheta}$, $r_{\varphi,\vartheta}$	initial and current principal radii of curvature
R, r	initial and current distance between a middle surface point and axis of bellows
$U_{r,z}$	radial and axial displacements of middle surface points

e_φ, e_ϑ	principal elongations of the middle surface
$\varepsilon^{\pm}_\varphi$, $\varepsilon^{\pm}_\vartheta$	principal strains in working layers
$\kappa_{\varphi,\vartheta}$	increments of middle surface curvatures
$\sigma^{\pm}_{\varphi,\vartheta}$	principal stresses in working layers
$N_{\varphi,\vartheta}$, $M_{\varphi,\vartheta}$	normal force and moment resultants
Q_φ	shearing force resultant (the reaction)
p_n, p_e	internal, normal pressure; elastic carrying capacity of bellows under pure pressure
R_m	bellows mean radius
N, N_e	overall axial force; elastic carrying capacity of bellows subject to axial force
Δ	overall axial deflection
R_Φ	convolution radius
H_u	thickness of the equivalent uniform wall of bellows
$\rho = R_\Phi / R_m$	ratio of convolution to mean radii
$\lambda = H_u R_m / R_\Phi^2$	toroidal shell parameter.

REDUCED SET OF STATE EQUATIONS (PERFECTLY ELASTOPLASTIC MODEL)

Equilibrium equations for toroidal shells combined with geometrical relations, generalized stress definitions and constitutive equations for perfectly elastoplastic, incompressible material, as described by the associated flow rule and the Huber-Mises-Hencky yield condition, lead to the following reduced set of six nonlinear, coupled differential equations, linear with respect to the first derivatives of increments of state functions (cf. SKOCZEŃ et al.[1992a]):

$$
\begin{aligned}
&(\delta\varphi)' = R_\varphi \delta\kappa_\varphi\\
&(\delta U_r)' = -R_\varphi[\delta e_\varphi \sin\varphi + (1+e_\varphi)\delta\varphi\cos\varphi]\\
&(\delta U_z)' = R_\varphi[\delta e_\varphi \cos\varphi - (1+e_\varphi)\delta\varphi\sin\varphi]\\
&(\delta N_\varphi)' = -\{R'\delta N_\varphi + R_\varphi[N_\vartheta\delta\varphi\cos\varphi + \delta N_\vartheta \sin\varphi]\}/R - Q_\varphi\delta\varphi' - \varphi'\delta Q_\varphi\\
&(\delta Q_\varphi)' = \{R_\varphi[\delta N_\vartheta\cos\varphi - N_\vartheta\delta\varphi\sin\varphi] - R'\delta Q_\varphi\}/R + N_\varphi\delta\varphi' + \varphi'\delta N_\varphi\\
&\qquad\qquad - R_\varphi[(1+e_\vartheta)p_n\delta e_\varphi + (1+e_\varphi)p_n\delta e_\vartheta + (1+e_\varphi)(1+e_\vartheta)\delta p_n]\\
&(\delta M_\varphi)' = -\{R'\delta M_\varphi + R_\varphi[M_\vartheta\delta\varphi\cos\varphi + \delta M_\vartheta\sin\varphi]\}/R - R_\varphi[Q_\varphi\delta e_\varphi + (1+e_\varphi)\delta Q_\varphi]
\end{aligned}
\qquad (1)
$$

where

$$\delta M_\vartheta=[(1-\frac{h}{r_\varphi})\delta\sigma_\vartheta^- -(1+\frac{h}{r_\varphi})\delta\sigma_\vartheta^+](1+e_\varphi)\frac{ht}{2}+M_\vartheta\frac{\delta e_\varphi}{(1+e_\varphi)}+(N_\vartheta+\frac{M_\vartheta}{r_\varphi})h^2\frac{\delta r_\varphi}{r_\varphi^2}$$

$$\delta e_\varphi=\{\delta N_\varphi+(\frac{1}{r_\vartheta}-\frac{1}{R_\varphi})\delta M_\vartheta+[M_\varphi(\frac{1}{r_\vartheta}-\frac{1}{R_\varphi})-N_\varphi]\frac{\delta e_\vartheta}{(1+e_\vartheta)}$$

$$-[N_\varphi h^2(\frac{2}{r_\vartheta}-\frac{1}{R_\varphi})+M_\varphi]\frac{\delta r_\vartheta}{r_\vartheta^2}\}/[Et(1+e_\vartheta)]-[(1-\frac{h}{R_\varphi})\delta\sigma_\vartheta^- +(1+\frac{h}{R_\varphi})\delta\sigma_\vartheta^+]/(4E)$$

$$+[\delta\lambda^-(2\sigma_\varphi^- -\sigma_\vartheta^-)(1+\frac{h}{R_\varphi})+\delta\lambda^+(2\sigma_\varphi^+ -\sigma_\vartheta^+)(1-\frac{h}{R_\varphi})]/2$$

$$\delta N_\vartheta=[(1-\frac{h}{r_\varphi})\delta\sigma_\vartheta^- +(1+\frac{h}{r_\varphi})\delta\sigma_\vartheta^+](1+e_\varphi)\frac{t}{2}+N_\vartheta\frac{\delta e_\varphi}{(1+e_\varphi)}+(M_\vartheta+N_\vartheta\frac{h^2}{r_\varphi})\frac{\delta r_\varphi}{r_\varphi^2}$$

$$\delta\kappa_\varphi=\{(\frac{1}{R_\varphi}-\frac{1}{r_\vartheta})\delta N_\varphi-\frac{\delta M_\varphi}{h^2}+[N_\varphi(\frac{1}{r_\vartheta}-\frac{1}{R_\varphi})+\frac{M_\varphi}{h^2}]\frac{\delta e_\vartheta}{(1+e_\vartheta)}$$

$$[N_\varphi+M_\varphi(\frac{2}{r_\vartheta}-\frac{1}{R_\varphi})]\frac{\delta r_\vartheta}{r_\vartheta^2}\}/[Et(1+e_\vartheta)]-\{(\frac{1}{R_\varphi}-\frac{1}{r_\varphi})\delta N_\vartheta$$

$$-\frac{\delta M_\vartheta}{h^2}+[N_\vartheta(\frac{1}{r_\varphi}-\frac{1}{R_\varphi})+\frac{M_\vartheta}{h^2}]\frac{\delta e_\varphi}{(1+e_\varphi)}+[N_\vartheta+M_\vartheta(\frac{2}{r_\varphi}-\frac{1}{R_\varphi})]\frac{\delta r_\varphi}{r_\varphi^2}\}/[2Et(1+e_\varphi)]$$

$$+[\delta\lambda^-(2\sigma_\varphi^- -\sigma_\vartheta^-)(1+\frac{h}{R_\varphi})-\delta\lambda^+(2\sigma_\varphi^+ -\sigma_\vartheta^+)(1-\frac{h}{R_\varphi})]/(2h). \tag{2}$$

In the above, $\delta\lambda^\pm$ are the nonnegative scalar factors which may be determined from the yield condition. The reduced set of equations supplemented with the boundary conditions constitute a two-point boundary value problem. Each of the rotationally symmetric bellows segments has the same geometry, is subjected to the same loadings and remains under the same support conditions. Hence, a single bellows segment is analysed (Fig.1).

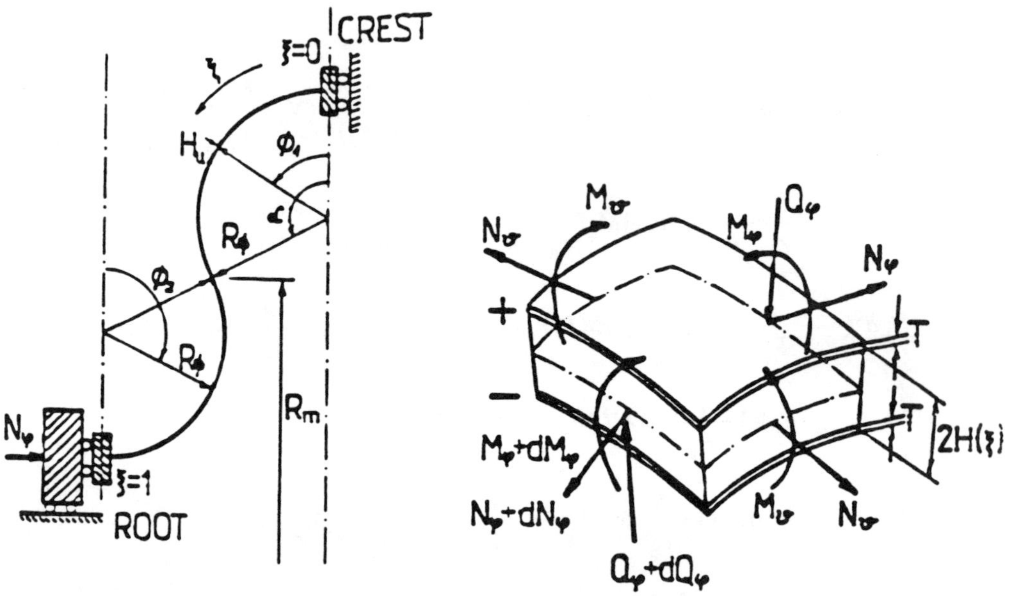

Fig. 1. Half of the S-type bellows pitch and notational conventions for generalized stresses.

Equations (1) are integrated between the crest and the root of the segment, where the conditions of symmetry are satisfied. Continuity conditions are applied at the point where the curvature of the meridian changes from positive to negative. In order to solve the two-point boundary value problem of type 3x3, the shooting method is combined with the two-step integration technique based on the Runge-Kutta IV procedure, and completed with the Newton-Raphson algorithm.

LARGE STRAINS FINITE ELEMENT ANALYSIS (PLASTIC HARDENING MODEL)

The numerical simulation of large cyclic deformations of bellows at finite strains is performed by using the ANSYS 5.0 code. The hypoelastic characterization of the elastic response is employed. The formulation is restricted to small elastic strains, but it allows for arbitrarily large plastic strains. The essential elements of the approach to large deformations of elastoplastic structures at finite strains are given below.

Let the deformation map for the body of interest be described as follows:

$$\phi:\Omega_0\subset\mathbb{R}^3\times\mathbb{I}\rightarrow\Omega^*\subset\mathbb{R}^3 \tag{3}$$

The current position vector **x** of a given point of the body is a function of the position vector **X** in the reference state and time t:

$$x=\phi(X,t) \tag{4}$$

The deformation gradient is defined by the formula:

$$F=\frac{\partial\phi(X,t)}{\partial X} \tag{5}$$

Applying the polar decomposition of F one obtains the rotation tensor **R** and the right stretch tensor **U**: **F=RU**. The material logarithmic strain tensor **E** is determined through the spectral decomposition of **U**:

$$E=lnU=\sum_{I=1}^{3}\ln(\lambda_I)N_I\otimes N_I \tag{6}$$

where

$$\lambda_I\ ,\ (I=1,2,3)\ ;\ N_I\ ,\ (I=1,2,3) \tag{7}$$

denote the principal stretches (eigenvalues of **U**) and the Lagrangian triad (eigenvectors of **U**).

The advantage of the logarithmic (Hencky) strain measure consists in the fact that the incompressibility constraint in finite strain regime is expressed in the same form as for small strains. It leads to a possibility of extension of the standard theories of plasticity (for small strains) to the large strain range. The results presented in the paper are based on the bilinear kinematic hardening material model. The isoparametric 8-node axisymmetric plane elements were applied. Relatively smooth distribution of strains through thickness was achieved by applying 5 elements in the direction normal to the midsurface of the shell.

SHAKEDOWN AND INADAPTATION MECHANISMS OF PERFECTLY PLASTIC BELLOWS

For the shakedown analysis, we assume S-type bellows subject to sustained internal pressure and variable repeated axial force. Furthermore, if for certain pairs of the pressure and axial force magnitudes **(p,N)** the structural behaviour, after some plastic deformation during the initial load cycles, eventually becomes elastic, we say that shakedown takes place. The permanent deflection measured at the most exerted point (root) stabilizes due to the cycle-by-cycle decreasing plastic energy dissipation. This means that the total amount of plastic energy dissipated during the process of cyclic elastoplastic deformation must be finite at each point of the structure (first local criterion):

$$\int_0^t \sigma_{ij}\frac{\partial\varepsilon_{ij}^p}{\partial t}d\tau < \infty \tag{8}$$

Another local functional, measured at a given point of the structure, the Odqvist parameter expressing length of the trajectory in the plastic strains space, may equivalently be applied as the shakedown criterion (second local criterion). If shakedown takes place, the Odqvist parameter:

$$\int_0^t \sqrt{\frac{\partial \varepsilon_{ij}^p}{\partial t}\frac{\partial \varepsilon_{ij}^p}{\partial t}}\,d\tau < C \tag{9}$$

must be bounded at each point of the structure. The cycle-by-cycle calculated Odqvist parameter increments are a monotonic function of the number of cycles and eventually, after a finite number of cycles, approach zero. The equivalent shakedown criteria on the level of cross- section (semiintegral) or body (global) may be formulated as well (cf. KOENIG [1987]). In what follows we shall introduce usefull distinction: the shakedown phenomenon is called elastic if the dissipation of plastic energy eventually stops; on the other hand, if after transient ratchetting the energy dissipation stabilizes on a certain nonzero level and reversed plasticity occurs we say that plastic shakedown takes place.

An example of elastic shakedown is presented in Fig. 2 for the following data:

$$\rho=0.1\ ,\ \lambda=1.0\ ,\ \frac{p_n}{p_e}=0.4\ ,\ \frac{N}{N_e}=1.05$$

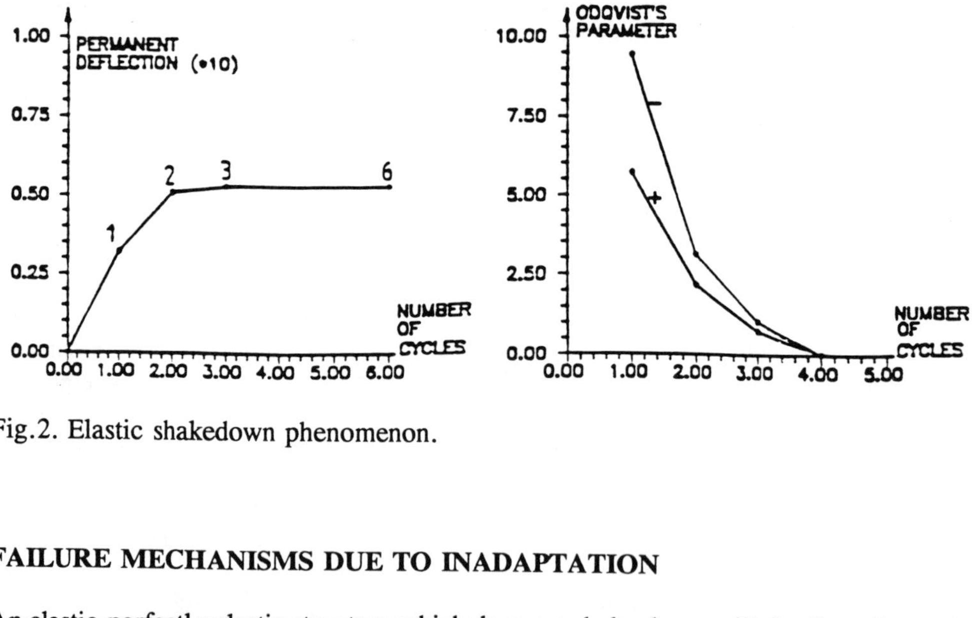

Fig.2. Elastic shakedown phenomenon.

FAILURE MECHANISMS DUE TO INADAPTATION

An elastic-perfectly plastic structure which does not shake down will finally collapse due to an inadaptation process. Generally, the mechanism of inadaptation is based on the alternating plasticity and/or incremental collapse (ratcheting) phenomena. The plastic shakedown or ratcheting are usually considered on the level of the whole structure. However, both mechanisms are associated with a specific development of plastic strain fields at the root and the crest of a bellows segment. The two-point substitutive wall section enables us to analyse plastic deformation at the cross-section at level. We distinguish between the alternating plasticity, the incremental, and the mixed (nonsymmetric alternating plasticity) mechanisms which may occur at the cross-section of the wall. As a result of rapid development of plastic strains within one or both layers of the sandwich section, due to the elastic-perfectly plastic material model, failure may occur due to the formation of a jump of slope in the normal to the midsurface. A more general classification of possible discontinuity modes in sandwich section has been presented by SKRZYPEK & ŻYCZKOWSKI [1983]. This is an incremental mechanism which leads to formation of a discontinuity line within the shell. A mixed mechanism at the cross-section at level means that in one layer there is incremental (progressive) deformation, and in the other layer there is the alternating plasticity phenomenon.

In most papers concerning the problem of failure mechanisms of shells due to cyclic loadings, the analysis is performed at the global level. From the point of view of the force-controlled process of cyclic loading of the elastic-perfectly plastic bellows, two basic collapse modes may be distinguished at the global level: the ultimate load mechanism and the formation of a discontinuity line. The latter is observed for deflection-controlled processes also. These collapse modes are associated with nonsymmetric alternating plasticity or incremental effects at the cross-section at level. For lower pressure the failure takes place at the root, for higher pressure at the crest, of the bellows.

ULTIMATE LOAD MECHANISM - FORCE CONTROLLED AXIAL CYCLES

This mechanism for cyclic loading may be regarded as analogous to the effect of reaching the maximal load for quasistatically loaded bellows. The presumed axial force amplitude cannot be reached as a result of the localization of plastic flow at the crest and the root of the segment (Fig. 3). Plastic strain increments at the root increase cycle-by-cycle, both at the inner and the outer layers (Fig. 4). Simultaneous expansion of the plastic zone around the root, and intensive yielding, lead to decrease of the carrying capacity after a limited number of cycles. From the point of view of the force-controlled process this phenomenon must be regarded as a collapse. The reverse point 12 (Fig. 3), which terminates the constant force amplitude process, corresponds to the ultimate load effect. Possible continuation of the process of cyclic loading from the point of ultimate load yields decrease of the carrying capacity with each subsequent loading cycle, for positive axial load increments. The residual deflection increments diminish and the expansion of the plastic zone around the root stops. On the other hand, a considerable accumulation of the plastic strain increments in the layers is observed (Fig. 7). Eventually, we face a smooth transition from the ultimate load mechanism to the formation of a plastic discontinuity line. The calculations were performed for the following data:

$$\rho=0.1\ ,\ \lambda=1.0\ ,\ p_n=0.0\ ,\ \frac{N}{N_e}=1.08$$

Evolution of the meridional strains at the root for each layer (Fig. 5) indicates that the nonsymmetric alternating plasticity predominates in the inner layer, whereas the incremental effect takes place in the outer one. This mixed type inadaptation mechanism is observed at the root. Evolution of the Odqvist parameter increments as well as the unlimited increase of permanent deflection are presented in Fig. 6.

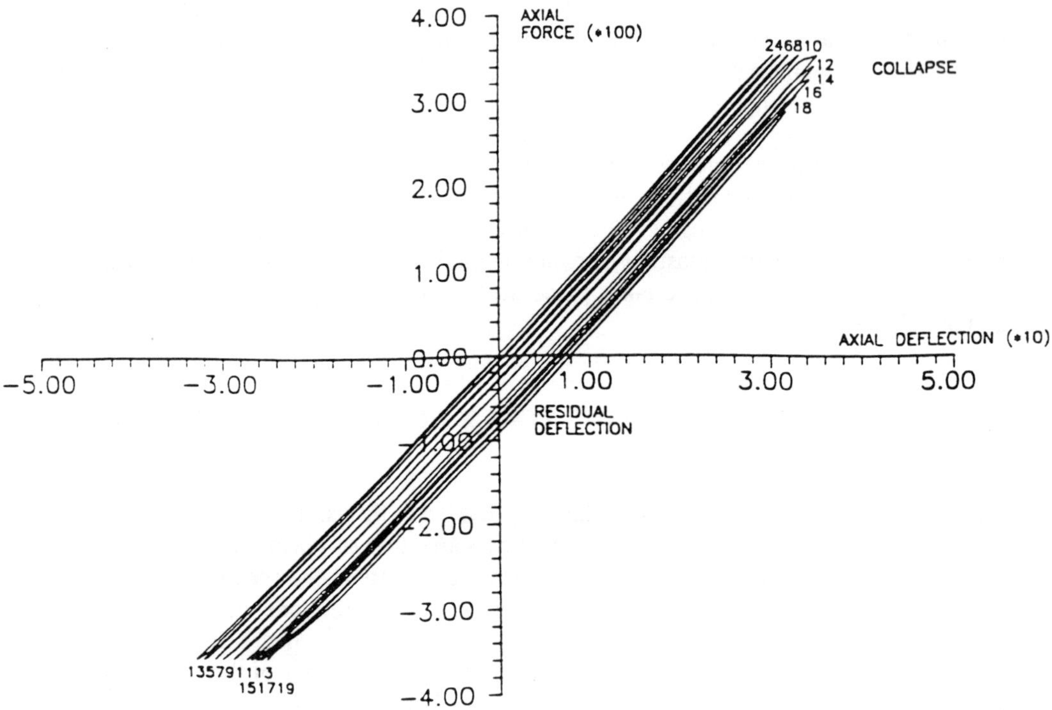

Fig. 3. Ultimate load mechanism.

It is worth pointing out that we can observethe effect of ratcheting (cycle-by-cycle increase of the residual deflection) in the absence of internal pressure. This phenomenon is caused by the anisosensitivity of S-type bellows for tension and compression, which may be detected only in the framework of the large deflections (geometrically nonlinear) theory. As a result of the fact that the bellows flexibility is different under tension and compression, the plastic strain fields develop in a different way, though the axial loading cycle is symmetric. Hence, after a full cycle a residual permanent axial deflection is observed.

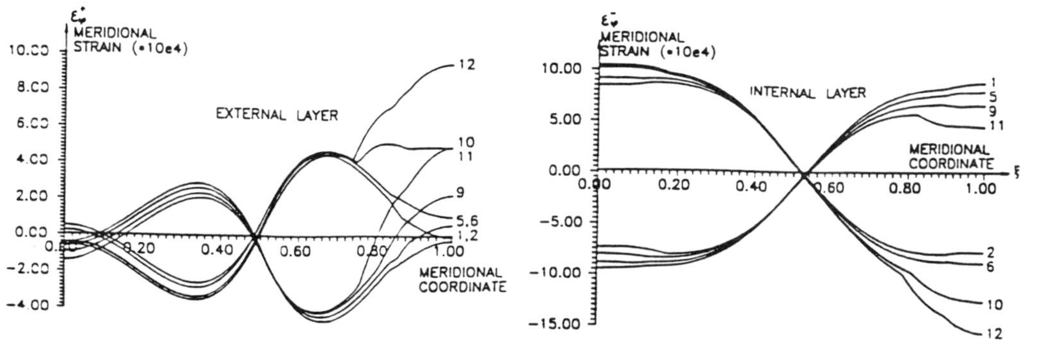

Fig. 4. Evolution of plastic strains due to ultimate load mechanism.

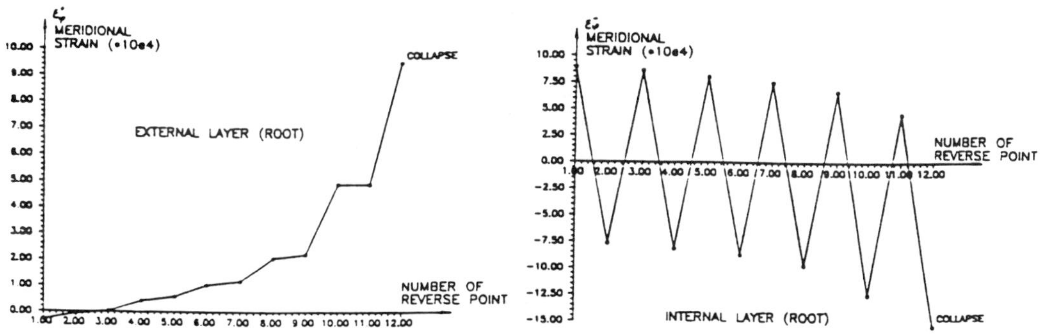

Fig. 5. Mixed inadaptation mechanism which develops at the root.

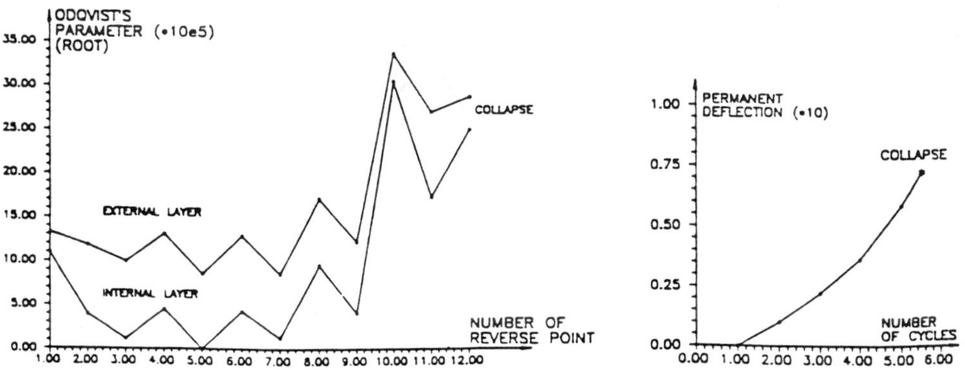

Fig. 6. Evolution of the Odqvist parameter at the root. Increase of permanent deflection.

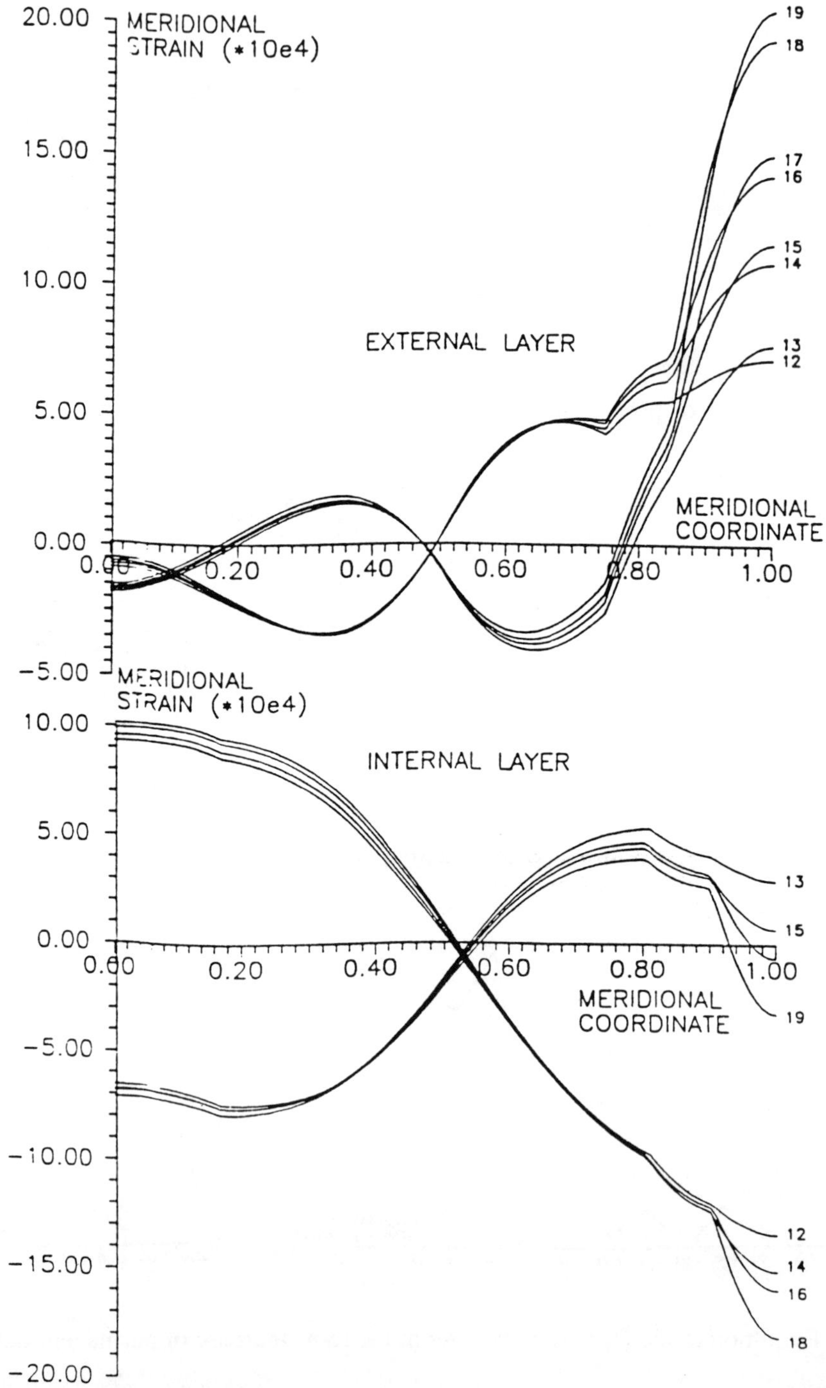

Fig. 7. Accumulation of plastic strains due to the extended process of cyclic loading.

FORMATION OF A DISCONTINUITY LINE - SUSTAINED PRESSURE AND FORCE-CONTROLLED AXIAL CYCLES

The effect of higher magnitudes of pressure on the failure mechanism of bellows is significant. The meridional strain distribution within the shell changes, and plastic strains concentrate around the crest of the segment. Switch of the failure point from root to crest of the segment appears to be a characteristic feature of higher magnitudes of sustained pressure. Another feature of this mechanism that plastic strain increments accumulate within a narrow zone of approximately constant length. An example of formation of a discontinuity line has been calculated for the following data:

$$\rho = 0.1 \ , \ \lambda = 1.0 \ , \ \frac{p_n}{p_e} = 0.5 \ , \ \frac{N}{N_e} = 1.06$$

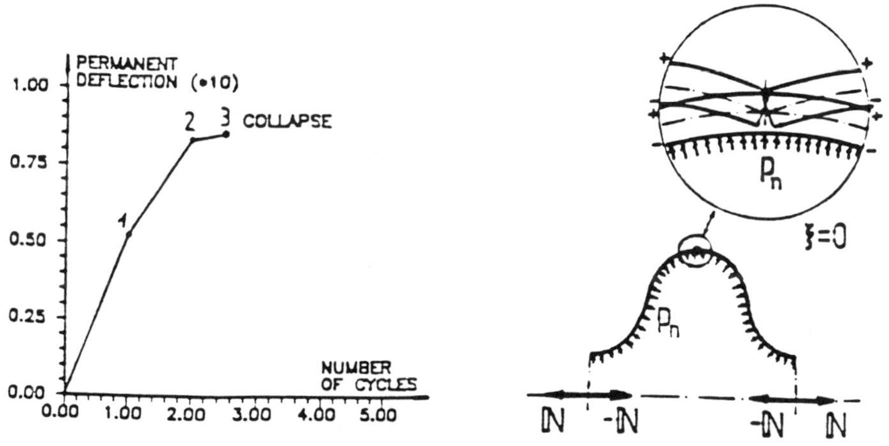

Fig. 8. Residual collapse deflection. Formation of a discontinuity line at the crest.

Strong concentration of tensile strains at the inner layer around the crest eventually results in a jump in the slope of the normal to the midsurface (discontinuity of order 0) of the shell (Fig. 8). Failure is associated with a predominance of the incremental effect (progressive deformation) within the shell wall.

FORMATION OF A DISCONTINUITY LINE - DEFLECTION-CONTROLLED AXIAL CYCLES

A similar failure mechanism of elastic-perfectly plastic S-type bellows was observed in the case of a deflection-controlled symmetric axial cycle. The calculations were performed for the following data:

$$\rho=0.1\ ,\ \lambda=1.0\ ,\ p_n=0.0\ ,\ \frac{\Delta}{\Delta_e}=1.05$$

After several cycles, the previously assumed amplitude of deflection was not reached (Fig.9), due to formation of plastic discontinuity at the root. The hysteresis loop grows cycle-by-cycle. This means that the plastic

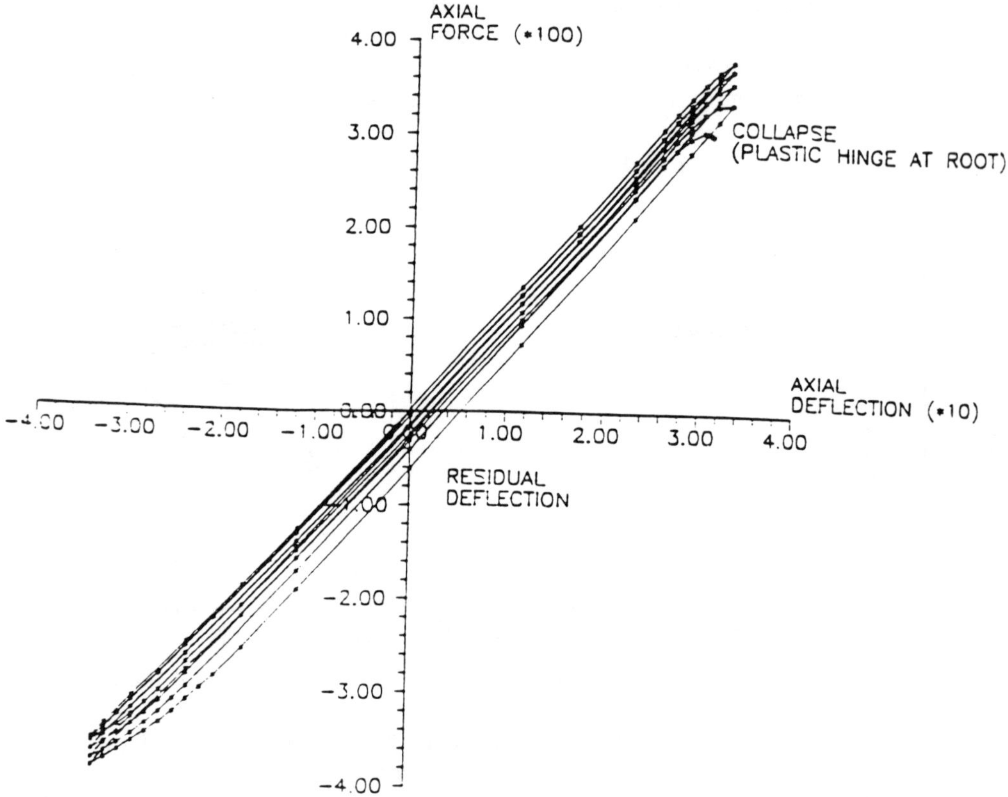

Fig. 9. Deflection-controlled process of cyclic loading of S-type bellows.

energy dissipation within a cycle is an increasing function of the number of cycles. Another interesting effect is that due to anisosensitivity of the shell (different tensile and compression stiffnesses), a significant increase of the residual axial force in the absence of sustained load (internal pressure) takes place. This effect cannot be detected with small displacements/rotations theory. The evolution of the meridional strains at the root as well as the cycle-by-cycle evolution of the Odqvist parameter increments are presented in Fig. 10.

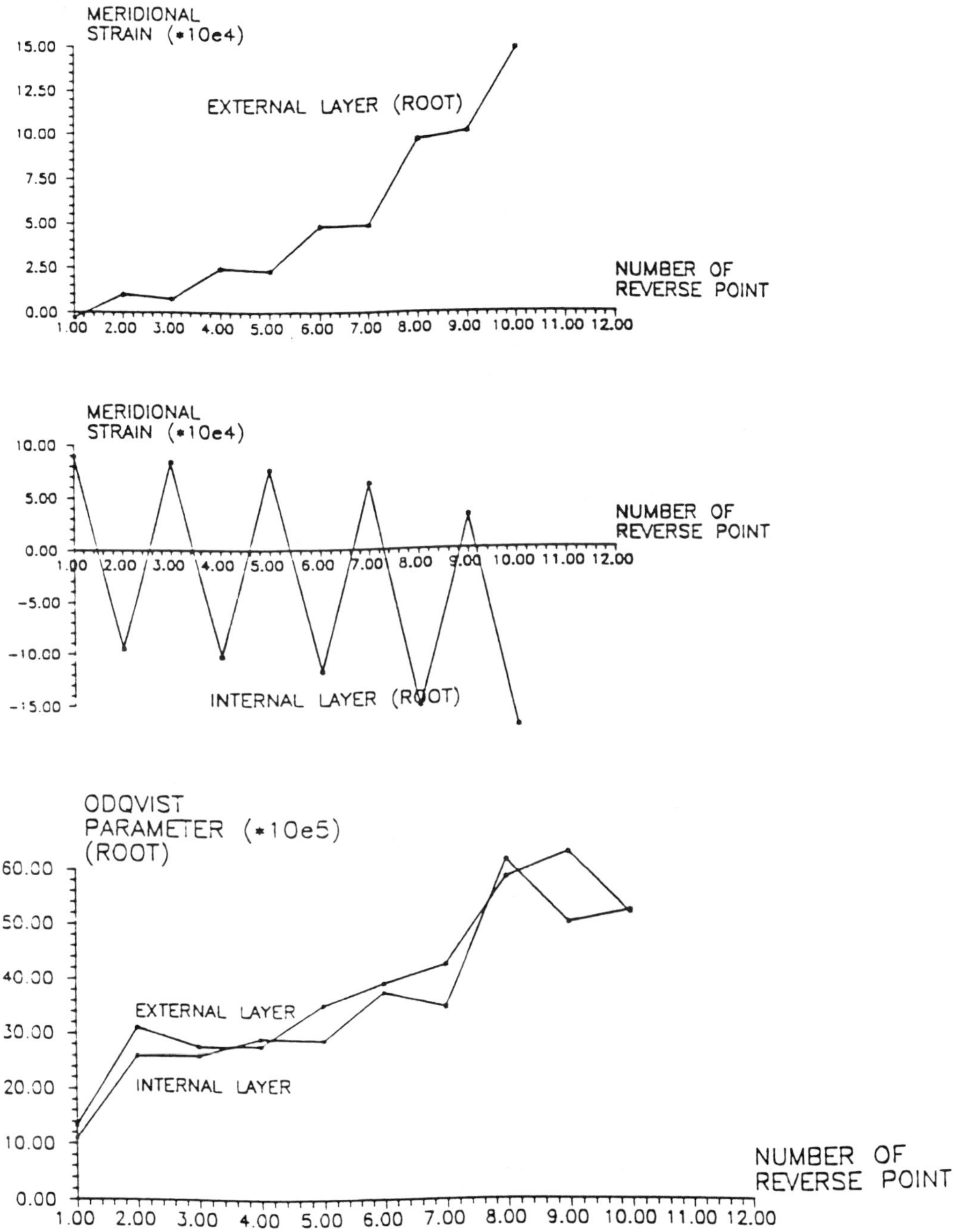

Fig. 10. Evolution of the meridional strains and the Odqvist parameter increments at the root.

SHAKEDOWN AND INADAPTATION DOMAINS - FORCE-CONTROLLED AXIAL CYCLES

Various response domains for the elastic-perfectly plastic S-type bellows subjected to sustained pressure and force-controlled cyclic axial loading are presented in Fig. 11. The domain of pure elastic response of the structure is bounded by the elastic carrying capacity curve (ECC). For lower pressure magnitudes (when compared to the elastic carrying capacity for pure pressure loading), yielding starts at the root: the failure of the shell occurs here also. For medium pressure, yielding starts at the crest, where failure of the structure eventually takes place. As pressure increases, the point of initial yielding gradually moves towards the doscontinuity in the meridional curvature. The two outer curves were determined for the quasistatically loaded structure in the framework of the limit analysis. LCC denotes the load carrying capacity whereas DCC stands for the formation of plastic discontinuity lines. The region between the initial yielding curve and both the curves obtained due to the limit analysis, contains all possible pairs of pressure and axial force that lead to elastoplastic response of the structure. This domain is divided into two subdomains: shakedown and inadaptation domains. The elastic response domain and the shakedown domain constitute the region of safe work of the structure. Both these failure mechanisms are restricted to the inadaptation domain. The ultimate load mechanism and the formation of a plastic discontinuity line are based on the plastic strain localization at the root or crest of the bellows.

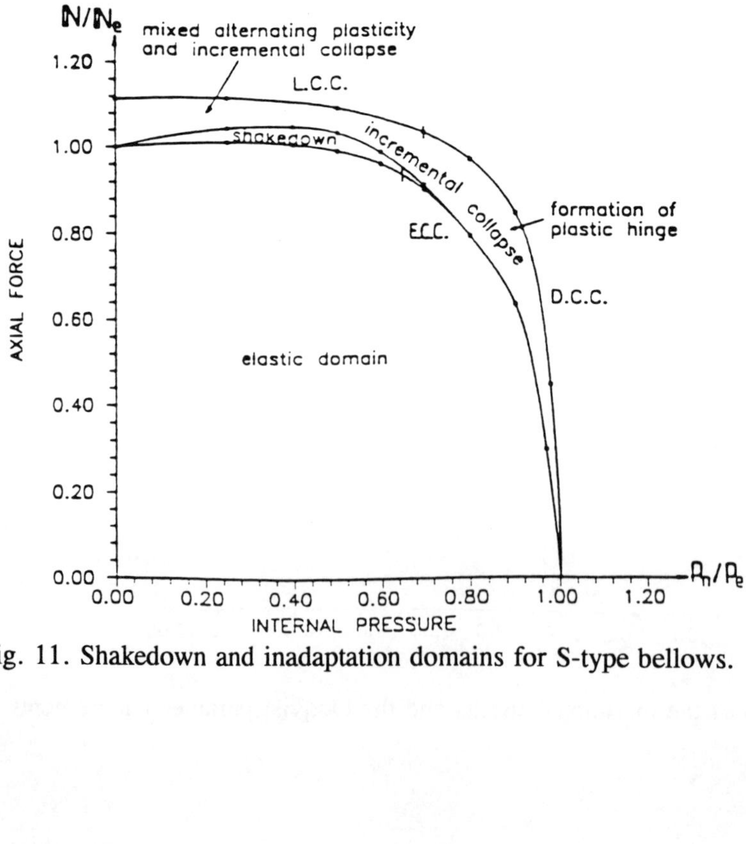

Fig. 11. Shakedown and inadaptation domains for S-type bellows.

Note that the adaptation and inadaptation domains are comparatively narrow with respect to the elastic domain as a result of the assumed two-point substitutive cross section of the shell wall. Development of the plastic strain fields in the direction normal to the midsurface (through thickness) is neglected.

RATCHETING IN U-TYPE BELLOWS - EFFECT OF A STIFF RING AT ROOT

The effect of the plastic strain localization at the root and crest of the U-type bellows segment has been confirmed by YAMASHITA et al. [1989]. A substitutive rectangular beam model for ratcheting proposed by the authors is based on the assumption that the distributions of bending moment along the beam length due to distributed force and to lateral deflection are analogous to distributions of the meridional bending moments along the shell convolute. The ratcheting mechanism consists of formation of a plastic hinge at one end of the beam in the initial half-cycle and another plastic hinge at the other end of the beam in the second half-cycle.

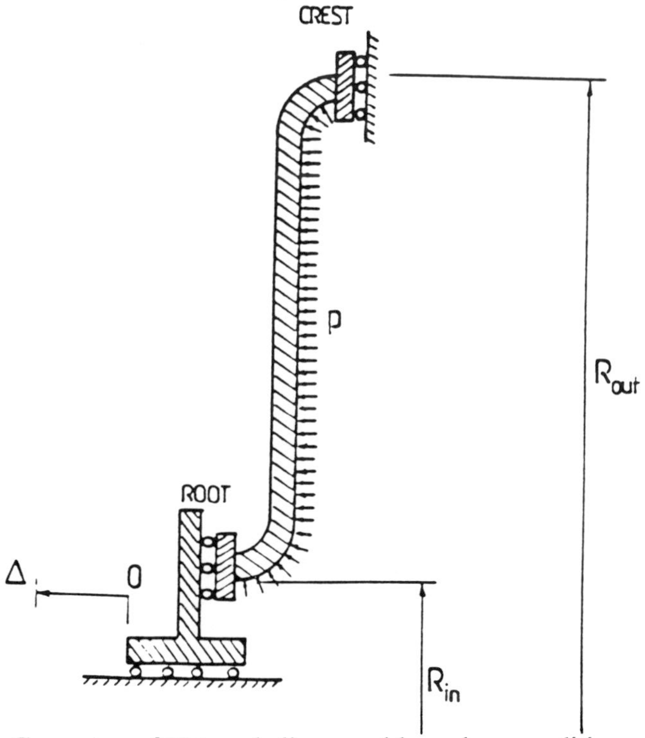

Fig. 12. Geometry of U-type bellows and boundary conditions.

As a result of accumulation of plastic strains at both ends of the beam, the total residual deflection increases with each cycle and ratcheting takes place. The ratcheting effect is strictly associated with the existence of the distributed force on the beam (sustained pressure in the case of bellows), and cannot be detected if the magnitude of the sustained load is equal to zero. The effect of anisosensitivity of U-type bellows may be neglected due to the much higher flexibility than for S-type bellows. Hence, application of the geometrically linear theory for small deflections (cf. YAMASHITA et al. [1989]) appeares to be justified. Generally, analysis of the progressive deformation based on the elastic-perfectly plastic material model and the geometrically linear theory leads to the so called infinite ratcheting. When the plastic strain increments accumulated within each cycle are constant, linear ratcheting effect takes place. Experiments do not confirm these results. On the other hand, deflections imposed upon a single segment are very often large and cannot be considered small. An accumulation of plastic strains may lead to significant magnitudes of plastic strain intensity, and in consequence, to necessity of formulating the problem in terms of finite strains. An example of such behaviour is given below.

A U-type bellows of thickness 3 mm subjected to internal pressure of 2 MPa and to axial cyclic deflection 0 to 1 mm at the root was analysed. Half of the bellows pitch is presented in Fig. 12. The total length of pitch is equal to 4 mm. The finite element model was based on ANSYS 5 code. The mesh was refined around the root and the crest of the shell, where localization of plastic strains is expected. Two general cases of boundary conditions were taken into account. In the first, the conditions of symmetry were assumed at the root and crest (Fig. 11). In the second, the

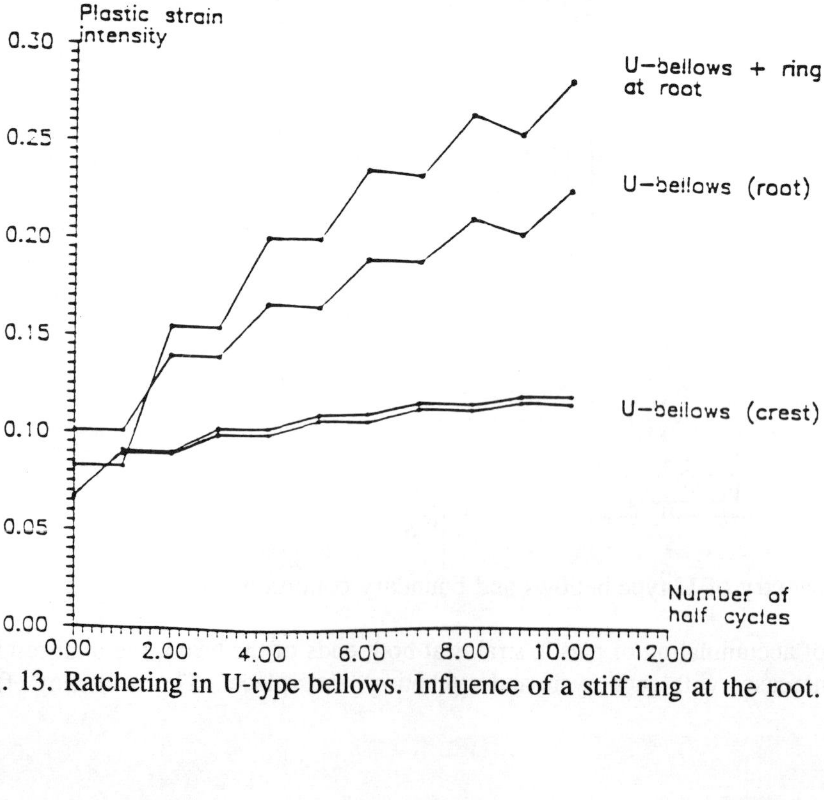

Fig. 13. Ratcheting in U-type bellows. Influence of a stiff ring at the root.

radial displacement of the shell at the root has been eliminated. The latter constraint is associated with the existence of a stiff ring at the root, installed by manufacturers in order to prevent the bellows from bulging. We aim at determining the influence of the ring on the progressive deformation of the bellows. For both the boundary conditions, five nonsymmetric cycles of axial deflection under constant level of pressure were performed. The analysis was based on the large deformation approach. An application of the stiff ring at the root forms much stiffer support condition than in the case without the ring. The results of calculations are presented in Fig. 13. The material data, obtained from the uniaxial tensile tests for thin (0.3 mm) sheets of stainless steel 316 L, were as follows:
E/H=116.5 (Young's modulus to hardening parameter ratio), E/SY=561.8 (Young's modulus to yield stress ratio). Initially (in the first half-cycle), plastic strain intensity at the root is lower for the shell with the ring than for the shell without the ring. Gradually, in the subsequent cycles, the accumulation of plastic strains at the root of the U-type bellows with the ring turns out to be larger than in the case of bellows without the constraint. The difference in accumulation of plastic strains at the crest of the bellows for both types of boundary conditions is negligible. The state of saturation after finite (transient) ratcheting for the shell with the ring may be reached at the higher level of plastic strain intensity and the number of cycles to failure may be considerably reduced. On the other hand, maximum deflection after 5 cycles for the bellows with the ring at the root is lower than in the case of the shell without the ring (Fig. 14).

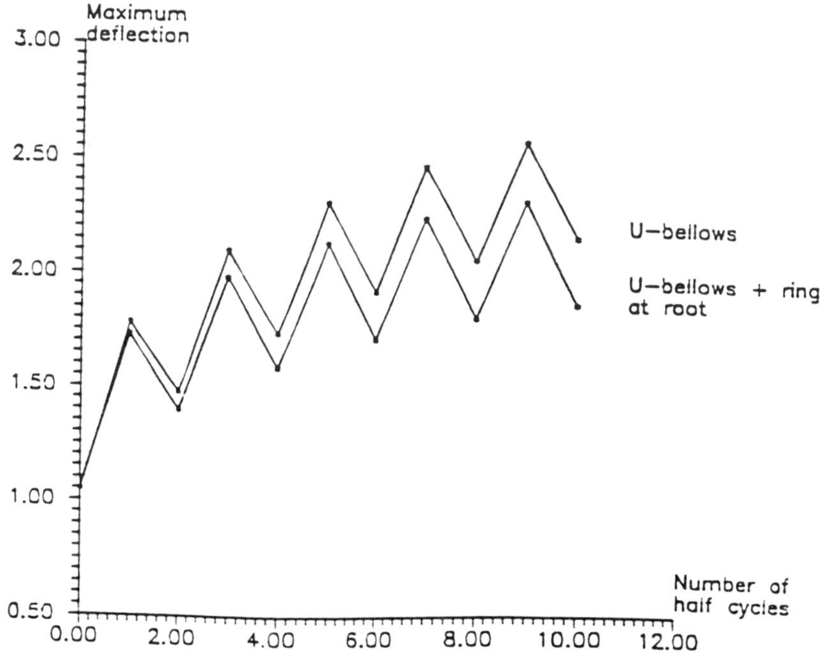

Fig. 14. Effect of a stiff ring at the root on the maximum deflection of U-type bellows.

CONCLUSIONS

There is a large variety of profiles used in the design of bellows expansion joints. Toroidal bellows (Ω,C,S-shaped) are usually much stiffer than U-type or nested bellows. As a result, toroidal bellows are strongly anisosensitive against axial loads. In order to describe properly the effect of ratcheting in anisosensitive shells, a geometrically nonlinear approach should be applied. This leads to detecting the effect of progressive deformation (ratcheting) even in the absence of sustained load. Note that elastic shakedown may occur after a certain number of cycles, and not necessarely in the first cycle as is usually assumed in small deflection theory.

The failure of elastic-perfectly plastic S-type bellows under force-controlled cycling is due either to the ultimate load mechanism or due to the formation of a discontinuity line. Deflection-controlled cycling leads here to the formation of a discontinuity line after a finite number of cycles.

Even in such a flexible structure as a U-type bellows, the accumulation of plastic strains within a few cycles may lead to large magnitudes of plastic strain intensity. In this case, the finite strain approach is fully justified, especially for the effect of transient ratcheting (which leads to reversed plasticity), because the failure criterion is based on the level of plastic strains reached after saturation.

REFERENCES

1967 BREE, J., "Elastic-plastic behaviour of thin tubes subjected to internal pressure and intermittent high-heat fluxes with application to fast-nuclear-reactor fuel elements", J. Strain Analysis, 2, 226.

1989 PONTER, A.R.S. and CARTER, K.F., "Upper bound methods for use in the design and assessment of axisymmetric thin shells subjected to cyclic thermal loading", Nuc. Eng. Design, 116, 239.

1992 JIANG, W. and LECKIE, F.A., "A direct method for the shakedown analysis of structures under sustained and cyclic loads", J. Appl. Mech., 59, 251.

1974 HAMADA, M. and TANAKA, M., "A consideration on the low-cycle fatigue-life of bellows (the fatigue life under completely-reversed-deflection cycles)", Bull. JSME, 17, 41.

1978 POROWSKI, J.S. and O'DONNELL, W.J., "Upper bounds for creep ratchetting in convoluted bellows", ASME PVP-PB-029, 15.

1989 YAMASHITA, T., TSUKIMORI, K., NAKAMURA, M., IWATA, K. and IMAZU, A., "Ratchetting mechanism of bellows and evaluation method", O-arai Engineering Center, Power Reactor and Nuclear Fuel Development Corporation, Ibaraki-ken, Japan.

1992a SKOCZEŃ, B., SKRZYPEK, J. and BIELSKI, J., "Shakedown and inadaptation mechanisms of bellows subject to constant pressure and cyclic axial force", Mech. Struct. & Mach., 20, 119.

1992b SKOCZEŃ, B. and SKRZYPEK, J., "Application of the equivalent column concept to the stability of axially compressed bellows", Int. J. Mech. Sci., 34, 901.

1987 KOENIG, J.A., Shakedown of elastic-plastic structures, Elsevier, Amsterdam/ Oxford/ New York/ Tokyo.

1983 SKRZYPEK, J. and ŻYCZKOWSKI, M., "Termination of processes of finite plastic deformations of incomplete toroidal shells", Sol. Mech. Arch., 8, 39.

1989 BIELSKI, J. and SKRZYPEK, J., "Failure modes of elastic-plastic curved tubes under external pressure with in-plane bending", Int. J. Mech. Sci., 31, 435.

1992 KOHNKE, P., "ANSYS user's manual for revision 5.0", Swanson Analysis Systems, Inc., P.O.Box 65, Johnson Road, Houston, PA 15342-0065.

1981 REISSNER, E., "On finite pure bending of curved tubes", Int. J. Solids Struct., 17, 839.

MIN-MAX APPROACH TO SHAKEDOWN AND LIMIT LOAD ANALYSIS FOR ELASTIC PERFECTLY PLASTIC AND KINEMATIC HARDENING MATERIALS

Janisław ZWOLIŃSKI
Institute of Aeronautics and Applied Mechanics
Warsaw University of Technology
ul. Nowowiejska 24, 00-665 Warsaw, Poland

ABSTRACT

The paper deals with the shakedown and limit load analysis of perfectly plastic and kinematic hardening materials. The numerical approach is based on the finite element method (FEM).
A solution procedure for elastic-plastic structures subjected to variable loads was developed using min-max formulation of the shakedown problem.
An IBM PC program MINIRES was prepared and applied to many practical problems encountered in power plants, oil refineries and other installations.
Two practical examples: an axial pipe compensator (Fig.2) and a pipe elbow (Fig.3) presented in the paper illustrate the effect of kinematic hardening on the shakedown behaviour.

INTRODUCTION

The conclusions of the Melan theorem on shakedown of structures under time-variable loads allow us to present a method of calculation of limit loads, to specify a maximal range of load ensuring the occurrence of shakedown, and to determine plastic strain which occurs under these circumstances. The method concentrates on minimization of the maximal reduced stress by optimal selection of the residual stress so that:

$$F = F\left[\sigma_{ij}^{e}, \sigma_{ij}^{res}\left(\varepsilon_{ij}^{p}\right)\right] = F(\boldsymbol{x},t) = \max_{(\boldsymbol{x},t)}\left[\sigma_{red}\left[\sigma_{ij}(\boldsymbol{x},t) = \sigma_{ij}^{e}(\boldsymbol{x},t) + \sigma_{ij}^{res}(\boldsymbol{x})\right]\right] \Rightarrow \min_{\left(\varepsilon_{ij}^{p}\right)} \quad (1)$$

Z. Mróz et al. (eds.), *Inelastic Behaviour of Structures under Variable Loads*, 363–380.

where: σ_{ij}^{e} - stresses resulting from an elastic solution of the problem,

σ_{ij}^{res}- residual stresses, time-independent, satisfying the condition of equilibrium and the free boundary conditions,

σ_{red}- reduced stress (also called the effective stress, or equivalent stress) according to Huber-Mises hypothesis,

ε_{ij}^{p} - plastic strain,

t - time,

$\boldsymbol{x}$ - Cartesian coordinates,

In the case of a variable yield stress $\sigma_Y(\boldsymbol{x},t)$, the maximal value of the function

$$F(\boldsymbol{x},t) = \sigma_{red}(\boldsymbol{x},t) / \sigma_Y(\boldsymbol{x},t) \tag{2}$$

should be minimized.

The Melan theorem states that the shakedown of a structure, subjected to time-variable load within specified limits, will occur when a field of statically admissible, time-independent, residual stress is found such that the sum of this stress and of the elastic stress resulting from the structure response to external loads will not violate the plasticity condition. Thus, it is possible to optimize the field of residual stresses in a structure in order to maximize the admissible load. For calculation, it is much better to fix the range of load, and to minimize the maximal reduced stress by optimal selection of the residual stress field. Physically it means finding a minimal value for yield stress of the material of the analyzed structure. This yield stress will also ensure the shakedown for the range of load in question. An extensive discussion of the min-max approach was presented by PYCKO & MRÓZ [1992].

Upon completing the calculations it suffices to normalize the calculated values with respect to the actual yield stress. Thus the range of admissible load for a particular structure is obtained.

Two cases can be distinguished in the method. In the first case the residual stresses $\sigma_{ij}^{res}(\boldsymbol{x})$ are sought directly among statically admissible stresses. This method of direct determination of the residual stress field allows us to calculate the values of shakedown load or limit load, but gives no information about plastic strains. In the second case, the residual stresses are specified in terms of plastic strains by the relation:

$$\sigma_{ij}^{res}(\boldsymbol{x}) = \int_{\Omega} G_{ij}^{kl}(\boldsymbol{x},\xi)\varepsilon_{kl}^{p}(\xi)d\Omega, \tag{3}$$

where $G_{ij}^{kl}(\boldsymbol{x},\xi)$ is the Green's function.

In this paper, we shall concentrate solely on the second case, of residual stresses generated by plastic strains.

BASIC RELATIONS

We shall consider two models of the material, namely
1) elastic-perfectly plastic (Fig.2b),
2) elastic-plastic kinematically hardening (Fig.2c).

The finite element method (FEM) is applied, using constant strain elements. The total strain is decomposed into elastic, plastic and residual strain components, so that

$$\{\varepsilon\} = \{\varepsilon^e\} + \{\varepsilon^p\} + \{\varepsilon^{res}\}, \tag{4}$$

where $\{\varepsilon^{res}\}$ is related by Hooke's law to the residual stress $\{\sigma^{res}\}$, while the vector of resultant stresses equals the sum of vectors of elastic and residual stresses:

$$\{\sigma\} = \{\sigma^e\} + \{\sigma^{res}\}. \tag{5}$$

The values of the components of vectors of displacements, elastic strains and stresses $\{q^e\}$, $\{\varepsilon^e\}$, $\{\sigma^e\}$ will be obtained from the solution of the elastic problem.

Let us briefly restate the fundamental finite element relations. We have, for the linearly elastic element

$$\{\varepsilon\}_m = [B]_m \{q\}_m \tag{6}$$

$$\{\sigma\}_m = [C]_m \{\varepsilon\}_m \tag{7}$$

$$[k]_m = [B]_m^T [C]_m [B]_m \tag{8}$$

where $[B]_m$ is the strain-displacement matrix, $[C]_m$ is the elasticity matrix, $\{q\}_m$ is the nodal displacement vector, and $[k]_m$ is the stiffness matrix. The subscript m denotes the element number.. The global stiffness matrix $[k]$ is specified as follows

$$[k] = \sum_{m=1}^{M} [\pi_1]_m^T [k]_m [\pi_1]_m \tag{9}$$

where the matrix $[\pi_1]_m$ provides the relation between the global $\{q\}$ and the element nodal displacement vectors, thus

$$\{q\}_m = [\pi_1]_m \{q\} \tag{10}$$

The boundary -value problem for an elastic body is thus reduced to the set of equations

$$[k]\{q\} = \{Q\} \tag{11}$$

where $\{Q\}$ is the generalized nodal force.

The relation between residual stresses and conjugate strains is identical to that for an elastic problem, namely

$$\left\{\sigma^{res}\right\}_m = [C]_m \left\{\varepsilon^{res}\right\}_m \tag{12}$$

On the other hand, the relation between the residual stress $\left\{\sigma^{res}\right\}$ and the plastic strain $\left\{\varepsilon^{p}\right\}$ can be presented in the form

$$\left\{\sigma^{res}\right\} = [A]\left\{\varepsilon^{p}\right\}. \tag{13}$$

where the residual stress matrix $[A]$ corresponds to the Green function $G_{ij}^{kl}(\boldsymbol{x},\xi)$ for the continuum formulation. This matrix can be generated from the stationary condition of the elastic potential energy and is obtained in the form

$$[A] = \sum_{m=1}^{M} [\pi_2]_m^T [C]_m \left([B]_m [\pi_1]_m [R] - [\pi_2]_m\right), \tag{14}$$

where

$$[R] = [k]^{-1}\left(\sum_{m=1}^{M}\int_{\Omega}[B]_i^T [C]_i d\Omega\right). \tag{15}$$

and $[\pi_2]_m$ is the matrix providing relations between global and element strains and stresses, so that

$$\left\{\varepsilon^{p}\right\}_m = [\pi_2]_m \left\{\varepsilon^{p}\right\} \qquad \left\{\sigma^{p}\right\}_m = [\pi_2]_m \left\{\sigma^{p}\right\} \tag{16}$$

The matrix $[A]$ was derived by MAIER [1975] for beam structures, and later the derivation was generalised by KÖNIG [1987], and BIELAWSKI & ZWOLIŃSKI [1986].

Let us note that the matrix $[A]$ provides the relation between residual stress and initial strain, such as plastic, thermal or phase tranformation strain. So, we may generally write

$$\left\{\sigma^{res}\right\} = [A]\left\{\varepsilon^{in}\right\} \tag{17}$$

where $\left\{\varepsilon^{in}\right\}$ is the initial strain.

Let us consider the constant strain element for which the yield function expressed in terms of the reduced stress is introduced

$$f_m = \left(\sigma_{red}^2\right)_m, \quad m = 1,2,\ldots,M. \tag{18}$$

Obviously, the value of f_m depends on the elastic and residual stress state, thus

$$f_m = \sigma_{red}^2\left(\sigma_{ij}\right)_m = \sigma_{red}^2\left[\left(\sigma_{ij}^{e}\right)_m + \left(\sigma_{ij}^{res}\right)_m\right] \tag{19}$$

Here f_m is associated with the Huber-Mises yield condition. The minimization of function

f_m is performed in the iterative way described by BIELAWSKI & ZWOLIÑSKI [1986].

Assume that the elastic solution was specified and the elements of maximal values of f_m are known. The elements of maximal reduced stress are assumed to flow plastically with the plastic strain following the gradient vector normal to the yield surface. Consider the step $t-1$ of the calculation process for which the plastic strain distribution is known. The residual and total stresses then are

$$\left\{\sigma^{res}\right\}_{t-1} = [A]\left\{\varepsilon^{p}\right\}_{t-1} \tag{20}$$

$$\{\sigma\}_{t-1} = \left\{\sigma^{e}\right\} + \left\{\sigma^{res}\right\}_{t-1} \tag{21}$$

and the value of $(f_m)_{t-1}$ is specified for all elements.

Consider now the minimization process of the maximal function $f^{\max}_{K(t-1)}$ reached in at least one of the elements. Assume the plastic strain increment to be specified as follows

$$\left\{\Delta\varepsilon^{p}\right\}_{t} = x_t\left\{\bar{\varepsilon}^{p}\right\}, \tag{22}$$

where $\left\{\bar{\varepsilon}^{p}\right\}$ is assumed to be known, for instance, following the gradient $\partial f_m/\partial\sigma_{ij}$ and x_t is to be determined.

The plastic strain vector and total stress components are now determined by the relations

$$\left\{\varepsilon^{p}\right\}_{t} = \left\{\varepsilon^{p}\right\}_{t-1} + \left\{\Delta\varepsilon^{p}\right\}_{t}, \tag{23}$$

$$\{\sigma\}_{t} = \{\sigma\}_{t-1} + \{\Delta\sigma\}_{t}. \tag{24}$$

and we can write

$$\{\Delta\sigma\}_{t} = \left\{\Delta\sigma^{res}\right\}_{t} = x_t\left\{\bar{\sigma}^{res}\right\}_{t}. \tag{25}$$

where $\left\{\Delta\sigma^{res}\right\}$ is the residual stress increment associated with the specified plastic strain $\left\{\bar{\varepsilon}^{p}\right\}$, so that

$$\left\{\bar{\sigma}^{res}\right\}_{t} = [A]\left\{\bar{\varepsilon}^{p}\right\}_{t}. \tag{26}$$

Substituting these relations into the yield function, we obtain the quadratic equation with respect to x_t, namely

$$f_m(x_t) = a_m x_t^2 + b_m x_t + c_m, \tag{27}$$

where:

$$a_m = \left[(\overline{\sigma}_{ii})_t (\overline{\sigma}_{ii})_t - (\overline{\sigma}_{ii})_t (\overline{\sigma}_{jj})_t + 3(\overline{\sigma}_{ij})_t (\overline{\sigma}_{ij})_t \right]_m,$$

$$c_m = \left[(\sigma_{ii})_{t-1} (\sigma_{ii})_{t-1} - (\sigma_{ii})_{t-1} (\sigma_{jj})_{t-1} + 3(\sigma_{ij})_{t-1} (\sigma_{ij})_{t-1} \right]_m, \qquad (28)$$

$$b_m = 2\left[(\overline{\sigma}_{ii})_t (\sigma_{ii})_{t-1} - (\overline{\sigma}_{ii})_t (\sigma_{jj})_{t-1} + 3(\overline{\sigma}_{ij})_t (\sigma_{ij})_{t-1} \right]_m .$$

and the function $f(x_t)$ is represented in the plane f_m, x_t as a parabola.

Let us now identify the maximal parabola at $x_t = 0$ so that the following inequality holds

$$f_K(x_t = 0) > f_m(x_t = 0) \qquad m = 1,2,\ldots,K-1,K+1,\ldots,M, \qquad (29)$$

Consider now the typical situation presented in Fig.i.

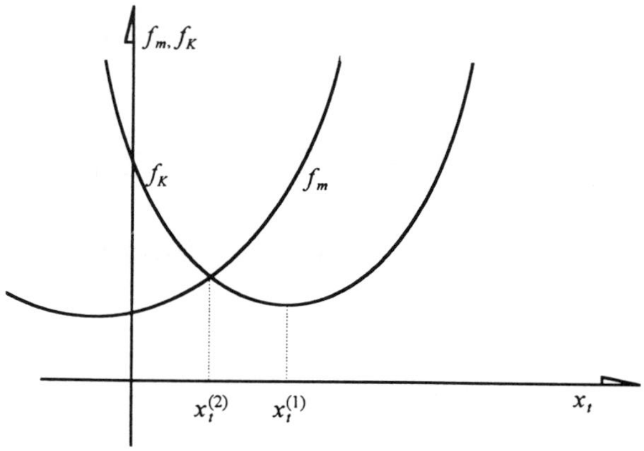

Fig.1.

Moving down along the parabola f_K, one may reach a minimum at $x_t^{(1)}$ or at a point of intersection $x_t^{(2)}$ with another parabola f_m. The first value $x_t^{(1)}$ is specified by the relation

$$\frac{\partial f_K(x_t)}{\partial x_t} = 0 \qquad \text{or} \qquad 2a_K x_t^{(1)} + b_K = 0 \qquad (30)$$

The intersection value $x_t^{(2)}$ is obtained from the equation

$$f_K\left(x_t^{(2)}\right) = f_m\left(x_t^{(2)}\right) \qquad m = 1,2,\ldots,K-1,K+1,\ldots,M \qquad (31)$$

Obviously, the smaller value of x_t is assumed in order to scale the plastic strain increment, specified by (22). When x_t is determined, the stresses are calculated from the relation

$$\{\sigma\}_t = \{\sigma\}_{t-1} + x_t\{\overline{\sigma}\}_t \tag{32}$$

and the next iteration can be initiated.
Consider now the case where at the iteration step t there are N elements reaching the same value of f_m. The stress state in these elements is then represented by one yield function. The plastic strain increments are then assumed to be generated by the rule

$$\left\{\Delta\varepsilon^p\right\}_t^{(n)} = x_t^{(n)}\left\{\overline{\varepsilon}^{p(n)}\right\} \qquad n = 1,\dots,N \tag{33}$$

where $\left\{\overline{\varepsilon}^{p(n)}\right\}$ is assumed to follow the normal direction to the yield surface f_m, thus

$$\overline{\varepsilon}^{p(m)} = \partial f_m / \partial\sigma^{(m)} \tag{34}$$

and the multipliers are to be determined from the condition of equal descent of f_m for all elements $n = 1,\dots,N$. This condition requires the elements to remain on the same yield surface in the next iterative step. Let us write this condition in the form

$$\left[\frac{\partial f_m}{\partial\varepsilon_j^p}\right]^T \left\{\Delta\varepsilon_j^p\right\} = \left[\frac{\partial f_m}{\partial\sigma^{(n)}}\right]^T [A]^{(n)} \left[\frac{\partial f_m}{\partial\sigma^{(n)}}\right]\left[\frac{\partial f_m}{\partial\sigma^{(n)}}\right] x_t^{(n)} = const. \quad n = 1,2,\dots,N \tag{35}$$

which provides $N-1$ equations to specify the multipliers $x_t^{(m)}$. The remaining N-th equation is then obtained from the condition of intersection with the consecutive parabola, see Eqs. (30) and (31).
The method presented is useful for elastic-perfectly plastic and kinematic hardening materials. In the case of kinematic hardening there is no ratcheting response and the asymptotic state is either elastic or alternating plasticity. When the structure is subjected to an alternating load which does not violate the yield condition, and a steady load, then elastic shake down always occurs after a permanent plastic strain is accumulated in the structure. The present procedure can be applied to this case. The minimized value of the reduced stress corresponds to the minimal yield stress σ_{mY} in a hardened state, necessary for the shakedown of a structure.

MAIN RESULTS

Following the algorithm outlined in the previous section, an IBM PC program called MINIRES was developed by Bielawski and Zwoliñski, applicable to plane stress or strain and to axisymmetric problems. This program was verified and applied to many technical problems associated with limit load, shake down and cyclic plasticity. Let us mention some major applications of this program:

i) axial bellow compensators,

ii) pipe elbows of circular and elliptic cross-sections, some possessing longitudinal U-notches,
iii) high pressure, multilayer vessels (pressures up to 2000 MPa) with initial prestress and of different shapes,
iv) a pipe flange connection,
v) a steam turbine compressor rotor under axial, centrifugal and thermal loading,
vi) nozzles of spherical and ellipsoidal pressure vessels heads,
vii) cylindrical rolling barrels, the study of shakedown of a contact hardened layer,
viii) limit load and shakedown of pin joints,

In most cases the structure element was subjected to imposed loading or pressure, kinematical constraints or loading, also with the yield limit $\sigma_Y = \sigma_Y(T)$. Some of the results obtained from the 2D analysis were then verified by incremental 3D solutions for cyclic load programs.

Now let us discuss several examples which are representative for the problems discussed above.

1. Axial pipe compensator (Fig.2a)

Consider the bellow compensator, installed in a pipe system in order to compensate for axial thermal dilatation. The compensator shape and dimensions are shown in Fig.2a. The loading is composed of two independent components, namely

i) temperature-induced axial elongation Δl (specified by the non-dimensional parameter $\overline{\kappa} = \Delta l/l = \Delta T/T$). This elongation is regarded as a uniform displacement boundary condition. It is assumed that $\overline{\kappa}$ may vary within limits $-\overline{\kappa}_a < \overline{\kappa} < +\overline{\kappa}_a$ so it is regarded as alternating loading. It is convenient to normalize $\overline{\kappa}_a$ with respect to the maximal elastic amplitude $\overline{\kappa}_{ea}$ corresponding to onset of yielding, thus $\overline{\overline{\kappa}} = \overline{\kappa}_a/\overline{\kappa}_{ea}$, and $\sigma_{red}(\kappa_{ea}) = \sigma_{oY}$,

ii) internal pressure p specified by the non-dimensional pressure parameter $\overline{p} = p/p_L$, where p_L denotes the limit pressure determined by limit analysis. It is assumed that $\overline{p}$ does not vary in time , i.e. $\overline{p} = const$.

The following geometric and material parameters were assumed in the analysis: $R_o / T / H / h = 400 / 100 / 100 / 10$ and $E/\sigma_{oY} = 200000/200$, $\nu = 0.3$.

The limit pressure specified by the numerical analysis for the assumed yield limit $\sigma_{oY} = 200\,MPa$ was found to be equal to $p_L = 9.75\,MPa$. The corresponding maximal pressure at the elasticity limit was $p_e = 4.88\,MPa$, i.e. $\overline{p}_e \cong 0.5$. Similarly, the maximal value of an elastic axial elongation amplitude was $\overline{\kappa}_{ea} = 0.011$, that is about 1% of the compensator length. This is a rather small value, because the compensator is thick-walled. Figs. 2b, c, present the perfectly plastic and kinematic hardening stress-strain diagrams. Note that kinematic hardening is associated with a nonlinear monotonically increasing curve. Fig. 2d presents the interaction (Bree-type) diagram in the plane $\overline{p}, \overline{\overline{\kappa}}_a$. The domains of elasticity, shakedown, alternating plasticity, and ratcheting are shown in the diagram. Fig. 1e presents the respective interaction diagram for the kinematic-hardening model. It is seen that the ratcheting domain disappears due to of back stress development. Further, the minimal value of the yield back stress required to eliminate ratcheting is shown in Fig. 2e by plotting the total plastic stress $\overline{\sigma}_{mY}$ (specified in Fig. 1e), being the sum of initial yield stress σ_{oY} and the back stress α in the uniaxial tension. The curve $\overline{\sigma}_{mY}(\overline{p})$ in Fig. 1e was obtained for the axial amplitude $\overline{\overline{\kappa}}_a = 1$. Fig. 2f presents the distribution of the reduced or effective plastic strain ε^{p}_{red} at the shake down state for $\overline{p} = 0.7, \overline{\overline{\kappa}}_a = 1$. Finally, Fig. 1g provides the distribution of the reduced (or effective) stress $\overline{\sigma}_{red} = \sigma_{red}/\sigma_{oY}$ for $\overline{p} = 0.7$ and $\overline{\overline{\kappa}} = -1$, that is for the compressed bellow.

2. Pipe elbow

Let us consider the pipe elbow, Fig. 3a, with the specified dimension ratios $R_k / R_Y / h = 300/26.75/3.5$ and the material parameters $E/\sigma_{oY} = 200000/200$, $\nu = 0.3$. The elbow is subjected to two independent proportional loads:

i) kinematically imposed bending in the elbow plane. The curvature variation is specified by the non-dimensional curvature parameter

$$\overline{\kappa} = \left(\frac{1}{R'_k} - \frac{1}{R_k} \right) / \frac{1}{R_k} = \frac{\Delta\varphi}{\varphi_o}$$

where R_k and R'_k are the initial and instantaneous curvature radii of the elbow.

ii) internal pressure p, specified by the non-dimensional parameter $\overline{p} = p/p_L$, where p_L denotes the limit elbow pressure. It is assumed that $\overline{p} = const$ and the curvature variation occurs within the range $-\overline{\kappa}_a < \overline{\kappa} < +\overline{\kappa}_a$. It is convenient to introduce the normalized curvature $\overline{\overline{\kappa}}_a = \overline{\kappa}_a / \overline{\kappa}_{ea}$, where $\overline{\kappa}_{ea}$ is the maximal elastic curvature at the onset of yielding $\sigma^{max}_{red}(\overline{\kappa} = \pm\overline{\kappa}_{ea}, \overline{p} = 0) = \sigma_{oY}$. The limit elbow pressure for $\sigma_{oY} = 200\,MPa$ is $p_L = 29.1\,MPa$. The corresponding limit pressure of a straight pipe $(R_k \to \infty)$ is $p^p_L = 30.3\,MPa$. The elastic pressure value is $\overline{p}_e = 0.877$ and the maximal value of elastic curvature amplitude is $\overline{\kappa}_{ea} = 0.0336$. It should be noted that the elbow flexure was calculated by using the program for axisymmetric solution of a toroidal closed shell by inducing the initial strain $\left|\varepsilon^{in}_{\varphi}\right| = \overline{\kappa}_a = const$.

Fig. 3 presents the results of numerical analysis. In Fig. 3b the interaction (Bree-type) diagram is shown in the plane $\overline{p}, \overline{\overline{\kappa}}$ for the perfectly plastic model, and in Fig. 3c the respective diagram is presented for the kinematically hardening material. As in the previous case, there is no ratcheting domain in Fig. 3c, and the required value of back stress is shown by plotting the required value of $\overline{\sigma}_{mY}$ for $\overline{\overline{\kappa}} = 1$. Fig. 3d presents the plastic domain within the elbow cross-section for $\overline{p} = 0.25, \overline{\overline{\kappa}}_a = 1$ and Fig. 3e shows the values of reduced plastic strains at the state of shakedown.

3. High pressure flange pipe connection.

Fig. 4a presents the flange connection design (external diameter $D_z = 0.32m$, pipe diameter $D_{wew} = 0.046m$) and Figs. 4b-f illustrate the details of the solution.

The connection is assumed to be loaded by

i) internal pressure $0 < p < p_o = 275\,MPa$

ii) initial screw prestrain corresponding to the total prestress force $P = 1.7 p_o \pi D_{wew}^2 / 4$

iii) thermal strains associated with stationary temperature distribution $T(x)$ varying within the range $200^0 C < T < 300^o C$.

The 3D incremental analysis was carried out by assuming the pressure to vary cyclically between 0 and p_o with fixed screw prestress and temperature. Two loading cycles were numerically calculated, accounting also for unilateral contact problem between flange-gasket and flange-screws. In fact, the contact area after screw prestress is different from that occurring after imposed pressure. Figures 4e,f present the results obtained by applying the MINIRES system, assuming the axisymmetric stress state and "smearing" the screw prestress. For the yield stress $\sigma_{oY} = 1060\,MPa$ (maraging steel) the results obtained by 3D and 2D models are similar.

4. The steam turbine rotor, (Fig. 5.)

Let us finally consider a portion of a steam turbine rotor, composed of a blade disk and labyrinth sealing. The analysis was performed by considering the turbine start-up with two loading systems.

i) the transient temperature field $T(x,t)$ generated by steam action will account for heat transfer conditions; the reduction of yield stress due to temperature field was accounted for,

ii) centrifugal forces corresponding to disk rotation, $n = 3000 rpm$.

The shakedown problem was analyzed at a set of discrete instants $t_i, i = 1,2 \ldots 100$, for which residual stress fields and plastic strain distribution were determined.

Fig. 5b presents the temperature field at $t_i = 13\,\text{sec.}$ after start-up, Fig. 5c shows the accumulated plastic strains at the sealing notch root at the shakedown state. Fig.4d presents the normalized reduced stress distribution.

ACKNOWLEDGEMENT

The work was partially supported by the Scientific Research Council (KBN) under the Research Project 30978-91-01.

REFERENCES

1975 MAIER, G., "Mathematical programming method in analysis of elastic-plastic structures" (in Polish), Arch. Inż. Ląd., 31, 387.

1986 BIELAWSKI, G., "An optimization of residual stresses for determination of shakedown and limit load" (Msc Thesis, Warsaw University of Technology, supervisor ZWOLIŃSKI J.).

1987 ZWOLIŃSKI,J., BIELAWSKI, G., "An optimal selection of residual stress for shakedown and limit load analysis", (in Polish). Proc. Conf. Comp. Meth. Struct. Mech., pp.459-462, Jadwisin.

1987 KÖNIG, J.A., "Shake down of elastic-plastic structures", PWN, Warszawa, Elsevier, Amsterdam-Oxford-N.York-Tokyo.

1988 BIELAWSKI, G., OPALIŃSKI, W., ZWOLIŃSKI, J., "Shakedown of steam turbine rotor at start-up" (Stress report for Institute of Energy, Warsaw).

1988 BIELAWSKI, G., OPALIŃSKI, W., ZWOLIŃSKI, J., "Elastic - plastic analysis of pipe flange" (Stress report for ENERGOPROJEKT Warsaw).

1991 ZWOLIŃSKI, J., BIELAWSKI, G., "Optimizational approach to the elasto-plastic analysis", EUROMECH 1st. European Solid.Mech.Conf., München.

1992 PYCKO, S., MRÓZ, Z., "Alternative approach to shakedown as a solution of min-max problem", Acta Mech. 93, 205 - 222, Springer-Verlag.

1993 ZAGRAJEK, T., ZWOLIŃSKI, J., PAWLICKI, J., "Shakedown and limit load analysis of high pressure pipe flange using axisymmetric and 3-D FEM". (Stress report for Petrochemia Płock SA).

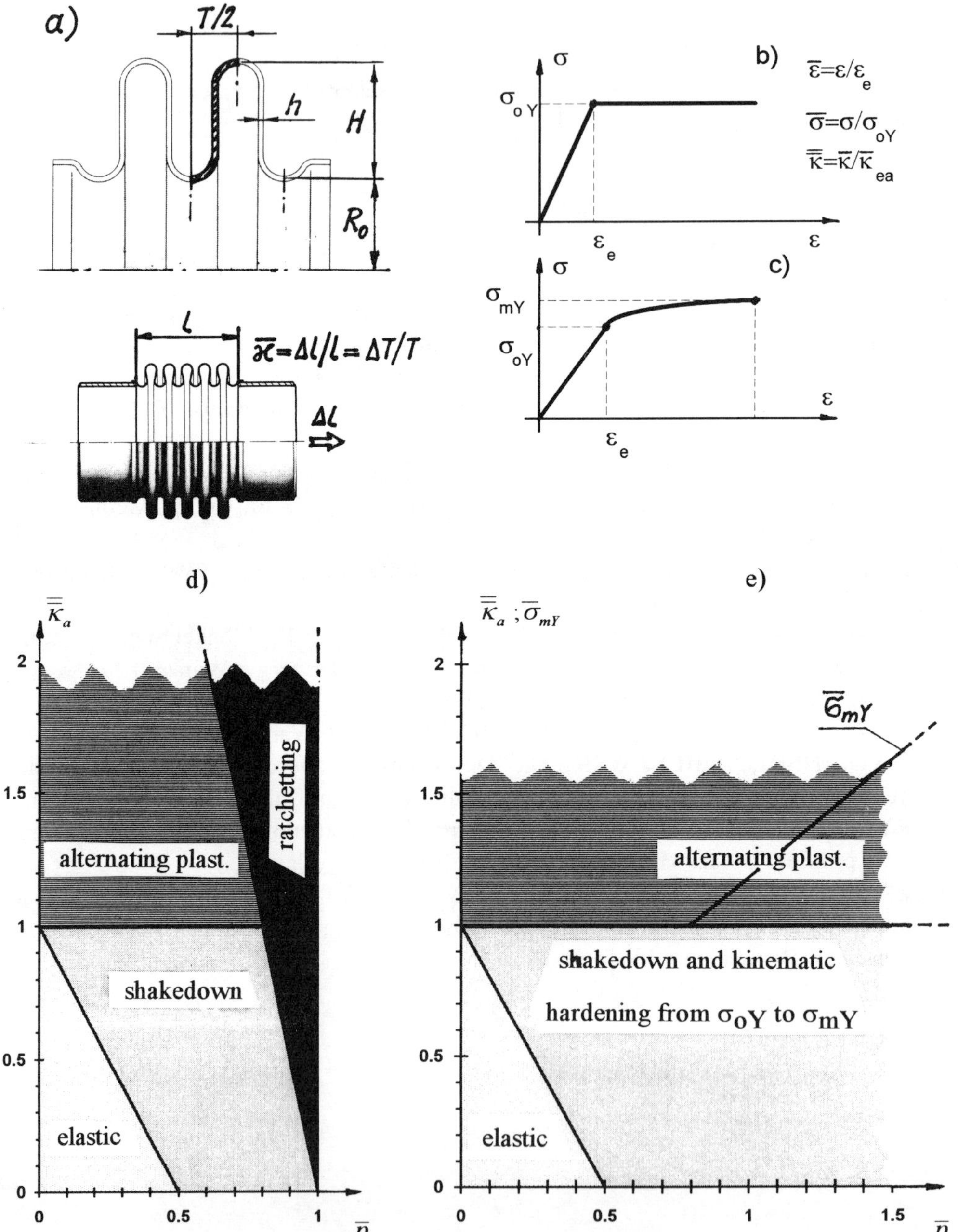

Fig. 2. The axial pipe compensator: 2D shakedown analysis.
a) Compensator geometry ($R_0 / T / H / h = 400 / 100 / 100 / 10$).
b) Elastic-perfectly plastic material.
c) Elastic-kinematic hardening material.
d) Results for elastic-perfectly plastic model (Bree-type diagram).
e) Results for elastic-kinematic hardening model

f)

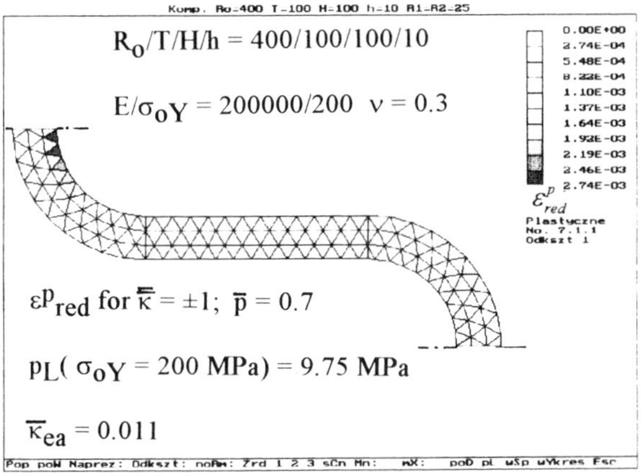

g)

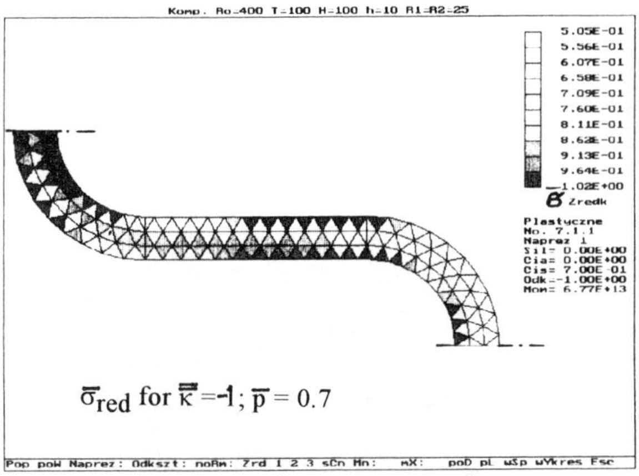

Fig. 2. The axial pipe compensator: 2D shakedown analysis (continued).
f) Cross section of the compensator bellow. Plastic strain for
$\overline{p} = 0.7; \overline{\overline{\kappa}}_a = 1.$
g) Reduced stress for $\overline{p} = 0.7; \overline{\overline{\kappa}} = -1.$

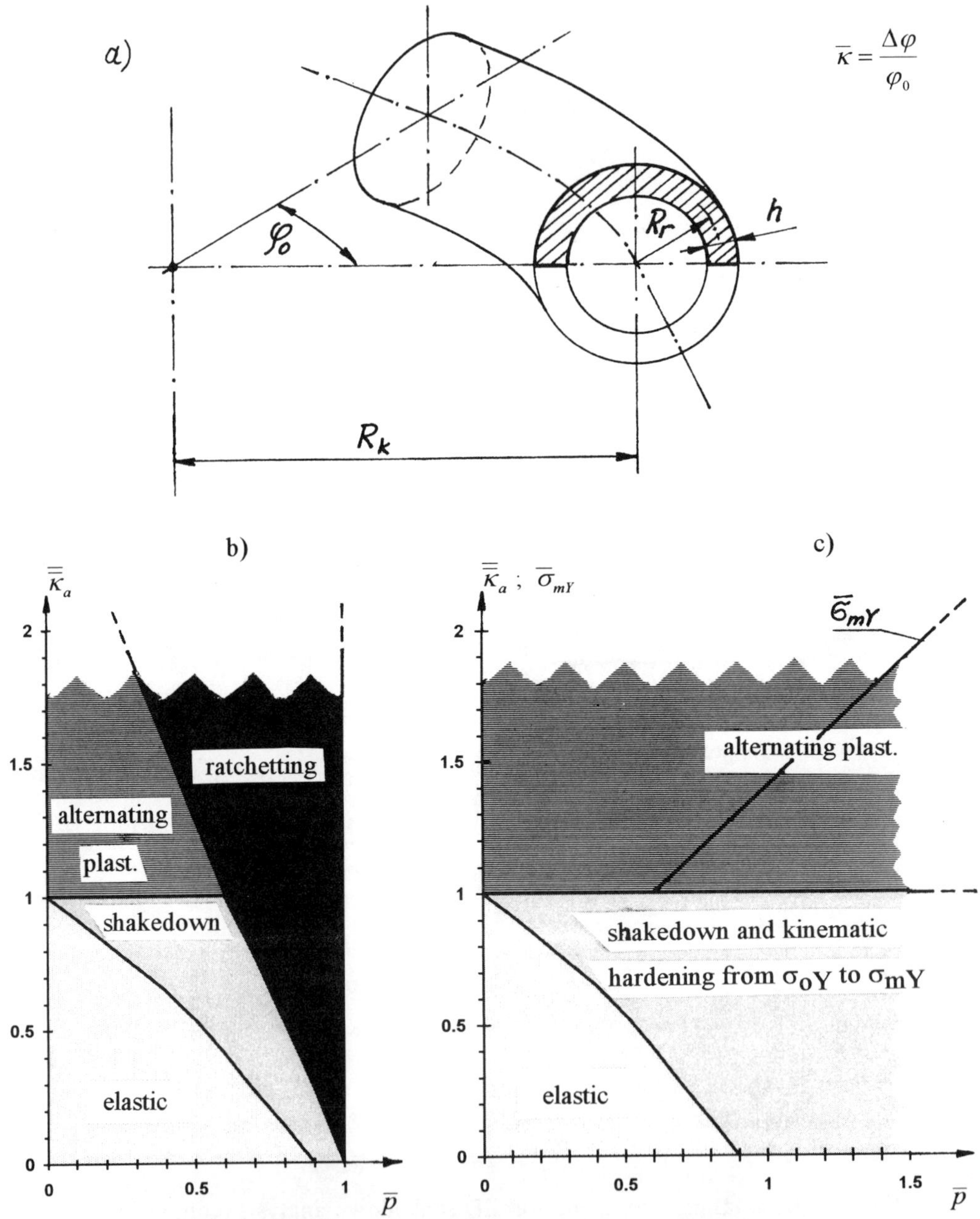

Fig. 3. The pipe elbow. Results of 2D shakedown analysis.
a) Pipe elbow geometry ($R_k / R_r / h = 300 / 26.75 / 3.5$).
b) Results for elastic-perfectly plastic model (Bree-type diagram).
c) Results for elastic-kinematic hardening model

d)

e)

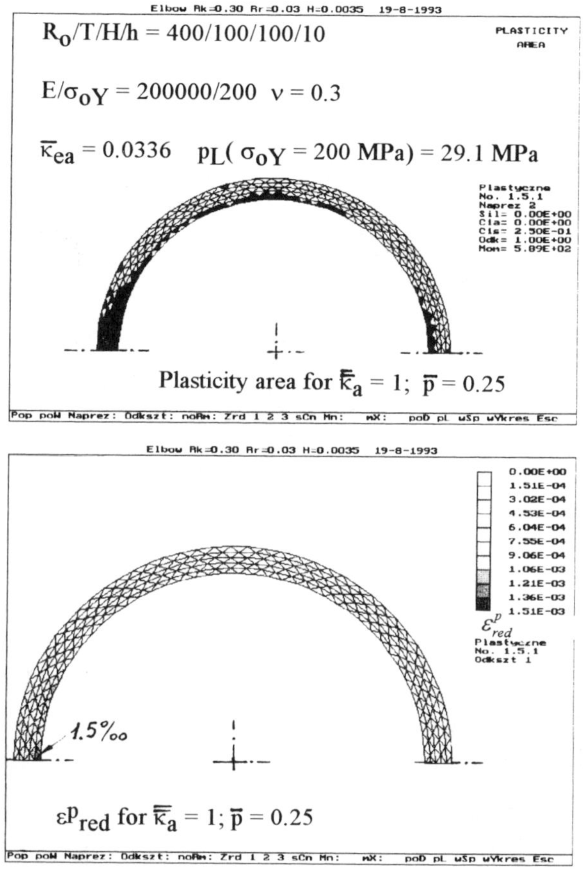

Fig. 3. The pipe elbow. Results of 2D shakedown analysis (continued).
d) Cross section of the pipe elbow in half view. Plasticity area for $\bar{p} = 0.7; \bar{\bar{\kappa}}_a = 1$.
e) Reduced plastic strain after shakedown.

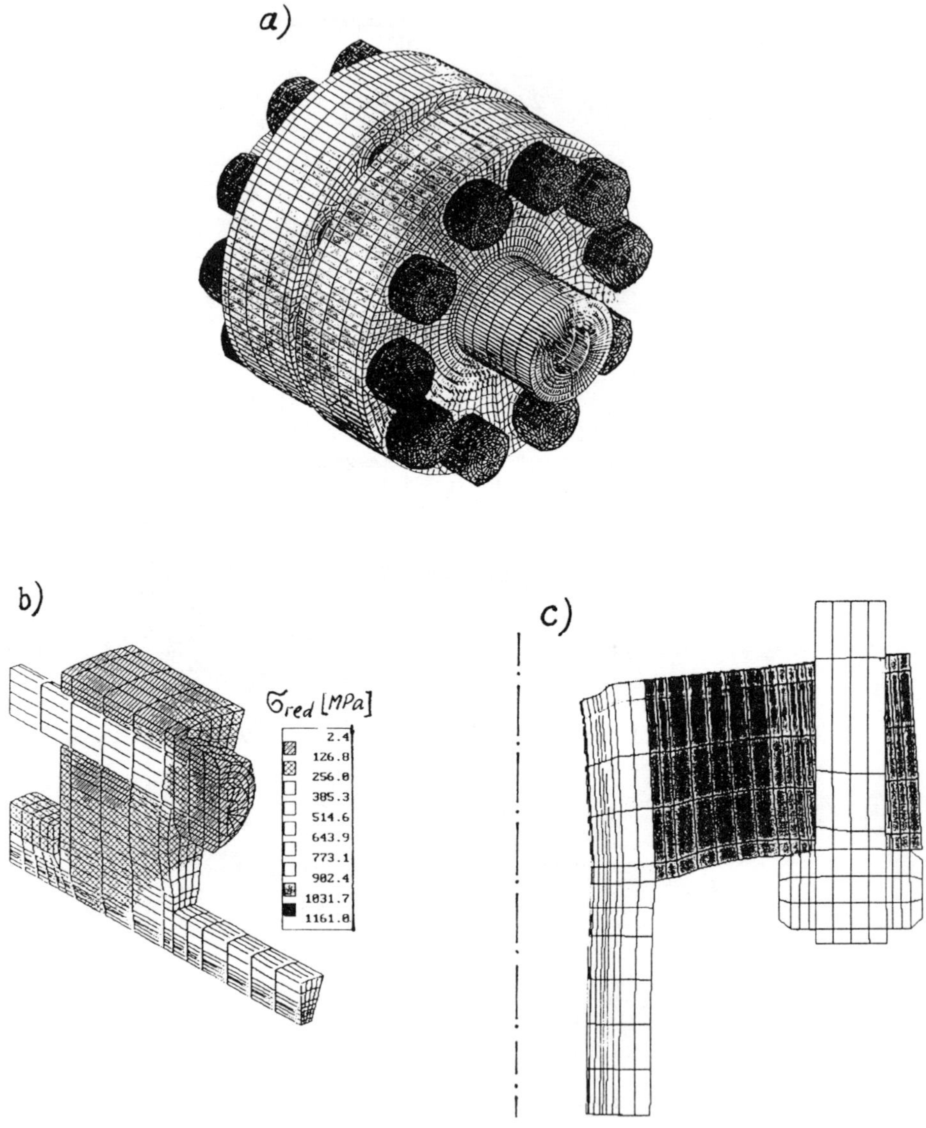

Fig. 4. High pressure (p =275 MPa) pipe-flange connection.
Elastic-plastic 3D stress analysis.
a) Pipe-flange connection: 3D view.
b) Reduced stress distribution.
c) Enlarged deformation.

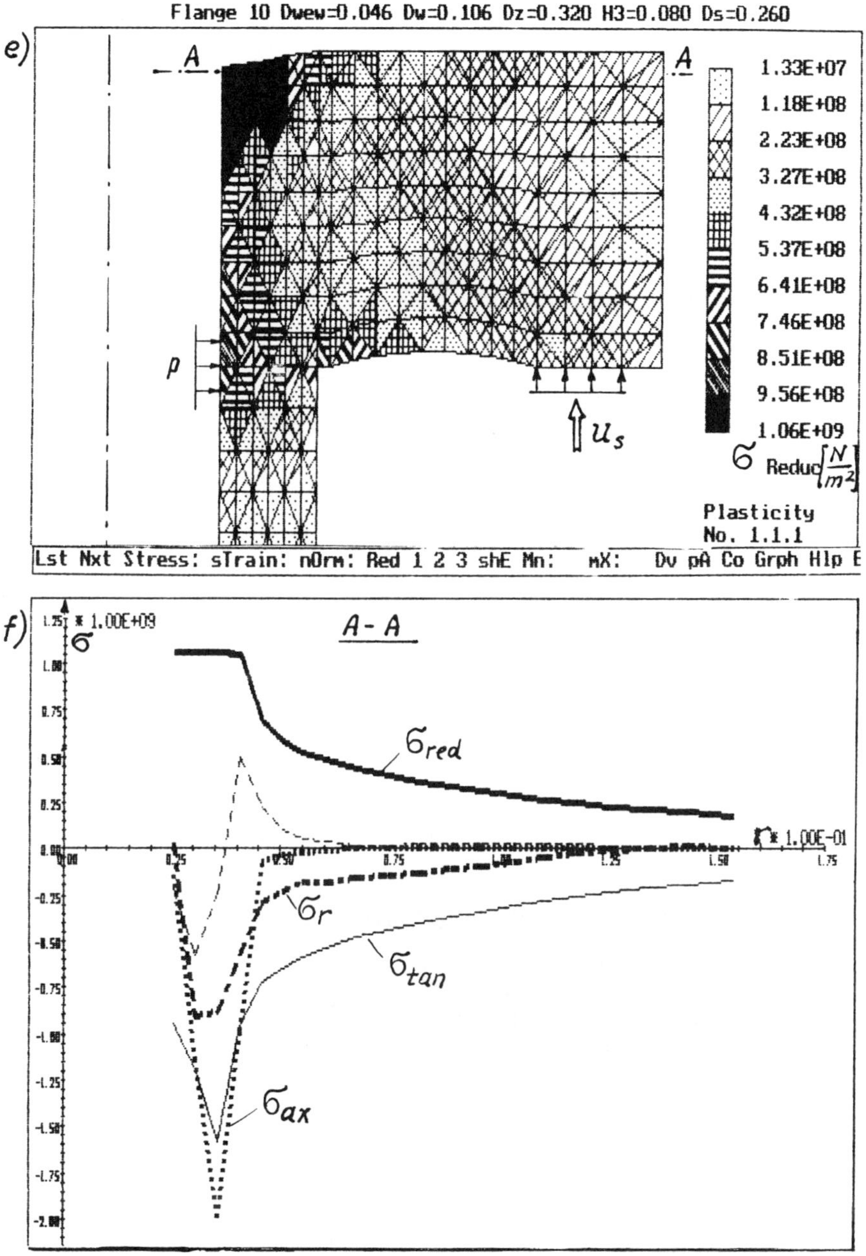

Fig. 4. High pressure (p =275 MPa) pipe-flange connection (continued).
Elastic-plastic 3D stress analysis.
d) Reduced stress distribution after shakedown in flange cross section.
e) Diagram of stress components in the cross section A-A.

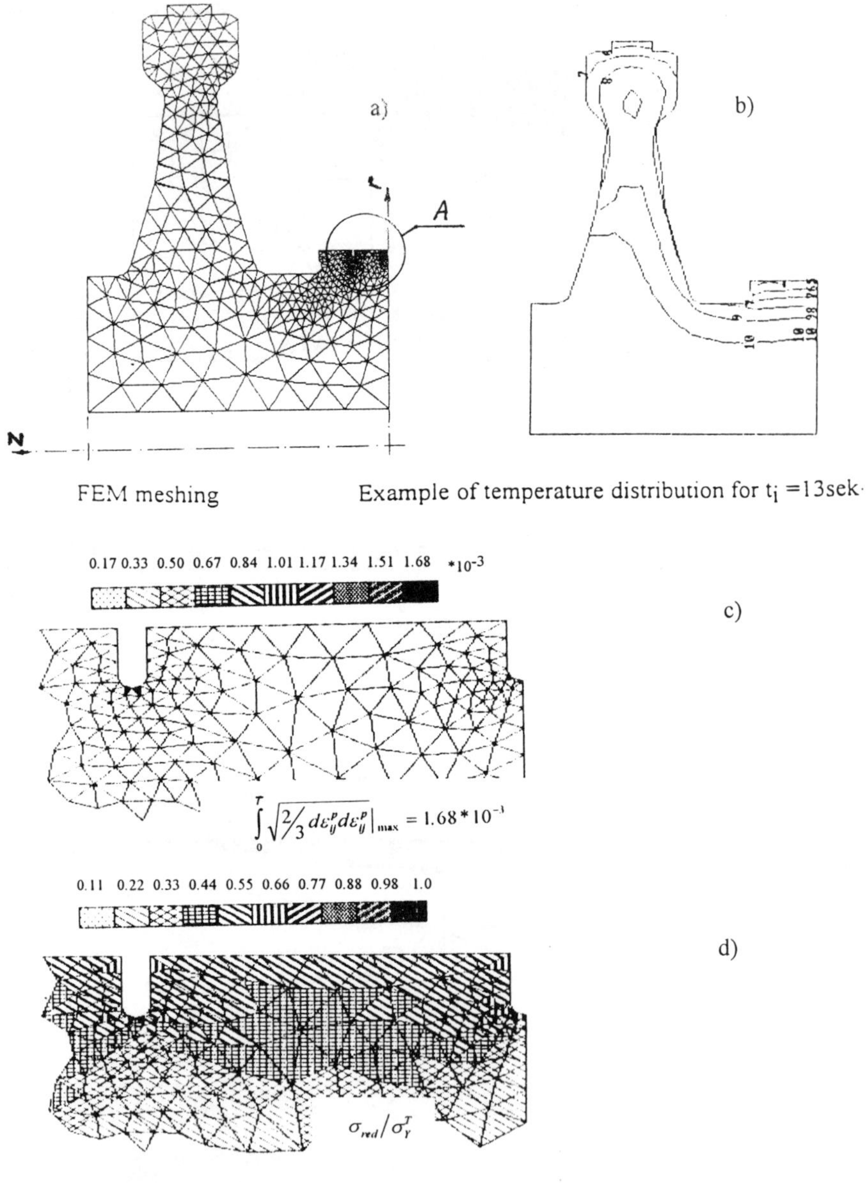

Fig. 5. The steam turbine rotor part.
Results of the 2D shakedown analysis
a) FEM meshing.
b) Example of temperature distribution for t_i = 13 sec.
c) Detail A, accumulated plastic strain after shakedown process.
d) Detail A, reduced stress distribution.

AN INDIRECT INCREMENTAL METHOD FOR SHAKEDOWN ANALYSIS BASED ON THE MIN-MAX APPROACH

Sławomir PYCKO
Institute of Fundamental Technological Research
ul. Świętokrzyska 21, 00-049 Warsaw, Poland

Abstract: A numerical algorithm based on the Newton-Raphson iterative scheme is proposed for a determination of limit analysis and shakedown load multipliers. A fictitious tangent matrix for variable loading is indirectly derived from the min-max formulation of the shakedown problem. It reduces the complex analysis of structures under variable loading to a certain equivalent fictitious incremental problem. A fast convergence of the procedure is illustrated on examples of space frames.

INTRODUCTION

Many engineering structures are exposed to variable operating conditions causing inelastic deformations. Analyses of these structures can be carried out using nonlinear finite step procedures or simplified solution methods. The former, though delivering complete information about the behaviour of the structure, are very costly and require specification of the load history, which is often unknown. The latter, though providing only a limited knowledge concerning safety loads or bounds on deformation, can be extensively used at the level of design, where incomplete information about the operating conditions is available.

Shakedown theory based on statical and kinematical theorems (cf KOITER [1960]) is a source of many simplified methods used for determination of the shakedown load multiplier for an elastic-plastic structure subjected to variable loading. The most common methods are based on a mathematical programming formulation (see e.g. ĈYRAS [1969], BELYTSCHKO [1972], MAIER [1973], HUNG & PALGEN 1979, MORELLE [1986]). More efficient methods of solution were recently developed that take into account limitation of the number of variables and constraints (see e.g. GENNA [1988], ORKISZ et al. [1990], STEIN et al. [1992]). The main difficulties

Z. Mróz et al. (eds.), *Inelastic Behaviour of Structures under Variable Loads*, 381–398.

affecting the effectiveness of these methods are related to the existence of a large number of inequality constraints. Other methods of analysis are based on specific iterative solution schemes (see e.g. BUCKTCHORPE & WHITE [1991], TOULIOS *et. al.* [1991]) and on subsequent modified elastic analyses (see e.g. ZARKA et al. [1988]). The latter require a linear kinematic hardening model for the material.

A min-max formulation of the shakedown problem which recently appeared in the literature (e.g. ZWOLIŃSKI & BIELAWSKI [1987], PYCKO & MRÓZ [1992], ZWOLIŃSKI [1994]), seems to be an alternative approach with respect to the mathematical programming. The main advantage of such an approach results from the lack of inequality constraints in the formulation that gives rise to the development of new efficient numerical procedures. The present contribution is a continuation of papers by PYCKO [1993a] and JANAS *et al.* [1995], in order to meet the requirements of the algorithm.

STRUCTURAL DESCRIPTION

Consider a discrete elastic-perfectly plastic structure. Denote by $\bar{\mathbf{Q}}$ and $\bar{\mathbf{q}}$ global vectors assembling all components of local stresses, $\mathbf{Q}$, and strains, $\mathbf{q}$, respectively, defined for Gauss points $\boldsymbol{\xi}_G$ of the finite element. Describe by $\bar{\mathbf{R}}$ and $\bar{\mathbf{r}}$ the global vectors of external loads and generalized displacements, respectively, defined in all nodal points $\boldsymbol{\xi}_N$ of the structure. These quantities have to be chosen in such a way that the internal and external work increments are equal and independent of the form of its description

$$\bar{\mathbf{Q}}\delta\bar{\mathbf{q}} = \bar{\mathbf{R}}\delta\bar{\mathbf{r}}. \tag{1}$$

Within the small displacement theory, the global vector of generalized elastic stresses $\bar{\mathbf{Q}}^E$ is a linear function of external loads applied to the nodal points of the discretized structure. The vectors of stresses, $\bar{\mathbf{Q}}$, and elastic stresses, $\bar{\mathbf{Q}}^E$, are assumed to be in equilibrium with the same external loads $\bar{\mathbf{R}}$. Therefore, the residual stress vector being in the equilibrium with vanishing loading is defined as

$$\bar{\mathbf{Q}}^{res} = \bar{\mathbf{Q}} - \bar{\mathbf{Q}}^E. \tag{2}$$

The vector of total strains is assumed to be decomposed into elastic and plastic parts $\bar{\mathbf{q}}^e$ and $\bar{\mathbf{q}}^p$, so that

$$\bar{\mathbf{q}} = \bar{\mathbf{q}}^e + \bar{\mathbf{q}}^p. \tag{3}$$

The stress-elastic strain relation is governed by the linear Hooke's law with $\bar{\mathbf{E}}$ being the symmetric elasticity matrix

$$\bar{\mathbf{Q}} = \bar{\mathbf{E}}\bar{\mathbf{q}}^e. \tag{4}$$

Within the framework of this description it is assumed that the vector of residual stresses is linearly related to the vector of plastic strains

$$\bar{\mathbf{Q}}^{res} = \bar{\mathbf{Z}}\bar{\mathbf{q}}^{p}. \tag{5}$$

Here $\bar{\mathbf{Z}}$ is called the influence matrix; it depends on the geometry of the structure, the configuration of a surface on which kinematical boundary conditions are imposed and on the kind of discretization. The form of the matrix is described elsewhere (see, e.g. MAIER [1969], COMI *et al.* [1992], ZWOLIŃSKI [1994]) and is assumed to be known.

Due to the elastic-plastic model of the material adopted for the analysis, the response of the structure at a given Gauss point $\boldsymbol{\xi}_G$ can be considered as elastic, elastic-plastic, or purely plastic. Therefore, elastic and yield loci F^E and F^P are introduced in the space of local generalized stresses (vectors without a bar)

$$F^E(\mathbf{Q}, \mathbf{k}) = 0, \qquad F^P(\mathbf{Q}, \mathbf{k}) = 0, \tag{6}$$

to distinguish the behaviour of the cross-section. Here, $\mathbf{k}$ denotes collectively the plastic moduli of the cross-section defined for the given Gauss point. Purely elastic response means that the stresses vary inside the elastic locus, whereas purely plastic behaviour is represented by the stress vector attaining the yield surface. In the latter case a local vector of plastic strain rates can be specified as follows

$$\dot{\mathbf{q}}^p = \dot{\lambda}\frac{\partial F^P}{\partial \mathbf{Q}}, \qquad \dot{F}^P \le 0, \qquad \dot{\lambda} \ge 0, \qquad \dot{\lambda}\dot{F}^P = 0. \tag{7}$$

Here, partial derivatives of a scalar by a vector are considered as a column vector. The function $F^P(.)$ can be arbitrarily nonhomogeneous with respect to the stress components. The scalar λ denotes the plastic multiplier. Relations (7) represent associated plastic flow rule and loading-unloading conditions.

Depending on the type of structure considered, the elastic locus can be different or identical to the yield locus. Moreover, it is assumed that the elastic locus is an even function of generalized stresses, i.e.

$$F^E(\mathbf{Q}, \mathbf{k}) = F^E(-\mathbf{Q}, \mathbf{k}). \tag{8}$$

The description introduced in this paper is fairly general. It can be employed for the analysis of both continuum and surface structures.

Variable loads can be described, without any loss of generality, by means of a finite number of $l = 1, \ldots, n_l$ independent systems of loads varying in time according to loading parameters $\beta_{(l)}(t)$ (see e.g. KÖNIG [1987])

$$\bar{\mathbf{R}}(t) = \sum_{l=1}^{n_l} \beta_{(l)}(t)\bar{\mathbf{R}}^{(l)}. \tag{9}$$

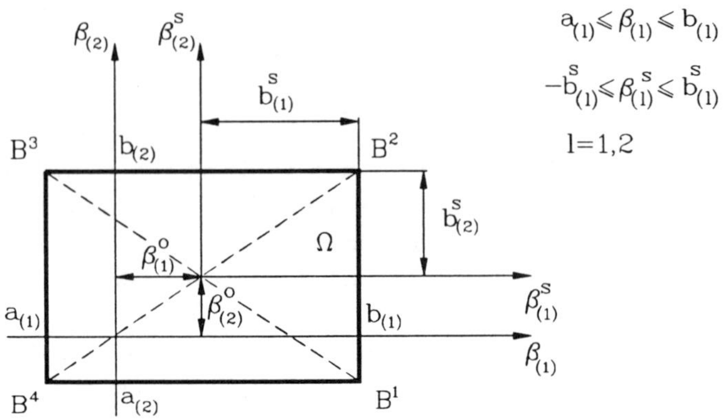

Figure 1: Load domains for total and symmetrical portion of loads

where $\bar{\mathbf{R}}^{(l)}$ denotes the lth time-independent system of load configuration applied to nodal points $\boldsymbol{\xi}_N$. Loading parameters $\beta_{(l)}$ are limited by a load polyhedron with vertices $\mathcal{B}^k$, $k = 1, \ldots, n_k$ (Fig. 1) in the n_l dimensional space. A domain of variation Ω of these parameters is described by

$$a_{(l)} \leq \beta_{(l)}(t) \leq b_{(l)}, \tag{10}$$

where $a_{(l)}$, $b_{(l)}$ are lower and upper limits of the loading parameter $\beta_{(l)}(t)$.

It is convenient to split the loads into parts that are constant and symmetrically varying in time (PYCKO & MRÓZ [1992]) according to the following formulae (see Fig. 1)

$$\begin{gathered} \beta_{(l)}(t) = \beta^o_{(l)} + \beta^s_{(l}(t), \\ \beta^o_{(l)} = \tfrac{a_{(l)}+b_{(l)}}{2}, \qquad -b^s_{(l)} \leq \beta^s_{(l)}(t) \leq b^s_{(l)}, \quad b^s_{(l)} = \tfrac{b_{(l)}-a_{(l)}}{2}. \end{gathered} \tag{11}$$

In the present considerations, the load domain Ω represented by the vertices of the polyhedron $\mathcal{B}^k$ is assumed to be fixed. A statement, simplifying operations with variable loading, quoted below is due to KÖNIG [1977]. It results from the convexity of the yield condition and from the linearity of the equilibrium equations:
If a structure shakes down in a cyclic loading path going through all the vertices $\mathcal{B}^1, \ldots, \mathcal{B}^{n_k}$ of the hyper-polyhedron domain Ω, then the structure shakes down also in any loading path contained within the domain Ω.

Hence, instead of dealing with any load variation inside the domain Ω, it is sufficient to consider only all the vertices $\mathcal{B}^k$ of the load polyhedron Ω.

SHAKEDOWN MULTIPLIER

Theoretical background concerning a min-max formulation was given by PYCKO & MRÓZ [1992]. Some considerations from this paper will be recalled and adopted here to make the presentation self-consistent.

Let us introduce a definition of an inverse multiplier η. This multiplier can be determined for a given statically admissible residual stress vector $\mathbf{Q}^{res}$ at each Gauss point $\boldsymbol{\xi}_G$ of the structure.

Definition 1
The elastic or plastic inverse multipliers η^E, η^P, respectively, associated with a point $\boldsymbol{\xi}_G$ are defined by the formulae

$$F^I\left(\frac{\mathbf{Q}^E(\boldsymbol{\xi}_G,\mathcal{B}^k)+\mathbf{Q}^{res}(\boldsymbol{\xi}_G,\bar{\mathbf{q}}^p)}{\eta^I},\mathbf{k}(\boldsymbol{\xi}_G)\right)=0,\qquad I=E,P. \tag{12}$$

In Eq.(12) the ratio of the total stress vector $\mathbf{Q}=\mathbf{Q}^E+\mathbf{Q}^{res}$ and the inverse multiplier η^I means a vector that remains on the elastic or the yield surfaces $F^I(.)=0$, I=E,P.

After solving the implicit equation (12) with respect to η^I we arrive at a modified function $\mathcal{F}^I$ in the form

$$\eta^I(\boldsymbol{\xi}_G,\mathcal{B}^k,\bar{\mathbf{Q}}^{res})=\mathcal{F}^I\left(\mathbf{Q}^E(\boldsymbol{\xi}_G,\mathcal{B}^k)+\mathbf{Q}^{res}(\boldsymbol{\xi}_G,\bar{\mathbf{q}}^p),\mathbf{k}(\boldsymbol{\xi}_G)\right),\qquad I=E,P. \tag{13}$$

Definition 2
The statical inverse (alternative) multiplier η^{aI} for a given statically admissible time-independent residual stress field $\mathbf{Q}^{res}$ is defined as the maximum of all inverse multipliers η^I over every nodal point $\boldsymbol{\xi}_G$ and load vertex $\mathcal{B}^k$, $k=1,\ldots,n_k$

$$\eta^{aI}(\bar{\mathbf{Q}}^{res})=\max_{\boldsymbol{\xi}_G,\mathcal{B}^k}\eta^I(\boldsymbol{\xi}_G,\mathcal{B}^k,\mathbf{Q}^{res}),\qquad I=E,P. \tag{14}$$

Shakedown analysis based on the min-max formulation consists of solutions of two fundamental problems (PYCKO & MRÓZ [1992]):

1. Determine the inverse shakedown multiplier η^{inc} with respect to the incremental collapse

$$\eta^{inc}=\min_{\bar{\mathbf{Q}}^{res}}\eta^{aP}(\bar{\mathbf{Q}}^{res})=\min_{\bar{\mathbf{Q}}^{res}}\max_{\boldsymbol{\xi}_G,\mathcal{B}^k}\mathcal{F}^P\left(\mathbf{Q}^E(\boldsymbol{\xi}_G,\mathcal{B}^k)+\mathbf{Q}^{res}(\boldsymbol{\xi}_G,\bar{\mathbf{q}}^p),\mathbf{k}(\boldsymbol{\xi}_G)\right); \tag{15}$$

2. Determine the inverse shakedown multiplier η^{alt} with respect to the alternating plasticity

$$\eta^{alt} = \eta^{aE}(\bar{\mathbf{0}}) = \max_{\boldsymbol{\xi}_G, \mathcal{B}^k} \mathcal{F}^E \left(\mathbf{Q}^{Es}(\boldsymbol{\xi}_G, \mathcal{B}^k), \mathbf{k}(\boldsymbol{\xi}_G)\right). \tag{16}$$

Here, $\mathbf{Q}^{Es}$ denotes elastic stresses obtained for a symmetrically variable part of loading shown in Fig. 1 and represented by parameters $\beta^s_{(l)}$ (11) (9)

$$\mathbf{Q}^{Es}(\boldsymbol{\xi}_G, \mathcal{B}^k) \equiv \mathbf{Q}^E(\boldsymbol{\xi}, \bar{\mathbf{R}}^s), \qquad \bar{\mathbf{R}}^s(\mathcal{B}^k) = \sum_{l=1}^{n_l} \beta^s_{(l)}(\mathcal{B}^k)\bar{\mathbf{R}}^{(l)}. \tag{17}$$

In view of what precedes, the shakedown inverse multiplier enforcing adaptation of the structure is defined as the supremum of inverse multipliers against incremental collapse and alternating plasticity

$$\eta^{SD} = \sup(\eta^{inc}, \eta^{alt}). \tag{18}$$

On the other hand, the load shakedown multiplier μ^{SD} enlarging the reference load domain Ω to the shakedown load domain Ω^{SD}, is expressed as

$$\mu^{SD} = \frac{1}{\eta^{SD}}, \qquad \Omega^{SD} = \mu^{SD}\Omega. \tag{19}$$

The residual stress state, which ensures adaptation of the body to the loading represented by the multiplier μ^{SD}, is equal to

$$\bar{\mathbf{Q}}^{res}_{SD} = \mu^{SD}\bar{\mathbf{Q}}^{res}_i = \frac{\bar{\mathbf{Q}}^{res}_i}{\eta^{SD}}. \tag{20}$$

Several explanatory comments of these formulations are enclosed below:

1. For a continuum, the elastic locus is identical to the yield locus $F^E = F^P$, and has to be an even homogeneous function of stresses of degree one. In the case when alternating plasticity (low cycle fatigue) is the actual mechanism of inadaptation, the solutions of the optimisation problems (15) and (16) give the same inverse shakedown multipliers $\eta^{inc} = \eta^{alt}$.

2. For a cross-section with the elastic locus different from the yield locus, the inverse multiplier with respect to the alternating plasticity can be obtained only from the problem (16). The proof of the statement (16) was given by PYCKO & MRÓZ [1992] for the even elastic locus (8) and results from considerations resting on continuum mechanics.

3. The problem (16) is a particular case of a more general min-max problem described by Eq. (15). Moreover, its solution requires only an elastic analysis and hence, is very efficient. According to (18) and (19) the inverse multiplier

η^{alt} provides an upper bound $\mu^{alt} = 1/\eta^{alt}$ to the shakedown multiplier μ^{SD}. The occurrence of stresses $\mathbf{Q}^{Es}$ in the formulation (16) is a consequence of the property that the time-independent set of loads, described by multipliers $\beta^o_{(l)}$, does not affect the alternating plasticity condition which is dependent only on the symmetrically varying loads specified by $\beta^s_{(l)}(t)$ (17) (PYCKO & MRÓZ [1992], POLIZZOTTO [1993]).

4. The assumption concerning superposition of elastic and plastic generalized strains (3) which is acceptable only for continuum and for sandwich cross-sections of the structure, excludes from the analysis pseudo-residual stresses self-equilibrated inside the cross-section. This situation does not affect the accuracy of the inverse shakedown multiplier with respect to the incremental collapse (PYCKO [1993b]). In fact, the residual stresses Q^{res} occurring in the formulation (15) have to be self-equilibrated only at the level of the whole structure. Though pseudo-residual stresses do not occur in explicit form in Eq. (15), they play a crucial role in determination of the yield locus F^P.

A geometrical interpretation of the min-max problem (15) can be found in the paper by JANAS et al. [1994].

PROBLEM FORMULATION

The problem (16) is a particular case of the more general problem (15), therefore, only the latter will be considered here in detail.

Let us express the residual stress vector $\mathbf{Q}^{res}$ at a given Gauss point $\boldsymbol{\xi}_G$ by means of the global plastic strain vector $\bar{\mathbf{q}}^p$, according to the formula (5). Then, the min-max problem (15) is directly transformed to

$$\eta^{inc} = \min_{\bar{\mathbf{q}}^p} \max_{\boldsymbol{\xi}_G, \mathcal{B}^k} \eta^P(\boldsymbol{\xi}_G, \mathcal{B}^k, \bar{\mathbf{q}}^p), \tag{21}$$

where

$$\eta^P(\boldsymbol{\xi}_G, \mathcal{B}^k, \bar{\mathbf{q}}^p) = \mathcal{F}^P\left(\mathbf{Q}^E(\boldsymbol{\xi}_G, \mathcal{B}^k) + \mathbf{Q}^{res}(\boldsymbol{\xi}_G, \bar{\mathbf{q}}^p), \mathbf{k}(\boldsymbol{\xi}_G)\right). \tag{22}$$

After determination of the inverse multiplier η^{inc}, the shakedown load multiplier against incremental collapse is obtained as $\mu^{inc} = 1/\eta^{inc}$. A few iterative methods of solution of this problem (see ZWOLIŃSKI & BIELAWSKI [1987], PYCKO [1993b], JANAS *et al.* [1995]) have appeared recently in the literature. All of them have a common feature, that the early iterations, for relatively large structures, are very fast and contribute very significantly to the solution of the shakedown load multiplier whereas the last iterations are time-consuming and provide only a minor improvement to the final result. Therefore, the almost exact determination of the shakedown multiplier, though possible, is not practically justified.

On the other hand, in engineering practice, it is often more important to know whether the structure is or is not able to shake down to the required load domain. The answer to this question is a subject of the present paper.

Assume that the intensity of the variable loads is described by the load multiplier μ. The statical inverse multiplier corresponding to this load multiplier is obtained as $\eta^{\mu} = 1/\mu$. Therefore, the problem under consideration is equivalent to the problem formulated below

If for an assigned value of the statical inverse multiplier $\eta^{\mu} = 1/\mu$ there exists such a field of plastic strains $\bar{\mathbf{q}}^p$ which satisfies the following optimisation problem

$$\eta^{\mu}(\bar{\mathbf{q}}^p) = \max_{\boldsymbol{\xi}_G, \mathcal{B}^k} \mathcal{F}^P \left(\mathbf{Q}^E(\boldsymbol{\xi}_G, \mathcal{B}^k) + \mathbf{Q}^{res}(\boldsymbol{\xi}_G, \bar{\mathbf{q}}_i^p), \mathbf{k}(\boldsymbol{\xi}_G) \right). \tag{23}$$

then the structure will shake down to the load represented by the multiplier $\mu = 1/\eta^{\mu}$.

It is seen that the problem (23) is only apparently different from that formulated by (21). In fact, for η^{μ} equal to η^{inc}, the field of plastic strains satisfying (23) is also the optimal field minimizing the maximal value of η^P specified by (21).

NUMERICAL ALGORITHM

Let us consider the ith step of the iterative procedure. Global vectors of elastic stresses $\bar{\mathbf{Q}}^E$, plastic strains $\bar{\mathbf{q}}^p$, and residual stresses $\bar{\mathbf{Q}}^{res}$, are known at the beginning of the iteration step. Therefore, the actual value of the statical inverse multiplier $\eta_i^{aP}(\bar{\mathbf{q}}_i^p)$ for the given plastic strains can be determined from the maximization problem performed over all Gauss points $\boldsymbol{\xi}_G$ and all vertices $\mathcal{B}^k$ of the load polyhedron.

$$\eta_i^{aP}(\bar{\mathbf{q}}_i^p) = \max_{\boldsymbol{\xi}_G, \mathcal{B}^k} \mathcal{F}^P \left(\mathbf{Q}^E(\boldsymbol{\xi}_G, \mathcal{B}^k) + \mathbf{Q}^{res}(\boldsymbol{\xi}_G, \bar{\mathbf{q}}_i^p), \mathbf{k}(\boldsymbol{\xi}_G) \right), \tag{24}$$

Our aim, according to (23), is to find such an increment of plastic strains $\Delta\bar{\mathbf{q}}_i^p$ for which the following holds

$$\eta^{\mu} = \eta_i^{aP}(\bar{\mathbf{q}}_i^p + \Delta\bar{\mathbf{q}}_i^p) < \eta_i^{aP}(\bar{\mathbf{q}}_i^p), \tag{25}$$

where

$$\eta_i^{aP}(\bar{\mathbf{q}}_i^p + \Delta\bar{\mathbf{q}}_i^p) = \max_{\boldsymbol{\xi}_G, \mathcal{B}^k} \mathcal{F}^P \left(\mathbf{Q}^E(\boldsymbol{\xi}_G, \mathcal{B}^k) + \mathbf{Q}^{res}(\boldsymbol{\xi}_G, \bar{\mathbf{q}}_i^p + \Delta\bar{\mathbf{q}}_i^p)), \mathbf{k}(\boldsymbol{\xi}_G) \right). \tag{26}$$

Each iteration of the solution procedure considered below consists of two basic subproblems. The first one is related to the maximization process (24). In this part all Gauss points $\boldsymbol{\xi}_i^{(v)}$ and corresponding vertices of the load polyhedron $\mathrm{B}_i^{(v)}$ $v = 1, ...n_v$ which maximize the function $\mathcal{F}^P$ for the given plastic strain vector $\bar{\mathbf{q}}_i^p$, are determined as:

$$(\boldsymbol{\xi}_i^{(v)}, \mathcal{B}_i^{(v)}): \qquad \eta_i^{aP}(\bar{\mathbf{q}}_i^p) = \eta_i^{(v)}(\mathbf{Q}^{(v)}) \tag{27}$$

where $\eta_i^{(v)}$ is defined as

$$\eta_i^{(v)}(\mathbf{Q}^{(v)}) \equiv \mathcal{F}^P \left(\mathbf{Q}^E(\boldsymbol{\xi}_i^{(v)}, \mathcal{B}_i^{(v)}) + \mathbf{Q}^{res}(\boldsymbol{\xi}_i^{(v)}, \bar{\mathbf{q}}_i^p), \mathbf{k}(\boldsymbol{\xi}_i^{(v)}) \right). \tag{28}$$

Let us restrict our analysis only to these pairs $(\boldsymbol{\xi}_i^{(v)}, \mathcal{B}_i^{(v)})$, $v = 1, \ldots, n_v$ which were determined in the first subproblem. This leads to a considerable reduction in the size of the problem This reduction concerns not only the number of Gauss points, but also the number of load vertices. Now, each Gauss point $\boldsymbol{\xi}_i^{(v)}$ is associated with only one load vertex $\mathcal{B}^{(v)}$ which maximizes the function $\mathcal{F}^P$ for given plastic strain vector $\bar{\mathbf{q}}_i^p$.

Let us pass, now, to the solution of the problem (25), (26) and expand $\eta_i^{aP}(\bar{\mathbf{q}}_i^p + \Delta\bar{\mathbf{q}}_i^p)$, which is the function of plastic strains, into the Taylor series with respect to $\bar{\mathbf{q}}_i^p$

$$\eta_i^{aP}(\bar{\mathbf{q}}_i^p + \Delta\bar{\mathbf{q}}_i^p) = \eta_i^{aP}(\bar{\mathbf{q}}_i^p) + \frac{\partial \eta_i^{aP}(\bar{\mathbf{q}}_i^p)}{\partial \bar{\mathbf{q}}^{p^T}} \Delta\bar{\mathbf{q}}_i^p +, \ldots, +. \tag{29}$$

The index ^{T} in Eq.(29) denotes transposition of a vector or a matrix.

Substituting $\eta_i^{aP}(\bar{\mathbf{q}}_i^p)$ by $\eta_i^{(v)}(\mathbf{Q}_i^{(v)})$ according to (27) and noting that $\eta_i^{(v)}$ can be expressed as a function of stresses $\mathbf{Q}_i$ by Eq. (28), we obtain n_v nonlinear equations

$$\eta_i^{aP}(\bar{\mathbf{q}}_i^p + \Delta\bar{\mathbf{q}}_i^p) = \eta_i^{(v)}(\mathbf{Q}_i^{(v)}) + \frac{\partial \eta_i^{(v)}}{\partial \mathbf{Q}_i^{(v)T}} \frac{\partial \mathbf{Q}_i^{(v)}}{\partial \bar{\mathbf{q}}_i^{p^T}} \Delta\bar{\mathbf{q}}_i^p +, \ldots, +. \tag{30}$$

$$v = 1, \ldots, n_v.$$

The number of equations corresponds to the number of Gauss points $\boldsymbol{\xi}_i^{(v)}$ in which the stress vector defined as

$$\mathbf{Q}_i^{(v)} = \mathbf{Q}_i^E(\boldsymbol{\xi}_i^{(v)}, \mathcal{B}_i^{(v)}) + \mathbf{Q}_i^{res}(\boldsymbol{\xi}_i^{(v)}, \bar{\mathbf{q}}_i^p) \tag{31}$$

maximizes the function $\mathcal{F}^P$. Assume that plastic strain increments can occur only at these points and that they can be expressed by the normality rule to the yield surface. Therefore, the set of n_v nonlinear equations with n_v unknown increments of plastic multipliers $\Delta\lambda_i^{(w)}$ is obtained from (30)

$$\eta_i^{aP}(\bar{\mathbf{q}}_i^p + \Delta\bar{\mathbf{q}}_i^p) = \eta_i^{(v)}(\mathbf{Q}_i^{(v)}) + \sum_w^{n_v} D_i^{vw} \Delta\lambda_i^{(w)} + \ldots +, \tag{32}$$

$$v,\ w = 1, \ldots, n_v.$$

with components of the matrix $\mathbf{D}$ expressed as

$$D^{vw} = \frac{\partial \eta_i^{(v)}}{\partial \mathbf{Q}_i^{(v)T}} \frac{\partial \mathbf{Q}_i^{(v)res}}{\partial \mathbf{q}_i^{(w)p^T}} \frac{\partial F_i^{(w)P}}{\partial \mathbf{Q}_i^{(w)}}. \tag{33}$$

We focus on the square matrix $\mathbf{D}$, called the *fictitious tangent matrix* for variable repeated loading. The dimension of the matrix is n_v and corresponds to the number of pairs $(\boldsymbol{\xi}_i^{(v)}, \mathcal{B}_i^{(v)})$ for which the function $\mathcal{F}^P$ attains its maximum value on the ith iteration. Determination of the fictitious tangent matrix $\mathbf{D}$ becomes the main goal of the second subproblem.

A similar expression concerning matrix $\mathbf{D}$ was obtained in the paper by JANAS *et al.* [1994], where the whole min-max problem (21) was solved. The procedure was as follows: the problem (32) was linearized, by neglecting all higher-order terms, to a system of algebraic equations. Next, the increment $\Delta\lambda^{(w)}$ was determined as a function of $\eta_i^{aP}(\bar{\mathbf{q}}_i^p + \Delta\bar{\mathbf{q}}_i^p)$. Having the direction of $\Delta\lambda^{(w)}$ fixed, its length on the i-th iterative step was found from the solution of a unidimensional nonlinear optimisation problem

$$\eta_i^{inc} = \min_{\alpha} \eta_i^{aP}(\bar{\mathbf{q}}_i^p + \alpha \times \Delta\bar{\mathbf{q}}_i^p). \tag{34}$$

In the present paper, instead of looking for the optimal shakedown multiplier η_i^{inc}, we consider a slightly different problem concerning shakedown of the structure to an *a priori* specified load domain described by the multiplier μ or inverse multiplier $\eta^\mu = 1/\mu$ (25). This leads directly to the system of nonlinear equations (32), (33) which can be easily solved by employing the well known Newton-Raphson iterative scheme. Therefore, the solution of the problem (26) is given in the following iterative algorithm:

1. The first subproblem: determine all pairs $(\boldsymbol{\xi}_i^{(v)}, \mathcal{B}_i^{(v)})$ for which the following holds

$$\text{find} \quad (\boldsymbol{\xi}_i^{(v)}, \mathcal{B}_i^{(v)}) \quad \text{such that} \quad \eta_i^{(v)}(\mathbf{Q}_i^{(v)}) = \eta_i^{aP}(\bar{\mathbf{q}}_i^p), \quad v = 1, \ldots, n_v \tag{35}$$

where

$$\eta_i^{aP}(\bar{\mathbf{q}}_i^p) = \max_{\boldsymbol{\xi}_G, \mathcal{B}^k} \mathcal{F}^P\left(\mathbf{Q}_i^E(\boldsymbol{\xi}_G, \mathcal{B}^k) + \mathbf{Q}_i^{res}(\boldsymbol{\xi}_G, \bar{\mathbf{q}}_i^p), \mathbf{k}(\boldsymbol{\xi}_G)\right). \tag{36}$$

2. The second subproblem: for given pairs $(\boldsymbol{\xi}_i^{(v)}, \mathcal{B}_i^{(v)})$, determine the fictitious tangent matrix $\mathbf{D}_i$ for variable loading

$$D_i^{vw} = \frac{\partial \eta_i^{(v)}}{\partial \mathbf{Q}_i^{(v)T}} \frac{\partial \mathbf{Q}_i^{(v)res}}{\partial \mathbf{q}_i^{(w)pT}} \frac{\partial F^{(w)P}}{\partial \mathbf{Q}_i^{(w)}}. \tag{37}$$

3. Iterative scheme: for the assumed value of $\eta^\mu = \eta_i^{aP}(\bar{\mathbf{q}}_i^p + \Delta\bar{\mathbf{q}}_i^p)$ solve the system of linear equations

$$\left\{ \begin{array}{c} \Delta\lambda_{i+1}^{(1)} \\ \vdots \\ \Delta\lambda_{i+1}^{(n_v)} \end{array} \right\} = \begin{bmatrix} D_i^{11} & \ldots & D_i^{1n_v} \\ \vdots & \vdots & \vdots \\ D_i^{n_v 1} & \ldots & D_i^{n_v n_v} \end{bmatrix}^{-1} \left\{ \begin{array}{c} \eta^\mu - \eta_i^{(1)} \\ \vdots \\ \eta^\mu - \eta_i^{(n_v)} \end{array} \right\} \tag{38}$$

4. Data updating

$$\Delta \mathbf{q}^{p(v)}_{(i+1)} = \Delta \lambda^{(v)}_{(i+1)} \frac{\partial F^P_i}{\partial \mathbf{Q}^{(v)}_i}, \tag{39}$$

$$\mathbf{Q}_{(i+1)}(\boldsymbol{\xi}_G, \mathcal{B}^k) = \mathbf{Q}^E(\boldsymbol{\xi}_G, \mathcal{B}^k) + \mathbf{Q}^{res}_{(i+1)}(\boldsymbol{\xi}_G, \bar{\mathbf{q}}^p_{(i)} + \Delta \mathbf{q}^{p(v)}_{(i+1)}), \tag{40}$$

The iterative process starts with vanishing plastic strains and residual stresses, i.e. the first iteration is determined for the elastic structure

$$\bar{\mathbf{q}}^p_1 = \mathbf{0}, \qquad \mathbf{Q}_1 = \mathbf{Q}^E. \tag{41}$$

The iterative procedure is continued unless the norm of the r.h.s. vector in Eq. (38) is smaller than the assumed positive constant ϵ

$$\left\| \begin{Bmatrix} \eta^\mu - \eta^{(1)}_i \\ \vdots \\ \eta^\mu - \eta^{(n_v)}_i \end{Bmatrix} \right\| < \epsilon \tag{42}$$

In some situations it may happen that the procedure is divergent. There are two possible situations leading to divergence. First, the assumed value of η^μ is smaller than η^{inc}, which means that shakedown is impossible. Second, the assumed value of η^μ is considerably smaller then the multiplier η^{aPE}, determined for the first iteration in the absence of residual stresses (elastic state). In this case it is necessary to divide our analysis into more steps. It should be noted that when η^μ tends to η^{inc}, the determinant of the matrix $\mathbf{D}$ tends to zero.

To complete the procedure, the form of the matrix $\mathbf{D}$ defined by (33) is specified

$$D^{vw} = \left\lfloor \frac{\partial \eta^{(v)}}{\partial Q_1}, \dots, \frac{\partial \eta^{(v)}}{\partial Q_{n_r}} \right\rfloor \begin{bmatrix} Z^{vw}_{11} & \dots & Z^{vw}_{1n_r} \\ \vdots & \vdots & \vdots \\ Z^{vw}_{n_r 1} & \dots & Z^{vw}_{n_r n_r} \end{bmatrix} \begin{Bmatrix} \frac{\partial F^{(v)P}}{\partial Q_1} \\ \vdots \\ \frac{\partial F^{(v)P}}{\partial Q_{n_r}} \end{Bmatrix}, \tag{43}$$

where the matrix $\mathbf{Z}$ relates vectors of residual stresses $\mathbf{Q}^{res}$ and plastic strains $\mathbf{q}^p$

$$\begin{Bmatrix} Q^{(v)res}_1 \\ \vdots \\ Q^{(v)res}_{n_r} \end{Bmatrix} = \sum_w \begin{bmatrix} Z^{vw}_{11} & \dots & Z^{vw}_{1n_r} \\ \vdots & \vdots & \vdots \\ Z^{vw}_{n_r 1} & \dots & Z^{vw}_{n_r n_r} \end{bmatrix} \begin{Bmatrix} q^{(w)p}_1 \\ \vdots \\ q^{(w)p}_{n_r} \end{Bmatrix}. \tag{44}$$

Notice that matrix $\mathbf{Z}$ is a large square matrix with a dimension equal to the number of stress components in all Gauss points. It depends only on the geometry of the structure, the configuration of the surface on which kinematic boundary conditions are imposed and the discretization techniques. Therefore, it can be determined at the beginning of the numerical procedure and, used in the determination of the matrix

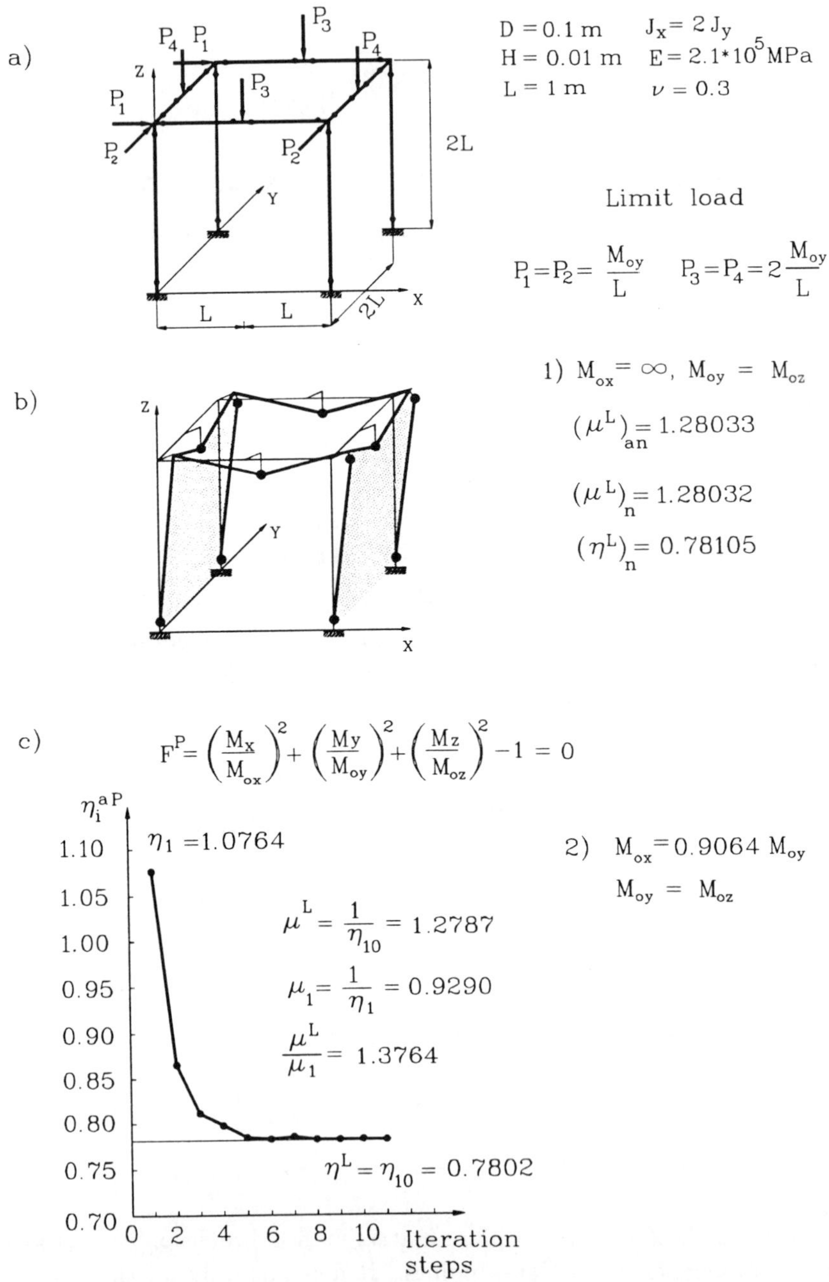

Figure 2: One storey frame under constant horizontal and vertical loading. a) Collapse mechanism and corresponding limit load multiplier. b) Convergence rate of iterative procedure for one step of load decrement.

D at each step of the iteration. On the other hand, the fictitious tangent matrix **D** which is used in the solution of the system of linear equations (38) has a dimension equal to the number of Gauss points n_v where the function $\mathcal{F}^P$ attains its maximum. This size changes from one iteration to another, but cannot exceed the number of Gauss points where a plastic failure mechanism will develop. So, the size of this matrix is relatively small in comparison to the number of degrees of freedom for the whole structure.

EXAMPLES

The iterative algorithm scheme derived in the previous Section is used here for a solution of limit analysis and shakedown problems. Efficiency of the algorithm is shown on 3D frames with tubular cross-section. The yield criterion depending on two bending moments M_y, M_z and torsion M_x is assumed to be

$$F^P = \left(\frac{M_x}{M_{ox}}\right)^2 + \left(\frac{M_y}{M_{oy}}\right)^2 + \left(\frac{M_z}{M_{oz}}\right)^2 - 1 = 0, \tag{45}$$

with $M_{oy} = M_{oz}$, and $M_{ox} = 0.9064 M_{oy}$ being cross-sectional plastic bending and torsional moments, respectively, when acting separately.

This kind of structure was chosen in order to avoid influence of other effects, not directly related to the algorithm (e.g. finite element discretization), on the convergence of the solution. The efficiency of the iterative method is demonstrated on two examples for one large step of load decrement from elastic to ultimate η^L or shakedown inverse multiplier η^{inc} with respect to incremental collapse. The latter are known to be determined as $\eta^{L,\,inc} = 1/\mu^{L,\,inc}$.

Example 1

Let us consider a one storey space frame shown in Fig.2a subjected to horizontal P_1, P_2, and vertical P_3, P_4 sets of forces. Limit analysis study is first carried out for the following proportions of forces $P_1 : P_2 : P_3 : P_4 = 1 : 1 : 2 : 2$. Neglecting the influence of torsional moment on yielding, by assuming $M_{ox} = \infty$, we obtain the ultimate load multiplier $(\mu^L)_{an} = 1.28033$ analytically. It corresponds to the plastic mechanism shown in Fig. 2b. (see e.g. JANAS *et al.* [1995]). On the other hand, the numerical procedure proposed in the paper appears to be convergent up to the load multiplier $(\mu^L)_n = 1.28032$ which corresponds to $\eta^L = 1/\mu^L = 0.78105$. The relative error between the numerical and the analytical results is negligibly small and is equal to 0.0008%.

Consider now the yield condition in the three dimensional space of bending and torsional moments (45). The limit load multiplier obtained from the numerical procedure is equal to $(\mu^L)_n = 1.2787$. With one large step of load decrement, it was found that after 10 iterations the convergence to the value $\eta^L = 1/\mu^L = 0.7802$ was

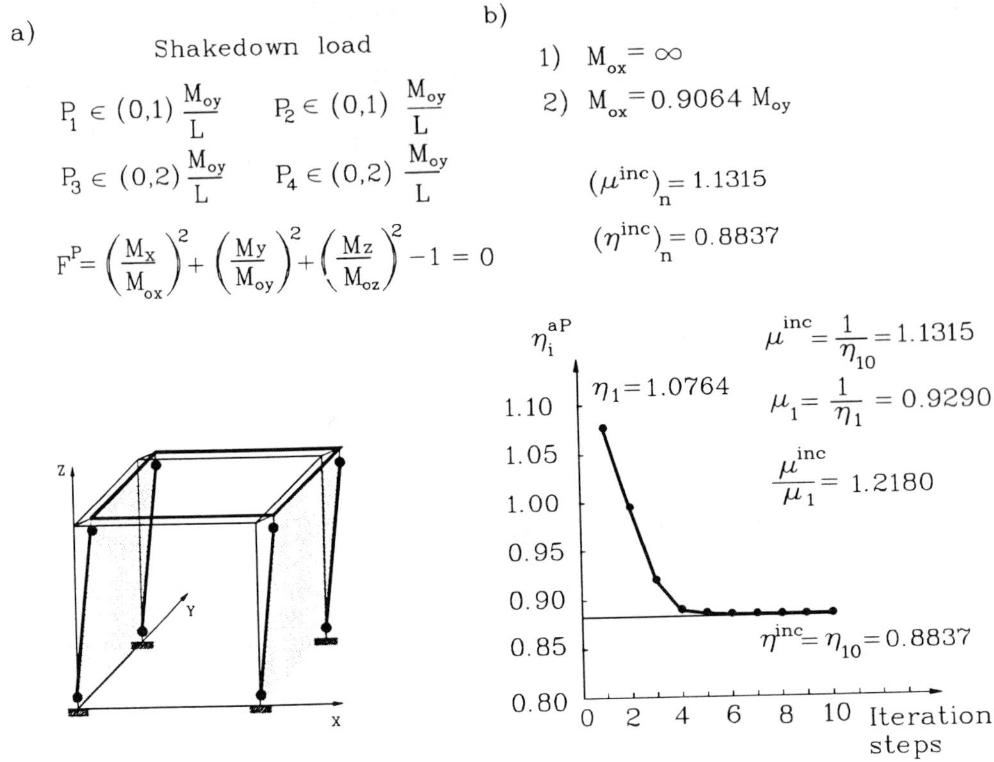

Figure 3: One storey frame under variable horizontal and vertical loading. a) Incremental collapse mechanism and shakedown multiplier. b) Convergence rate of iterative procedure for one step of load decrement.

achieved (Fig.2c).

Shakedown analysis is carried out for four independently varying sets of loading $P_1 \in (0,1)M_{oy}/L$, $P_2 \in (0,1)M_{oy}/L$, $P_3 \in (0,2)M_{oy}/L$, $P_4 \in (0,2)M_{oy}/L$, (Fig. 2a and Fig. 3a), which gives $n_k = 16$ vertices $\mathcal{B}^k$ of the load polyhedron in the four dimensional space of loading. The shakedown multiplier with respect to incremental collapse is equal to $(\mu^{inc})_n = 1.1315$ and corresponds to $(\eta^{inc})_n = 0.8837$. The convergence rate for the one step of load decrement is shown in Fig.3b. The same result is obtained for both cases of yield locus with and without torsion.

Example 2

Let us consider a rectangular clamped plane grid, shown in Fig. 4a, subjected to constant load $P_1 = P_3 = 2P$, and $P_2 = P_4 = 3P$, $P = M_{oy}/L$. It consists of four spans in the 'x' direction and six spans in the 'y' direction. The upper bound limit load multiplier determined analytically for collapse mechanism shown in Fig. 4a is

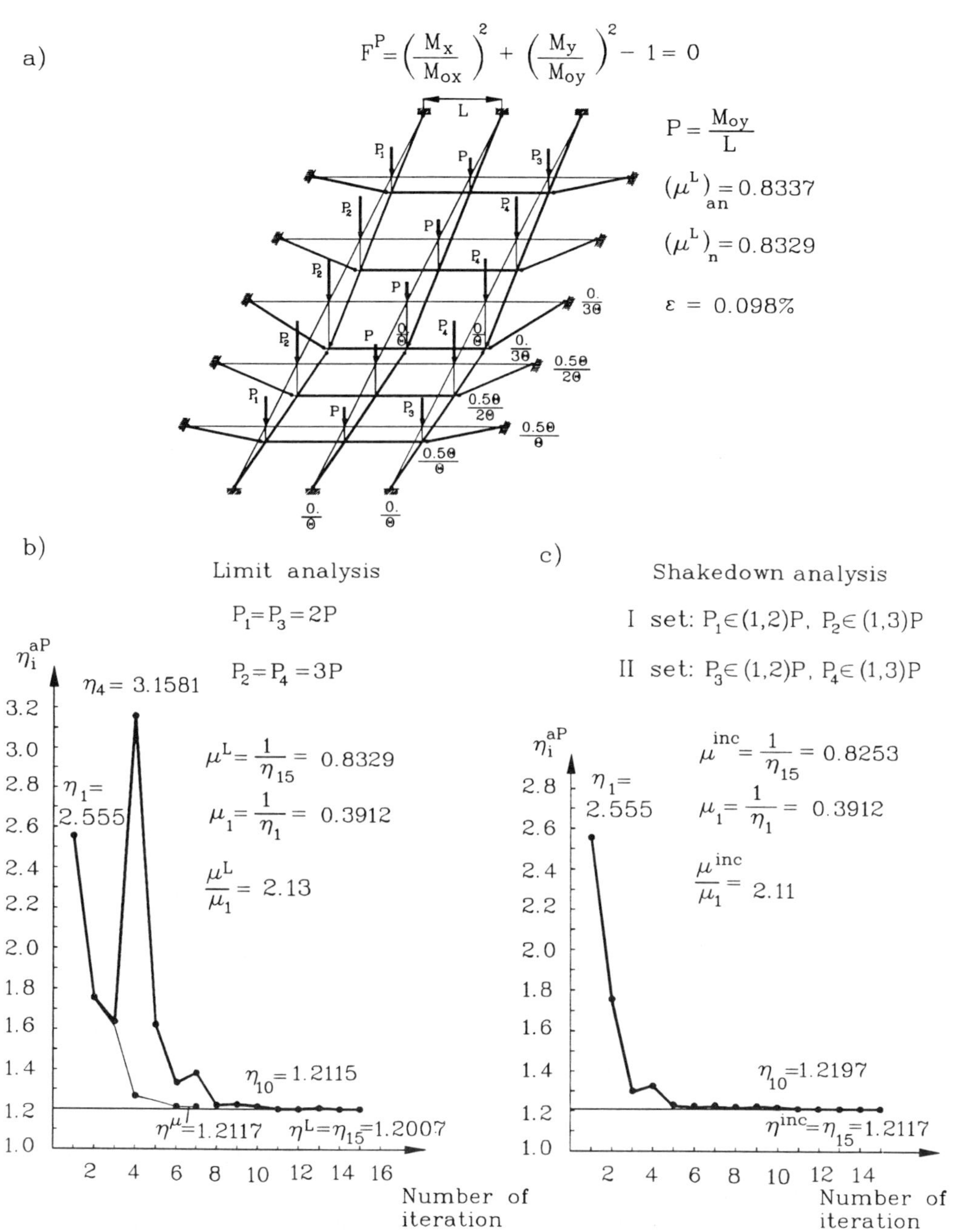

Figure 4: Plane grid under vertical constant set of loading. a) Collapse mechanism and corresponding limit load multiplier, b) Convergence rate of iterative procedure for one step of load decrement.

$(\mu^L)_{an} = 0.8337$. The numerical procedure shows the convergence up to the load multiplier $(\mu^L)_n = 0.8329$ ($\eta^L = 1.2007$) which gives 0.098% of the relative error between the numerical and the upper bound limit multipliers. The convergence rate for one step of load decrement (the ratio of limit and elastic inverse multipliers is $\eta^L/\eta^1 = 0.47$ or $\mu^L/\mu^1 = 2.13$) is marked by the thick line in Fig. 4b. Notice, that the fourth iteration gives the value $\eta_4 = 3.1581$ significantly different from others. This is due to the plastic unloading taking place in 6 nodes in which increments of plastic multipliers $\Delta\boldsymbol{\lambda}$ are negative. These nodes are regarded as the elastic ones in this iteration.

On the other hand the thin line in Fig. 4b represents the iterative process carried out for the value $\eta^\mu = 1.2117$ ($\mu = 0.8253$) the same as in the case of shakedown analysis (see below). The convergence is very high and is obtained after 6 iterations. This is a consequence of the fact that the tangent matrix $\mathbf{D}$ at the last iterative steps is nonsingular for the proportional load smaller than the limit load.

Shakedown analysis is performed for two independently varying sets of loads P_1, P_2 and P_3, P_4, respectively. Their limits of variation $P_1, P_3 \in (1,2)P$ and $P_2, P_4 \in (1,3)P$ are expressed with respect to $P = M_{oy}/L$. Fig. 4c shows up the convergence of the solution to the value $\eta^{inc} = 1.2117$ achieved after 15 iterations for one step of load decrement. At the last iterative steps the determinant of fictitious stiffness matrix for variable loading $\mathbf{D}$ tends to zero. The shakedown multiplier $\mu^{inc} = 1/\eta^{inc} = 0.8253$ against incremental collapse is only slightly slower than the limit load multiplier $(\mu^L)_n = 0.8329$ for proportional loads. This results in a similarity of collapse modes for shakedown and limit loads, however, the former is obtained as the accumulation of plastic strains after the whole load cycle.

CONCLUDING REMARKS

The min-max formulation (ZWOLIŃSKI & BIELAWSKI [1987], PYCKO & MRÓZ [1992]) was employed in order to derive the iterative solution procedure for limit analysis and shakedown problems. The presented method differs significantly from those based on mathematical programming. The main advantage of the method concerns its iterative character based on the well known Newton-Raphson scheme. Therefore, the solution of the problem exhibits very good convergence rate which makes the computational effort, even for variable loading, quite reasonable. Such an approach makes the procedure similar to the incremental time-stepping methods extensively used in the elastic-plastic analysis of structures. However, both mentioned methods are substantially distinct. The crucial point of the proposed method is a determination of the matrix $\mathbf{D}$. It can be interpreted as a fictitious tangent elasto-plastic matrix for variable loading. Its dimension is, however, much smaller than the dimension of the tangent stiffness matrix used in classic incremental plasticity. Determination of the matrix becomes the most important and, undoubtedly, the most costly part of the procedure. But this cost seems to be worthwhile, especially for the analysis of structures under variable loading. It is also important to mention that the approach can profit from many algorithms developed in the area of the Finite Element Method.

ACKNOWLEDGEMENT
The work was supported by the State Committee for Scientific Research (KBN) under the Research Project 3 0978 91 01.

REFERENCES

1960 KOITER, W.T., General theorems for elastic plastic solids, In *Progress in solid mechanics*, SNEDDON, I.N. and HILL, R.(Eds.), 165-221, North-Holland, Amsterdam.

1969 MAIER, G., Shakedown theory in perfect elasto-plasticity with associated and non-associated flow laws; a finite element linear programming approach, *Meccanica*, **4**, (3), 250 - 260.

1969 ĈYRAS, A.A., *Linear programming method of elastic-plastic systems* (in Russian), Stroizdat, Leningrad.

1972 BELYTSCHKO, T., Plane stress shakedown analysis by finite elements. *Int. J. Mech. Sci.*, **19**, 619–625.

1973 MAIER, G. A shakedown matrix theory allowing for hardening and second order geometric effects, In *Foundations of Plasticity*, SAWCZUK, A. (Ed.), 417–433, Noordhoff Int. Publishing Co., Leyden.

1977 KÖNIG, J.A., A new method of shakedown structural analysis. In *Symposium Franco-Polanais: Problemes Non-Lineaires de Mecanique*, 333–347, Cracovie.

1979 DANG, HUNG NGUYEN and PALGEN, L., Shakedown analysis by displacement method and equilibrium finite element, In *Proc. SMIRT 5 Conference*, **L**3/3, Berlin.

1986 MORELLE, P., Numerical shakedown analysis of axisymmetric sandwich shells. An upper bound formulation, *Int.J.Num.Meth.Engng.*, **23**, 2071–2088.

1987 KÖNIG, J.A., *Shakedown of elastic-plastic structures,*. Elsevier & PWN-Polish Scientific Publishers, Warsaw.

1987 ZWOLIŃSKI, J. and BIELAWSKI, G., An optimal selection of residual stress for determination of limit and shakedown multipliers (in Polish), In *Proc. of Conf. on Comp.Meth.Struct.Mech.*, 459–462, Jadwisin.

1988 BORKOWSKI, A., *Analysis of skeletal structural systems in the elastic and plastic range*, PWN, Elsevier, Warsaw.

1988 GENNA, F., A nonlinear inequality, finite element approach to the direct computation of shakedown load safety factors, *Int.J.Mech.Sci.*, **30**, 769–789.

1988 ZARKA, J.,FRELAT, J., INGLEBERT, G., and KASMAI-NAVIDI, P., *A new approach to inelastic analysis of structures*,Martinus Nijhoff Publishers, Dordrecht.

1990 ORKISZ, J., ORRINGER, O., HOŁOWIŃSKI, M., PAZDANOWSKI, M., and CECOT, W., Discrete analysis of actual residual stress resulting from cyclic loadings, *Comp.Struct.*, **35**, 397–412.

1991 BUCKTCHORPE, D.E., and WHITE, P.S. Shakedown based design assesments of LM/BR structures, In *Proc. SMIRT 11 Conference*, **E**, 197.

1991 TOULIOS M., WHITE P.S. and BUCKTCHORPE D.E., Simplified design methods by shakedown and an algorithm for FE implementation, In *Applied Solid Mechanics-4*, PONTER, A.R.S. and COCKS, A.C.E. (Eds.), Elsevier Applied Science, London.

1992 COMI, C., MAIER, G., and PEREGO, U., Generalized variable finite element modeling and extremum theorems in stepwise holonomic elastoplasticity with internal variables, *Comp.Meth.Appl.Mech.Eng.*, **96**, 213–237.

1992 PYCKO, S. and MRÓZ, Z., Alternative approach to shakedown as a solution of min-max problem. *Acta Mechanica*, **93**, 205–222.

1992 STEIN, E., ZHANG, G., and KÖNIG, J.A., Shakedown with nonlinear strainhardening including structural computation using finite element method, *Int. J.Plasticity*, **8**, 1–31.

1993 POLIZZOTTO, C., On the conditions to prevent plastic shakedown of structures: part I–theory, *Trans. ASME J.Appl. Mech.*, **60**,15–19.

1993a PYCKO, S., Method of determination of steady plastic cycles under cyclic loads, In *Proc. SMIRT 12 Conference*, KUSSMAUL, K. (Ed.), Elsevier Science Publisher, North-Holland.

1993b PYCKO, S., Variable loading and imposed displacements in the theory of shakedown, IFTR Reports 31.

1994 ZWOLIŃSKI, J., Min-max approach to shakedown and limit load analysis for elastic perfectly plastic and kinematic hardening materials, In *Inelastic behaviour of structures under variable loads*, MRÓZ, Z., WEICHERT, D., DOROSZ, S. (Eds.), Kluwer Academic Publishers BU (in print).

1995 JANAS, M., PYCKO, S., and ZWOLIŃSKI, J., A min-max procedure for the shakedown analysis of skeletal structures, *Int. J. Mech. Sci.* (in print).

SLACKENED SYSTEMS UNDER VARIABLE LOADS

Andrzej GAWĘCKI, Piotr KRÜGER
Institute of Building Structures
Poznań Technical University
ul. Piotrowo 5, 60-965 Poznañ, Poland

Abstract: The work concerns the problem of slackened systems, i.e. structural systems with gaps at connections, subjected to variable loads. Considerations have been mainly limited to systems assembled of structural elements made of elastic-perfectly plastic and rigid-perfectly plastic materials. Particular attention has been paid to problems of cyclic loading and elastic shakedown of structures with uniparameter contact conditions. Numerous examples, presented in the work, provide a lot of observations and allow the formulation of important conclusions. The examples concern simple slackened structural systems under cyclic loads (beams and frames with rotation constraints at hinges; trusses with longitudinal gaps at the end parts of bars). The general mathematical model of slackened-elastic-plastic systems and theoretical problems connected with loading, unloading and shakedown have also been included.

INTRODUCTION

The problem of slackened systems, i.e. structures with gaps (clearances) at connections of structural elements, is not quite new, nevertheless, very few works on this problem are known. Numerous illustrative examples, important observations and general theory of such systems, as well as discussions of the results which have been obtained, may be found in several works by the first author (cf. GAWĘCKI [1983, 1985, 1986, 1987, 1988, 1989, 1992]. Fundamental theoretical problems of slackened systems made of elastic-plastic material, together with some theorems concerning the problems of their analysis and synthesis, have been presented. Particular attention has been paid to the problem of uniqueness of solutions. The mathematical model proposed can also describe the behaviour of locking-elastic-plastic structures and all "friction-less" mechanical systems with unilateral constraints. Theoretical considerations have been carried out according to the following fundamental assumptions:

- the system is assembled of deformable (linear elastic-plastic) elements and of indestructible and undeformable connecting elements of very small dimensions,
- relative motions (due to the presence of gaps at the connections) of the structural element and connecting element are permitted,

Z. Mróz et al. (eds.), Inelastic Behaviour of Structures under Variable Loads, 399–417.

- displacements and clearances are so small that the application of the geometrically linear theory is justified,
- all friction effects are neglected,
- only quasi-static processes are considered,
- the *ideal structure* (i.e. the reference structure without gaps) is geometrically stable.

The relative displacements of structural elements and connecting elements play the role of concentrated *clearance strains* which appear within *a clearance region*. The clearance region represents a closed set in the clearance strain space and is bounded by *a clearance surface*, which corresponds to the locking surface (cf. PRAGER [1957]). Thus, as in the theory of plasticity, the concept of *a generalized clearance hinge* can be introduced. Note that the clearance surface can be constructed for any connections with clearances that can occur in real structures. However, the theory is restricted to a convex clearance region.

The most important results which have been obtained for slackened systems are:

A. The presence of gaps at structural connections does not affect the ultimate limit load in comparison with the reference (ideal) elastic-perfectly plastic structure without gaps. However, the ultimate limit load is usually attained in a step-wise way where *sublimit clearance-plastic flow mechanisms* of lower load multiplier can occur.

B. Clearance "dissipation" is always equal to zero.

C. The elastic response region in the load space can be non-convex.

D. The dimensions of gaps (*clearance moduli*) can be chosen in such a way that displacements, stresses and global strains are identical with those of the reference elastic-perfectly plastic structure subjected to the same external loads. In particular, for a given load path the limit load can be always reached in an elastic way, even in the case of initial loading.

E. A suitable choice of clearance moduli can lead to an increase in the critical load of the elastic slackened structure in comparison to the corresponding elastic structure without gaps.

Slackened systems, even for a proportional loading, demonstrate a lot of unexpected effects, particularly in the case when plastic deformations occur. For variable loads the behaviour of slackened systems becomes much more complicated. Theoretical problems of elastic shakedown appear to be especially difficult. This remark relates also to numerical problems due to the presence of unilateral constraints that requires non-standard and time-consuming incremental procedures.

The main aim of the work is to present the results of a numerical study devoted to problems of cyclic loading and shakedown of slackened structures made of elastic-perfectly plastic and rigid-perfectly plastic materials. The work is supplemented by the mathematical model and theoretical considerations concerning problems of proportional loading and unloading as well as shakedown of slackened systems.

MATHEMATICAL MODEL OF SLACKENED SYSTEMS

In order to construct a consistent FEM-oriented theory, the piece-wise approximation and notation, introduced by MAIER (cf., for examples [1975]), has been used. The complete systems of relations which describes the mathematical model of slackened-elastic-plastic system can be presented as follows (see GAWĘCKI [1987,1992]):

$$
\begin{array}{lll}
\mathbf{Cu} - \boldsymbol{\varepsilon} = \mathbf{0} & \mathbf{g} = \mathbf{M}^T\boldsymbol{\varepsilon}_L - \mathbf{l} \le \mathbf{0} & \dot{\boldsymbol{\lambda}}^T\mathbf{f} = 0 \\
\mathbf{C}^T\boldsymbol{\sigma} - \mathbf{p} = \mathbf{0} & \boldsymbol{\sigma} = \mathbf{M}\boldsymbol{\psi} & \dot{\boldsymbol{\lambda}}^T\dot{\mathbf{f}} = 0 \\
\mathbf{f} = \mathbf{N}^T\boldsymbol{\sigma} - \mathbf{H}\boldsymbol{\lambda} - \mathbf{k} \le \mathbf{0} & \boldsymbol{\psi} \ge \mathbf{0} & \boldsymbol{\psi}^T\mathbf{g} = 0 \\
\dot{\boldsymbol{\varepsilon}} = \mathbf{N}\dot{\boldsymbol{\lambda}} & \boldsymbol{\varepsilon} = \boldsymbol{\varepsilon}_L + \boldsymbol{\varepsilon}_E + \boldsymbol{\varepsilon}_P & \boldsymbol{\psi}^T\dot{\mathbf{g}} = 0 \\
\dot{\boldsymbol{\lambda}} \ge \mathbf{0} & \boldsymbol{\varepsilon}_E = \mathbf{E}^{-1}\boldsymbol{\sigma} &
\end{array} \qquad (1)
$$

In these relations, $\boldsymbol{\sigma}$ and $\boldsymbol{\varepsilon}$ are supervectors that collect stresses (or generalized stresses) and strains (or generalized strains), respectively, of all the structural elements; **p** and **u** denote the respective supervectors of loads (or generalized loads) and displacements (or generalized displacements) of all the connecting elements; **C** is the geometric compatibility matrix that depends only on geometry and boundary conditions; **E** is a strictly positive, definite matrix of elasticity. The yield condition is assumed to be piece-wise-linear in the form of matrix inequalities ($\mathbf{f} \le \mathbf{0}$), where matrix **N** collects all the external normals to the yield hyperpolyhedron; **H** is a square matrix of linear plastic hardening; $\dot{\boldsymbol{\lambda}}$ represents a vector of plastic strain multiplier rates; and components of **k** denote the distances from the respective sides to the origin. The clearance region is approximated by the hyperpolyhedron in the clearance strain space and **M** collects all the external normals, while **l** contains the distances from the individual sides to the origin. The normality rule for the stresses, resulting from friction-less behaviour, is expressed by the supervector $\boldsymbol{\psi}$ of stress multipliers. The orthogonality conditions express the unilateral constraints imposed on stress and strain states. Since the problem is considered in the framework of small deformations, the total strain $\boldsymbol{\varepsilon}$ can be assumed to be the sum of clearance $\boldsymbol{\varepsilon}_L$, elastic $\boldsymbol{\varepsilon}_E$ and plastic $\boldsymbol{\varepsilon}_P$ components.

Relations (1) can be treated as Kuhn-Tucker conditions and, after some modifications, they are equivalent to the saddle point problem or to the corresponding dual extremum principles for both the holonomic and incremental analyses (cf. GAWĘCKI [1987,1992]).

It should be pointed out that the mathematical model (1) describes a relatively wide class of materials and structure behaviour, namely: elastic, elastic-plastic with or without linear plastic hardening, rigid-plastic, locking (slackened)-rigid-plastic, locking (slackened)-elastic, locking (slackened)-elastic-plastic with or without linear plastic hardening.

SOME REMARKS ON LOADING AND UNLOADING

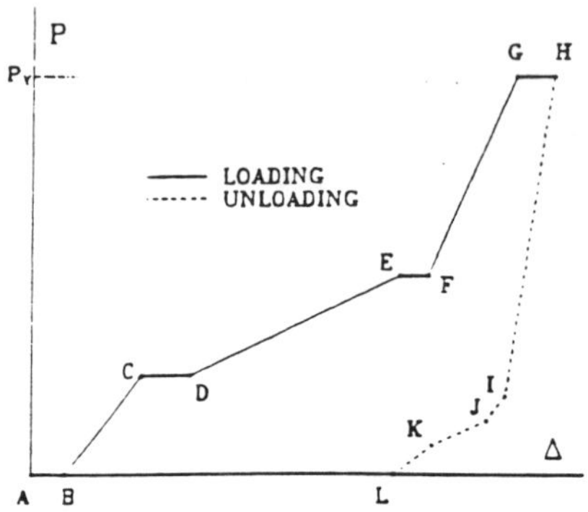

Fig. 1. Typical P-Δ diagram for a slackened-elastic-perfectly plastic structure

In order to illustrate some characteristic features of slackened-elastic-perfectly plastic structures, a typical P-Δ diagram for loading and unloading is presented in Fig. 1. The structure is loaded up to the limit load P_Y and, after some plastic deformation, unloaded. On horizontal segments of the P-Δ diagram (i.e. AB, CD, EF, GH), the structure behaves like a mechanism. The segment AB corresponds to a *clearance mechanism* that occurs if the structure is geometrically unstable due to the presence of gaps. This mechanism stops when the system converts into a non-zero stiffness structure (a so called *original structure*), depending on the applied load vector. From the point B up to the point C the stiffness of the structure is different from zero and deformations of the structure are elastic or elastic-plastic. At the point C *a clearance-plastic mechanism* begins and a *sublimit flow* develops. From the point D the mechanism converts again into a non-zero stiffness structure and elastic or elastic-plastic deformations develop. The segment EF represents the second clearance-plastic mechanism and the segment FG corresponds again to elastic or elastic-plastic deformations of the structure. At the point G the limit load is reached and the segment GH represents the purely *plastic flow mechanism*. The unloading curve (indicated by the dashed line) is far from that commonly observed in elastic-plastic structures. In spite of purely linear elastic deformations, the changes of the structure stiffness are noted. It occurs either due to the loss of existing contacts or due to the appearance of new contacts at slackened connections.

Let us consider the problem of unloading using the energy approach. At a load level **p**, the elastic energy of a current mechanical state of the structure W_E is equal to $W^* + W_{Eo}$, where W^* denotes the recovery energy and W_{Eo} is the energy stored in the structure after unloading. With the compatibility and equilibrium equations, the recovery energy can be expressed as (GAWĘCKI [1993]):

$$W^* = \frac{1}{2}\mathbf{p}^T(\mathbf{u}-\mathbf{u}_0) - \frac{1}{2}(\boldsymbol{\sigma}-\boldsymbol{\sigma}_0)^T(\boldsymbol{\varepsilon}_P - \boldsymbol{\varepsilon}_{P0} + \boldsymbol{\varepsilon}_L - \boldsymbol{\varepsilon}_{L0}) \ , \tag{2}$$

where subscript "o" refers to the state after unloading ($\mathbf{p}_0 = \mathbf{0}$). Taking into account the convexity of clearance region, one obtains

$$\frac{1}{2}\mathbf{p}^T(\mathbf{u}-\mathbf{u}_0)-\frac{1}{2}\boldsymbol{\sigma}^T(\boldsymbol{\varepsilon}_L-\boldsymbol{\varepsilon}_{L0})-\frac{1}{2}(\boldsymbol{\sigma}+\boldsymbol{\sigma}_0)^T(\boldsymbol{\varepsilon}_P-\boldsymbol{\varepsilon}_{P0})\leq W^*$$

$$\tag{3}$$

$$W^*\leq\frac{1}{2}\mathbf{p}^T(\mathbf{u}-\mathbf{u}_0)+\frac{1}{2}\boldsymbol{\sigma}_0^T(\boldsymbol{\varepsilon}_{L0}+\boldsymbol{\varepsilon}_L)-\frac{1}{2}(\boldsymbol{\sigma}+\boldsymbol{\sigma}_0)^T(\boldsymbol{\varepsilon}_P-\boldsymbol{\varepsilon}_{P0}).$$

If reverse plasticity does not take place (i.e. $\boldsymbol{\varepsilon}_P = \boldsymbol{\varepsilon}_{P0}$), and the clearance region is convex, Ineq. (3) takes the form

$$\frac{1}{2}\mathbf{p}^T(\mathbf{u}-\mathbf{u}_0)-\frac{1}{2}\boldsymbol{\sigma}^T(\boldsymbol{\varepsilon}_L-\boldsymbol{\varepsilon}_{L0})\leq W^*\leq\frac{1}{2}\mathbf{p}^T(\mathbf{u}-\mathbf{u}_0)+\frac{1}{2}\boldsymbol{\sigma}_0^T(\boldsymbol{\varepsilon}_{L0}-\boldsymbol{\varepsilon}_L) \tag{4}$$

From Ineq. (4) it follows that the case of slackened structures is rather complicated and the unloading can be non-linear. The elastic energy stored in the slackened structure after unloading is often equal to zero, and consequently the residual stresses vanish. Ineq. (4) explains the nontypical behaviour of slackened structures during unloading, discussed in the example presented. On the other hand, in common elastic-plastic structures the residual stresses are usually different from zero, the elastic stored energy is equal to $\mathbf{p}^T(\mathbf{u}-\mathbf{u}_0)/2$ and the path of unloading is linear. This conclusion results directly from Ineq. (4).

CYCLIC LOADING AND SHAKEDOWN OF SLACKENED STRUCTURES

Let us consider the problems of cyclic loading and shakedown of slackened structures made of elastic-perfectly plastic and rigid-perfectly plastic materials. Because the clearance strains can be treated as time-dependent distortions and the residual stresses in rigid plastic material cannot be uniquely determined, some modification should be made to the classical definition of shakedown, namely:

"A slackened structure shakes down if after a finite number of load cycles a time-independent *plastic strain field* forms and the stress states do not violate the yield condition".

Elastic-perfectly plastic materials

As it was mentioned, the problem of shakedown of slackened structures is very complicated. Since clearance distortions are always load-dependent, even in the case of adaptation, the classical approach (cf. KÖNIG&KLEIBER [1978], KÖNIG [1987]) cannot be used. Moreover, to the authors' knowledge, there is no useful theorem concerning the problem of shakedown of slackened systems. That is why we deal with the shakedown problems, restricting ourselves only to cyclic loading analyses.

In order to explain the theoretical difficulties connected with the shakedown analysis we recall the standard approach to determine the bounds of plastic dissipation (cf., for example KÖNIG [1987]). Let us introduce the strictly positive definite quadratic form of energy of plastic deformations as a function of conventional time "t":

$$Y(t) = \frac{1}{2}[\sigma(t) - \sigma'(t)]^T [\varepsilon_E(t) - \varepsilon'_E(t)] > 0 \tag{5}$$

where the primed symbols denote the stress and strain states for the structure which shakes down under prescribed load program. The remaining stresses and strains correspond to the same structure subjected to the same load program, but during the initial cycles of load when plastic deformations still occur. After differentiation with respect to time one obtains

$$\begin{aligned} \dot{Y}(t) &= [\sigma(t) - \sigma'(t)]^T [\dot{\varepsilon}_E(t) - \dot{\varepsilon}'_E(t)] = \\ &-[\sigma(t) - \sigma'(t)]^T \dot{\varepsilon}_P(t) - [\sigma(t) - \sigma'(t)]^T [\dot{\varepsilon}_L(t) - \dot{\varepsilon}'_L(t)] \ . \end{aligned} \tag{6}$$

If $\varepsilon_L \equiv \mathbf{0}$, then

$$\dot{Y}(t) = -[\sigma(t) - \sigma'(t)]^T \dot{\varepsilon}_P(t) \le 0 \ . \tag{7}$$

Ineq. (7) implies the boundedness of both the total and local energy dissipated. In the case of slackened structures the sign of the second term in Ineq. (6) cannot be uniquely determined, unless the difference $\dot{\varepsilon}_L(t) - \dot{\varepsilon}'_L(t)$ is kinematically admissible. Then this term vanishes and the function Y(t) is monotically descreasing in time. Up to now, the proof of such a property of clearance of strain rates in slackened systems does not exist.

The numerical experience of the authors suggests that the theoretical difficulties come from non-convexity of elastic response regions of slackened structures. The shapes and dimensions of elastic loci appear to be very sensitive to the changes of plastic deformations This problem will be illustrated by examples in the next sections.

Rigid-perfectly plastic materials

If the system is assembled of rigid-perfectly plastic structural elements, the external load $\mathbf{p} = \mu\mathbf{p}_o$ is limited by the yield condition at critical cross-sections (μ denotes here the scalar load multiplier), and for sufficiently large value of $\mu = \mu_L$ the limit load problem should be solved. For slackened structures, the solution of limit load problem appears to be somewhat different from that of *ideal* rigid-perfectly plastic structures. It has been shown (cf. GAWĘCKI [1988]) that if the clearance region is convex and bounded by a closed clearance surface, then the ultimate collapse load and the corresponding plastic flow mechanism are identical to those of the reference ideal rigid-perfectly plastic structure (without clearances). The ultimate state is usually attained in a step-wise way where sublimit clearance-plastic flow mechanisms can occur before reaching the purely plastic one. It has been also derived that the limit load problem can be resolved into a set of consecutive linear programming sub-problems. Several examples of original structure, sublimit and limit load problems have been presented by KRÜGER&GAWĘCKI [1990]. From this work it follows that a *sublimit load region* in the load space, as a sum of convex regions for various types of the structure, can be non-convex. The behaviour of slackened structures during the load processes exhibits repeated changes of the clearance-plastic mechanism in sublimit states. In particular, one can observe a case in which the different clearance-plastic sublimit mechanisms correspond to the same value of the load multiplier. This results from the fact that the change of mechanism corresponds to the change of clearance mechanism only, associated with a "non-dissipative" motion.

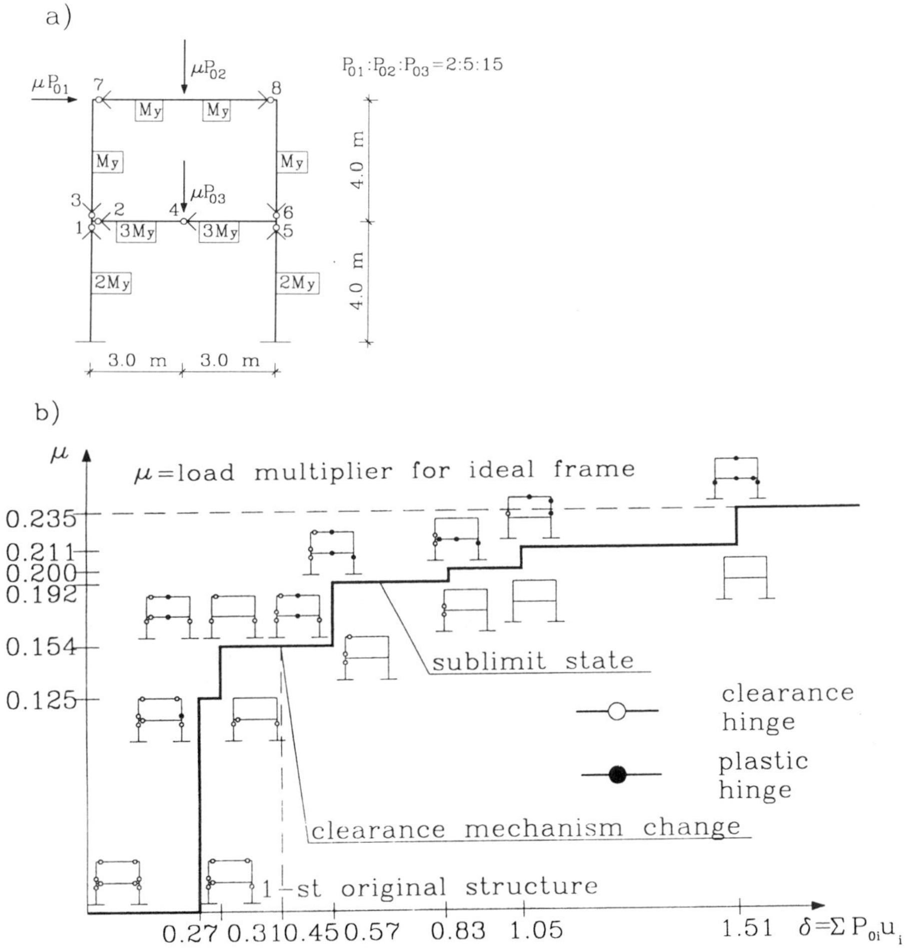

Fig. 2. Limit load of a two-storey slackened frame

In order to form an opinion on the complexity of the limit load problem, we present the results of calculation carried out for the two-storey slackened frame of Fig. 2a. The structure slackening consists in the presence of constraints imposed on mutual rotations at hinges. The corresponding μ–δ step-wise relation is presented diagrammatically in Fig. 2b (δ denotes the weighted displacement of energy dimension: $\delta = \Sigma P_{0i} u_i$). The characteristic point of the non-dissipative rotation of the adjoining hinges "1","2" and "3" of the connecting plate is also indicated. Note that the ultimate limit load is equal to that obtained by BORKOWSKI [1985] for the same frame without gaps at connections.

In the case of variable loads, the current shape and dimensions of the sublimit region depend on the plastic deformations accumulated in the previous sublimit load states. Thus, we can formulate a new problem of *shakedown of slackened-rigid-perfectly plastic structures*. The sufficient condition for adaptation is the formation of time-independent plastic strains, such that the corresponding sublimit region contains the prescribed load domain. A link between this formulation and classical one is clearly seen. The sublimit region and steady plastic deformations correspond to the elastic region and the residual stresses, respectively. The concept of residual stresses is useless because the stress state in rigid structures cannot be determined. Up to now, there are no theorems concerning this problem and therefore we restrict ourselves to presenting some illustrative examples.

The incremental analysis consists in the determination of effects of load changes connected with plastic deformations and time-dependent clearance strains. At the current step of load, two cases can occur: either the load increment induces plastic strains, or the load does not induce plastic strains. In the first case, a new original structure and sublimit state (or states) have to be determined. In the second case, the problem consists in the determination of a new original structure only.

In cyclic loading, the fundamental role is played by the shape and the dimension of sublimit region which demonstrate *step-wise changes* when the plastic deformations develop. The boundary of the sublimit region corresponds to the load for which the clearance-plastic mechanism begins. Thus, the slackened-rigid-perfectly plastic structure shakes down if the domain of load variations is contained within the current sublimit region.

EXAMPLES

Cyclic loading and shakedown of a beam with rotation constraints at the support

In order to illustrate the behaviour of slackened systems subjected to cyclic loading we present P-Δ diagrams for a simply supported beam with constraints imposed on the angle of rotation ϕ at the right support: $-\phi^- \le \phi \le \phi^+$; $\phi^- \ge 0$, $\phi^+ \ge 0$ (see Fig. 3a). Computations have been carried out for the beam of ideal I-cross-section with the depth $h = 16$ cm and the moment of inertia $J = 1000$ cm^4. The material of the beam is linear elastic-perfectly plastic with Young's modulus $E = 200$ GPa and the yield stress

σ_p = 240 MPa. The fully plastic bending moment of the cross section equals M_p=30 kNm and the limit load of the beam fully fixed at the right support (i.e. $\phi^-=0$, $\phi^+=0$) ισ P_Y = 75 kN. The beam is loaded by the variable concentrated force P ($P_Y \leq P \leq P_Y$), according to the program : 0, P_Y,, -P_Y,, P_Y,, -P_Y, etc.
It turns out that the shape of lines, representing the particular cycles of loading, depends only on the sum of limit rotations (clearance moduli) ϕ^- and ϕ^+, independently of their proportions. Figures 3b-3h show the influence of the angle $\psi = \phi^- + \phi^+$ on the P-Δ diagrams. The ranges of piece-wise linear elastic response of the beam are indicated (thick solid lines): $\Delta P_E = \max(P^+ + P^-)$, $P^+ > 0$, $P^- > 0$, where $-P^- \leq P_E \leq P^+$. The thin solid lines in these figures represent the cyclic loading path in the P-Δ space. The dotted lines indicate the P-Δ relation for the beam of ψ = 0 and the dashed lines describe the behaviour of the beam at the beginning stage of cyclic loading. The relation $\Delta P_E(\psi)$ is presented diagrammatically in Fig. 4. It is clearly seen that the presence of gaps at the

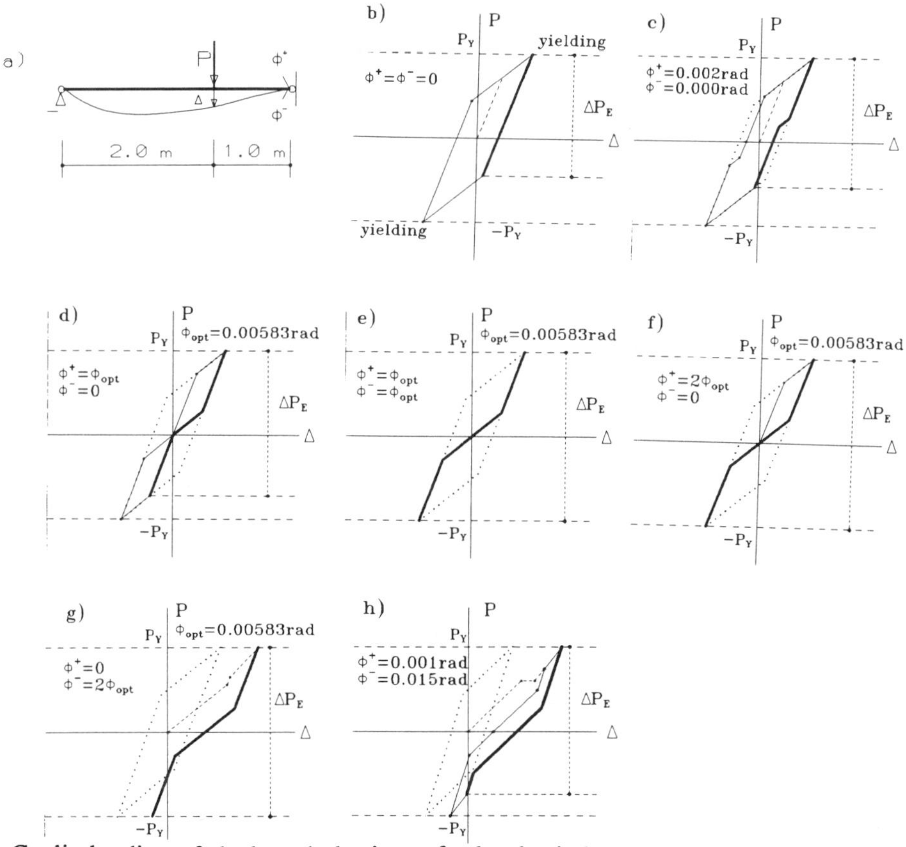

Fig. 3. Cyclic loading of slackened-elastic-perfectly plastic beam with increasing rotation constraints at the right support

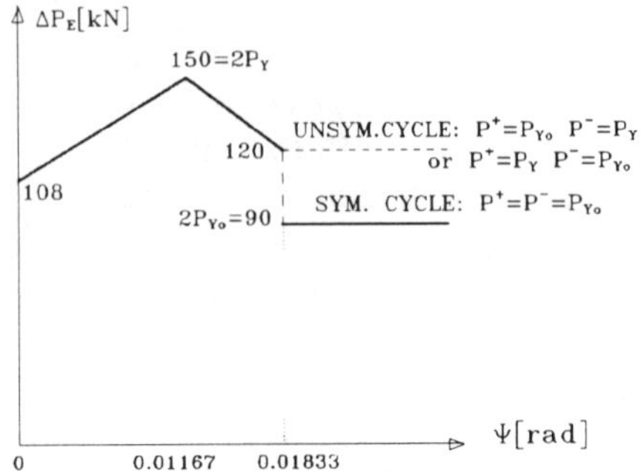

Fig. 4. Shakedown loads versus sum of limit rotations, ψ

connection usually gives a greater value of the shakedown load $\Delta P_E(\psi)$ than that of the fully fixed beam at the right support, where $\psi = 0$. For *small* values of ϕ^- and ϕ^+ an increase of ψ is profitable and plastic dissipation occurs only at the right support. If $\psi = \psi_{opt} = 2\phi_{opt} = 0.01167$ rad, then the shakedown load reaches its maximum ($\Delta P_E = 2P_Y = 150$ kN) and the plastic dissipation is bounded, whereas for $\phi^- = \phi^+ = \phi_{opt}$, the load cycle corresponds to pure elastic behaviour and the plastic dissipation is equal to zero. For *medium* values of ψ ($\psi_{opt} < \psi < 0.01833$ rad), ΔP_E decreases, however it is still larger than that of the beam fixed at the right support. Large values of ψ ($\psi > 0.01833$ rad) lead to a qualitative change of the behaviour of the slackened beam. For symmetric load cycles a *discontinuity of the shakedown load* is observed. The elastic strength suddenly decreases from $\Delta P_E = 120$ kN to $\Delta P_E = 2P_{Yo} = 90$ kN. The latter value corresponds to the pin-ended beam at both supports. At the singular point, where $\psi = \psi_{CR} = 0.01833$ rad, the shakedown load is extremely sensitive to changes of ψ.
This very simple example illustrates some difficulties which can arise in the analysis of shakedown of slackened structures. The results obtained allow us to conclude that the fundamental question is to answer whether the clearance moduli of the slackened system are small, medium or large.
From the analysis of the P-Δ diagrams, it follows that for a wide range of clearance moduli, according to the Coffin-Miner formula, a significant increase in the critical number of load cycles to failure can be predicted. It should be also stressed that strongly non-linear behaviour of slackened systems occurs even in the range of small values of limit rotations and deflections. For the beam considered in the example, the maximum deflection reaches L/360, where L denotes the length of the beam.

Incremental collapse and shakedown of two-span beam with rotation constraints

A two-span beam of uniform ideal I-cross sections (cf. Fig. 5a) is loaded by P_1 and P_2 varying according to the following load region: $0 \le P_1 \le Q$, $0 \le P_2 \le Q$. The cyclic loading proceeds according to the points 1-2-3-4 indicated in Figs. 5b,c. The maximum value of Q for which the structure behaves elastically is denoted by Q_S. It appears that the shakedown load depends on the location of rotation gaps. If the constrained rotation is introduced only at the central support (point "3") the shakedown load corresponds to that of the ideal structure (Q_S=50.53 kN) and the deflection of the mid-span, the point "2", reaches a stabilized value. Quite different behaviour of the beam is observed for the rotation gap introduced at point "2". Then the shakedown load (Q_S = 47.60 kN) is less then that of the elastic-perfectly plastic beam with fully constrained rotation. The modification of elastic loci and yield polygon are presented in Figs. 5b,c. The deflection of the point "4" versus conventional "time" (measured here as the sum of loads along the load path) is visualized in Fig. 5b and Fig. 5c.

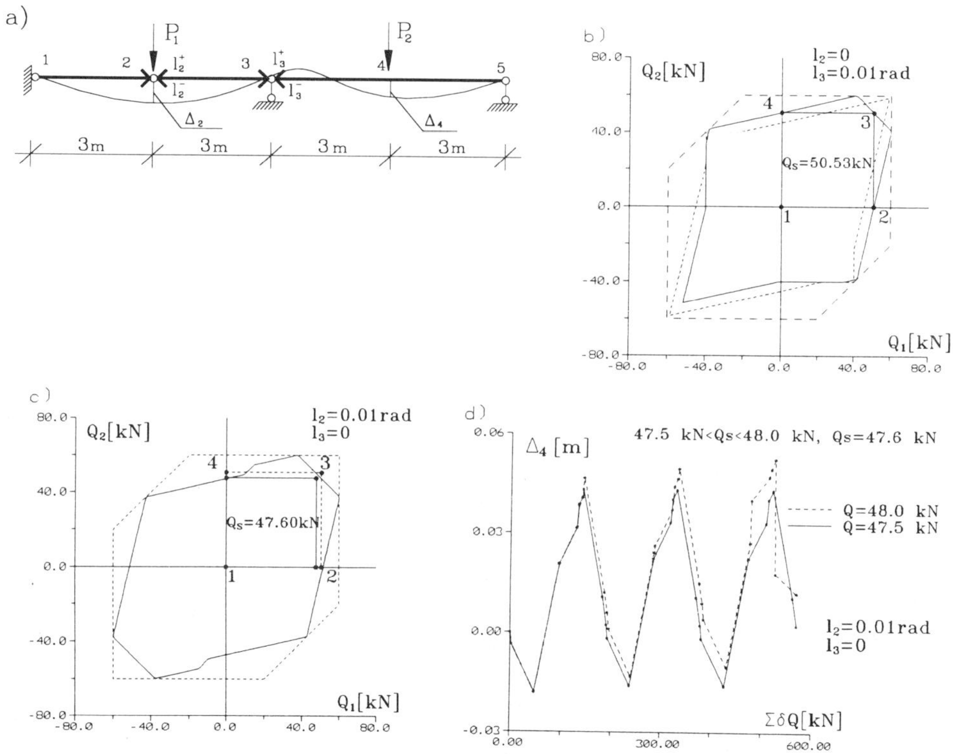

Fig. 5. Shakedown and inadaptation due to incremental collapse of slackened beam

Optimal slackening of the beam

The simple beam presented in Fig. 6a is optimally slackened by a suitable choice of limit rotations at both supports. The absolute maximum of symmetric alternate load for the ideal I-cross section is equal to the yield load of the corresponding beam fully fixed at both supports. The beam is subjected to loads that can vary within the limits: $-Q \leq P_1 \leq Q$, $-Q \leq P_2 \leq Q$. In this case it can be expected that the alternating plasticity leads to restriction of extremum loads. As a result of optimal slackening, we obtain a significant increase in the shakedown load (Q_S = 60 kN) as compared with the result obtained for a fully fixed beam whose shakedown load is equal to 40 kN. So, we can conclude that the shakedown load reaches the absolute maximum for the optimal slackened beam. This profitable result is obtained due to a non-convexity of the elastic region which has been formed after one cycle of the load. The optimal value of clearance moduli depends only on the sum of limit rotations introduced at each support. The optimal clearance moduli can be determined using the theorem derived by GAWĘCKI [1987]. The shakedown domains together with the initial and final elastic loci as well as the yield polygon are presented in Fig. 6b.

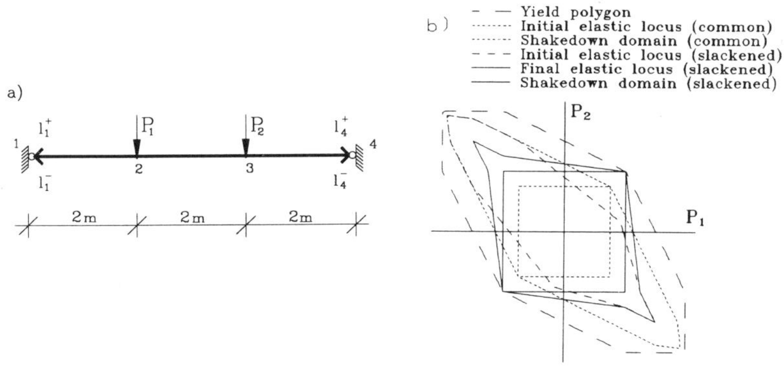

Fig. 6. Shakedown of the optimally slackened beam under alternate loads.

Cyclic loading of a slackened-elastic-perfectly plastic truss

The example concerns the truss presented in Fig. 7a. We intend to examine the influence of clearance moduli on the elastic regions and on the ability to adapt of the truss during cyclic loading.

First, consider the case of proportional loading for P_1 / P_2 = 0.2 and optimally distributed longitudinal gaps at the ends of the elements. The optimal gaps determined from the solution of the elastic-perfectly plastic truss are: $l_4^+ = 1.4312$ cm and $l_6^- = 1.0585$ cm. Final results are presented in Fig. 7b. Note that the elastic region for the slackened truss is non-convex and the limit load can be reached in an elastic way.

In Fig. 7c, the modification of elastic loci in the case of adaptation of the truss subjected to cyclic load is shown. The clearance constraints correspond to $l_4^- = l_5^+ = 0.58327$ cm.

Figure 7d shows the increase in shakedown load of the same slackened truss as in Fig. 7c, in comparison with the ideal truss without gaps. This increase reaches 10%. Thus, the presence of gaps leads to an increase in shakedown load even when the loads are limited by the incremental collapse mechanism.

The last case corresponds to the situation when the shakedown load of the slackened truss is less then that of the truss without gaps. The assumed clearance moduli are: $l_1^- = l_1^+ = l_5^- = l_5^+ = 1$ cm, $l_2^- = 2$ cm, $l_2^+ = 3$ cm, $l_6^- = l_6^+ = 1.5$ cm and the corresponding decrease in shakedown load is equal to 30%. The yield polygon, and the elastic loci for the common and slackened trusses and load programs are illustrated in Fig. 7e.

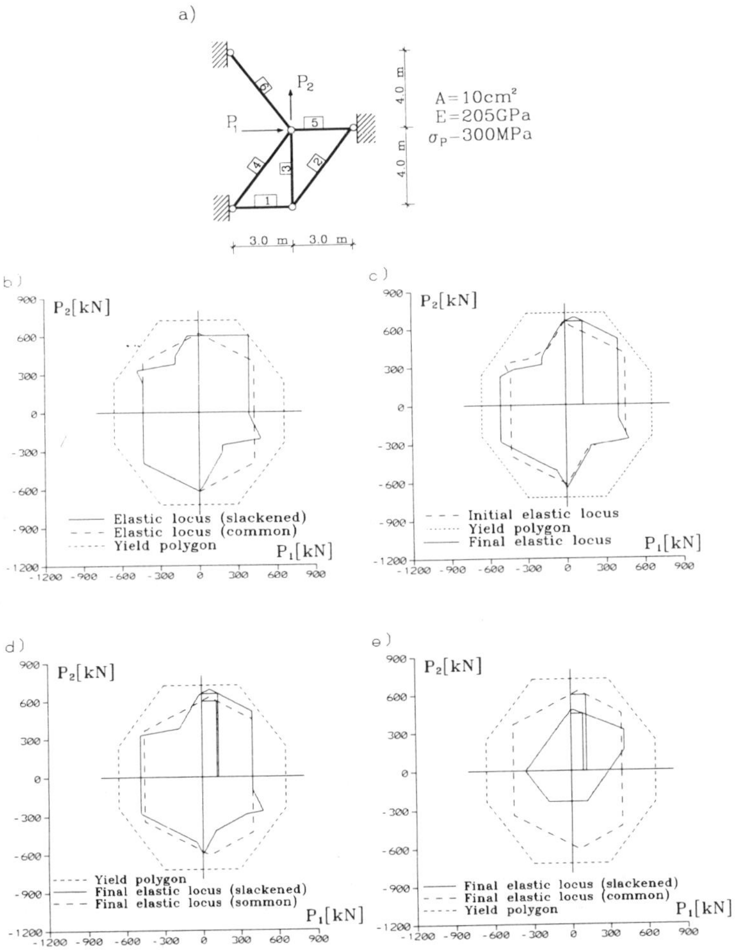

Fig. 7. Proportional loading and shakedown of slackened truss.

Cyclic loading and inadaptation of slackened-rigid-perfectly plastic portal frame

Consider the slackened-rigid-perfectly plastic portal frame with rotation constraints subjected to horizontal and vertical loads (cf. Fig. 8a). The mechanical model of slackened frame together with the details of connections is explained in Fig. 8b. The cyclic loading of the frame proceeds according to the "counter clockwise" and "clockwise" along the boundary of the rectangular load path. The load program, the initial sublimit locus and the yield polygon are presented in Fig. 8c.

In the beginning, consider the case of cyclic loading according to the counter clockwise path. The first modification of the sublimit region occurs at the point $A_{1,}$, when the plastic hinge at the node II develops (see Figs. 9a,c). The corresponding sublimit polygon is shown in Fig. 9b.

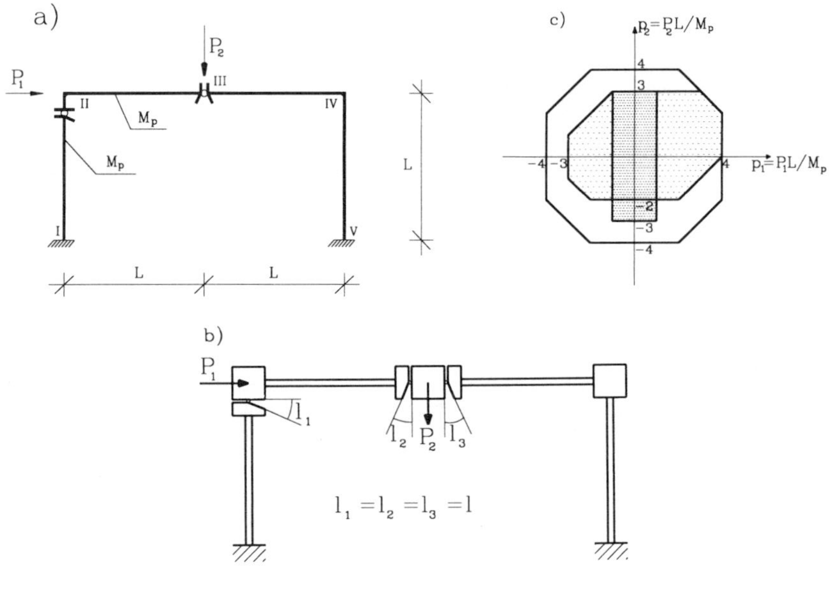

Fig. 8. Slackened-rigid-perfectly plastic portal frame, yield and sublimit polygons, and load program

From this picture it follows that in the initial stage of the second cycle a new plastification appears at the point B. These deformations induce the modification of the sublimit region to that presented in Fig. 9a (shaded area). From now on, we observe the stabilization of limit cycle in which succeeding plastifications occur at the point A (Fig. 9a) and at the point B (Fig. 9b). The structure will collapse according to the alternating plasticity. The plastic deformations are located only at the node IV (see Fig. 9c). So, during the limit cycle, the sublimit regions take one of the non-convex forms visualized in Figs. 9a,b. The deflection of the point III as a function of conventional time, corresponding to the path "length" measured in the load space, is shown in Fig. 9d.

The second case corresponds to cyclic loading in a the clockwise direction (cf. Fig. 10a). The behaviour of the structure is similar as in the first case. The limit cycle stabilizes in the second loop and plastic deformations develop also at the node IV. However, the sublimit regions appear to be convex in this case.

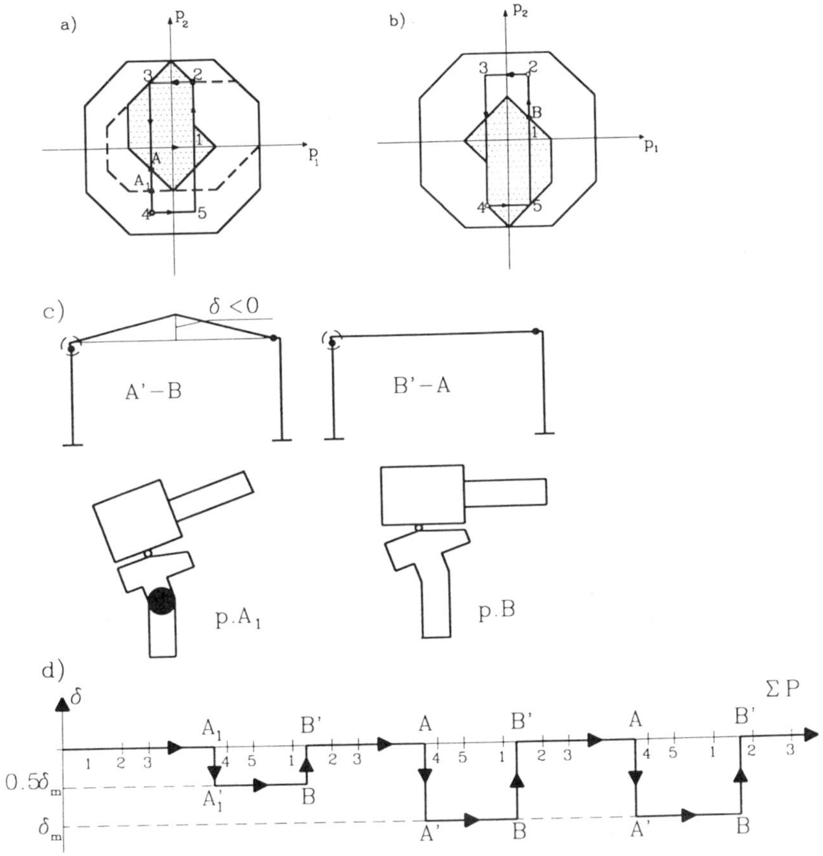

Fig. 9. Modifications of sublimit regions, plastic deformations of the frame, changes of deflection during cyclic loading (first case)

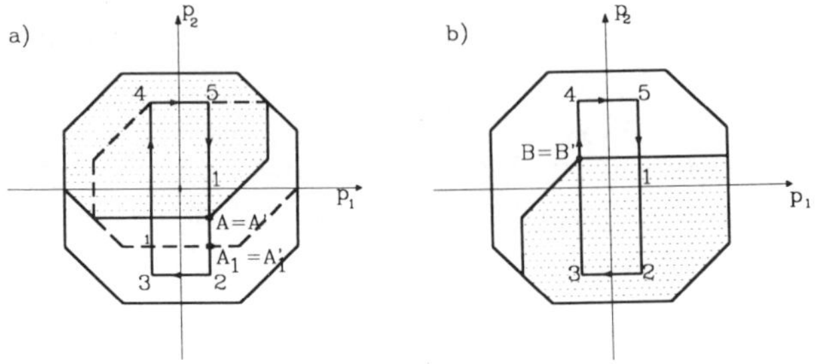

Fig. 10. Modifications of sublimit regions (second case)

Cyclic loading and inadaptation of slackened-rigid-perfectly plastic beam

Consider the two-span beam with symmetrical rotation constraints (0.01 rad) introduced at the point II (cf. Fig. 11a). The beam is loaded by vertical forces P_1 and P_2. The load program coincides with the sides of a square of dimensions 50 kN. From Figs. 11b,c it is seen that the beam will collapse according to the incremental mechanism. In each cycle the combined mechanism occurs (the point A on the load path). Then we observe the beam mechanism (the point B on the load path) in the left span of the beam.

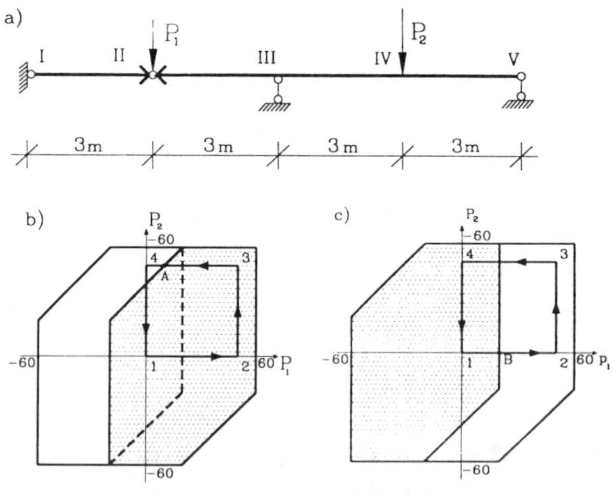

Fig. 11. Cyclic loading of slackened-rigid-perfectly plastic beam, modifications of sublimit loci

Cyclic loading and shakedown of slackened-rigid-perfectly plastic beam

Consider again the two-span beam with symmetrical rotation constraints (0.01 rad) assuming that gaps are located at the support III. The clock-wise path of loading corresponds to the rectangular sides (1-2-3-4) of dimensions 40 kN×50 kN (cf. Fig. 12). It turns out that concentrated plastic deformations appear only in the first cycle of loading at the point IV. During further loading cycles, the beam does not exhibit any plastic deformations and the load program is contained within the stabilized sublimit region (see Fig. 12b,c). Thus, the adaptation of the beam is observed. It should be mentioned that for the counter clockwise cycle of loading the beam also shakes down and the corresponding modifications of sublimit loci are illustrated in Figs. 12d,e. So, we can anticipate that the adaptation of the slackened beam does not depend on the path of loading.

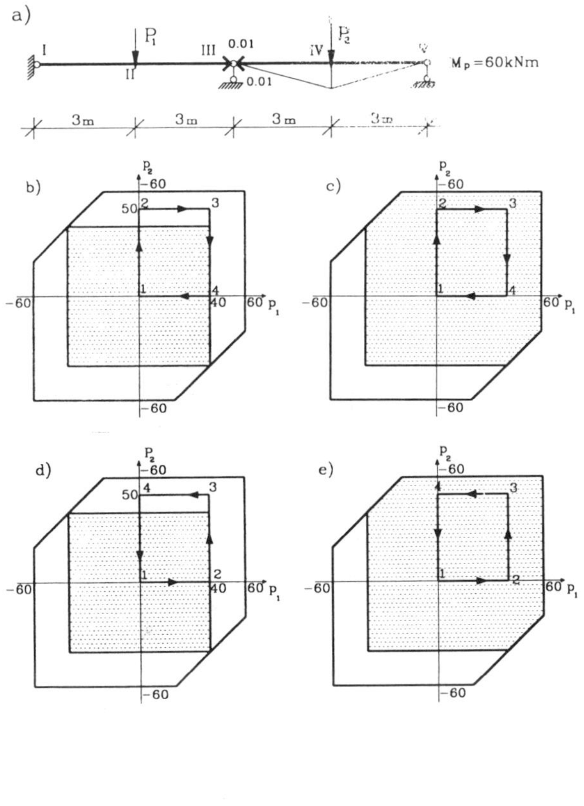

Fig. 12. Cyclic loading and shakedown of slackened-rigid-plastic beam, modifications of sublimit loci

CONCLUSIONS

1. The shape and dimension of elastic or sublimit loci, constructed in the external load space, play an important role in the problems of cyclic loading and shakedown of slackened systems. The presence of gaps implies a non-linear behaviour of the structure before the appearance of plastic deformations. It is known that such a property can lead to a non-convex elastic response of the structure. The non-convexity of elastic and sublimit regions seems to be the main source of difficulties appearing in the construction of theorems about shakedown for slackened systems.
2. The numerical examples allow us to suppose that elastic and sublimit regions are star-shaped (i.e. only one point of intersection of radius-vector and boundary of a region can occur).
3. A new problem of *shakedown of slackened-rigid-perfectly plastic structures can be formulated.* The sufficient condition for adaptation of a structure consists in the formation of time-independent plastic strains such that the corresponding sublimit region contains the prescribed load domain. A link between this formulation and the classic one is clearly seen. The sublimit region and steady plastic deformations correspond to the elastic region and the residual stresses, respectively.
4. Taking into considerations the presence of clearances, a modification of the classic definition of shakedown should be made, namely:

"A slackened structure shakes down if after a finite number of load cycles *a time-independent plastic strain field* forms and the stress states do not violate the yield condition".

5. The presence of gaps can induce an increase or decrease in shakedown load of the slackened-elastic-perfectly plastic structure. The shakedown load strongly depends on the location of gaps and clearance moduli. The optimization of them can lead to a considerable increase in shakedown load with respect to either plastic fatigue or incremental collapse.
6. From the numerical experience of the authors it follows that the shakedown load is independent of the load path and initial distortions.
7. It has been shown that small gaps are beneficial, but it is not profitable to have them too large. In the latter case the shakedown load can exhibit a strong discontinuity for a certain singular value of the clearance modulus.
8. The qualitative modifications of the shape and relatively large shifting of sublimit regions appear to be characteristic features of slackened-rigid-perfectly plastic structures under variable loads. This remark also applies also to elastic regions for slackened structures made of the elastic-perfectly plastic material.
9. The examples included in the work refer only to beam structures. Nevertheless, the FEM-oriented approach allows one to solve more complex systems (plates, shells, continuum media). Moreover, passing into a micro scale and taking into consideration the results obtained for slackened systems, we can suppose that the non-associated flow rule and the non-convexity of yield region result from the presence of internal gaps in the material.

10. The problem of slackened systems is still insufficiently clear and a lot of problems remain open, especially when the variable loads are applied.

REFERENCES

1957 PRAGER, W., "Ideal locking materials", Trans.Soc.Rheol., 1, 169.

1975 MAIER, G., "Mathematical programming methods in analysis of elastic-plastic structures" (in Polish), Arch. In¿. L¹d., 31, 387.

1978 KÖNIG, J.A. and KLEIBER M., "On a new method of shakedown analysis", Bull. Polon. Acad. , Sci. Ser. Sci. Tech., 26, 165.

1983 GAWĘCKI A. ,"Nonlinear effects of joint clearances on behaviour of bar structures", 24-th Polish Solid Mech. Conf., Jachranka (Poland)

1985 BORKOWSKI, A., "Analysis of Skeletal Structural Systems in the Elastic and Plastic Ranges" (in Polish), 1-st ed., PWN, Warszawa-Poznañ, pp.200.

1985 GAWĘCKI, A., "On an elastic-plastic beam with the rotation constraints at the support", Engg. Trans., 33, 423.

1986 GAWĘCKI, A., "Elastic-plastic beams and frames with unilateral boundary conditions", J.Struct. Mech., 14, 53.

1987 KÖNIG, J.A., "Shakedown of Elastic-Plastic Structures", PWN-Elsevier, Warsaw, pp.267.

1987 GAWĘCKI, A. ,"Elastic-Plastic Bar Structures with Clearances" (in Polish), Poznań Univ. Ed.,ser. "Rozprawy" ,185, Poznañ pp.99.

1988 GAWĘCKI, A. ,"On plastic flow mechanisms in the perfectly plastic-slackened structures", Arch. Mech. 40, 653.

1989 GAWĘCKI, A. ,"Matrix theory of elastic-plastic-slackened structures", Bull. Polon. Acad. Sci. ser.sci.Tech., 37, 23.

1990 KRÜGER, P. and GAWĘCKI, A., "Numerical study on perfectly plastic-slackened systems", 28-th Polish Solid Mechanics Conference, Kozubnik (Poland)

1992 GAWĘCKI, A. ,"Elasto-plasticity of slackened systems", Arch. Mech., 44, 363.

1993 GAWĘCKI, A., "Bounds on energy in discrete deformable systems", Arch. Mech. 45, 439.

PLASTIC ANALYSIS AND DESIGN OF SKELETAL STRUCTURES ACCOUNTING FOR THEIR SENSITIVITY TO RATCHETTING

Hosain Mohamed H. BONDOK,
Faculty of Civil Engineering, Zagazig University, Zagazig, Egypt
Marek JANAS,
Institute of Fundamental Technological Research,
ul. Świętokrzyska 21, 00-049 Warsaw, Poland;

Abstract: A step-by-step linearized FE procedure is proposed for elastic-perfectly plastic analysis of structures accounting for geometry changes due to the appearance of an incremental (or immediate) collapse mechanism. Spatial skeletal structures are considered, with biaxial bending, torsion and axial forces taken into account. A linear programming formulation of the shakedown and limit analysis problems is adopted and the post-yield behaviour approach is used. The structure configuration is modified according to the kinematics of the collapse mode corresponding to the preceding step of the analysis. In this way, evolution of the shakedown load domain and of the limit load surface, following the structure deformation, is studied. The corresponding load multipliers obtained at a chosen level of acceptable deformation permit to determine the structure sensitivity to deformations due to the initiation of the immediate or incremental collapse process. Depending on this sensitivity, a modification of the nominal safety factor may be proposed. The case studies revealed some interesting features of the spatial structures behaviour during the advanced ratcheting process.

INTRODUCTION

The safety margin of the structure depends on the consequences of its collapse, i. e., on its behaviour in the post-yield or adaptation phase. That is why an analysis concerning these phases is of practical interest and, in the case of structures

Z. Mróz et al. (eds.), Inelastic Behaviour of Structures under Variable Loads, 419–432.

undergoing exceptional (single or repeated) loading, becomes necessary. Although accounting for physical and/or geometrical non-linearities results in a serious complication of the computation process, numerical computation facilities available today permit the analysis of large-scale systems, including even very sophisticated constitutive laws, large deformations and complex boundary conditions. Unfortunately, all the procedures aiming at non-linear problems have an important weak point: their dependence on the loading history. In many important cases, and probably in the majority of cases of civil engineering structures, the loading history is unknown; only a domain for an arbitrary variation of loads may be specified. This fact makes it very difficult and expensive or frequently impossible to apply a direct approach needed for a realistic analysis and design.

Here, our attention is focused on the classes of structure response that do not depend upon the history preceding the loading situation considered. The above concerns the cases of unconstrained plastic flow (incipient immediate collapse), adaptation (shakedown load) and, of course, linear elastic behaviour. All these classes may be treated as particular cases in the framework of the shakedown analysis.

Determination of the surfaces of limit loads and of the shakedown domains in the space of load parameters is already a standard task described in many monographs (e.g., COHN and MAIER [1979], MASSONNET and SAVE [1978], HORNE [1979], KÖNIG [1987], SAWCZUK [1989]) and enters into advanced courses of structural mechanics. This concerns, of course, the geometrically linear approach. Taking into account geometrical non-linearities in the framework of the shakedown theory was proposed by MAIER [1973], WEICHERT [1990], TRITSCH and WEICHERT [1993] and STUMPF [1993]. These proposals are either oriented, first of all, toward elastic pre-shakedown deformations or needed a path dependent analysis. Their applicability to large-scale structures seems to be still limited, especially when a generalised variable description is needed. It should be remarked that in the general case of geometrically non-linear formulation, the existence of history-independent shakedown load domains is disputable.

It seems that a possible way of overcoming these difficulties is the use of an approach known as the post-yield analysis, which may be applied both to the immediate collapse by an unconstrained plastic flow and to the incremental collapse by ratcheting. At each step of the deformation process a geometrically linear problem is solved, with the configuration of the structure modified using a collapse mechanism determined at the preceding step. In spite of the obviously approximate character of this approach (pre-collapse deformations are neglected), it may give reasonably accurate results for structures whose elastic deformability is small.

The post-yield approach was probably proposed first by ONAT [1955] and extensively developed, e.g., for frames (HORNE [1979], DUSZEK and SAWCZUK [1976]) and plates (SAWCZUK and WINNICKI [1965], JANAS [1968]). The same idea may be useful in the shakedown analysis. After the first proposals by DAVIES [1967], this approach was applied by KÖNIG [1980] to study instability of the inadaptation process and was developed, together with SIEMASZKO [1985], into an effective method for plane frames. Currently this approach is used to deal, in a unified way, with all the history-independent types of spatial skeletal structures response.

FORMULATION AND PROCEDURE

The loading process is represented, in a standard way, by a finite number m of scalar parameters $\beta_i(t)$ independently varying in time, and by a spatial distribution (configuration) of "unit" body forces and surface tractions $\mathbf{t}_i(x)$ corresponding to these parameters:

$$\mathbf{t}(x,t) = \sum \beta_i(t)\mathbf{t}_i(x) \tag{1}$$

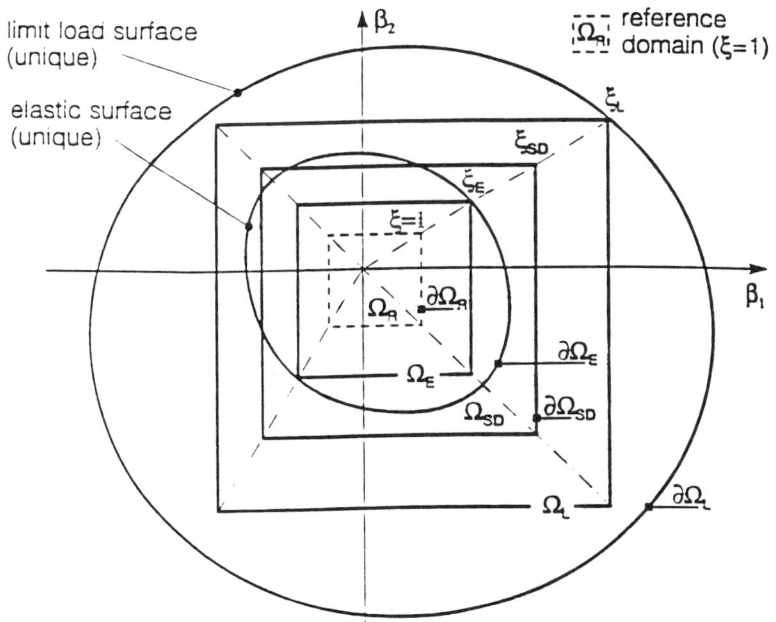

Fig. 1. Load domains in the space of load parameters.

The loading history represents a path in an m-dimensional space of these parameters. Domains of plastic (Ω_L) , shakedown (Ω_{SD}), and purely elastic (Ω_E) responses are enclosed inside the corresponding convex surfaces $\partial\Omega$ (Fig. 1). The limit-load $(\partial\Omega_L)$ and the elastic-limit $(\partial\Omega_E)$ surfaces are unique, whereas the shakedown surface $(\partial\Omega_{SD})$ is determined depending on the choice of the domain shape. A reference domain $\mathbf{\Omega}_R$ (2) defining this shape should be chosen and the shakedown domain will be its homothetic expansion by a multiplier ξ_{SD}

$$\gamma_i^- \leq \beta_{iR}(t) \leq \gamma_i^+ \tag{2}$$

$$\xi_{SD}\gamma_i^- \leq \beta_i(t) \leq \xi_{SD}\gamma_i^+ \tag{3}$$

Determining the shakedown load means determining the intensity factor ξ_{SD} corresponding to a given shape of the domain. To permit specification of the corresponding multipliers concerning the limit load ξ_L and at the elastic limit ξ_E, domains homothetic to the shakedown (and reference) domain inscribed into the limit-load surface $\partial\mathbf{\Omega}_L$ the elastic limit surface $\partial\mathbf{\Omega}_E$, respectively, are looked for (Fig. 1).

The lower-bound (Melan) and the upper-bound (Koiter) theorems for the shakedown load multiplier ξ_{SD}, when applied to structures discretized into lumped compliance finite element may be presented using Maier classical formulation (MAIER [1969]) as a dual pair of linear programming problems:

$$\begin{aligned} \xi_{SD} &= \max\{\xi \mid \xi\mathbf{d} + \mathbf{N}^T\mathbf{s}^r \leq \mathbf{k}, \quad \mathbf{C}^T\mathbf{s}^r = 0\}, \\ \xi_{SD} &= \min\{\mathbf{k}^T\underline{\lambda} \mid \mathbf{d}^T\underline{\lambda} = 1, \quad \mathbf{Cu} = \mathbf{N}\underline{\lambda} \quad \underline{\lambda} \geq \mathbf{0}\}. \end{aligned} \tag{4}$$

Here $\mathbf{s}$ is a global vector of generalized stresses with four components (two bending moments, a torque, and an axial force) at each nodal cross-section, and the matrix $\mathbf{N}$ reassembles piece-wise yield criteria for all elements:

$$\mathbf{f} = \mathbf{N}^T\mathbf{s} - \mathbf{k} \leq \mathbf{0}, \tag{5}$$

with $\mathbf{k}$ being the vector of plastic moduli. The associated flow rule integrated over a cycle reads

$$\mathbf{Cu} = \mathbf{N}\underline{\lambda}, \tag{6}$$

with $\mathbf{C}, \mathbf{u}$ and $\underline{\lambda}$ standing for a compatibility matrix, the vector of nodal displacements and the vector of plastic flow intensity, respectively. A reduced envelope of elastic stresses

$$\mathbf{d} = \max_{\Omega_R} \mathbf{N}^T\mathbf{s}^e \tag{7}$$

corresponds to loads $\mathbf{p}$ varying inside the reference domain Ω_R. Indices e and r denote elastic and self-equilibrated residual stresses, respectively.

If an elastic-limit multiplier ξ_L has been looked for, the solution is directly obtained from (4) in the absence of residual stresses. The yield criterion (5) should be replaced by an elastic-limit criterion corresponding to the first plasticity appearing in the cross-section.

The limit analysis problem may be represented also using Eq. (4), with the envelope **d** reduced to values corresponding to vertices of the reference domain. However, it is more convenient to use a classic formulation of the limit analysis theorems (COHN and MAIER [1979]):

$$\xi_L = \max\left\{\xi \mid \mathbf{C}^T\mathbf{s} = \xi\mathbf{p}, \quad \mathbf{N}^T\mathbf{s} \le \mathbf{k}\right\}, \underline{\dot{\lambda}}$$
$$\xi_L = \min\left\{\mathbf{k}^T\underline{\dot{\lambda}} \mid \mathbf{p}^T\dot{\mathbf{u}} = 1, \quad \mathbf{C}\dot{\mathbf{u}} = \mathbf{N}\underline{\dot{\lambda}}, \quad \underline{\dot{\lambda}} \ge \mathbf{0}\right\}. \tag{8}$$

To solve the above problems for space skeletal structures a PC program SDLAS has been prepared (BONDOK, JANAS, and SIEMASZKO [1993]). It uses elastic solutions needed for determination of the envelope **d** derived from the classic SAP program.

Solution of the problems (5) or (8) gives the load multiplier and the corresponding collapse mode (incremental or immediate, respectively). According to the post-yield approach, the above formulation may be applied to the structure deformed according to the corresponding collapse mode. Subsequent deformed geometries are determined by the vectors of nodal coordinates

$$\mathbf{x}^{\delta+1} = \mathbf{x}^{\delta} + \mu^{\delta}\mathbf{u}^{\delta} \tag{9}$$

with the normalizing factor μ determined from the condition that an increment of the maximum displacement is equal to a chosen deformation step $\boldsymbol{\Delta}$. The program SDLAS determines the current collapse mode and automatically modifies the structure geometry for the analysis of the next step.

Rigorously speaking, the above procedure does not apply for inadaptation by alternating plasticity. There is no collapse mode in this case and the procedure stops after the first step. The alternating plasticity may be considered, following PYCKO and MRÓZ [1992], as an elastic-limit problem, with the envelope **d** corresponding to a reduced reference load domain

$$\xi_{SDA}\left(\gamma_i^+, \gamma_i^-\right) = \xi_E\left(\pm\gamma_i^{'}\right) \quad \gamma_i^{'} = \frac{1}{2}\left(\gamma_i^+ - \gamma_i^-\right) \tag{10}$$

The load multiplier corresponding to the alternating plasticity ξ_{SDA} may be determined at each step of an incremental inadaptation process, checking if the latter is more dangerous.

Any piece-wise linear yield criterion (5) may be used, but in the actual version of the procedure, a so-called bisectrix approximation (GRIERSON [1977]) is applied. Vertices of the yield polyhedron are on axes and on the hyper-bisectrix in each octant of the 4-dimensional space of generalized stresses. Finding vertices on axes (plastic moduli) is elementary; those on the bisectrices are determined as approximations of non-linear yield criteria known from case studies or found by the limit analysis procedure applied to the yielding of the cross-section. In normalized coordinates, vertices on axes are, (0,1,0,0),... and in each octan (±d,±d,±d,±d); for PWL surfaces inscribed in exact ones this coordinate varies between $d = 0.46$ and $d=0.77$, depending on the shape of the cross-section and the number of the generalized variables involved (GRIERSON [1977]). To permit a better PWL approximation of the exact yield surface obtained from the FE limit analysis of the cross-section, the octant points may be chosen outside the hyper-bisectrices, with the coordinates $d_i(d_1, d_2, d_3, d_3)$ appropriately chosen.

The piece-wise linear formulation may appear time-consuming if a more exact approximation of the yield surface is needed. In this case a min-max formulation (JANAS, PYCKO and ZWOLIŃSKI [1994], PYCKO [1994]) permitting the use of non-linear yield criteria will be probably more competitive. However, if linearized yield criteria are used, it will be possible to prepare a catalogue of plastic properties of structural cross-sections containing plastic moduli under separately acting cross-sectional forces and coordinates of the octant points; the latter may be chosen not only depending on the shape of the cross-section but also on the type of the loading situation and the geometry of the whole structure.

EXAMPLE

We shall report here some results from a quantitative study concerning a simple 4-leg clamped frame with a hollow-box cross-section (Fig. 2) under a system of three independently varying sets of loads. Different geometry data and different cross-section properties were considered. A geometrical softening effect under downward loads is shown in Fig. 3.

The octant PWL approximation of the yield criterion for $d = 0.5$ (polyhedron inscribed into the exact surface, see GRIERSON [1977]) is compared with a lower-bound (pyramid) and with an upper-bound (cube) approximations. The latter is disqualified by the results from Fig. 3.

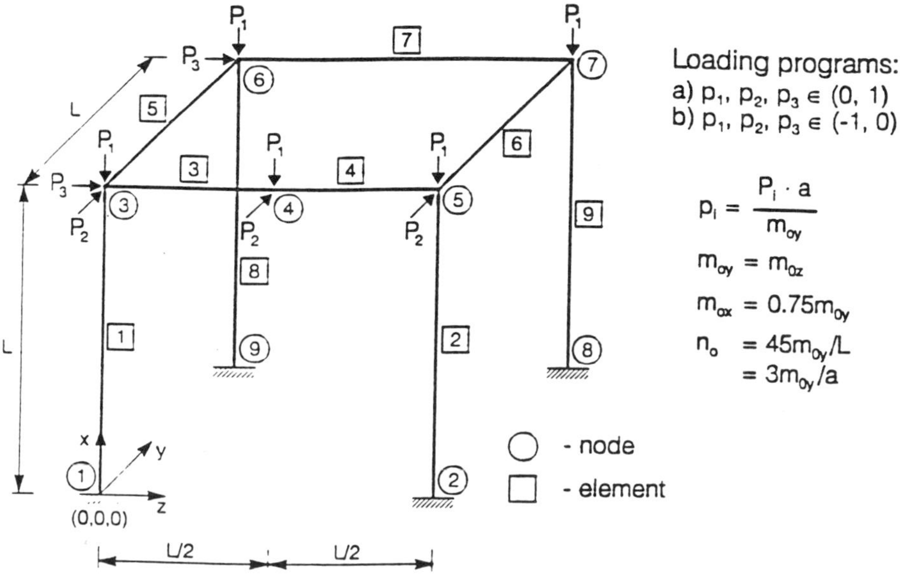

Fig. 2. Frame geometry and loads.

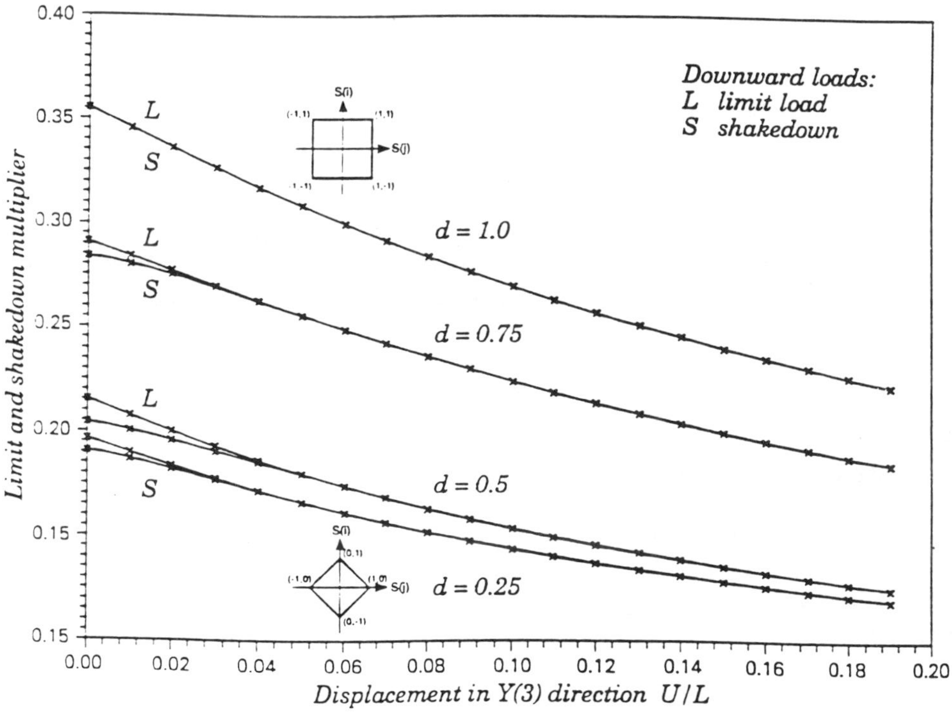

Fig. 3. Comparison of results for different yield criteria approximation.

SENSITIVITY TO GEOMETRICAL EFFECTS AND A MODIFIED SAFETY FACTOR

The post-yield analysis consists of a step-by-step linearization of a non-linear problem, with the shakedown or limit-analysis procedure sequentially applied to deformed configurations. If the analysis is limited to the first two steps it will correspond to a finite-difference formulation of a sensitivity problem. A measure of the structure response (a limit or a shakedown load multiplier ξ_L and ξ_{SD} , respectively) is considered as a function of another measure of the response: a maximum displacement rate $\max|\mathbf{u}|$ in the current collapse (immediate or incremental) mechanism taken as an independent value. This measure is treated as an imperfection of the ideal (i.e. initial) configuration.

Load multipliers determined at two deformation levels are needed to describe the secant sensitivity

$$\nu_k = \frac{\xi_k^f - \xi_k^i}{|\mathbf{u}^*|^f - |\mathbf{u}^*|^i} \qquad k = L \quad \text{or} \quad SD \tag{11}$$

The undeformed $(\delta = 0)$ configuration is usually chosen as the initial one (i), but it may also include some supposed initial imperfections. The scaling factor μ^f to describe the final configuration following (9) should be chosen in such a way that the maximal displacement is equal to the value chosen as admissible

$$\max\left(\mu^f |\dot{\mathbf{u}}^*| - |\mathbf{u}^*|_{\mathrm{adm}}\right) = 0 \tag{12}$$

The asterisk (*) means that the corresponding vectors may contain, in certain cases, only some components considered to be of special importance. The standard way is to adopt as the control value in (11) the generic parameter of the SDLAS program, $\max|\mathbf{u}|$.

Components of the vector $\mathbf{u}_{\mathrm{adm}}$ may be defined, e.g., as a fraction of the span or the height of the structure, following the experience from incremental analysis of selected prior-to-collapse histories. It should be remarked that the rules of selecting admissible values are different for geometrical hardening $\left(\xi_k^f > \xi_k^i\right)$ and for softening $\left(\xi_k^f < \xi_k^i\right)$. For hardening, the increase in the load multipliers corresponds simply to the maximum deformation tolerable from the service point of view. For softening, a decrease in the multipliers will depend on prior-to-collapse deformation and/or on supposed initial imperfections.

The resulting value μ^f may be far larger than the value μ^δ corresponding to a single Δ accepted in (9). Therefore, the values ξ^f should be calculated with a sequence of Δ steps from δ^C to δ^f. However, it is sufficient to verify if the collapse modes at the $i = 0$ and f configurations were the same. The sequential procedure will be needed only in the rare cases, when the collapse mode changes before the final configuration is attained.
If we design a structure to satisfy a safety factor S required by the code, with respect to the collapse load of the undeformed structure, its real safety level will be different, depending on the phenomenon of geometrical hardening or softening. To maintain the real security level required by the code, we can either apply the code factor to the collapse load determined for the structure deformed up to a certain control level, or perform the analysis for the undeformed structure, only using a modified safety factor S'. The code factor S will be increased or decreased depending on the geometrical softening or hardening, respectively. In the simplest case, with the initial secant sensitivity taken with respect to the undeformed structure, we shall have

$$S_k' = S_k\left(1 - \frac{\xi_k^f - \xi_k^o}{\xi_k^o}\alpha\right), \tag{13}$$

with α being a penalty factor amplifying or attenuating the influence of the geometrical effects, and depending, e.g., on the uncertainty level. Here it will be assumed that $\alpha = 1$. Values of the factor S' may be determined for some benchmark geometries and loading programs and then applied to selected class of design situations.

The above procedure is illustrated with a frame shown in Fig. 4a under upward and downward loads. The corresponding limit and shakedown load multipliers are plotted against the sway displacement in Fig 4b. The values ξ_k^f taken from this figure at an allowable sway displacement equal to L:150, give the modified safety factor S' as follows:
- for upward loads:

$$S_L' = 0{,}930\ S_L \qquad S_{SD}' = 0{,}983\ S_{SD}$$

- for downward loads:

$$S_L' = 1{,}062\ S_L \qquad S_{SD}' = 1{,}045\ S_{SD}$$

When a class of different load configurations should be considered, it may be advantageous to use envelopes for the load multipliers for the undeformed structure indicating a level of its sensitivity to geometrical effects. An example of such a display is given in Fig. 5. Corrections to the code factors S are proportional to the ordinates traced along the sides of the envelopes. For the sake of the simplicity of the presentation, a loading program corresponding to only two-dimensional collapse modes is considered. Such a representation was used, e.g., in DUSZEK and SAWCZUK [1976] to indicate the tangent stability of the immediate collapse mode.

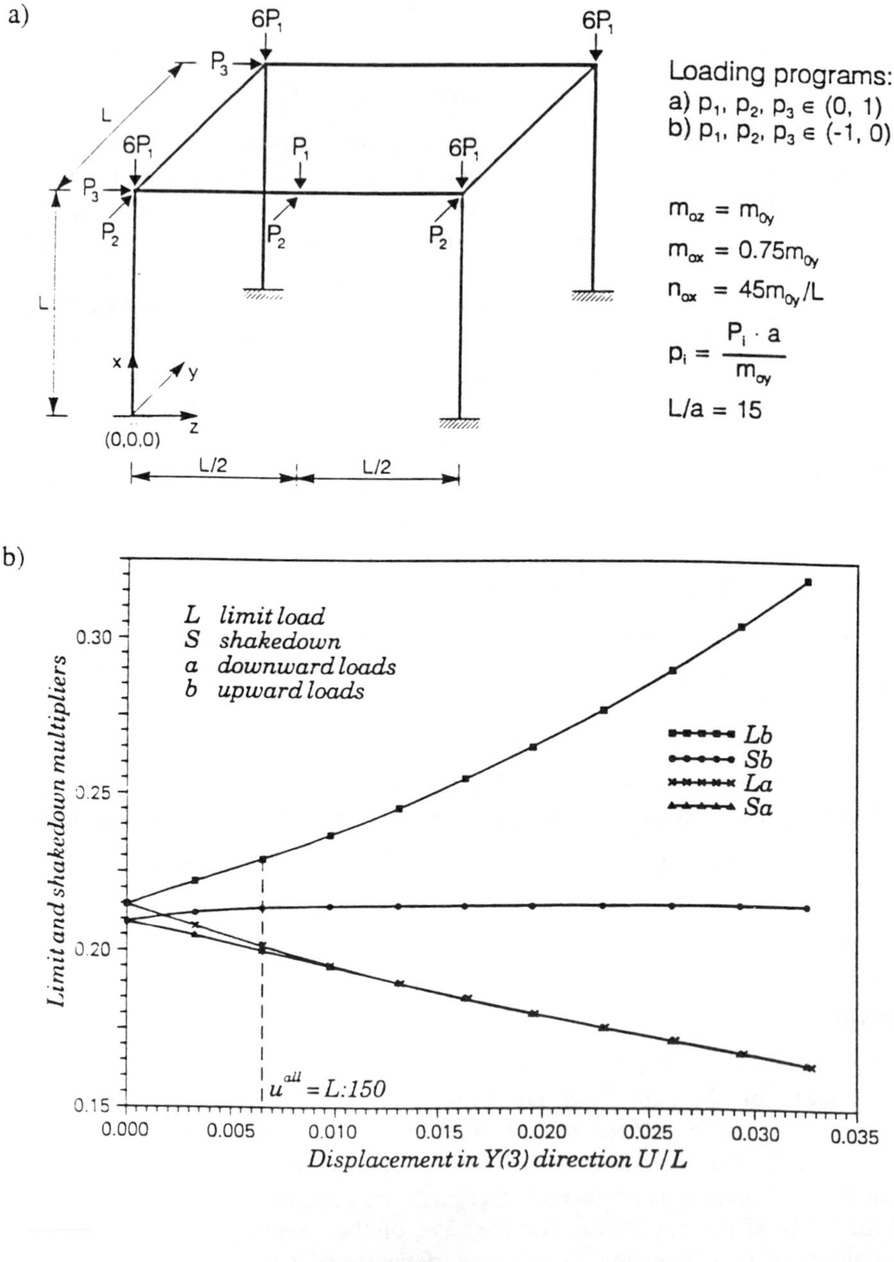

Fig. 4. Stable and unstable behaviour of a space frame

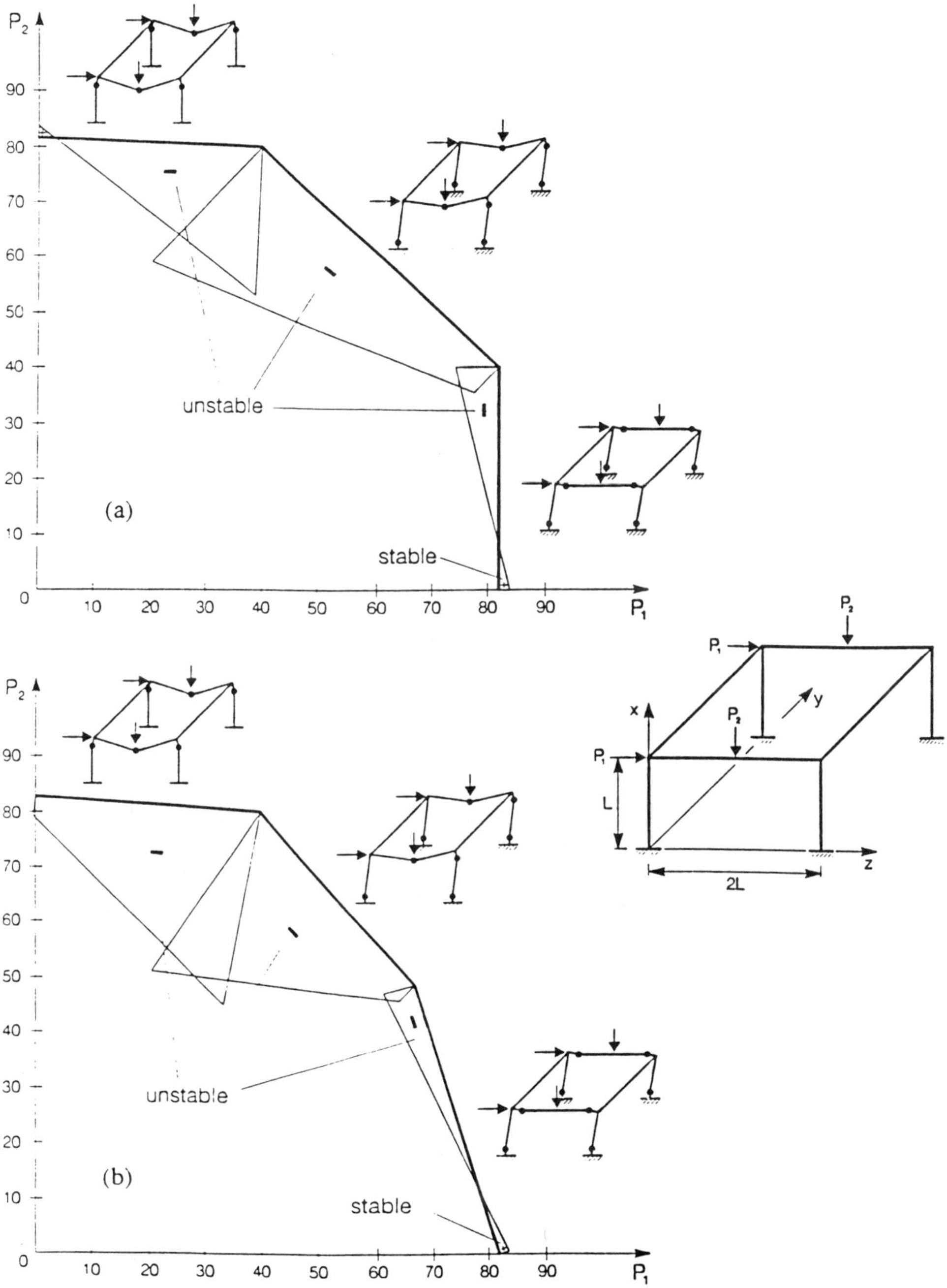

Fig. 5. Envelopes for: (a) Limit load multiplier, (b) Shakedown load multiplier

FINAL REMARKS

The procedure that has been implemented gives a unified analysis of all history-independent responses of the structure. It permits an approximate but simple analysis of the inadaptation and post-yielding processes. The above concerns also cases when an incremental analysis is not feasible because the loading history is unknown. All this can be done using standard FE procedures and data from commercial codes.

Geometry changes taken into account are due exclusively to compatible plastic deformations (collapse mechanisms). Therefore, the approach may concern post-collapse rather than prior-to-collapse processes. To account for elasticity of more compliant systems, the method should be combined with proposals aimed at elastic non-linearities MAIER [1972], TRITSCH and WEICHERT [1993], STUMPF [1993].

Since the procedure permits current modifications of the structure geometry, it may also be used to assess safety of deformed/damaged structures, with the configuration changes obtained from in-situ measurements.

The two-step procedure permitting determination of the secant sensitivity seems to be a more reliable way for estimation of the real safety of the structure than the verification of the initial stability of the collapse process (MAIER [1973], DUSZEK [1973]). As it appears in case studies (BONDOK [1993]), sometimes a process initially unstable may stabilize quickly and vice-versa. Our approach deals with an average sensitivity concerning a considered domain of deformation.

The case studies (BONDOK [1993]) revealed an interesting feature concerning the structural behavior at large displacements. The limit load multipliers appeared to be more sensitive to geometry changes than the shakedown multipliers. The latter are interior to the former and, therefore, at geometrical hardening the two corresponding curves are divergent, whereas at softening they converge. In many cases (cf. Fig. 3) the shakedown curve soon joins the limit-load curve. This effect seems to be more visible when the structure is more complex and torsion-resistant. We can interpret this phenomenon as an ability of the structure to improve its geometry, provided the complexity of the geometry leaves a sufficient freedom for that. That corresponds, in a manner, to a plastic redistribution of stress in continuous media.

ACKNOWLEDGEMENT

The work was supported partially by the Scientific Research Council (KBN) under the Research Project 30978-91-01.

REFERENCES

1955 ONAT, E.T., On certain second-order effects in limit design of frames, J.Aeronaut.Sci., **22**, 681-684.

1965 SAWCZUK, A., WINNICKI, L., Plastic behavior of simply supported concrete plates at moderately large deflections, Int.J.Solids Struct., **1**, 97-110.

1967 DAVIES J.M., Collapse and shakedown loads of plane frames, J. Struct. Div., Proc. ASCE, **93**, 35-50.

1968 JANAS, M., Large plastic deformations of reinforced concrete slabs, Int. J. Solids Struct., **4**, 61-74.

1969 MAIER, G., Shakedown theory in perfect elastoplasticity with associated and non associated flow laws, Meccanica, **4**, 250-260.

1972 MAIER, G., A shakedown matrix theory allowing for work hardening and second-order geometric effects, Foundation of Plasticity (Ed. A. Sawczuk) 417-433, Pergamon-Press, Oxford.

1973 DUSZEK, M.K., On stability of rigid-plastic at the yield-point load, Bull.Ac.Pol.Sci., **IV 21**, 79-87.

1976 DUSZEK, M.K., and SAWCZUK, A., Stable and unstable states of rigid-plastic frames at the yield point load. J.Struct.Mech., **4**, 33-47.

1976 MASSONNET, CH., SAVE, A., Calcul plastique des constructions, V.I., Nelissen, Liege.

1977 GRIERSON, D.E., ABDEL-BASSET, S.B., Plastic analysis under combined stresses, J.Eng.Mech. Proc ASCE, **103**, 837-854.

1979 COHN, M.Z., MAIER, G. (Editors), Engineering Plasticity by Mathematical Programming, Pergamon Press, London.

1979 HORNE, M.R., Plastic Theory of Structures, Pergamon Press, London 1979.

1980 KÖNIG, J.A., On stability of the incremental collapse, Arch. Inż. Lądowej, **26**, 219-229.

1985 SIEMASZKO, A., KÖNIG, J.A., Analysis of stability of incremental collapse of skeletal structures, J.Struct.Mech., **13**, 301-324.

1987 KÖNIG, J.A., Shakedown of Elastic-Plastic Structures, PWN-Elsevier, Warsaw - Amsterdam.

1989 SAWCZUK, A., Mechanics and Plasticity of Structures, PWN-Ellis Harwood Warsaw - Chichester.

1990 WEICHERT, D., Advances in the geometrically nonlinear shakedown theory, Inelastic Solids and Structures (Eds. Kleiber, M., König, J.A.), 489-502, Pineridge Press.

1992 PYCKO, S., MRÓZ Z., Alternative approach to shakedown as a solution of a min-max problem, Acta Mechanica, **93**, 202-222.

1993 BONDOK, H.M., Geometry changes in the elastic-plastic analysis of steel spatial skeletal structures, IFTR Reports, 37, (PhD Thesis).

1993 BONDOK, H.M., JANAS, M., and SIEMASZKO, A., Numerical program for post-yield and inadaptation analysis of space skeletal structures. Proceeding of the XI Polish Conf. on Computer Methods in Mechanics, Ed. W. Gilewski, T.U. of Kielce, 129-138.

1993 TRITSCH, J.B., WEICHERT, D., Shakedown of elastic-plastic structures at large deformations: a comparative study of static shakedown theorems, ZAMM, **73**, T767.

1993 STUMPF, H., Theoretical and computational aspects in the shakedown analysis of finite elastoplasticity, Int. J. Plasticity, **9**, 583-602.

1994 JANAS, M., PYCKO, S., ZWOLINSKI, J., A min-max procedure for the shakedown analysis of skeletal structures, to be published in Int. J. Mechanical Sciences.

1994 PYCKO, S., An indirect incremental method for the shakedown analysis based on a min-max approach, In: Inelastic Response of Structure under Variable Loads, Ed. Z. Mróz, D. Weichert, S. Dorosz, Kluwer 1994.

SHAKEDOWN OF RAIL CORRUGATIONS

Wiktor GAMBIN
Institute of Fundamental Technological Research
Polish Academy of Sciences
ul. Świętokrzyska 21, 00-049 Warsaw, Poland

Abstract: A model of the elastic-plastic behaviour of a rail under moving and cyclically changing loads is proposed. Simple analytical expressions enable us to estimate the changes of the limit shear stress and residual stresses under the running rail surface after repeated passages of a wheel. It is shown that magnitudes of local hardening and residual stresses stabilize after a few passages of the load. The places of local hardening and peaks of residual stresses do not move along rails and they correspond to the positions of rail corrugations.

INTRODUCTION

Corrugations are wave-like patterns of wavelengths 35-85 mm and depth of up to 1 mm on the running surface of rails. They occur near stations where traction and braking are heavy, creating an environmental noise, and damage to track components. To eliminate them, many authors have looked for mechanisms for their formation. MAIR [1980] found that track resonant frequencies are close to those of observed corrugation frequencies. Assuming that vibrations of track can give loads sufficiently high to cause plastic deformation of the rail steel, BOGACZ [1992] showed that plastic yield played a significant role in the process of formation of corrugations.

The aim of the paper is to give a simple analytical model of the process. The model will be based on the elastic-plastic effects connected with cyclic overloads of rails. Our investigations will be focused on a thin wheel/rail contact layer under the running surface of a rail. Two quantities determine the behaviour of the layer material: the current limit shear stress and the longitudinal components of current residual stresses. Due to dynamic overloads, the limit shear stress, uniform in new rails, increases periodically along the running surface of the exploited rails.

Z. Mróz et al. (eds.), Inelastic Behaviour of Structures under Variable Loads, 433–447.

To obtain a reasonable description of rail corrugation formation, the following problems will be considered and presented in turn: an initial residual stress field in a new rail; wheel/rail contact forces - normal and tangential; the elastic stress field caused by these forces; redistributions of the limit shear stresses and residual stresses in the rails after successive passages of moving loads. Two loads are considered: loads moving with constant amplitude; loads moving with amplitude cyclically changing in time. The proposed model of rail behaviour enables us to determine the distributions of the critical shear stresses and the longitudinal residual stresses. Simple analytical expressions allow us to estimate the extreme values of these quantities.

ELASTIC STATE OF A RAIL

In a new rail, there is a residual stress field due to rail straightening in the manufacturing process. One can observe the tensile longitudinal component of residual stresses R_0 under the running surface of roller-straightened rails (Fig. 1). In a working rail, the initial residual stress field changes. After a number of load passages, a compressive longitudinal component of residual stresses R appears on the running surface of rails. It will be shown later that the distribution of the peaks in this residual stress is the same as the distribution of the smaller hardening places. These effects are complementary, but have different natures. The local hardening in the new rail is limited to the material in a thin layer under the running surface of a rail. Changes of the residual stress field due to the passages of the loads affect the whole cross-section of the rail. A new rail, with the initial residual stress R_0, is subjected to wheel/rail contact forces in the utilization process. Distributions of corresponding normal and tangent loads are given in the paper by BRZOZOWSKI, BOGACZ & POPP [1990]. A typical rail load is shown in Fig. 2. Note the complex character of the tangent load. The quantity $2l$ denotes the width of the contact zone. For a vertical load of 200 kN caused by a wheel of radius 900 mm, the parameter l has the value of 10 mm. The distribution of tangent forces depends on the conditions of the wheel motion. A static wheel/rail interaction generates the tangent load shown in Fig. 3 as $\tau''(x_2)$. Under the running surface of rails, such a load produces tensile longitudinal residual stresses during loading and compressive ones after unloading. This is *the ideal rolling contact* (see JOHNSON [1985]). For a train running with a constant speed, the distribution of tangent forces is very similar to this and the resultant force is almost zero. For heavy traction or braking of a train, the distribution of tangent forces is similar to that shown in Fig. 3 as $\tau'(x_2)$. The direction of the resultant force is the same as the direction of the motion, for braking, and opposite when there is traction. This is *the ideal sliding contact.*

We can assume that the total load on the rails is the superposition of three components: a) parabolic normal forces $p(x_2)$, b) parabolic tangent forces due to the sliding friction $\tau'(x_2)$ and c) tangent forces due to the rolling friction $\tau''(x_2)$, which are described by the fifth degree polynomial (Fig. 3):

$$p(x_2) = \frac{p_0}{l^2} (l^2 - x_2^2) \tag{1}$$

$$\tau(x_2) = \alpha \, \tau'(x_2) + (1 - \alpha) \, \tau''(x_2); \quad 0 \leq \alpha \leq 1 \tag{2}$$

where

$$\tau'(x_2) = \frac{\tau_0}{l^2} (l^2 - x_2^2) \tag{3}$$

$$\tau''(x_2) = \frac{\tau_0}{b} x_2 (l^2 - x_2^2) (x_2^2 - a^2) \tag{4}$$

for

$$a^2 = \frac{3l^2 - 5c^2}{l^2 - 3c^2} c^2 \tag{5}$$

$$b^2 = (l^2 - c^2) (c^2 - a^2) c \tag{6}$$

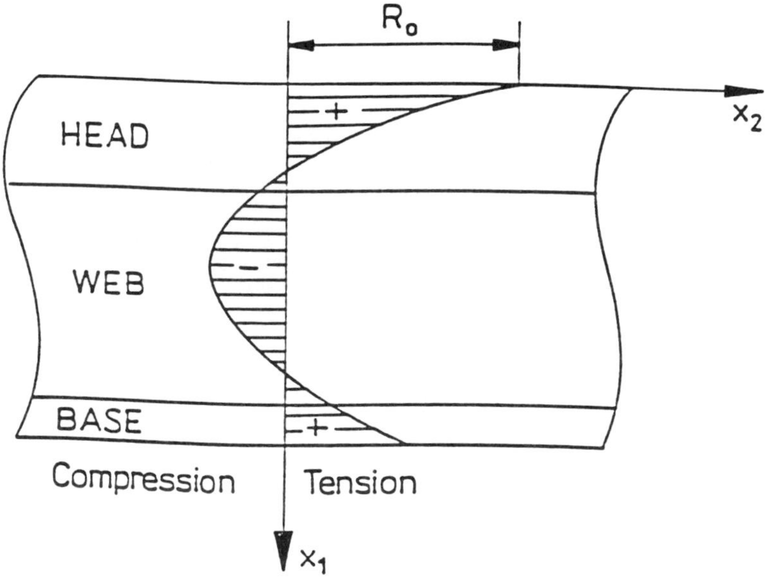

Fig. 1. Longitudinal residual stress in roller-straightened rail.

Here, p_0 is the amplitude of the normal forces, and τ_0 the amplitude of the tangent forces, both for sliding friction and for rolling friction. The quantity $2c$ (for $c < \sqrt{3}/3$) denotes the distance between extremes of the function $\tau''(x_2)$ and depends on wheel/rail friction conditions. It is assumed that $c = l/10$. The ratio:

$$\kappa = \frac{\tau_0}{p_0} \tag{7}$$

describes the friction conditions and it takes values the interval $0.1 \leq \kappa \leq 0.4$. The resultant normal and tangent forces are denoted by P and T, respectively.

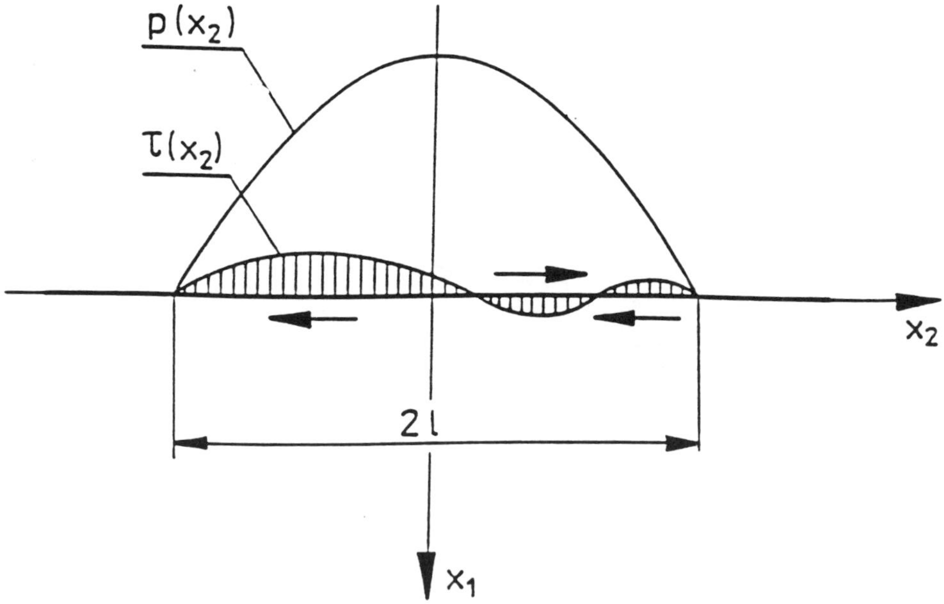

Fig. 2. Typical load of rail; $p(x_2)$ normal forces, $\tau(x_2)$ tangent forces (their directions are shown by arrows).

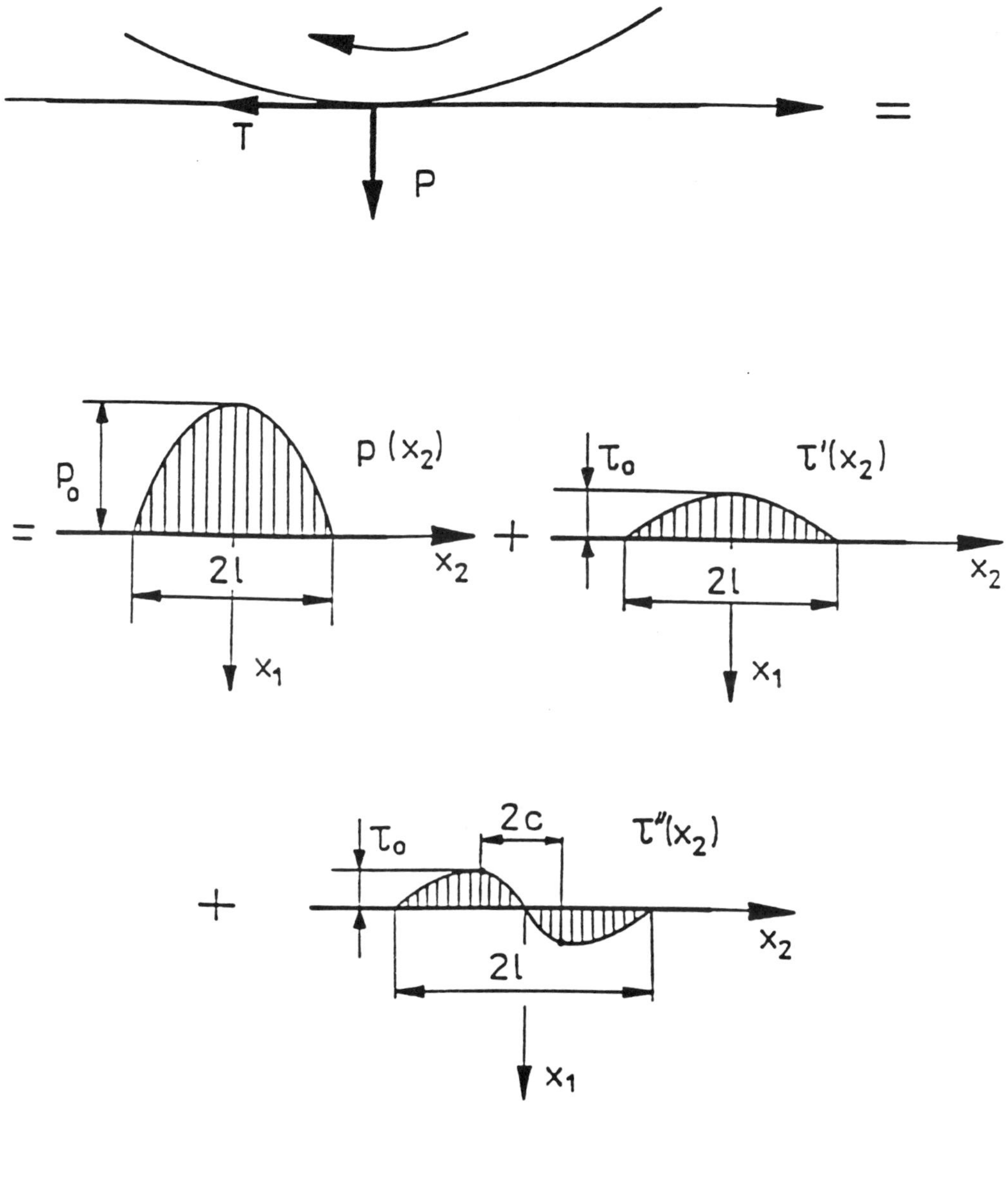

Fig. 3 Assumed decomposition of rail loads.

To find an onset of the plastic yield and to obtain a residual stress distribution in the rail, it is necessary to know the elastic stress state under the applied load. GAMBIN [1994] gives the stress field in the elastic half-space loaded by these normal and tangential forces. It is shown that both the onset of the plastic yield and the value of residual stresses R are determined by the value of $\sigma_{22}(x_1, x_2)$ for $x_1 = x_2 = 0$, denoted by σ_{max}:

$$\sigma_{max} = (A - 1)\, p_0 \tag{8}$$

where

$$A = \kappa\,(1 - \alpha)\,\frac{1}{\pi}\,\frac{l}{l - c}\,\ln\left(\frac{l}{c}\right)^4 \tag{9}$$

To introduce compressive residual stresses R in a new rail, σ_{max} should be a tensile stress. This is possible when the parameter A is large enough; A takes its highest value for ideal rolling contact ($\alpha = 0$) and a strong friction between wheel and rail ($\kappa = 0.4$).

PLASTIC YIELD UNDER MOVING LOAD WITH CONSTANT AMPLITUDE

Consider the moving load of rail shown in Fig. 2. Let the amplitudes p_{max} and $\tau_{max} = \kappa\, p_{max}$ be large enough to cause plastic yield of the loaded zone. Introduce the dimensionless coordinates:

$$r = \frac{x_1}{l} \qquad\qquad s = \frac{x_2 - \dot{x}_2}{l} \tag{10}$$

where $\dot{x}_2$ is the position of the load zone center. The applied forces, normal and tangential, are functions of $\dot{x}_2$ and s. For fixed $\dot{x}_2$, the applied forces are functions of the coordinate s:

$$p(s) = p_{max}(1 - s^2) \tag{11}$$

$$\tau(s) = \tau_{max}\left[\alpha\,(1 - s^2) + (1 - \alpha)\,\frac{s}{\kappa_1}\,(1 - s^2)\,(s^2 - \kappa_2)\right] \tag{12}$$

where:

$$\kappa_1 = \frac{c}{l}\left(1 - \frac{c^2}{l^2}\right)\left(1 - \frac{a^2}{l^2}\right) \qquad \kappa_2 = \frac{a^2}{l^2} \tag{13}$$

The quantity a is defined by (5).

The proposed analysis of plastic effects in rails is only a static one. It is assumed that the dynamic behaviour of running trains is modelled by moving loads acting on the rails. Inertial effects in the plastic zone concentrated in the thin plastic layer under the rail running surface are neglected. The only restriction imposed on the behaviour of the plastic layer is the following:

The shear stress component does not change through the layer thickness, i.e:

$$\sigma_{rs} = \sigma_{rs}(s) \tag{14}$$

Introducing the boundary conditions:

$$\sigma_{rr}(0,s) = -p(s) \qquad \sigma_{rs}(s) = -\tau(s) \tag{15}$$

and the restrictions (14) into the equilibrium equations:

$$\begin{cases} \sigma_{rr,r} + \sigma_{rs,s} = 0 \\ \sigma_{ss,s} + \sigma_{rs,r} = 0 \end{cases} \tag{16}$$

one can obtain the following solution:

$$\begin{cases} \sigma_{rr}(r,s) = -p(s) + \dfrac{d\tau(s)}{ds}\, r \\ \sigma_{ss}(r,s) = C_1 = \text{const} \\ \sigma_{rs}(s) = -\tau(s) \end{cases} \tag{17}$$

The value C_1 can be determined from the elastic solution. Let us denote by $\bar{p}$ and $\bar{\tau} = \bar{p}/\kappa$ the amplitudes of $p(s)$ and $\tau(s)$, respectively, under which the onset of plastic yield occurs. Let $\bar{\sigma}_{max}$ will be the corresponding value of $\sigma_{ss}(0,0)$ in the elastic state. Then:

$$C_1 = \bar{\sigma}_{max} + R \tag{18}$$

where R is the longitudinal component of a current residual stress at the point where the plastic yield occurs. Taking into account (8), one can finally write:

$$C_1 = (A - 1)\,\bar{p} + R \tag{19}$$

In the plastic zone, the following Huber-Mises yield criterion should be satisfied:

$$(\sigma_{ss} - \sigma_{rr})^2 + 4\,\sigma_{rs}^2 = 4\,K^2 \tag{20}$$

The quantity K is an unknown, current shear stress limit.

To determine $\bar{p}$, R and K, consider the plastic yield onset in a new rail after *the first* passing of the load. It takes place on the running surface of the rail ($r = 0$), at the point $s = 0$. Then $\bar{p}$, R and K take values $\bar{p}_1$, R_0 and K_0, respectively, where the values R_0 and K_0 are known. According to (17), the stress state at the point $s = 0$ is the following:

$$\begin{cases} \sigma_{rr}(0,0) = -\bar{p}_1 \\ \sigma_{ss}(0,0) = (A - 1)\,\bar{p}_1 + R_0 \\ \sigma_{rs}(0) = -\,\bar{p}_1\,\kappa \end{cases} \tag{21}$$

Introducing (21) into (20), with K equal K_0, one can determine the amplitude of the normal forces which causes the plastic yield:

$$\bar{p}_1 = \frac{-AR_0 + \sqrt{A^2R_0^2 + (A^2 + 4\alpha^2)(4K^2 - R_0^2)}}{A^2 + 4\alpha^2} \tag{22}$$

We conclude that the stress field in the plastic zone on the running surface of the new rail after *the first* passing of the load is:

$$\begin{cases} \sigma_{rr}(s,0) = p_{max}^2\,(s - 1) \\ \sigma_{ss}(s,0) = \bar{p}_1\,(A - 1) + R_0 \\ \sigma_{rs}(s) = p_{max}\,\kappa\,(s^2 - 1)\,\left[\alpha + (1 - \alpha)\,\frac{s}{\kappa_1}\,(s^2 - \kappa_2)\right] \end{cases} \tag{23}$$

for $p_{max} \geq \bar{p}_1$.

Introducing the solution (23) into the yield criterion (20), one can obtain a value of K in the new rail after *the first* passing of the load. Repeating the calculations, one can find the limit shear stress $K_{(i)}$ after *the i-th* passing of the load with amplitude $p^{(i)}_{max} \geq \bar{p}_{(i)}$:

$$K_{(i)} = \frac{1}{2}\left\{\left[p^{(i)}_{max} + (A - 1)\,\bar{p}_{(i)} + R_{(i-1)}\right]^2 + (2\,\frac{\alpha}{K}\,p^{(i)}_{max})^2\right\}^{1/2} \tag{24}$$

where:

$$\bar{p}_{(i)} = \frac{-AR_{(i-1)} + [A^2R^2_{(i-1)} + (A^2 + 4\alpha^2)(4K^2_{(i-1)} - R^2_{(i-1)})]^{1/2}}{A^2 + 4\alpha^2} \tag{25}$$

for $i = 1,2,3,...$ and $K_{(0)} = K_0$, $R_{(0)} = R_0$, $\bar{p}_{(1)} = \bar{p}_1$.

After *the i-th* passing of the load $p^{(i)}_{max} \geq \bar{p}_{(i)}$, residual stresses can be determined from the general rule:

$$R_{(i)} = \sigma^{(i)}_{ss} - \bar{\sigma}^{(i)}_{max} \tag{26}$$

where $\sigma^{(i)}_{ss}$ and $\bar{\sigma}^{(i)}_{max}$ are the appropriate stress components in plastic and elastic state, respectively. According to (23_2) and (8), one can write:

$$\sigma^{(i)}_{ss} = \bar{p}_{(i)}(A - 1) + R_{(i-1)} \tag{27}$$

$$\bar{\sigma}^{(i)}_{max} = p^{(i)}_{max}\,(A - 1) \tag{28}$$

Then:

$$R_{(i)} = R_{(i-1)} + (A - 1)(\bar{p}_{(i)} - p^{(i)}_{max}) \tag{29}$$

The rules (24-25) and (29) fully describe the problem for a load moving with a constant amplitude. Notice that our approach reduces the analysis of plastic effects to *statically determinate analysis*, i.e. the analysis expressed in terms of stress, without using any stress-strain relations. The only relation describing the plastic behaviour of rails is the Huber-Mises yield condition expressed in terms of a current stress state and a current limit shear stress K. On the basis of the above analysis, kinematical effects may be described independently.

Let us apply these results to the analysis of plastic effects caused by typical wheel/rail interactions. According to data from BHP Steel International Group (MAIR & GROENHOUT [1980]), a maximum wheel load P is 300 kN, and residual stress in roller-straightened rail is $R_0 = 40$ MPa. Taking $l = 10$ mm, $c = 1$ mm, a wheel/rail contact area as 200 mm^2, and $\kappa = 0.4$, one can calculate amplitudes of normal contact forces p_{max} as 1125 MPa, for the resulting loads P equal to 300kN. Heavy traction or braking of the train, when $\alpha = 0$, is assumed.

Consider a series of successive passages of the same load P applied to a new rail. The rules give the following results, which are valid for i = 1,2,3,...:

$K_{(i)} = 648$ MPa, $R_{(i)} = -171$ MPa, when $K_0 = 300$ MPa,
$K_{(i)} = 671$ MPa, $R_{(i)} = -124$ MPa, when $K_0 = 400$ MPa,
$K_{(i)} = 694$ MPa, $R_{(i)} = -78$ MPa, when $K_0 = 500$ MPa.

Notice that for each case, K and R take their final values after the first passing of the load.

Sometimes, due to the used rail technology, the initial residual stress R_0 is much more than the assumed one. Let us consider the last series of passages for the increasing loads applied to a new rail with $R_0 = 120$ MPa. Then:

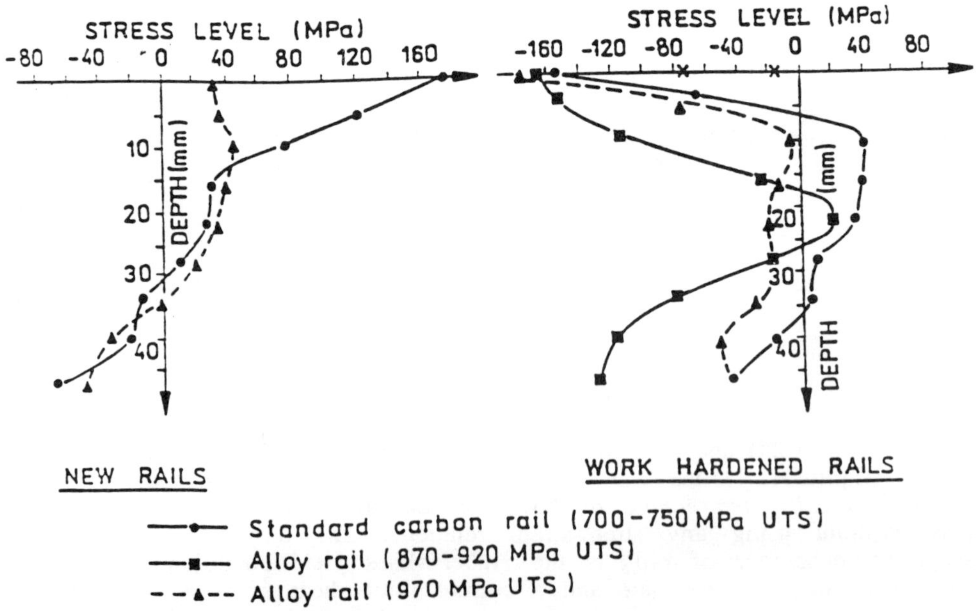

Fig. 4. Redistribution of longitudinal residual stress in rail (after MAIR & GROENHOUT [1980]).

$K_{(1)} = 678$ MPa, $R_{(1)} = -109$ MPa, when $K_0 = 300$ MPa,
$K_{(2)} = 702$ MPa, $R_{(2)} = -63$ MPa, when $K_0 = 400$ MPa,
$K_{(i)} = 725$ MPa, $R_{(i)} = -16$ MPa, when $K_0 = 500$ MPa.

for $i = 3,4,5,...$. Fig. 4 shows the results of tests for standard carbon and alloy rails from MAIR & GROENHOUT [1980]. Notice a qualitative agreement between the calculations and the tests.

PLASTIC YIELD UNDER MOVING LOAD WITH CYCLICALLY CHANGING AMPLITUDE

Our analysis should take into account the fact that the considered loads change cyclically during their motion. Because load oscillations described by smooth harmonic functions lead to very complex relations, we assume the step-wise load variation shown in Fig. 5. In this case, one can regard the load variation as a function of position of the contact zone center $\dot{x}_2$:

$$p_c(\dot{x}_2) = \begin{cases} p_{max} & \text{for } -L_1 + nL \leq \dot{x}_2 \leq L_1 + nL \\ p_{min} & \text{for } L_1 + nL \leq \dot{x}_2 \leq -L_1 + (n+1)L \end{cases} \tag{30}$$

where $n = 0, \mp 1, \mp 2, ...$, and L and L_1 are shown in Fig. 5.

Let us repeat the analysis. Let the values $\bar{p}$ and $\bar{\tau}$ correspond to the plastic yield onset. Consider the stress field components: $\sigma_{rr}(\dot{x}_2,s)$, $\sigma_{ss}(\dot{x}_2,s)$ and $\sigma_{rs}(\dot{x}_2,s)$, in the plastic zone on the running surface of a new rail. It is easy to see that after *the first* passage of the cyclically changing load, these components are given by the rules (23), where $p_c(\dot{x}_2)$ and $\tau_c(\dot{x}_2)$ are taken instead of p_{max} and τ_{max}, respectively. Cyclic oscillations of loads lead to nonuniform fields: $\bar{p}(\dot{x}_2)$, $K(\dot{x}_2)$ and $R(\dot{x}_2)$. For simplicity, the argument $\dot{x}_2$ will be omitted in further considerations. To obtain the fields, one can use the rules (24-25) and (29) with a certain modification. Notice that these rules, when applied to the cyclic loads, lead to a step-wise distribution of K an R. The real distributions of K and R are smooth at the ends of step intervals. Moreover, $\bar{p}$ and R in the rules should determine the constant C_1 in the rule (20). This constant was derived for the case of the initial plastic yield. Then, $\bar{p}$ and R must be taken *at the fronts of plastic zones*. However, the residual stresses R, generated by the thin plastic layers, are partially taken over by the adjacent elastic zones. R takes a value between that in the plastic zone and that in the elastic one. Let $\bar{p}^{f}_{(i)}$ and $R^{f}_{(i)}$ be the values of $\bar{p}_{(i)}$ and $R_{(i)}$ at the fronts of plastic zones. If the distance L is not too large, it is reasonable to assume that:

$$R^{f}_{(i)} = \frac{1}{2}\left[R^{f-}_{(i)} + R^{f+}_{(i)}\right] \tag{31}$$

where $R^{f-}_{(i)}$ and $R^{f+}_{(i)}$ are the values of $R_{(i)}$ at the points of discontinuity.

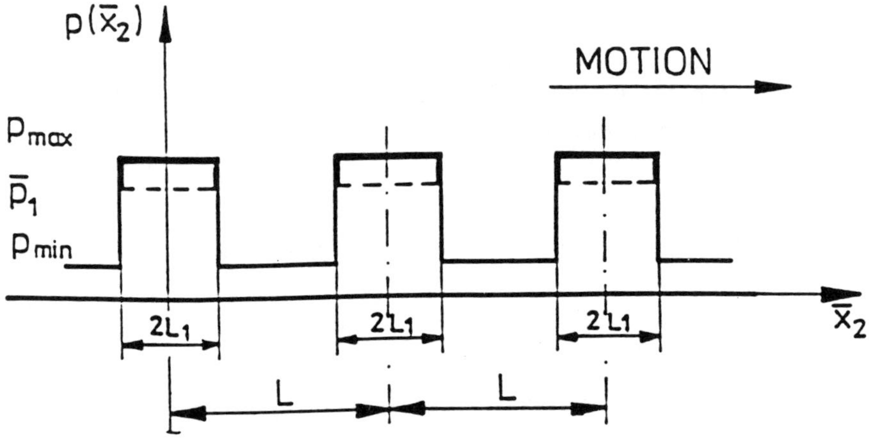

Fig. 5. Moving and oscillating load of rail.

Following the rules (24-25), one can obtain the limit of shear stress after *the i-th* passing of the load:

$$K_{(i)} = \frac{1}{2}\left\{\left[p^{(i)}_{max} + (A-1)\,\bar{p}^{f}_{(i)} + R^{f}_{(i-1)}\right]^2 + (2\frac{\alpha}{\kappa}p^{(i)}_{max})^2\right\}^{1/2} \tag{32}$$

in plastically deformed zones, while $p^{(i)}_{max} \geq \bar{p}_{(i)}$, and $K_{(i)} = K_{(i-1)}$ in the remaining zones of the rail surface layer. The quantity $\bar{p}^{f}_{(i)}$ is calculated from the relation:

$$\bar{p}^f_{(i)} = \frac{-AR^f_{(i-1)} + \left\{ A^2 \left(R^f_{(i-1)} \right)^2 + (A^2 + 4\alpha^2) \left[4K^2_{(i-1)} - \left(R^f_{(i-1)} \right)^2 \right] \right\}^{1/2}}{A^2 + 4\alpha^2} \tag{33}$$

The corresponding residual stresses may be estimated on the basis of the rule:

$$R_{(i)} = R^f_{(i-1)} + (A - 1) \left[\bar{p}^f_{(i)} - p^{(i)}_{max} \right] \tag{34}$$

Notice that the relations (31-34) fully describe the problem.

Now, let us consider a series of successive passages of the same cyclically changing load with amplitudes shifted by the distance $L_1 + L$ (see Fig. 6). Let us assume the same data as previously, i.e. the initial limit shear stress $K_0 = 300$ MPa, the initial residual stress $R_0 = 40$ MPa, $l = 10$ mm, $c = 1$ mm, the wheel/rail contact area 200 mm^2, and $\kappa = 0.4$. For the maximum load P equal to 250 kN ($p_{max} =$ 750 MPa), and $\alpha = 0$, the rules (31-34) lead to the following results (see also Fig. 6):

$K_{(1)}(\dot{x}_2) = 554$ MPa $\quad R_{(1)}(\dot{x}_2) = -114$ MPa $\quad$ for $(4n-1)L_1 \leq \dot{x}_2 \leq (4n+1)L_1$,

$K_{(1)}(\dot{x}_2) = 300$ MPa $\quad R_{(1)}(\dot{x}_2) = 40$ MPa $\quad$ for $(4n+1)L_1 \leq \dot{x}_2 \leq (4n+3)L_1$,

$K_{(2)}(\dot{x}_2) = 554$ MPa $\quad R_{(2)}(\dot{x}_2) = -114$ MPa $\quad$ for $(4n-1)L_1 \leq \dot{x}_2 \leq (4n+1)L_1$,

$K_{(2)}(\dot{x}_2) = 539$ MPa $\quad R_{(2)}(\dot{x}_2) = -145$ MPa $\quad$ for $(4n+1)L_1 \leq \dot{x}_2 \leq (4n+3)L_1$,

$K_{(3)}(\dot{x}_2) = 554$ MPa $\quad R_{(3)}(\dot{x}_2) = -114$ MPa $\quad$ for $(4n-1)L_1 \leq \dot{x}_2 \leq (4n+1)L_1$,

$K_{(3)}(\dot{x}_2) = 539$ MPa $\quad R_{(3)}(\dot{x}_2) = -145$ MPa $\quad$ for $(4n+1)L_1 \leq \dot{x}_2 \leq (4n+3)L_1$,

$K_{(i)}(\dot{x}_2) = 554$ MPa $\quad R_{(i)}(\dot{x}_2) = -114$ MPa $\quad$ for $(4n-1)L_1 \leq \dot{x}_2 \leq (4n+1)L_1$,

$K_{(i)}(\dot{x}_2) = 544$ MPa $\quad R_{(i)}(\dot{x}_2) = -133$ MPa $\quad$ for $(4n+1)L_1 \leq \dot{x}_2 \leq (4n+3)L_1$,

for i = 4, 5, 6,...

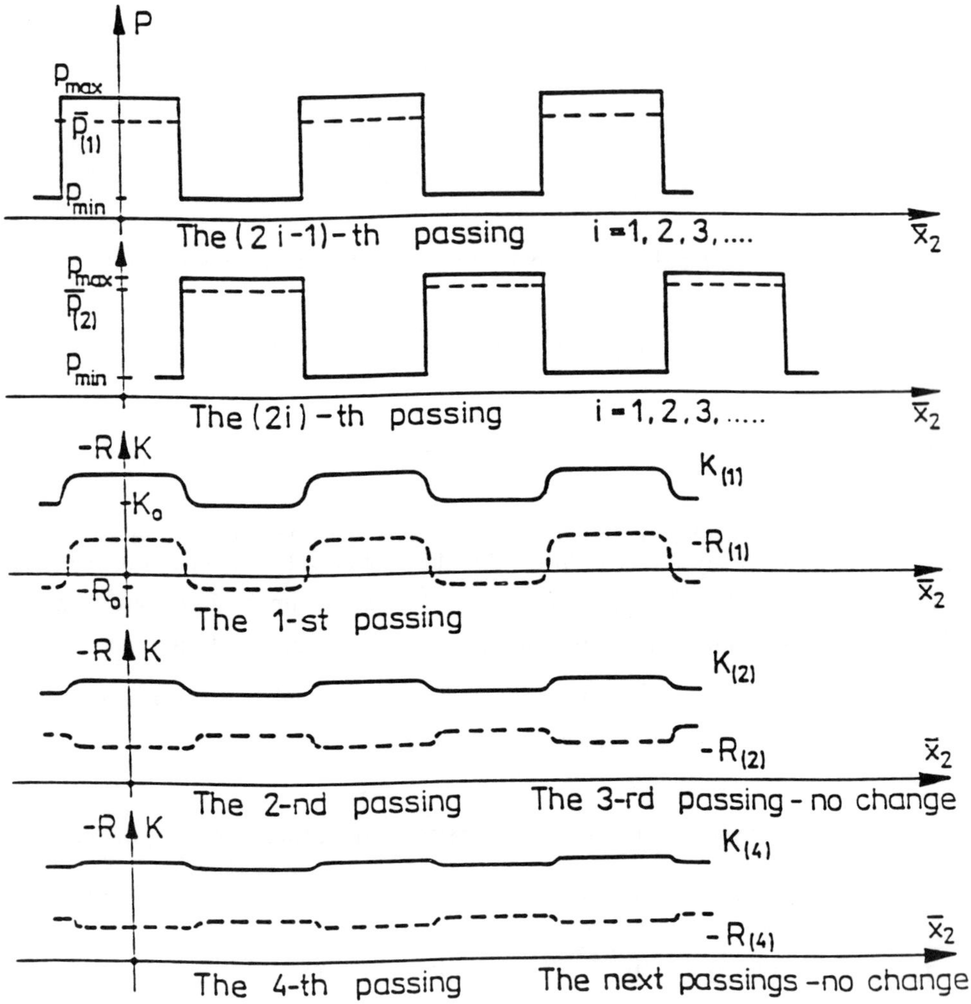

Fig. 6. Distribution of critical shear stress and residual stresses after successive passages of loads with shifted amplitudes.

The results of calculations show that after the second passage, the hardening of the plastically deformed zones is smaller than after the first one, but the introduced compressive stresses are higher. The situation does not change after the next passage. This means that the peaks of residual stresses are situated at the weaker places of the rail. The places of local hardening and highest residual stresses do not move along rails after successive passages of cyclically changing loads. Notice that after the third, fifth and next passages no plastic zones occur on the running surface of the rail. It means that the limit shear stress and the longitudinal residual stress take their ultimate values after a number of passages.

CONCLUSIONS

The presented analysis proves that elastic-plastic effects due to dynamic wheel/rail interactions are responsible for formation of rail corrugations. Cyclically changing moving loads cause a heterogeneity of distribution of both material hardening and residual stresses in rails. The magnitude of local hardening and residual stresses becomes stabilized after a few passages of the load. Places of lower hardening correspond with positions of residual stress peaks on the running surface of rails. These places do not move along rails after successive passing of the load and indicate positions of rail corrugations.

REFERENCES

1980 MAIR, R. I., "Natural frequency of rail tracks and its relationship to rail corrugation", in Rail Research Papers; Compendium of BHP Steel International Group, vol. **1**, Melbourne, 89-94.

1980 MAIR, R. I. & GROENHOUT, R., "The growth of transverse fatigue defects in head of railway rails", in Rail Research Papers; Compendium of BHP Steel International Group, vol. **1**, Melbourne, 170-186.

1985 JOHNSON, K. L., Contact Mechanics, Cambridge University Press,1985

1990 BRZOZOWSKI, M., BOGACZ, R. & POPP, K., "Zur Reibungsmodellierung beim Rollkontakt, *ZAMM* **70**, 6, T 678 - T 679.

1992 BOGACZ, R., "On residual stresses in rails and wheel/rail interaction"; in Residual Stresses in Rails, vol. **2**, Kluwer Academic Publishers, 87-100.

1994 GAMBIN, W., "Simplified model of rail corrugations formation", Engng. Trans. **42**, 1-2, 123-143.

STEADY CYCLIC STATE OF A STRUCTURE: METHODS OF ITS DIRECT DETERMINATION

D.A.GOKHFELD and O.S.SADAKOV
Cheliabinsk State Technical University, Cheliabinsk,Russia

ABSTRACT

Unlike the majority of publications in which the problem in question is considered within the framework of the shakedown theory and its specific mathematical methods, another approach is here developed. The latter is based on the structural (composite, layers) rheological model by which a more realistic description of deformation behavior of elastic-visco-plastic bodies can be obtained. In accordance with this model, the steady cyclic state of any structure made of an actual material is characterized, in the general case, by alternating inelastic deformation superposed on constant strain distribution which has been derived during the previous transient stage of the process. Methods of direct determination of the hysteresis loop parameters and of total (asymptotic) accumulated strain are discussed in the paper.

INTRODUCTION

Interest in the problem of strength and durability of structures subjected to cyclic actions of thermal fields and mechanical loads is due to requirements of modern technologies like nuclear power plants, aircraft engines, metallurgical, chemical and some other industries.
Let a structural component be subjected to intensive periodic external actions (mechanical loads,imposed displacements, temperature fields) which can be described by a set of parameters A_i:

$$A_i(t+t_c) = A_i(t) \tag{1}$$

(where t_c is the period). Its response (reactive forces and displacements, stresses and strains) will be designated by variables R_j. Specifying some reference time moment $t'<t_c$ we can trace the variation of the response during any cycle N (Fig.1):

$$R_j^c(t,N) \equiv R_j(t) - R_j(t_N), \tag{2}$$

$$t_N \le t \le t_{N+1}, \qquad t_N \equiv (N-1)t_c + t'. \tag{3}$$

Difference (2) will be termed the "cyclical" part of the response while parameter R_j magnitudes at moments t_N represent its "statical" part:

$$R_j^{st}(N) \equiv R_j(t_N). \tag{4}$$

Z. Mróz et al. (eds.), Inelastic Behaviour of Structures under Variable Loads, 449–461.

Each of the external action parameters A_i can be similarly divided into two analogous parts: $A_i(t,N)$ and $A_i\ (N)$. Under the regular loading conditions (1) they do not depend on cycle number N. It may be shown (GOKHFELD & SADAKOV [1984]) that the response of any structure made of cyclically stable material and subjected to cyclic loading gradually stabilizes to some steady cyclic state (SCS). Difficulties related to the kinetics analysis during thousands of loading cycles attracted attention to the methods of direct

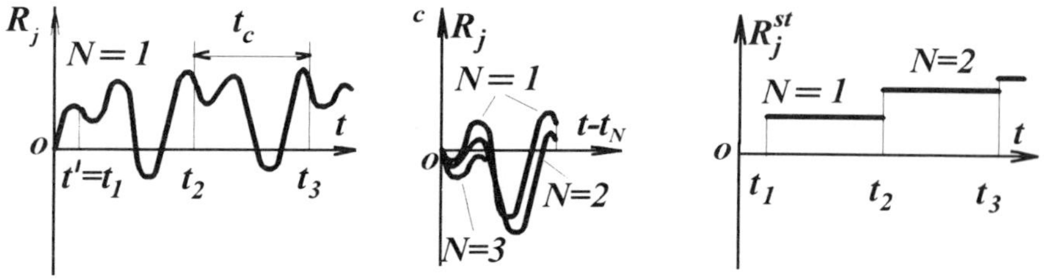

Fig 1. "Cyclical" and "statical" parts of response

determination of the asymptotic state of a structure subjected to cyclic actions of mechanical loads and temperature fields. It should be mentioned here that there is a wide group of investigations which are based on the model of perfect plasticity and may be considered as a generalization of the classical shakedown theory by which various inelastic processes can be embraced. The problem of existence and uniqueness of the inelastic steady state of a structure subjected to cyclic loading as well as available methods of its direct determination were studied by FREDERIC & ARMSTRONG [1966], MROZ [1971], HALPHEN [1975], GOKHFELD & CHERNIAVSKY [1980], PONTER & KARDENIZ [1985], POLIZOTTO et al. [1990] and other authors. According to the limit analysis, asymptotical behavior of any structural component subjected to cyclic loading beyond its shakedown limits is characterized by alternating plasticity and/or unceasing strain accumulation with constant increment per cycle. These processes represent both parts (cyclical and statical) of the response of the structure. As an example of another approach to the problem, some papers published by SHORR (e.g. [1952]) can be cited. Here structures made of perfectly viscous materials (i.e. at creep conditions) were considered. To obtain an asymptotic solution, the usual equations for the kinetics analysis were complemented with a relation by which closure of the stress cycle is stipulated.

The approach which is developed in this paper is based on the structural model of the elasto-visco-plastic solid. By this model (in the Western technical literature usually termed composite or layer model) a quite adequate description of the rheological properties of actual materials under various types of variable-repeated loading is provided (GOKHFELD & SADAKOV [1984, 1990, 1991]). Making use of the adopted basic version of the structural model some new possibilities have been found for direct determination of the hysteresis loop parameters and of total (asymptotic) accumulated strain.

STRUCTURAL MODEL

The rheological models of structural type are based on the concept of micro-inhomogeneity of actual materials. Unlike the more complicated "physical" models this inhomogeneity is reflected formally, as the structural one, similarly to the known Masing scheme. According to this scheme, any volume element is assumed to be a set of subelements (SEs) having equal strains (or strain tensors, in general), temperature and elastic moduli, but different rheological characteristics. The resulting self-equilibrated micro-stresses reproduce the role of the material memory in its deformation behavior.

The simplest basic version of the structural model is characterized by the following concepts (GOKHFELD & SADAKOV [1984]). Isotropic hardening of materials is not taken into account, i.e. the latter are supposed to have cyclically stable deformation behavior, initially, or after preliminary loading cycles. The subelements (SEs) are perfectly viscous, i.e. their creep rates $\dot{p}_k$ (k - the number of SE) depend on current individual stresses σ_k and temperature T only:

$$\dot{p}_k = \Phi_k(\sigma_k / E, T); \tag{5}$$

E - elasticity modulus, Φ_k - rheological functions (RF) of SEs.

The RFs assumed to be similar to RF of a material:

$$\dot{p}_k = \Phi(\sigma_k / (E z_k), T), \tag{6}$$

z_k is the similarity coefficient of the SE number k. It should be noted that regularities of both types of inelastic deformation, i.e. creep and plasticity, can be reflected (GOKHFELD & SADAKOV [1991]) by the model. The distribution of similarity coefficients determines the required shape of the stress-strain curve

$$\sigma = E\varphi(\varepsilon) \tag{7}$$

which corresponds to some basic values of strain rate $\dot{\varepsilon}_b$ and temperature T_b. Functions φ, RF (6) and $E(T)$ are the basic functions determining the model behavior. We shall use one more function

$$K(x) \equiv d\varphi(x)/dx, \tag{8}$$

which represents the relative tangent modulus of the basic stress-strain curve (7) at a given value of ε. It is worth noting that by making use of elastic strain $r = \sigma/E$ instead of stress σ, elastic modulus temperature dependence can be excluded from consideration. This results in the simplification of the non-isothermal deformation analysis. It is assumed that in any body element there is a group of perfectly-elastic SEs which can be represented by one such SE characterized by a proper relative weight. Hence unlimited anisotropic hardening of the simulated solid follows: its stress-strain diagrams at arbitrary strain rate and temperature have asymptotes which are characterized by non-zero tangent modulus and, accordingly, creep rate tends to zero independently of stress magnitude. Naturally, this implies that any type of failure and loss of stability is ignored.

On the other hand, it is adopted that, at any small value of stress, a little group of SEs which suffer inelastic deformation remains; i.e. there is no distinct boundary between elastic and inelastic deformation processes. As follows from the last two assumptions, only one type of asymptotic deformation behavior exists at the SCS for a structure subjected to cyclic loading: independently of the loading intensity it suffers alternating inelastic deformation superposed on some time-independent strain distribution accumulated during the preceding transient stage of the process. In other words, unlike the generalized shakedown theory, both the cyclical and statical responses of a structure become independent of cycle number N.

It is essential to note that stabilization of the hysteresis loops in actual structures usually finishes practically after a few loading cycles. On the other hand, strain accumulation (cyclic creep) may continue for a long time, especially at elevated temperatures (Fig.2).

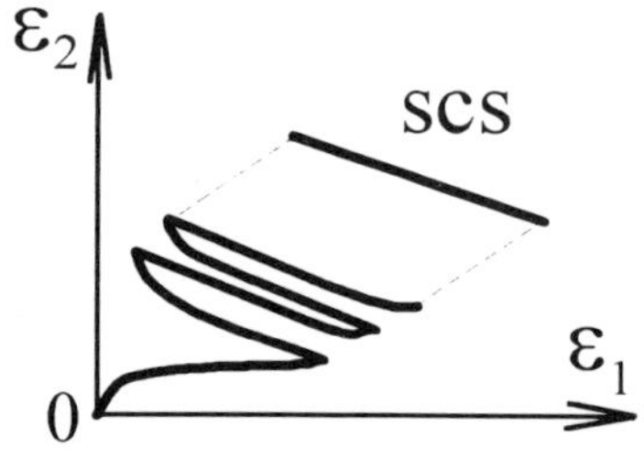

Fig.2. Stabilization of the stress-strain cycle

SIMILARITY PRINCIPLE

At proportional loading conditions, taking into account essential nonlinearity of RFs of alloys within their operating temperature ranges, we can reduce the SM relationships to a macroscopic state equation (GOKHFELD & SADAKOV [1984,1989,1990]), by which the current creep rate of a material can be determined. The latter proves to be dependent on two new state parameters and temperature. By introducing current differences

$$r_* \equiv r - r_\nu, \qquad \varepsilon_* \equiv \varepsilon - \varepsilon_\nu \tag{9}$$

(r-σ/E - elastic strain, ε - power strain, i.e. the total strain minus its thermal component, by index "ν" values of parameters at the moment of the last strain reverse are denoted) we can find the current parameter η from the parametric equation

$$r_* / \eta = \varphi(\varepsilon_* / \eta). \tag{10}$$

Here φ is one of the basic functions of the model. Then we can find the state parameter

$$s = \eta + s_\nu \tag{11}$$

(s_V is the value of parameter s at a reversal moment. It is known as well as r_V, e_V). When loading begins, the values r_V , e_V , s_V are equal to zero; then, until the first reverse occurs $r = r_V$, $\varepsilon = \varepsilon_V$, $s = \eta$. Creep rate is determined by the state equation:

$$\dot{p} \equiv dp / dt \equiv \varepsilon - \dot{r} = \Phi(r_b s, T)(1 - K(\varepsilon_* / \eta)) \qquad (12)$$

where Φ is the rheological function of material, K is derivative of φ (7), r_b is a constant of the simulated material

$$\Phi(r_b, T_b) \equiv \varepsilon_b. \qquad (13)$$

State equation (12) supplemented by the memory rules (which determine the conditions for the formation of the turning points and their erasure from the memory of material; see, for example, GOKHFELD & SADAKOV [1989]) has been termed the similarity principle (SP) as it specifies a set of similarity relations which characterize the deformation regularities. It permits us to describe both stress-strain diagrams at "rapid" (active) deformation and creep curves under isothermal as well as non-isothermal initial and cyclic loading. The reality of the predictions has been confirmed by results of experimental investigations.

It should be stressed that the hysteresis loop drift ("static" response) cannot be determined by the SP, as the cyclic creep phenomenon is excluded by the assumptions adopted. But additional analysis of the structural model relations has led to an equation by which final (limiting) location of the hysteresis loop can be determined (GOKHFELD & SADAKOV [1984]).

PHYSICAL EQUATIONS FOR SCS DETERMINATION

Equations which determine the specific relation between stresses and strains in a solid in its SCS will be termed "the physical (or determinative) equations". They can be derived from the corresponding experimental data directly, or obtained by a proper analysis of any rheological model of the considered medium. The similarity principle (SP) presents an example of a SCS physical equation of the second type; however, it is intended only for the determination of the cyclic part of the response of a material.

The possible cyclic loading programs for actual structural components are very diversified. The suitable methods and calculation procedures which can be derived for the main types of loading programs differ essentially from each other. This is related to the different SCS physical equations which should be used; thus, it seems to be reasonable to analyse them.

The proportional symmetrical isothermal loading cycle at constant absolute value of strain rate is the simplest loading program. Under these conditions the necessary physical equation can be obtained from the experimental data directly:

$$\sigma_a = f(\varepsilon_a) = E\varphi(\varepsilon_a) \tag{14}$$

(σ_a and ε_a are stress and strain amplitudes). This is the well known cyclic curve. If the mechanical properties of the material are indifferent to the strain rate sign, it can be asserted that there is no strain accumulation. Another situation appears for the asymmetrical loading cycle; two SCS physical equations are needed: $\sigma_a = \sigma_a(\varepsilon_a, \varepsilon_m)$ and $\sigma_m = \sigma_m(\varepsilon_a, \varepsilon_m)$,where σ_m, ε_m denote the average values of stress and strain. Experimental determination of these functions is rather time-consuming. But in accordance with the SM, the strain amplitude is independent of cycle asymmetry, and this is confirmed by experiments. Thus, the cyclic curve (14) can be used to determine the cyclical part of the response. But, for the corresponding statical part (σ_m, ε_m) determination, some analytical procedures seem to be appropriate. Especially for more complicated loading programs, analytical methods based on adequate rheological models represent the only way to solve the problem of SCS determination Refering to the SM analysis, relation (14) is valid under asymmetrical and symmetrical loading if $|\dot{\varepsilon}| = \dot{\varepsilon}_b, T = T_b$ only. But from the SP the possibility of the proper generalization follows:

$$\sigma_a = E\vartheta\varphi(\varepsilon_a / \vartheta) \tag{15}$$

where

$$\vartheta = \Phi^0(|\dot{\varepsilon}|, T) / \Phi^0(\dot{\varepsilon}_b, T_b) = \vartheta(\dot{\varepsilon}, T); \tag{16}$$

the values $|\dot{\varepsilon}|$,T here supposed to be constant; function $\Phi^0(x,T)$ is the inverse function of the RF in regard to the first argument: $\Phi^0(\Phi(x,T),T) = x$.
However, strain accumulation (i.e. the statical part of response) cannot be determined by the SP relations. The necessary results were obtained by analysis of the SEs deformation behavior when gradually approaching the asymptotic state (GOKHFELD & SADAKOV [1984]). In this way the following relation has been obtained:

$$\sigma_m = E\vartheta_o(\varphi((\varepsilon_m + \varepsilon_a) / \vartheta_o) - \varphi(\varepsilon_a / \vartheta_o)). \tag{17}$$

Here $\vartheta_o(T)$ presents the rheological characteristic determined by (16) which corresponds to strain rate which tends to zero. Accordingly, the stress-strain diagram at $\dot{\varepsilon} = \dot{\varepsilon}_o \to 0$ is

$$\sigma = E\vartheta_o\varphi(\varepsilon / \vartheta_o) \equiv E\varphi_o(\varepsilon) \tag{18}$$

Apparently, stress-strain diagrams of this type were termed by KREMPL [1984] as "equilibrium" ones. They can be got from the special experiments which include relaxation processes.
Corresponding analysis under non-isothermal conditions requires proper modification of the adopted definitions. Now the loading cycle should be termed a "symmetrical" one if there is no shift of the initial hysteresis loop. Accordingly, the cycle asymmetry is to be

there is no shift of the initial hysteresis loop. Accordingly, the cycle asymmetry is to be determined, taking into account actual temperature conditions. Thus, stress (or strain) opposite amplitudes may be not equal. As an example, let us consider a non-isothermal loading cycle (without hold-time periods) characterized by the following parameters: (ε_1 ,T_1) $\Leftrightarrow$ (ε_2, T_2), where numbers 1,2 correspond to the reversal points. The relation between peak-to-peak amplitudes Δr and $\Delta\varepsilon$ takes the form

$$\Delta r = m\varphi(\Delta\varepsilon / m), \tag{19}$$
$$m = \vartheta(\dot{\varepsilon}_1, T_1) + \vartheta(\dot{\varepsilon}_2, T_2). \tag{20}$$

Note that the corresponding relation between stress ($\sigma = E\varepsilon$) and strain peak-to-peak amplitudes proves to be, in general, dependent on the cycle asymmetry:

$$\Delta\sigma = \Delta(Er) = E_m \Delta r + r_m \Delta E. \tag{21}$$

The only situation when the last equation does not contain the average elastic strain r_{m} of the deformation cycle is when $\Delta E = 0$ (or practically negligible dependence of elastic modulus on temperature). To derive expressions for "static" response of the SM under non-isothermal cyclic loading, let us introduce some non-trivial definitions for the strain "amplitudes" and "mean" values:

$$\varepsilon_{ai} = (\varepsilon_i - \varepsilon_{3-i})\vartheta_i / m; \tag{22}$$
$$\varepsilon_m = \varepsilon_i - \varepsilon_{ai}, \qquad i = 1,2. \tag{23}$$

Here $\vartheta_i = \vartheta(\dot{\varepsilon}_i, T_i)$; $m = \vartheta_1 + \vartheta_2$ (see eq.(20)); by eq. (22) two "amplitudes" having opposite signs are determined. In an analogous way, parameters of the stress cycle r_{ai} , r_m are determined. All the definitions coincide with the traditional ones at $\vartheta_1 = \vartheta_2$. Note, that eq.(19) may be written in the form

$$r_{ai} = \vartheta_i \varphi(\varepsilon_{ai} / \vartheta_i), \tag{24}$$

if the function φ (i.e. $\varphi(-x) = -\varphi(x)$) is assumed to be odd. Making use of the introduced terms, we can write the SCS state equation by which statical response is determined in the form:

$$r_m = \vartheta_{oi}(\varphi(\varepsilon_i / \vartheta_{oi}) - \varphi(\varepsilon_{ai} / \vartheta_{oi})) \tag{25}$$

where the actual (significant) number i is due to the requirement:

$$\varepsilon_m \varepsilon_{ai} > 0. \tag{26}$$

Characteristics $\vartheta_{oi} = \vartheta(\dot{\varepsilon}_o, T_i)$ are as defined before. The equation generalizes eq.(17) for the non-isothermal conditions. The deformation cycle determined by equation $\varepsilon_m = \varepsilon_i - \varepsilon_{ai} = 0$ can be termed a "symmetrical" one; this means that such a type of non-isothermal loading cycle is not accompanied by cyclic creep. Direct SCS determination in some other, more complicated situations, will not be considered here in detail; we shall limit ourselves to some remarks only. For loading cycles including hold-time periods, to get the cyclic part of the response, the state equation (12) already known can be used. Naturally, numerical integration must be used. For non-proportional loading conditons, it is necessary to draw the reader's attention to a situation which is rather typical for many structural components: the stress trajectory under the SCS conditions is linear. However it does not pass through the coordinate origin. This means that the cyclical part of stress variation is proportional, while its vector (in the stress space) is not collinear with the vector of the static part. As follows from the analysis based on the SM (the results were confirmed by experiments: GOKHFELD & SADAKOV [1984, 1990]) this type of loading is often accompanied by intensive strain accumulation (In Fig.3, a plane in the deviatoric strain space is shown). It should be stressed that for proportional cyclic loading conditions, the ratchet mechanism may be of scleronomic type. Naturally, the process can be accelerated by creep.

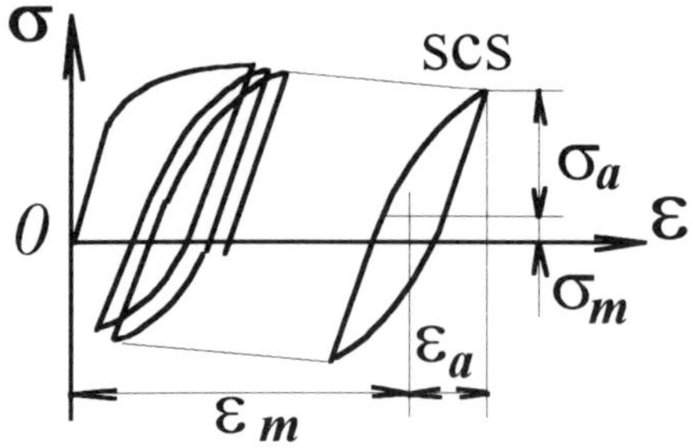

Fig.3. Stabilization of cycle at non-proportional loading (special case)

Relations between cyclic stresses and strains become similar to those of proportional loading (eqs. (15),(16) for isothermal conditions and (19), (20) for non-isothermal ones). Determination of the statical part of the response is more complicated, as the SCS physical equations for SEs of the structural model cannot be summarized as simply as for proportional loading conditions. The equations which are valid at non-proportional loading take too much space to be given here. In such cases, direct application of the structural model equations is necessary (see GOKHFELD & SADAKOV [1984]).

DIRECT DETERMINATION OF SCS OF STRUCTURES

The formulation of the problem is similar to that of the deformation kinetics analysis. Like the latter, it includes equilibrium and compatibility equations, but differs from it by specific physical relationships which were considered in the previous paragraph. The methods which can be used to solve the problem can be divided into two groups. The first one is related to structures which operate either under isothermal conditions ($\dot{E}=0$) or at non-isothermal ones if the elastic modulus variations during the loading cycle can be neglected. The solution can be divided into two parts to be performed in succession: the static part of the response of a structure can be determined after the cyclic response is obtained. The latter problem can be solved by using either the SP or the structural model general equations. It should be remembered that, unlike the SP, the initial model reflects strain accumulation and, therefore, the hysteresis loop does not become closed until total stabilization is reached. However, variations of the hysteresis loop shape practically cease after a few loading cycles.

For a temperature-dependent elastic modulus, separation of the problem into two parts becomes impossible, and the full set of equations must be solved simultaneously.

Another classification of the problem is due to the type of loading trajectory, proportional (radial) or non-proportional. Some specific type of non-proportional loading conditions is considered above. In this case determination of total strain accumulation (until the SCS is reached) is possible, but rather complicated, as the corresponding solution requires analysis of any SE deformation behavior. The parameters of the hysteresis loop can be found by using the procedure suggested for proportional loading conditions, as in this case the limit (SCS) loading trajectory is linear.

The simplifying assumptions related to the SCS direct determination lead to some new approaches to inelastic deformation analysis of structures. In this way, generalized rheological models can be formulated which exclude the necessity of tracing of deformation kinetics in many points of a structure. In particular, Neuber's method intended for stress concentration analysis belongs to such a type of model. The latter can be generalized to cyclic non-isothermal loading and cyclic loading with hold-time periods. Analogous macro-models intended for other typical cases can be suggested as well. Lastly, the possibility of constructing specialized nomographs should be checked. The idea is grounded on the similarity in deformation behavior of a group of structures made of materials which have RFs close to each other (or even coinciding) but which have different deformation curves involving the corresponding generalized parameters. Calculations executed for any structures belonging to the group can be easily converted for the others if their initial loading deformation curves are known. The results obtained in this way prove to be quite trustworthy.

Finally, some constraints which must be imposed on the suggested methods should be taken into account: synchronism in the appearance of turning points within the entire body is required. Otherwise, the equilibrium and compatibility equations cannot be formulated in stress (strain) amplitude and mean values. Usually this condition is satisfied, however some specific exclusions were revealed (KOITER [1952], GOKHFELD & CHERNIAVSKY [1980]).

THE EXAMPLE: SCS ANALYSIS FOR A SIMPLE TRUSS. COMPARISON WITH THE GENERALIZED SHAKEDOWN THEORY

Let a two-bar static-indeterminate truss be subjected to constant load while bar 2 suffers cyclic temperature changes. The latter are characterized by thermal strain peak-to-peak amplitude $\Delta\vartheta$ (Fig.4; symmetric bars 1 are adopted as a unit with total cross-section area A_1). Temperature dependence of the stress-strain diagrams will be ignored.
The compatibility equation for the truss has the form

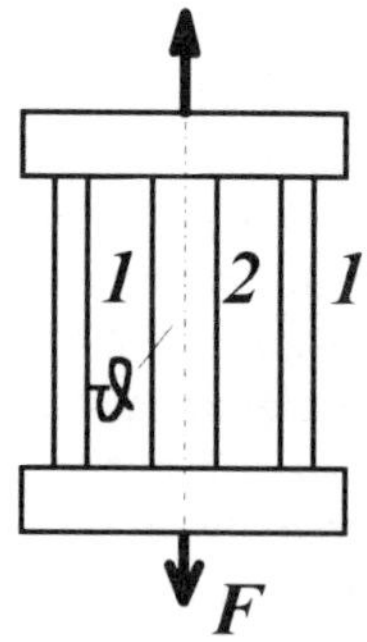

Fig.4. Two-bar truss

$$\varepsilon_1 = r_1 + p_1 = \varepsilon_1^+ = \varepsilon_2 = r_2 + p_2 + \vartheta_2 = \varepsilon_2^+ + \vartheta_2 \qquad (27)$$

where ε_i is total strain while r,p,ϑ are its elastic, plastic and thermal components;ε_i^+ is the power part of strain; i =1,2 - the bar number.
The equilibrium equation is

$$r_1 G_1 + r_2 G_2 = F, \qquad G_i = E_i A_i \qquad (28)$$

At perfectly elastic behavior the solution is as follows:

$$\begin{aligned} N_1 &= G(F/G_2 + \vartheta_2); \\ N_2 &= G(F/G_1 - \vartheta_2); \\ N_i &= \sigma_i A_i, \qquad G = (G_1^{-1} + G_2^{-1})^{-1}. \end{aligned} \qquad (29)$$

According to the shakedown theory the alternating plasticity condition follows directly from solution (29):

$$G\Delta\vartheta > 2\min(F_{o1}, F_{o2}), F_{oi} = \sigma_{si} A_i. \qquad (30)$$

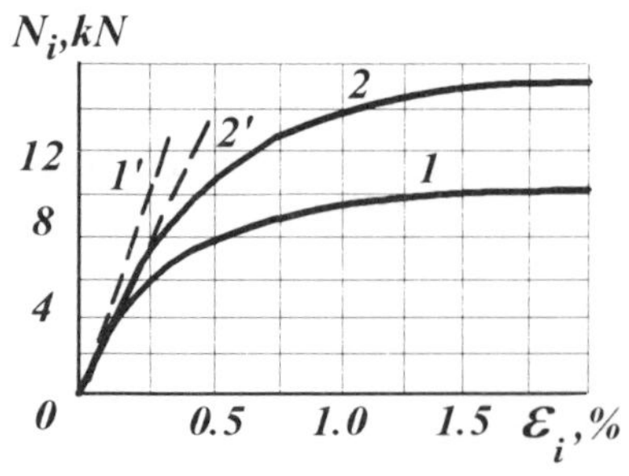

Fig.5. Load-strain curves for the bars

The possible strain accumulation (progressive collapse) mechanism is obvious: at any loading cycle both the bars in turn (i.e. alternatively) suffer equal plastic elongation. The corresponding condition can easily be obtained:

$$F > F_{o1} + F_{o2} - F_{\vartheta}, \qquad F_{\vartheta} \equiv G\Delta\vartheta. \qquad (31)$$

The shakedown diagram is shown by a dotted line in Fig.6.

Unlike the shakedown theory, the following SCS determination for the truss is based on actual load-strain curves given for the bars (Fig.5). Under the conditions considered, the problem can be divided into two parts. The cyclic response determination for the truss is reduced to a set of equations formulated in strain and stress amplitudes. Two of them (compatibility and equilibrium equations) follow directly from eqs.(27) and (28) while the third presents the cyclic diagrams for the materials the bars are made of:

$$\varepsilon_{a1} = \overset{+}{\varepsilon}_{a2} + \vartheta_a, \qquad \vartheta_a = \Delta\vartheta / 2;$$
$$\sigma_{a1} A_1 + \sigma_{a2} A_2 = 0; \sigma_{ai} = f(\varepsilon_{ai}). \tag{32}$$

The results for different values of $\Delta\vartheta$ (or of F) are illustrated in Fig.6 by horizontal lines marked by values of maximum inelastic strain double amplitude (for a bar 1). It should be remembered that static action (constant force F) has no effect on alternating plasticity parameters.

To determine the static part of the response of the structure, i.e. the hysteresis loop final location on the plane $\{N_i, \varepsilon_i\}$, the compatibility and equilibrium equations of the bars formulated in strains and stresses mean values are to be used together with the equation of the type (25):

$$\varepsilon_{m1} = \overset{+}{\varepsilon}_{m2} + \vartheta_m, \qquad \vartheta_m = \Delta\vartheta / 2;$$
$$N_{m1} + N_{m2} = F; \tag{33}$$
$$N_{mi} = (f_i((|\overset{+}{\varepsilon}_{mi}| + \varepsilon_{ai}) / \eta_{oi}) - f_i(\overset{+}{\varepsilon}_{ai} / \eta_{oi})) A_i \eta_{oi} sign \overset{+}{\varepsilon}_{mi}.$$

The set of equations (33) can be solved without any difficulties, and a proper graphical construction may be used. Total accumulated inelastic strain was determined making use of the diagrams given in Fig.5. Dependence of accumulated inelastic strain in bar 1 on loading parameters F and $\Delta\vartheta$ is illustrated in Fig.6 by dashed lines (scleronomic material, η_{oi}=1; there is no hysteresis loop drift at proportional cyclic loading). For another case ($\eta_{o1} = 0.7, \eta_{02} = 0.9$ are values characteristic for structural alloys at normal and elevated temperatures, accordingly) the results are represented by solid lines.

Comparing this with the shakedown diagram (dotted lines in Fig.6) we can see a good qualitative correspondence in the prognosis for both the unsafe states (alternating plasticity and progressive strain accumulation). For the quantitative correspondence, dependence of the shakedown domain on tolerance adopted for the proper approximation of the actual stress-strain diagram should be taken into account. Referring to Fig.6, we can note that accumulated strain at the loading parameters which correspond to the shakedown domain boundary exceed twice, or even more, the tolerance adopted for residual strain.

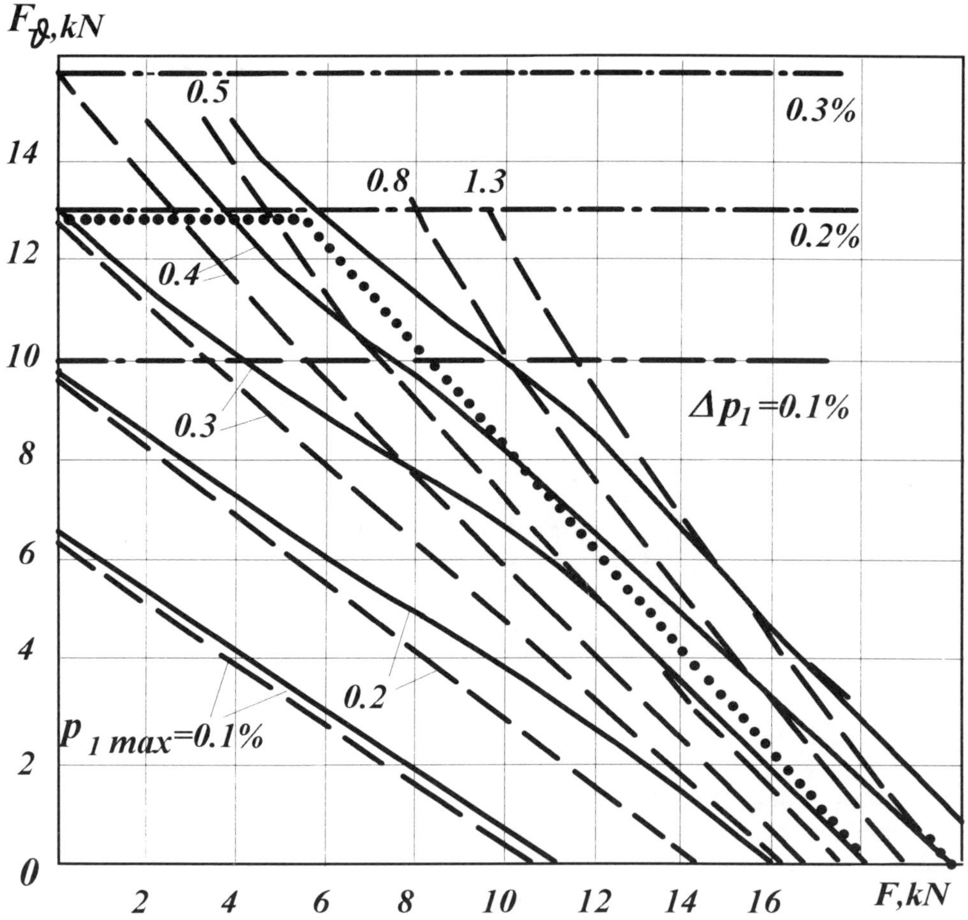

Fig.6. Two-bar truss SCS chart

This study has been carried out in the framework of a research project sponsored by the Russian Fund for Fundamental Research (grant 93-013-16779).

REFERENCES

1966 FREDERICK,C.O. & ARMSTRONG,P.J., Convergent internal stresses and steady cyclic state of stress. J. Strain Anal.,1,No.2,154.

1971 MROZ, Z., On the theory of steady plastic cycles in structures, Trans. 1st Intern. SMiRT Conf., Berlin, Paper L5/6.

1975 HALPHEN,B., Stress accommodation in elastic-perfectly plastic and viscoplastic structures, Mech. Res. Comm., 2, 5-6.

1980 GOKHFELD,D.A.& CHERNIAVSKY,O.F., Limit analysis of structures at thermal cycling. Sijthoff & Noordhoff, The Netherlands-USA.

1985 PONTER,A.R.S.& CARDENIZ,S., An extended shakedown theory for structures that suffer cyclic thermal loading, Part I: Theory, J. Appl. Mech., 52, pp.877-882.

1990 POLLIZOTTO,C., BORNIO,G.& FUESCHI,P., On the steady state response of elastic-perfectly plastic solids to cyclic loads, in: M. Kleiber & J.A König (eds), Inelastic Solids and Structures (Antoni Sawczuk Memorial Volume), Pineridge Press, Swansea, U.K., pp.473-478.

1987 KÖNIG,J.A., Shakedown of elasto-plastic structures, PWN-Elsevier, Warsaw-Amsterdam.

1964 SHORR, B.F. Cyclic creep of non-uniformly heated cylinders, in Russian, in: Thermal Stresses in Structural Members,4,Naukova Dumka, Kiev.

1984 GOKHFELD, D.A.& SADAKOV, O.S. Plasticity and creep of structural elements subjected to repeated loading, in Russian. Mashinostrojenije, Moscow.

1989 GOKHFELD, D.A., KONONOV, K.M., POROSHIN, V.B. & SADAKOV, O.S., Coupled mathematical models for cyclic inelastic deformation and damage accumulation processes Trans.10th Int. SMiRT Conf. Anaheim, Calif., USA, Paper L1/5.

1984 KREMPL,E. & KALLIANPUR,V.V., Some critical uniaxial experiments for viscoplasticity at room temperature, J. Mech. Phys. Solids. 32, No.4.

1990 GOKHFELD D.A., & SADAKOV, O.S., On the main approaches to strength and lifetime evaluation of structures under variable-repeated load and temperature, in: M. Kleiber & J.-A.König (eds.), Inelastic Solids and Structures (Antoni Sawczuk Memorial Volume), Pineridge Press, Swansea, U.K. pp. 415-432.

1952 KOITER, W.T., Some remarks on plastic shakedown theorems. Proc. 8th Intern. Congr. Appl. Mech., Istanbul,1,220.

An Improved Boundary Element Analysis for the Bending of a Thin Plate with a Crack

O. Mahrenholtz and Pin Lu*

Technische Universität Hamburg-Harburg, D-21072 Hamburg, Germany
* University of Science and Technology of China, Hefei, Anhui 230026, P.R. China

Abstract

In this paper, a modified fundamental solution (MFS) for a thin plate with a cutout is presented in terms of MUSKHELISHVILI'S complex variable method. As the MFS satisfy traction free conditions on the cutout surface, integrals along the cutout boundary can be avoided when they are used as the kernels of boundary integral equations; only outer boundary discretization is necessary in a numerical solution procedure. Numerical examples for the bending of a plate with a crack show that higher numerical accuracy of the results could be attained with relatively fewer discrete elements by using the MFS in boundary element analysis.

1 Introduction

In analysis of bending of a plate with a hole or a crack by the usual BEM, a large number of discrete elements near the crack tip or the hole boundary are required. This sometimes renders the method inefficient, and it has been shown (LU, 1990) that the fashion of discrete modelling for a crack has a considererable influence on the precision of the computed results. Therefore, if a set of special kernels is used in boundary integral equations which can remove the integrals along the crack surface, the above difficulties could be avoided effectively. Based on the consideration, some work has been done for two dimensional problems in boundary element analysis (MUKHERJEE, 1982)

In this paper, a modified fundamental solution for an infinite thin plate with an elliptical hole or a Griffith crack is derived by using MUSKHELISHVILI'S complex variable method, in which the traction free conditions on the cutout surface have been satisfied. Thus, the effect of the cutout on the stress and displacement fields is incorporated in the MFS. This means that the integrations along the cutout boundary are no longer necessary and the unknowns of the problems rely on the outer boundary of the plate only, which avoids the discretization of the cutout boundary in a numerical procedure.

Z. Mróz et al. (eds.), Inelastic Behaviour of Structures under Variable Loads, 463–472.

2 Modified Fundamental Solution

The fundamental solution v_1^0 of bending of a thin plate without cutout is (e.g. STERN, 1979)

$$v_1^0(P,Q) = \frac{1}{8\pi D}\, r^2 \,\ln r, \tag{1}$$

where r is the distance between source point P and field point Q and D is the bending stiffness. Eq.(1) can be represented by two complex variable functions as (MUKHERJEE, 1982)

$$v_1^0(z,z_0) = \frac{1}{8\pi D}\, Re[\, \bar{z}\, \phi^0(z,z_0) + \chi^0(z,z_0)\,] \tag{2}$$

with

$$\begin{aligned} \phi^0(z,z_0) &= (z-z_0)\ln(z-z_0), \\ \chi^0(z,z_0) &= -\bar{z}(z-z_0)\ln(z-z_0), \end{aligned} \tag{3}$$

where z and z_0 correspond to points P and Q in a complex plane, respectively. If the plate is weakened by a cutout, e.g. a hole or a crack, the fundamental solution can be written as

$$v_1(z,z_0) = \frac{1}{8\pi D}\, Re[\, \bar{z}\, \phi(z,z_0) + \chi(z,z_0)\,], \tag{4}$$

where

$$\begin{aligned} \phi(z,z_0) &= \phi^0(z,z_0) + \phi^*(z,z_0), \\ \chi(z,z_0) &= \chi^0(z,z_0) + \chi^*(z,z_0). \end{aligned} \tag{5}$$

In the above equations, $\phi^0(z,z_0)$ and $\chi^0(z,z_0)$ are known functions given by eq.(3), $\phi^*(z,z_0)$ and $\chi^*(z,z_0)$ are unknown modified complex functions which are analytic in the region outside the outcut, and can be determined according to the traction free condition of the cutout boundary. This condition has been given in the form of complex variable functions by SAVIN (1961):

$$\kappa\, \phi^*(\tau,z_0) + \overline{\tau \phi^{*\prime}(\tau,z_0)} + \overline{\psi^*(\tau,z_0)} = -[\, \kappa\, \phi^0(\tau,z_0) + \overline{\tau \phi^{0\prime}(\tau,z_0)} + \overline{\psi^0(\tau,z_0)}\,] \tag{6}$$

where $\kappa = -\frac{3+\nu}{1-\nu}$, ν is poisson's ratio, τ the value of z on the boundary of the cutout, the prime denotes differentiation with respect to the variable z, and $\psi(z,z_0) = \frac{d}{dz}[\chi(z,z_0)]$.

If a mapping function $z = \omega(\zeta)$ can be found, which transforms the region on and outside the cutout in the z-plane to a region on and inside a unit circle γ in the ζ-plane, and if

$$\begin{aligned} \phi^*(z, z_0) &= \phi^*(\omega(\zeta), z_0) = \phi_1^*(\zeta, z_0), & \psi^*(z, z_0) &= \psi^*(\omega(\zeta), z_0) = \psi_1^*(\zeta, z_0), \\ \phi^0(z, z_0) &= \phi^0(\omega(\zeta), z_0) = \phi_1^0(\zeta, z_0), & \psi^0(z, z_0) &= \psi^0(\omega(\zeta), z_0) = \psi_1^0(\zeta, z_0), \end{aligned} \tag{7}$$

eq.(6) can be written in the following form:

$$\kappa\, \phi_1^*(\sigma, z_0) + \frac{\omega(\sigma)}{\overline{\omega'(\sigma)}}\, \overline{\phi_1^{*'}(\sigma, z_0)} + \overline{\psi_1^*(\sigma, z_0)} = F(\sigma, z_0), \tag{8}$$

Here

$$F(\sigma, z_0) = -\left[\kappa\phi_1^0(\sigma, z_0) + \frac{\omega(\sigma)}{\overline{\omega'(\sigma)}}\, \overline{\phi_1^{0'}(\sigma, z_0)} + \overline{\psi_1^0(\sigma, z_0)}\right] \tag{9}$$

is known, σ is a point of ζ on the unit circle γ in the ζ-plane, and the prime denotes differation with respect to the complex variable ζ.
Multiplying eq.(8) and it conjuncate by the factor $\frac{1}{2\pi i}\frac{d\sigma}{\sigma+\zeta}$, respectively and integrating along the unit circle γ, one can obtain a pair of integral equations:

$$\kappa\phi_1^*(\zeta, z_0) = \frac{1}{2\pi i}\int_\gamma \frac{\omega(\sigma)}{\overline{\omega'(\sigma)}}\, \overline{\phi_1^{*'}(\sigma, z_0)}\, \frac{d\sigma}{\sigma-\zeta} = \frac{1}{2\pi i}\int_\gamma \frac{F(\sigma, z_0)}{\sigma-\zeta} d\sigma, \tag{10}$$

$$\psi_1^*(\zeta, z_0) + \frac{1}{2\pi i}\int_\gamma \frac{\overline{\omega(\sigma)}}{\omega'(\sigma)}\, \phi_1^{*'}(\sigma, z_0)\, \frac{d\sigma}{\sigma-\zeta} = \frac{1}{2\pi i}\int_\gamma \frac{\overline{F(\sigma, z_0)}}{\sigma-\zeta} d\sigma, \tag{11}$$

Solving $\phi_1^*(\zeta, z_0)$ and $\psi_1^*(\zeta, z_0)$ from eqs.(10) and (11), and substituting them into eq.(5), we can obtain the modified fundamental solution with the cutout.

If the cutout is an elliptical hole, the mapping function is

$$z = \omega(\zeta) = R\,(\frac{1}{\zeta} + c\,\zeta), \tag{12}$$

where $R = \frac{a+b}{2}$, $c = \frac{a-b}{a+b}$, a and b are the semi-major and semi-minor axes of the ellipse, respectively. Inserting eq.(12) into eqs.(10) and (11), and after proper manipulation, one can obtain

$$\begin{aligned} \phi_1^*(\zeta, z_0) &= \frac{1}{\kappa}\{-Rc(1+\kappa\ln Rc + \ln R)\zeta - (Rc\zeta - z_0)[\kappa\ln(\zeta - z_0) + \ln(\zeta - t_0)] \\ &\quad -\frac{R}{\zeta}\left[\kappa\ln\left(1-\frac{\zeta}{r_0}\right) + \ln\left(1-\frac{\zeta}{t_0}\right)\right]\}, \end{aligned} \tag{13}$$

$$\begin{aligned} \psi_1^*(\zeta, z_0) &= -R(1+\kappa\ln Rc + \ln Rc)\,\zeta - (R\zeta - \bar{z}_0)\,[\kappa\ln(\zeta - t_0) + ln(\zeta - r_0)] \\ &\quad -\frac{Rc}{\zeta}\left[\kappa\ln\left(1-\frac{\zeta}{t_0}\right) + \ln\left(1-\frac{\zeta}{r_0}\right)\right] - \frac{\zeta(\zeta^2+c)}{c\zeta^2-1}\phi_1^{*'}(\zeta, z_0) \end{aligned} \tag{14}$$

where

$r_{0,i} = \frac{z_0 \pm \sqrt{z_0^2 - 4R^2c}}{2Rc}$ are the roots of equation $c\zeta^2 - \frac{z_0\zeta}{R} + 1 = 0$,

$t_{0,i} = \frac{\bar{z}_0 \pm \sqrt{\bar{z}_0^2 - 4R^2c}}{2Rc}$ are the roots of equation $\zeta^2 - \frac{\bar{z}_0\zeta}{R} + c = 0$,

and

$$|r_0| \geq 1, \quad |t_0| \geq 1, \quad |r_i| < 1, \quad |t_i| < 1.$$

Therefore, the modified fundamental solution of the plate bending with an elliptical hole can be obtained by inserting eqs.(13) and (14) into eq.(4).
The mapping function for a Griffith crack with a lenght $2a$ can be obtained simply by letting $R = \frac{a}{2}$ and $c = 1$ in eq.(12). Substituting it into eqs.(10) and (11) and simplifying, we can write the complex varibale function ϕ and ψ for the crack problem as

$$\phi(z, z_0) = (z - z_0)\ln(z - z_0) + \frac{1}{\kappa}\tilde{\phi}^*(z, z_0), \tag{15}$$

$$\psi(z, z_0) = -\bar{z}_0 \ln(z - z_0) + \tilde{\phi}^*(z, \bar{z}_0) - \frac{z}{\kappa}\tilde{\phi}^{*\prime}(z, z_0), \tag{16}$$

Here

$$\begin{aligned} \tilde{\phi}^*(z, z_0) &= -(z - z_0)[\kappa \ln(\zeta - r_0) + \ln(\zeta - t_0)] + \frac{a}{2}[\kappa \ln(-r_0) + \ln(-t_0)]\frac{1}{\zeta} \\ &\quad -\frac{a}{2}[1 + (\kappa + 1)\ln\frac{a}{2}]\,\zeta, \end{aligned} \tag{17}$$

and $\tilde{\phi}^*(z, \bar{z}_0)$ can be obtained simply by replacing the complex values z_0, r_0 and t_0 with their conjugations, respectively, in eq.(17), and

$$\zeta = \frac{z \pm \sqrt{z^2 - a^2}}{a}, \qquad r_0 = \frac{z_0 \pm \sqrt{z_0^2 - a^2}}{a}, \qquad t_0 = \frac{\bar{z}_0 \pm \sqrt{\bar{z}_0^2 - a^2}}{a}$$

with

$$|\zeta| < 1, \qquad |r_0| \geq 1, \qquad |t_0| \geq 1.$$

Furthermore, eq.(16) implies

$$\begin{aligned} \chi(z, z_0) = \int \psi(z, z_0)\, dz &= -\bar{z}_0(z - z_0)\ln(z - z_0) - \frac{1}{\kappa}z\tilde{\phi}^*(z, z_0) \\ &\quad + \int \left[\tilde{\phi}^*(z, \bar{z}_0) + \frac{1}{\kappa}\tilde{\phi}^*(z, z_0)\right] dz. \end{aligned} \tag{18}$$

Substituting eqs.(15) and (18) into eq.(4), we can obtain the modified fundamental solution of the plate bending with a crack.

3 Boundary Integral Equations

As an application, the problem of plate bending with a crack is solved by the boundary element method, and the modified fundamental solution obtained before is used as the kernel in the boundary element analysis.

3.1 Direct Integral Formulation

Besides eq.(4), another fundamental solution for solving plate bending problems, which is used in direct integral equations, is

$$\begin{aligned} v_2(P,Q) &= \frac{\partial}{\partial n_P} v_1(P,Q) = \frac{\partial}{\partial n_P} v_1(z,z_0) \\ &= \frac{1}{8\pi D} Re \left\{ e^{i\theta} \left[\overline{\phi(z,z_0)} + \bar{z}\phi'(z,z_0) + \psi(z,z_0) \right] \right\}, \end{aligned} \tag{19}$$

Here $\frac{\partial}{\partial n_P}$ indicates the outward normal derivative at the source point P; θ is the angle between the outward normal direction at the point P and the positive direction of the x-axis; $\phi(z,z_0)$ and $\psi(z,z_0)$ are given by eqs.(15) and (16) for tzhe crack problem. When P is on the smooth boundary, the integral equations are (STERN, 1984)

$$\begin{aligned} \frac{1}{2} w(P) &= -\int_\Gamma [V_{n_Q}(v_1) w(Q) - M_{n_Q}(v_1) \frac{dw(Q)}{dn_Q} + \frac{dv_1(P,Q)}{dn_Q} M_{n_Q}(w) \\ &\quad - v_1(p,Q) V_{n_Q}(w)] dS_Q, \end{aligned} \tag{20}$$

$$\begin{aligned} \frac{1}{2} \frac{dw(P)}{dn_P} &= -\int_\Gamma [V_{n_Q}(v_2)\{w(Q) - w(P)\} - M_{n_Q}(v_2) \frac{dw(Q)}{dn_Q} \\ &\quad + \frac{dv_2(P,Q)}{dn_Q} M_{n_Q}(w) - v_2(P,Q) V_{n_Q}(w)] dS_Q, \end{aligned} \tag{21}$$

where Γ is the outer boundary of the plate, w is the deflection of the plate, $M_{n_Q}(\cdot)$ and $V_{n_Q}(\cdot)$ are bending moment and equivalent shearing operators acting on the boundary point Q, respectively.

Stress intensity factors at the crack tip for the plate bending with a crack placed on the x-axis are

$$\left\{ \begin{array}{c} K_I \\ K_{II} \end{array} \right\} = \lim_{\rho \to 0} \frac{6}{h^2} \sqrt{2\pi\rho} \left\{ \begin{array}{c} M_y(w) \\ M_{xy}(w) \end{array} \right\}_{\theta=0}, \tag{22}$$

where h is the thickness of the plate, ρ is the distance between the crack tip and the point outside the crack on the x-axis, in which bending moments $M_y(w)$ and $M_{xy}(w)$ are calculated. Inserting eq.(20) into eq.(22), we have

$$\begin{aligned} K_i &= \lim_{\rho \to 0} \frac{6}{h^2} \sqrt{2\pi\rho} \int_\Gamma [V_{n_Q}(K_i^*) w(Q) - M_{n_Q}(K_i^*) \frac{dw(Q)}{dn_Q} \\ &\quad + \frac{dK_i^*}{dn_Q} M_{n_Q}(w) - K_i^* V_{n_Q}(w)] dS_Q \,|_{\theta=0}, \quad i = I, II \end{aligned} \tag{23}$$

where the functions $K_I^*(P,Q)$ and $K_{II}^*(P,Q)$ are given in the Appendix.

3.2 Indirect Integral Formulations

As two boundary values are specified at any point on the boundary of a thin plate, we could choose a sum of two potentials to represent the solution $w(P)$ (BREBBIA, 1981):

$$w(P) = \int_\Gamma \left[v_1(P,Q) m(Q) + v_2^*(P,Q) k(Q) \right] dS_Q, \tag{24}$$

where $m(Q)$ and $k(Q)$ are the functions to be solved, and $v_2^*(P,Q) = \frac{\partial v_1(P,Q)}{\partial n_Q}$. As the point p in the region tends to the Point P on the smooth boundary, the following relations can be obtained (KITAHARA, 1985)

$$\begin{aligned}
w(P) &= \int_\Gamma [v_1(P,Q)m(Q) + v_2^*(P,Q)k(Q)]dS_Q, \\
\frac{dw(P)}{dn_P} &= \int_\Gamma \left[\frac{\partial v_1(P,Q)}{\partial n_P} m(Q) + \frac{\partial v_2^*(P,Q)}{\partial n_Q} k(Q)\right] dS_Q, \\
M_{n_P}(w) &= \int_\Gamma \left[M_{n_P}(v_1)m(Q) + M_{n_P}(v_2^*)k(Q)\right] dS_Q - \frac{1}{2}k(P), \\
V_{n_P}(w) &= \int_\Gamma Vn_P(v_1)m(Q)dS_Q + \frac{1}{2}m(P) + \lim_{p\to P}\int_\Gamma [V_{n_P}(v_2^*)k(Q)]\, dS_Q.
\end{aligned} \tag{25}$$

In terms of the boundary conditions of the plate, $m(Q)$ and $k(Q)$ can be determined by choosing two related equations in eq.(25). The stress intensity factors can be obtained as in eq.(22):

$$K_i = \lim_{\rho\to 0} \frac{6}{h^2}\sqrt{2\pi\rho}\int_\Gamma \left[K_i^*(P,Q)m(Q) + \frac{\partial K_i^*}{\partial n_Q}k(Q)\right] dS_Q|_{\theta=0}, \tag{26}$$

In addition, the stress intensity factors can also be determined by another formula (SIH, PARIS & ERDOGAN, 1962):

$$K_I - iK_{II} = -\frac{12\sqrt{\pi}D(3-\nu)}{h^2\sqrt{a}}\frac{1}{8\pi D}\int_\Gamma \left[\phi'(\zeta,z_0)m(Q) + \frac{\partial \phi_1'(\zeta,z_0)}{\partial n_Q}k(Q)\right] dS_Q|_{\zeta=1}, \tag{27}$$

where

$$\phi_1'(\zeta,z_0) = \frac{\partial}{\partial\zeta}[\phi_1^*(\zeta,z_0) + \phi_1^0(\zeta,z_0)]. \tag{28}$$

4 Numerical Treatments and Examples

We now illustrate the numerical procedure according to indirect integral equations.

As shown in Fig. 1, the outer boundary of a cracked plate is divided into N elements, in which constant interpolation functions are used for simplicity. Therefore, the functions $m(Q_j)$ and $k(Q_j)$ in each element j (with $j = 1, 2..., N$) are constant and equal to the values at the middle point Q_j of the element. According to the outer boundary conditions of the plate, one can select two corresponding integral equations from eq.(25). After proper numerical treatments, this lead to:

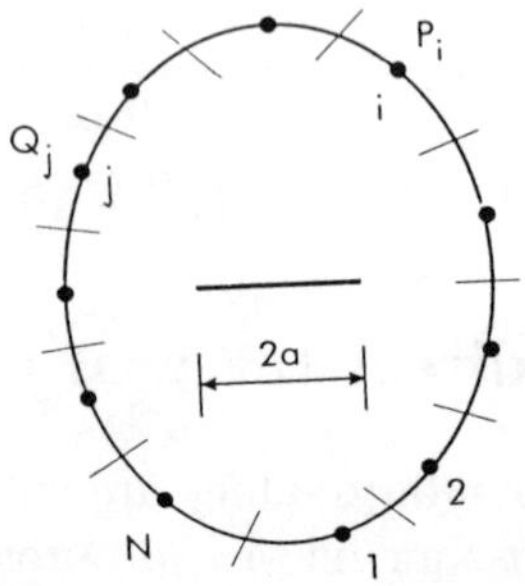

Fig. 1

$$\begin{aligned} B_1(P_i) &= \sum_{j=1}^{N}\{\alpha_1(P_i,Q_j)m(Q_j)+\alpha_2(P_i,Q_j)k(Q_j)\}, \\ B_2(P_i) &= \sum_{j=1}^{N}\{\beta_1(P_i,Q_j)m(Q_j)+\beta_2(P_i,Q_j)k(Q_j)\}, \qquad (i=1,2,...N) \end{aligned} \tag{29}$$

where $B_1(P_i)$ and $B_2(P_i)$ are two known boundary values at the point P_i. α_1, α_2, β_1 and β_2 are known integral values on the boundary j. For example, if the bending moment M_n and the equivalent shearing V_n are known on the outer boundary, eq.(29) can be expressed as

$$\begin{aligned} Mn_{P_i} &= \sum_{j=1}^{N}\left\{\left[\int_{\Gamma_j} M_{n_P}(v_1)ds_Q\right]m(Q_j)+\left[\int_{\Gamma} M_{n_P}(v_2^*)ds_Q\right]k(Q_j)\right\}-\frac{1}{2}k(P_i), \\ Vn_{P_i} &= \sum_{j=1}^{N}\left\{\left[\int_{\Gamma} V_{n_P}(v_1)ds_Q\right]m(Q_j)+\lim_{p\to P}\left[\int_{\Gamma_j} V_{n_P}(v_2^*)ds_Q\right]k(Q_j)\right\}+\frac{1}{2}m(P_i), \end{aligned} \tag{30}$$

where Γ_j is the length of the elements j. For each boundary point, one can obtain two corresponding discreted equations. This can be written in the matrix form as

$$\{B\} = [A]\,\{X\}, \tag{31}$$

where $\{B\}$ and $\{X\}$ are vectors of order $2N \times 1$, $[A]$ is a matrix of order $2N \times 2N$. The unknowns $m(Q_j)$ and $k(Q_j)$ (with $j = 1, 2, ..., N$) could be obtained by solving eq.(31), and the stress intensity factors therefore can be determined form eq.(26) or (27).

1. As an example, we first consider the problem of an infinite plate containing a crack of length $2a$, where a bending moment M_0 is applied all around the boundary of the plate at infinity, as shown in Fig. 2.

   In order to compare with K_I^∞, the stress intensity factor of the infinite plate with a center crack and uniform bending moment acting at infinity, we choose $a/L = a/W = 0.04$, and divide the outer boundary into eight constant elements (two elements in each side). The computed results according to formulas (26) and (27) are listed below

	eq.(26)	eq.(27)
$\frac{K_I}{K_I^\infty}$	0.99948	0.99891

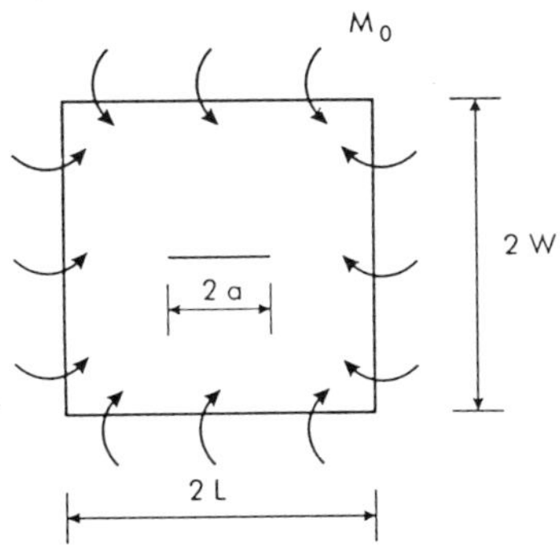

Fig. 2

The results show that higher numerical precision can be attained with very few elements. This indicates that the formulas obtained are correct and this special example can be well approximated by the modified fundamental solution.

2. Consider a rectangular plate containing a sharp crack of length $2a$, and subjected to uniform bending moment as shown in Fig. 3. The stress intensity factors are calculated for $a/W = 0.1$ and different values of a/L. The results are shown in the following table and compared with the results obtained by the finite element method:

K_I/K_I^∞ a/L	0.1	0.2	0.3	0.4
The authors	1.0053	1.049	1.0581	1.0983
CAAR (1981)	1.0058	1.024	1.058	1.104
No. of elements	16	16	16	20

It can be seen that the number of elements of outer boundary elements should be taken to be proportional to the variable a/L.

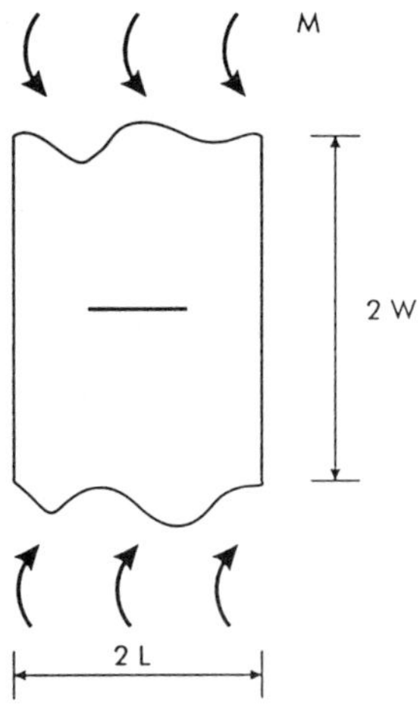

Fig. 3

5 Discussion

In this paper, we derive a kind of modified fundamental solution for a thin plate with a cutout, in which the effect of the cutout on the stress and displacements fields is incorporated. The discrete modelling of the cutout surface is no longer necessary if the modified kernels are used in BEM analysis. Numerical examples for the bending of plate with a crack show that higher computer accuracy can be attained with relatively fewer outer boundary elememts; some difficulties encountered by the standard BEM approach can be avoided in this way. The method is therefore suitable for solving plate bending problems with holes or cracks, which have high stress gradient near the crack tips or hole boundaries.

Appendix

Let

$$\tilde{F}(z,z_0)=\bar{z}\phi(z,z_0)+\chi(z,z_0),$$

Subsituting eqs.(15) and (18) into above formula, one can obtain

$$\tilde{F}(z,z_0)=\overline{(z,z_0)}I_1-\frac{1}{\kappa}(z-\bar{z})I_2+\int\left[I_3+\frac{1}{\kappa}I_2\right]dz,$$

where

$$\begin{aligned}I_1(z,z_0)&=(z-z_0)\ln(z-z_0),\\ I_2(z,z_0)&=\tilde{\phi}^*(z,z_0),\\ I_3(z,z_0)&=\tilde{\phi}^*(z,\bar{z}_0).\end{aligned}$$

Therefore

$$\begin{aligned}K_I^*(P,Q)&=M_y(v_1)=-\frac{1}{8\pi}Re\left[\frac{\partial^2\tilde{F}}{\partial x^2}+\nu\frac{\partial^2\tilde{F}}{\partial y^2}\right]\\ &=-\frac{1}{8\pi}Re\left\{(1-\nu)\left[-(\overline{z-z_0})\frac{\partial^2 I_1}{\partial z^2}+\frac{1}{\kappa}(z-\bar{z})\frac{\partial^2 I_2}{\partial z^2}-\frac{\partial I_3}{\partial z}\right]\right.\\ &\left.+2(1-\nu)\frac{\partial I_1}{\partial z}+\frac{3+\nu}{\kappa}\frac{\partial I_1}{\partial z}\right\}\end{aligned}$$

$$\begin{aligned}K_{II}^*(P,Q)&=M_{xy}(v_1)=-\frac{1}{8\pi}(1-\nu)Re\left[\frac{\partial^2\tilde{F}}{\partial x\partial y}\right]\\ &=\frac{1-\nu}{8\pi}Im\left[(\overline{z-z_0})\frac{\partial^2 I_1}{\partial z^2}-\frac{1}{\kappa}(z-\bar{z})\frac{\partial^2 I_2}{\partial z^2}+\frac{\partial I_3}{\partial z}-\frac{1}{\kappa}\frac{\partial I_2}{\partial z}\right],\end{aligned}$$

and $M_n(v_1)$, $V_n(v_1)$, $M_n(v_1^*)$ and $V_n(v_2^*)$ can also be obtained in a similar way which is not listed here.

References

1958 Savin, G.N., Stress Concentration Near Holes, Chinese translation, China Academic Publishers, Beijing.

1959 Muskhelishvili, N.I., Some Basic Problems of the Mathematical Theory of Elasticity, Chinese translation, China Academic Publishers, Beijing.

1962 Sih, G.C., Paris, P.C. Erdogan, F.,Crack-tip stress- intensity factors for plane extension and plate bending probelms, J. Appl. Mech. 29, pp. 306 - 312.

1979 Stern, M., A general boundary integral formulation for the numerical solution of plate bending problems, Int. J. Solid Structures, 15, pp. 765-782.

1979 Tottenham, H., The boundary element method for plates and shells, in: Developments in Boundary Element Mehtods - 1, P.K. Banerjee, R. Butterfield (Eds.), Applied Science Publishers Ltd., New York.

1981 Brebbia, C.A. (Ed.), Progress in Boundary Element Methods, Vol. 1, Pentech Press Ltd., London.

1981 China Academy of Aeronautical Science (CAAR) (Ed), Handbook of Stress Intensity Factors, China Academic Publishers, Beijing (in Chinese).

1982 Mukherjee, S., Boundary Element Methods in Creep and Fracture, Applied Science Publishers Ltd., New York.

1982 Stern, M., Plate bending problems, in: Boundary Element Techniques in Computer-Aided Engineering, C.A. Brebbia (Ed.), Martinus Nijhoff Publisher.

1982 Kitahara, M., Boundary Integral Methods in Eigenvalue Problems of Elastodynamics and Thin Plates, Elsevier Science Publisher, London.

1982 Lu, P., A boundary element solution for the crack problems, J. of China University of Science and Technology, Vol. 20, 4, 506.

STABILITY OF PAVEMENT STRUCTURES UNDER LONG TERM REPEATED LOADING

Lutfi RAAD, Transportation Research Center,
University of Alaska Fairbanks
Fairbanks, Alaska 99775, U.S.A.

Dieter WEICHERT, L.M.L. (EUDIL)-CNRS URA 1441,
Université des Sciences et Technologies de Lille
F-59655 Villeneuve d'Ascq, France

ABSTRACT

This paper addresses the stability of pavement structures under long term repeated loading using the shakedown theory. The accumulation of plastic strains in a given system may increase under repeated load applications, leading to incremental collapse, or plastic strains may cease to increase with time, resulting in a stable response or shakedown. An improved numerical method using finite element formulation coupled with an optimization technique is introduced. The method takes into consideration the variable elastic coefficients of both the coarse-grained and fine-grained layers in the pavement structure through the application of an extension of the classical shakedown theorem. The proposed method is used to illustrate the influence of layer properties on the shakedown behavior of general pavement systems.

OVERVIEW

Pavement structures are subjected to repeated traffic loads of varying magnitude and number of repetitions. Although pavement response parameters such as stresses, strains, and deflections could be determined with reasonable accuracy limits using finite element techniques, performance models fall short of predicting the stability of pavement systems under long term repeated loading. Of particular significance in this case is whether such systems will exhibit progressive accumulation of plastic strains or whether the accumulation of plastic strains will cease and a stable response or shakedown condition is attained.

The shakedown theory was originally developed by MELAN [1936]. According to this theory, a system will shakedown under repeated cyclic loads if a self-equilibrated residual stress field could be found such that equilibrium conditions, boundary conditions, and yield conditions are pointwise satisfied. It is assumed in this case that the material is elastic-perfectly plastic with convex yield surface and applicable normality condition and that viscous and inertia effects are negligible. This theory has been applied to discrete structures and plates (KÖNIG [1969], BELYTSCHKO [1972],

Z. Mróz et al. (eds.), Inelastic Behaviour of Structures under Variable Loads, 473–496.

MAIER & CORRADI [1974], ALWIS & GRUNDY [1980]). Although attempts have been made to apply this theory to general continua using various numerical algorithms (ABOUSTIT & REDDY [1980], PANDE et al. [1980], SHARP & BOOKER [1984], SHARP [1985]), these methods do not satisfy all constraints associated with the shakedown theory, specifically, equilibrium conditions, boundary conditions, and yield criteria by an arbitrary chosen residual stress field in the system under consideration. More recently, the shakedown theory has been applied numerically to pavement structures (SHARP [1985], RAAD et al. [1988]). In this case the stability of these pavements under long term repeated traffic loads have been addressed particularly since the question of increased accumulation of plastic strains has a significant influence on the long term performance of these structures. However, these numerical applications do not take into consideration the non-linear material properties of the base and subgrade layers of the pavement.

The Extended Shakedown Theorem

The extension of the classical shakedown theorem (MELAN [1936], KOITER [1960]) to include more sophisticated material laws in addition to linear elastic-perfectly plastic or linear elastic-linear hardening has been researched in the last years (e.g. MAIER [1987], WEICHERT & GROSS-WEEGE [1988], STEIN et al. [1990]). For certain class of materials with variable elastic coefficients that simulate soil-layer behavior in pavement structures, an extension of the shakedown theorem has been developed by WEICHERT & RAAD [1992]. In this case, it is assumed that the rate of elastic strains $\dot{\boldsymbol{\varepsilon}}^e$ is related to the rate of stress $\dot{\boldsymbol{\sigma}}$ by

$$\dot{\boldsymbol{\varepsilon}}^e = \mathbf{L}:\dot{\boldsymbol{\sigma}} \tag{1}$$

where $\mathbf{L}$ is the positive definite tensor of process dependent elastic coefficients with the usual symmetries. As evolution law for the following form is assumed

$$\mathbf{L} = \mathbf{L}^\circ + \mathbf{L}^d\, \mathbf{e}^{-\kappa} \tag{2}$$

where $\mathbf{L}^\circ$ and $\mathbf{L}^d$ are process-independent tensors obeying individually the same symmetry conditions as $\mathbf{L}$. $\mathbf{L}^\circ$ is positive definite, and $\mathbf{L}^d$ is positive definite for strengthening and negative definite for weakening materials. The exponent κ is a monotonically increasing function of the process with value zero for $t = 0$, where t is actual time. It is noted that the time measure need not coincide with physical time but is a parameter controlling the evolution of the plastic process. Therefore it is justifiable to use κ as a time parameter. In this case, the extended shakedown theorem could be stated as follows:

If under the preceding assumptions, for an elastic-plastic body B under variable loads, with finite volume V and sufficiently regular surface S consisting of the disjoint parts S_k and S_f, where displacements and forces are prescribed respectively, a time-independent state of residual stresses $\bar{\boldsymbol{\rho}}(\mathbf{x})$ exists such that the yield condition is satisfied with

$$F(\boldsymbol{\sigma}^{\circ} + \bar{\boldsymbol{\rho}}) < 0 \qquad (3)$$

with

$$\operatorname{div} \bar{\boldsymbol{\rho}} = 0 \ \text{ in } \ V \quad , \quad \mathbf{n}.\bar{\boldsymbol{\rho}} = 0 \ \text{ on } \ S_f \qquad (4)$$

where $\mathbf{n}$ denotes the unit outer normal vector on S, and with $\boldsymbol{\sigma}^{\circ}$ as state of stress in a purely elastic reference body B° with the same elastic (process dependent) properties and under the same loads as B, then the evolution of plastic strains in B is limited and B is said to shake down under the given loading history. The proof of this theorem is found in WEICHERT & RAAD [1992].

Applicability in Pavement Structures

The applicability of the extended shakedown theorem in pavements is limited by the assumption that a purely elastic reference solution is known for the specific class of materials with weakening or strengthening elastic coefficients. Such conditions are exhibited by pavement soil layers whereby the elastic coefficients are stress-dependent and vary with load repetitions in a strengthening or weakening mode depending on the type of soil, density, water content, and applied stress state. Experimental studies on a number of fine-grained cohesive soils and coarse-grained cohesionless soils under repeated-load/triaxial-testing conditions indicate that the elastic modulus, E_R, also known as the resilient modulus, defined as the ratio of the peak repeated deviator stress to the elastic or recoverable strain, reaches a limiting value with increasing number of deviator stress repetitions N. E_R, could increase or decrease with N but always reaches a limiting value, even under repeated stress conditions that lead to the incremental collapse of the soil specimen. In fact, the limiting E_R seems to be attained always prior to the occurrence of collapse or failure. Such behavior has been illustrated under controlled-stress and controlled- strain repeated load triaxial tests (RAAD & FIGUEROA [1980], WEICHERT & RAAD [1992]).
The limiting E_R for coarse-grained and fine-grained soils is given by the following (KNUTSON et al. [1977], THOMPSON & ROBNETT [1979], RAAD & FIGUEROA [1980]): For coarse-grained soils,

$$E_R = K\theta^{\eta} \qquad (5)$$

in which $\theta = \sigma_1 + \sigma_2 + \sigma_3$ with σ_1, σ_2, σ_3 as principal stresses and where K and η are material constants. For fine-grained soils,

$$E_R = G_2 - G_3[G_1 - (\sigma_1 - \sigma_3)] \quad ; \qquad G_1 > (\sigma_1 - \sigma_3) \qquad (6)$$

and

$$E_R = G_2 - G_4[(\sigma_1 - \sigma_3) - G_1] \quad ; \quad G_1 < (\sigma_1 - \sigma_3) \qquad (7)$$

in which $G_{i,}$, i = 1, 4, are material constants. Failure of soils under repeated stress states is governed by the Mohr-Coulomb yield criterion (RAAD & FIGUEROA [1980]), where

$$\sigma_1 = \sigma_3 \tan^2 (45 + \phi/2) + 2c \tan (45 + \phi/2) \qquad (8)$$

in which c and ϕ are the cohesion and friction of the soil material under consideration, and σ_1 and σ_3 are the major and minor principal stresses respectively. Note that in Equations 6-8 compressive stresses are assumed positive whereas tensile stresses are assumed negative.

In the application of the extended shakedown theorem to pavements, it is proposed to use the limiting values of elastic constants of the material associated with weakening or strengthening conditions for the purely elastic elastic solution. This is justified particularly if it is assumed that these limiting states are achieved prior to the occurrence of structural collapse or failure. Numerical applications to pavement structures have been illustrated by RAAD et al. [1989a, 1989b].

NUMERICAL FORMULATION

In the proposed method, the pavement is discretized into a series of rectangular elements, each with four external primary nodes. The displacement functions used are complete to the second degree and satisfy compatibility conditions. The material is assumed to be initially elastic-perfectly plastic with convex yield surface and applicable normality condition. A quasi-static analysis is used assuming negligible viscous and inertia effects. If stress states $\boldsymbol{\sigma}^{\circ}$, $\boldsymbol{\sigma}^{s}$, and $\boldsymbol{\sigma}^{a}$ correspond, respectively, to body forces $\boldsymbol{P}^{\circ}$, statically applied forces $\boldsymbol{f}^{s}$, and repeated loads $\boldsymbol{f}^{a}$, then the system will shake down provided a time-independent stress increment $\Delta\boldsymbol{\sigma}$ can be found such that equilibrium conditions, boundary conditions, and yield relations are satisfied. In this case, plane strain finite element analysis is used to determine the stresses in the system.

The determination of the shakedown load is then reduced to an optimization problem that could be stated as follows:

Minimize

$$\mathbf{Q} = -\boldsymbol{\alpha} + \sum_{i=1}^{NP} (\mathbf{R}_{xi})^2 + \sum_{i=1}^{NP} (\mathbf{R}_{yi})^2 \qquad (9)$$

subject to the following constraints:

$$\alpha > 0 \qquad (10)$$

$$Y(\boldsymbol{\sigma}) \leq 0 \tag{11}$$

$$\sigma_1 \geq -2c \tan(45 + \phi/2) \tag{12}$$

where

$$\boldsymbol{\sigma} = (\sigma_{ij})^{\circ} + (\sigma_{ij})^{s} + \alpha(\sigma_{ij})^{a} + \Delta\sigma_{ij} \tag{13}$$

such that $(\sigma_{ij})^{\circ}$, $(\sigma_{ij})^{s}$, and $(\sigma_{ij})^{a}$ are the stresses at the center of a given element due to $\boldsymbol{P}^{\circ}$, $\boldsymbol{f}^{s}$ and $\boldsymbol{f}^{a}$, respectively. $\Delta\sigma_{ij}$ is an arbitrary stress increment applied at the center of each element, NP is the number of nodal points, α denotes the load multiplier associated with repeated loads $\boldsymbol{f}^{a}$.
R_{xi}, R_{yi} are the resultant forces in the x and y directions at a nodal point due to $\Delta\sigma_{ij}$ with respect to a global set of coordinates (x, y) and Y denotes the Mohr-Coulomb yield criterion with failure occurring for $Y = 0$. Y is given by

$$Y = \sigma_1 - \sigma_3 \tan^2(45 + \phi/2) + 2c \tan(45 + \phi/2) \tag{14}$$

Minimizing Q subject to the indicated constraints would yield a maximum value for the load multiplier α, while satisfying equilibrium conditions, boundary conditions, and yield conditions in a weak sense. Because Q is quadratic with nonlinear constraints, quadratic optimization techniques are not feasible. Instead, a pattern search algorithm is developed based on the original work by HOOKE & JEEVES [1961]. The method could be summarized in the following steps.

1. Determine the stresses resulting from $\boldsymbol{P}^{\circ}$, $\boldsymbol{f}^{s}$, and initially applied repeated loads $\boldsymbol{f}^{a}$

2. Find a load multiplier α_{st} such that $\alpha_{st}\,\boldsymbol{f}^{a}$ would cause yielding of the critically stressed element in the system. This will shift the search to the vicinity of the region of interest.

3. The search starts by determining Q for α_{st} and a set of $\Delta\sigma_{ij}$ that satisfy the constraint conditions of eqn (10-12).

4. During a given exploratory sequence, the variable α is allowed one disturbance in the direction of decreasing Q. Each of the stress variables $\Delta\sigma_{ij}$ is allowed as many disturbances, each equal to its step size and in the same direction as long as the objective function Q decreases and the imposed constraints are satisfied. Otherwise the exploratory sequence is rated a failure.

5 A new search is initiated about the last base point determined in Step 4, using smaller step sizes. The algorithm terminates when the values of the step sizes are reduced to a certain preassigned value. In this case, the shakedown load will be equal to $\alpha_{st}\alpha\boldsymbol{f}^{a}$.

The above approach has been applied to multilayer pavement systems assuming elastic-perfectly plastic material behavior (RAAD et al. [1988]). However, as indicated previously, the elastic or resilient moduli of soil layers are stress-dependent. In this case, the *extended shakedown theorem* applies and the numerical solution should also satisfy the stress-dependent elastic moduli relations in addition to shakedown conditions. The shakedown capacity could then be obtained using a series of iterative steps. The assumption that the response under a given repeated state of stress at a given point in the pavement stabilizes and remains elastic so long as these stresses do not exceed the strength as defined by the Mohr-Coulomb yield criterion. Such behavior is illustrated schematically in Fig. 1, and is consistent with the *extended shakedown theorem*. A series of iterations using finite element analysis is conducted so that the stresses at the center of each element satisfy the stress-dependent elastic moduli relationship.

A new shakedown pressure acting on the pavement surface is then calculated by using the newly determined moduli at the center of elements and following the approach previously summarized in Steps 1 through 5. The shakedown pressure P_i for a given iterative step, i in this case, will be equal to $\alpha_{st}\alpha_t P_{i-1}$, where α_{st} is the load multiplier associated with initiation of failure, α_t is the maximum load multiplier obtained from eqn (9), and P_{i-1} is the shakedown pressure obtained for the previous iterative step $i-1$. The procedure is repeated until convergence is attained, whereby the shakedown pressure in two consecutive steps reaches essentially the same value, and hence shakedown conditions are satisfied simultaneously with the stress-dependent elastic moduli relations. The proposed method is shown in Fig. 2 and has good convergence properties (RAAD et al. [1989a]).

APPLICATIONS

The proposed numerical method has been applied to study 1) the shakedown behavior of a three layer pavement structure with particular emphasis of the granular base properties on the stability and fatigue of the pavement system and, 2) the shakedown behavior of a two-layer system consisting of a full-depth asphalt concrete layer over a subgrade with special emphasis on the influence of properties of the asphalt concrete on the response of the pavement. A more comprehensive set of results using shakedown analysis for these pavement systems have been published by RAAD et al. elsewhere (1989a, 1989b). The results presented below are selective and are used primarily to illustrate the application of the proposed numerical approach to practical pavement systems.

a) Three-Layer Pavements

The pavement considered consists of an asphalt concrete surface layer, and a granular base course layer overlying a clay subgrade. The properties of the pavement materials considered are summarized in Table 1. The variation of subgrade elastic modulus with applied repeated deviator stress is illustrated in Fig. 3.

Table 1. Pavement Materials Properties

Asphalt Concrete Layer	Granular Layer	Subgrade
$E = 3 \times 10^5$ psi $\upsilon_a = 0.40$ $\gamma_a = 140$ lb/cu ft $k_o = 0.50$ $\phi_a = 37°$ $c_a = 650$ psi	$E_R = 5000\theta^{0.5}$ (psi) $E_R = 2000\theta^{0.6}$ (psi) $\upsilon_b = 0.35$ $\gamma_b = 120$ lb/ cu ft	Soft, $c_s = 3$ psi, $\phi_s = 0°$ Stiff, $c_s = 12$ psi, $\phi_s = 0°$ $\upsilon_s = 0.47$ $\gamma_s = 110$ lb/ cu ft $k_o = 0.50$

Notes:

E	elastic modulus of asphalt concrete layer.
γ_a, υ_a, ϕ_a, c_a	density, Poisson's ratio, friction angle, and cohesion of asphalt concrete layer.
γ_b, υ_b	density and Poisson's ratio of granular base.
γ_s, υ_s, c_s, ϕ_s	density, Poisson's ratio, cohesion, and friction angle of subgrade.
k_o	coefficient of at rest earth pressure

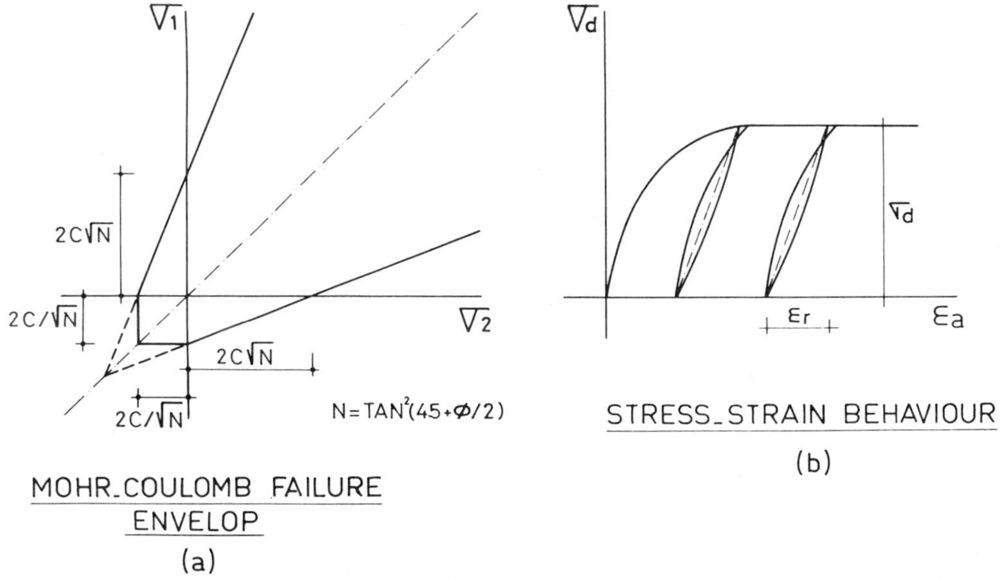

Figure 1 Schematic Representation of Stress-Strain Behavior and Yield for a Typical Soil under Repeated Loading.

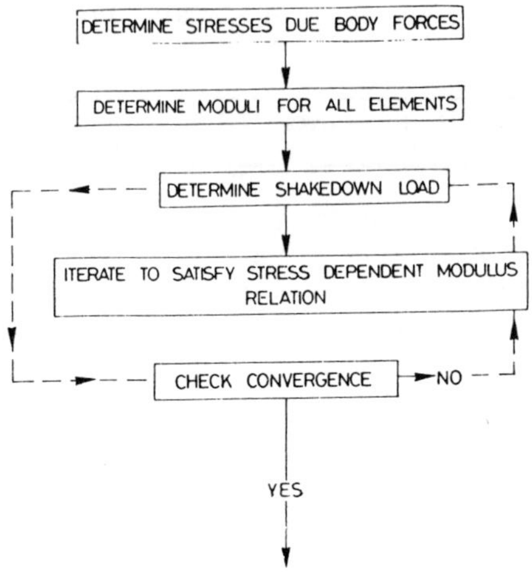

Figure 2 Flow Chart of the Proposed Numerical Method.

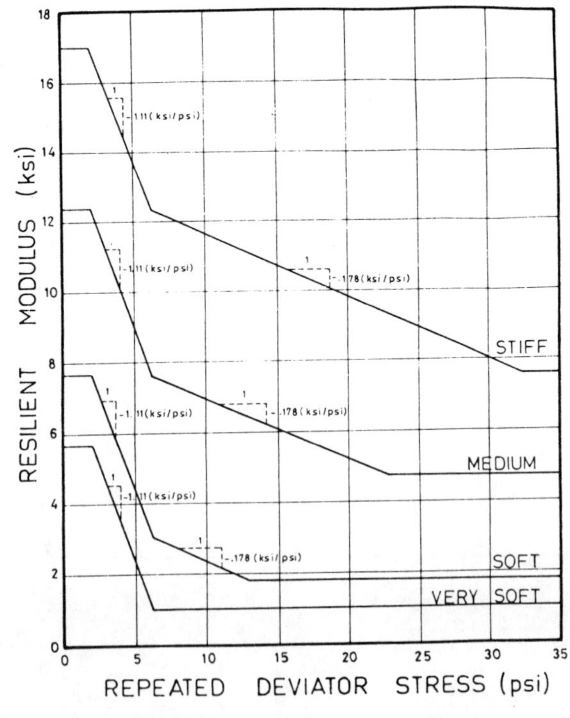

Figure 3 Resilient Properties of Subgrade Soils (25).

Influence of Cohesion and Friction of Granular Base

Analyses were performed to determine the influence of the strength of the granular base on the shakedown behavior of the pavement. According to the Mohr-Coulomb yield criterion, the strength of the granular base depends on the parameters of friction and cohesion as defined by the friction angle ϕ, and cohesion c, respectively. In this case, the pavement considered consists of a 3 in. asphalt concrete and a 12 in. granular base ($E_R = 5000\ \theta^{0.5}$) overlying soft or stiff clay subgrades was analyzed under an applied uniform surface pressure of radius equal to 6 in., assuming plane strains. Results are presented in Figures 4 and 5. For a given angle of friction, increasing the cohesion of the granular base would significantly increase the shakedown capacity of the pavement. Moreover, for values of cohesion less than about 5 psi, in the case of soft subgrade, and about 17 psi, in the case of stiff subgrade, an increase in the friction angle would enhance the shakedown capacity of the system. For lower cohesion, however, an increase in the angle of friction reduces shakedown capacity. It should be noted that, for a given base cohesion, an increase in the angle of friction would not only reduce the tensile strength of the base but would increase its shear strength. It seems, therefore, that for high values of cohesion, the relative effect on shakedown capacity associated with a decrease in the tensile strength of the base due to an increase in its friction angle surpasses the effect of the corresponding increase in its shear strength. This could probably be a result of increased mobilization of the shear resistance of the base prior to the initiation of tensile failure on its underside.

A comparison between the applied pressure on the pavement surface needed to initiate failure P_{st} and the shakedown load P_{SD} is presented in Fig. 6. Results show that the ratio P_{st}/P_{SD} is essentially a function of base cohesion. The variation of friction angle between 30° and 45° will not influence this ratio by more than 5%. It can be concluded that the shakedown pressure is larger than the surface pressure required to initiate failure in the base. It follows that the pavement system would shake down even though some elements of the granular base are in a state of yield. In this case, the yield zone is contained, and the pavement would still stabilize under long-term repeated loading. It is interesting to note that, for values of cohesion less than 5 psi, P_{st}/P_{SD} will increase and could reach 0.97 for a value of cohesion equal to 0.10 psi.

Fatigue and Shakedown Behavior

Prediction of fatigue and shakedown is required for the proper assessment of pavement performance. For example, a pavement designed to carry a number of load applications in fatigue could actually be carrying greater load than its shakedown capacity, and would, therefore, exhibit incremental collapse due to the continuous accumulation of plastic strains. Moreover, if the magnitude of the wheel loads exceeds the shakedown capacity of the pavement after application of maintenance and rehabilitation procedures, then the accumulation of plastic strains will continue and incremental collapse could occur.

Analyses were conducted to investigate the influence of strength and elastic (i.e. resilient) properties of the granular base on fatigue and shakedown predictions. Fatigue resistance of the asphalt concrete surface was determined using a fatigue failure

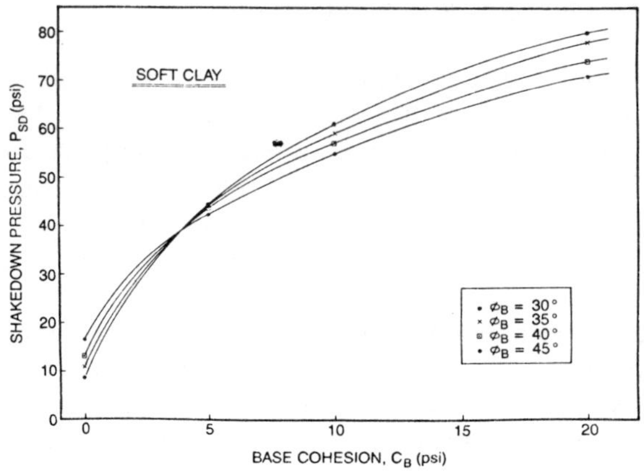

Figure 4 Influence of Base Cohesion and Friction on Shakedown for Soft Subgrade Conditions.

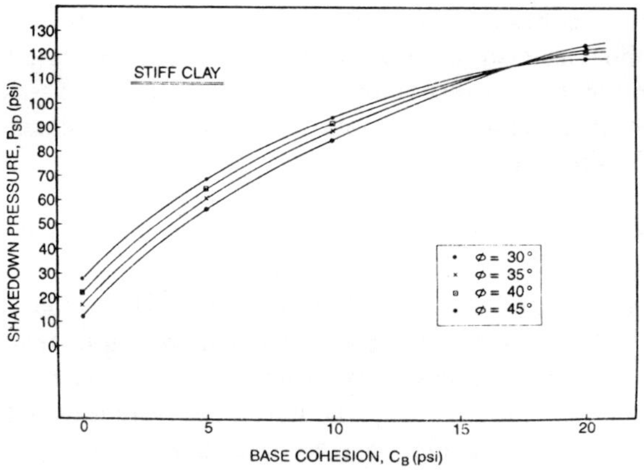

Figure 5 Influence of Base Cohesion and Friction on Shakedown for Stiff Subgrade Conditions.

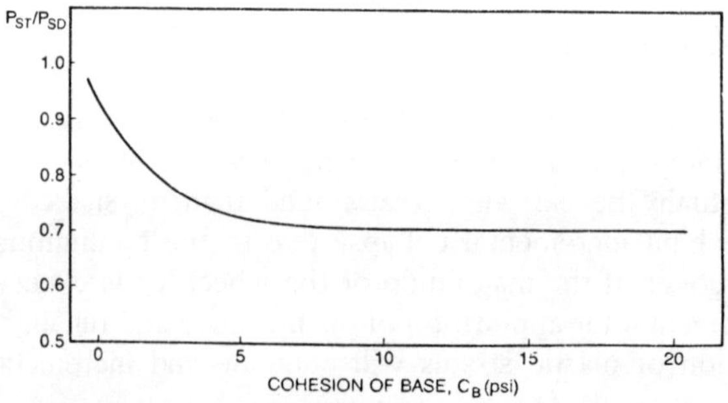

Figure 6 Variation of P_{st} / P_{SD} with Base Cohesion.

criterion proposed by MONISMITH et al. [1970]. This criterion expresses limiting tensile strains as a function of number of load repetitions required to cause fatigue failure for different moduli values of the asphalt concrete layer (Figure 7). The limiting tensile strains on the underside of the asphalt concrete layer were determined for plane strain loading conditions using the stress-dependent resilient properties of the granular base and subgrade.

The influence of cohesion of granular base ($E_R = 5000\theta^{0.5}$, $\phi = 40°$) on shakedown and fatigue behavior is illustrated in Figure 8. Fatigue resistance seems to be essentially unaffected by the cohesion of the granular base. On the other hand, shakedown behavior is significantly affected by base cohesion. For example, a reduction of base cohesion from 5 to 0.10 psi would reduce the shakedown capacity of a pavement with an 8 in. base from 50 psi to 15 psi. In this case, a pavement designed to satisfy acceptable fatigue and shakedown conditions could be unsafe with respect to shakedown, if loss in base cohesion reduces the shakedown limit below the magnitude of applied traffic loads. This could eventually cause sudden increase in the rate of accumulation of permanent deformations, thereby leading to incremental collapse.

Analyses were also conducted to compare fatigue and shakedown behavior of pavements with low cohesion-high friction granular base ($E_R=5000\theta^{0.5}$, $\phi=40°$, c=5 psi) with pavements having high cohesion-low friction granular base ($E_R=2000\theta^{0.6}$, $\phi= 20°$, c = 15 psi), assuming stiff subgrade conditions (Figures 9 and 10). Results indicate that pavements with low cohesion-high friction base exhibit better fatigue properties and lower shakedown capacity than pavements having high cohesion-low friction base. Moreover, in both cases, the increase in base thickness does not affect significantly the fatigue resistance of the asphalt concrete surface layer for large number of load applications. This agrees well with findings from other procedures (SHELL CO. [1978]). Pavement loading conditions represented by points below the shakedown curve would result in a stable pavement response, provided proper maintenance procedures are implemented as soon as fatigue cracking appears on the pavement surface.

Although pavements with high cohesion-low friction base have larger shakedown capacity, they would probably exhibit larger permanent deformations before shakedown and stability conditions are attained. Moreover, these bases are more susceptible to environmental changes (such as increase in base moisture and freezing-thawing), which would reduce base cohesion and lead to significant decrease in shakedown capacity.

b) Full-Depth Asphalt Concrete Pavements

Full-depth asphalt concrete pavements consist of an asphalt concrete layer over the subgrade. The proposed numerical method for shakedown analysis was used to investigate the behavior of these pavements, particularly since the properties of the asphalt concrete surface layer are significantly influenced by the material temperature.

Analyses were conducted for a variety of asphalt concrete temperature and for subgrade conditions corresponding to soft and stiff clay, respectively. The asphalt concrete stiffness or elastic modulus used is assumed to be temperature dependent. In addition,

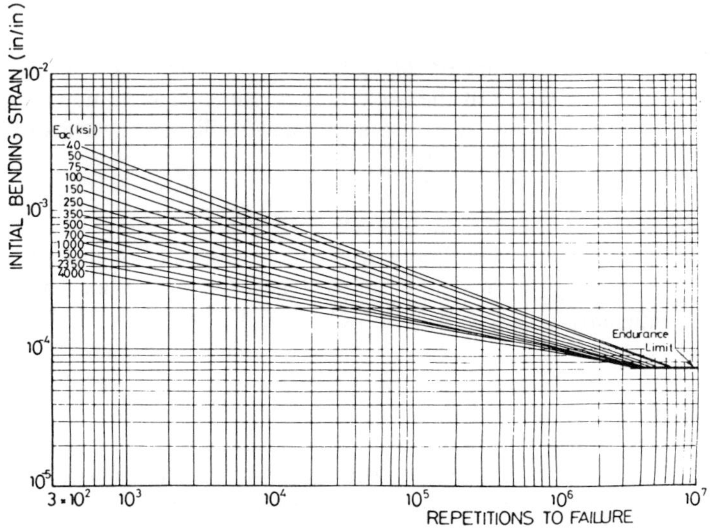

Figure 7 Fatigue Failure Criterion for Asphalt Concrete Mixtures (22).

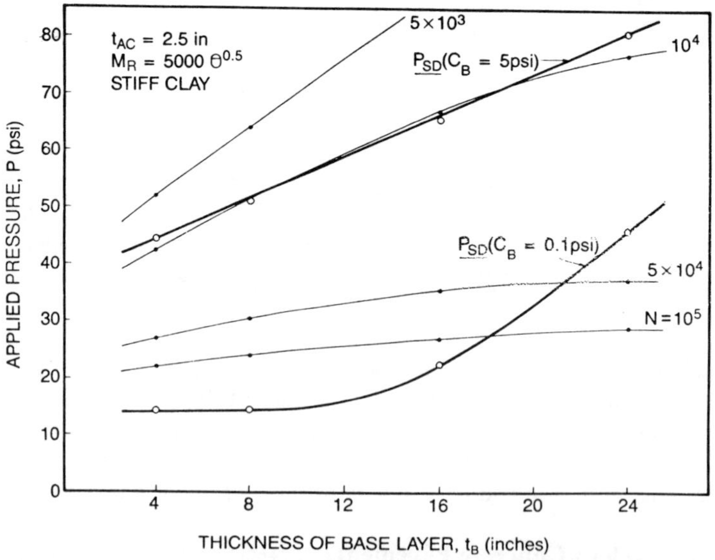

Figure 8 Influence of Base Cohesion on Fatigue and Shakedown.

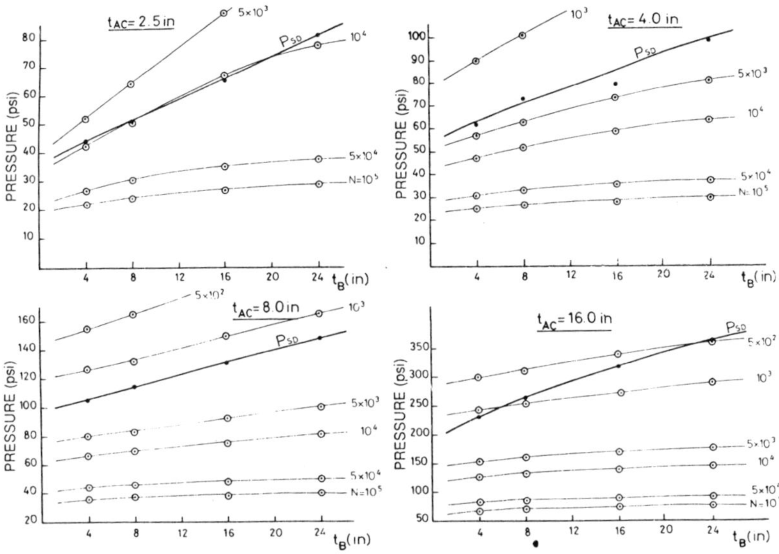

Figure 9 Fatigue and Shakedown Behavior for Pavements with a Low Cohesion-High Friction Base.

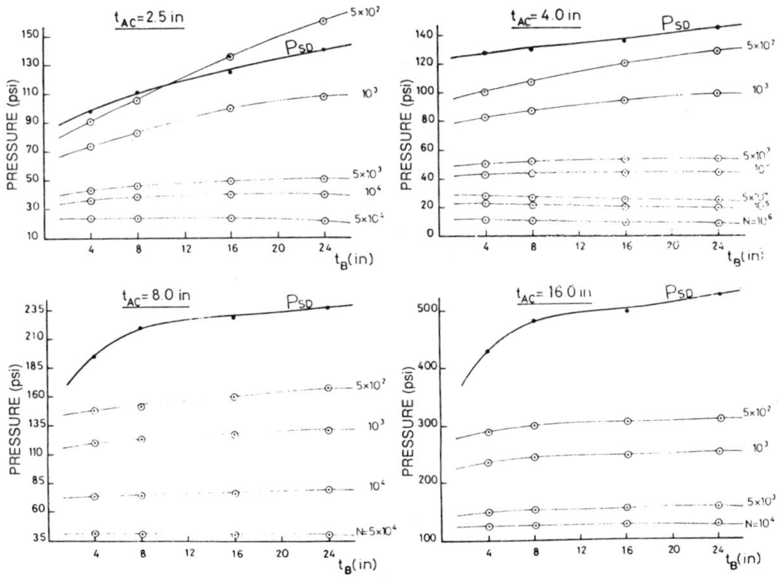

Figure 10 Fatigue and Shakedown Behavior for Pavements with a High Cohesion-Low Friction Base.

its friction and cohesion, as defined by the Mohr-Coulomb yield criterion, are also temperature dependent. The variation of stiffness and strength properties is illustrated in Fig. 11 and 12 (SALAM [1971]). The subgrade is assumed to have zero friction with cohesion equal to 3 psi and 12 psi for soft and stiff conditions, respectively. The variation of subgrade elastic modulus with repeated deviator stress is shown in Fig. 3. Plane strain analyses was used with applied wheel loads simulated by a uniform surface pressure having a radius equal to 6 in.

Shakedown Behavior

The shakedown load seems to decrease with increase in temperature of the asphalt concrete layer above 40°F. Such behavior reflects probably the predominant effect of the subgrade on shakedown. An increase in aspnalt concrete temperature would reduce the stiffness of the asphalt concrete layer and would, therefore, result in a larger transfer of applied load to the subgrade (Fig. 13 and 14). On the other hand, for values of asphalt concrete layer temperature under 40°F, shakedown behavior becomes influenced essentially by the asphalt concrete layer. A decrease in layer temperature in this case would be reflected in an increase in its stiffness and strength. The asphalt concrete layer would, therefore, carry a larger proportion of the applied load, which could result in a lower shakedown capacity, particularly for thinner surfaces and stiffer subgrades. It is also evident that the shakedown load increases with increasing subgrade strength, however, the beneficial effect of subgrade strength is less for thinner pavements and lower temperature. The shakedown capacity in this case is limited by the strength of the surface layer.

The influence of thickness of asphalt concrete layer on shakedown is shown in Fig. 15 and 16. An increase in thickness results in an increase in the shakedown load. This increase is greater for stiff subgrade support than for soft subgrade conditions. It should be noted that, in case of the stiff subgrade and for thicknesses of the asphalt concrete layer less than about 20 in., the shakedown load increases with increasing layer temperature above 10°F, but decreases as the temperature exceeds 40°F. On the other hand, for thicknesses the asphalt concrete layer larger than about 20 in., an increase in layer temperature would result in a decrease in shakedown capacity.

Shakedown Versus Fatigue

Analyses were also conducted to compare the shakedown and fatigue behavior of full-depth asphalt concrete pavements. The asphalt concrete was analyzed assuming layer temperature values of 10°F and 68°F for stiff and soft subgrade conditions. The fatigue criterion used is shown in Figure 7.

Results of analyses presented in Fig. 17 through 18 indicate a significant influence of subgrade conditions on pavement performance in terms of shakedown and fatigue. For the case of stiff subgrade, fatigue of the asphalt concrete layer seems to be the predominant distress mode, and the loads required to induce fatigue failure are smaller than the shakedown loads for all the pavement temperatures and thicknesses considered (Fig. 17 and 18). On the other hand, for the case of a soft subgrade, fatigue or shakedown could govern depending on pavement thickness, pavement temperature, and

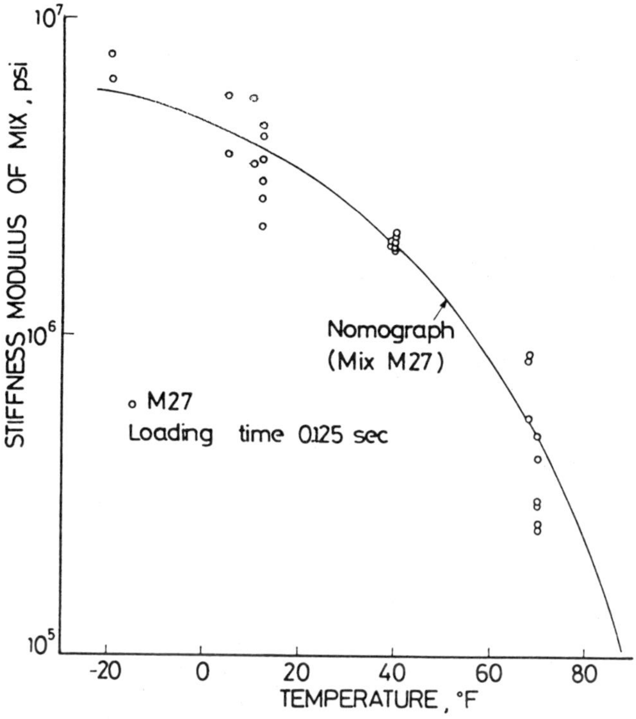

Figure 11 Variation of Mix Stiffness with Temperature

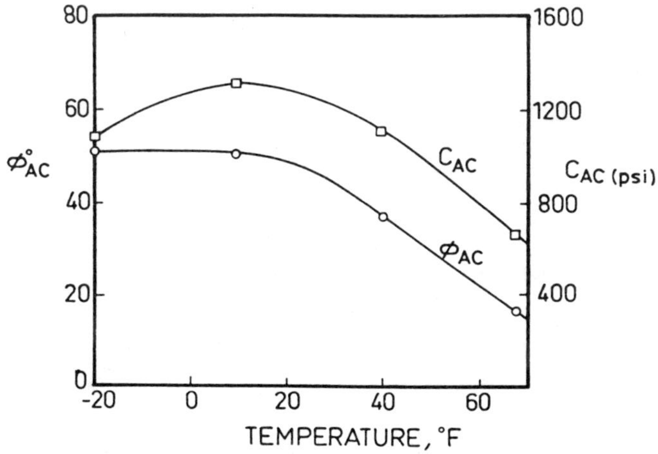

Figure 12 Variation of Cohesion c_a and Friction ϕ_a of Asphalt Concrete Mix with Temperature

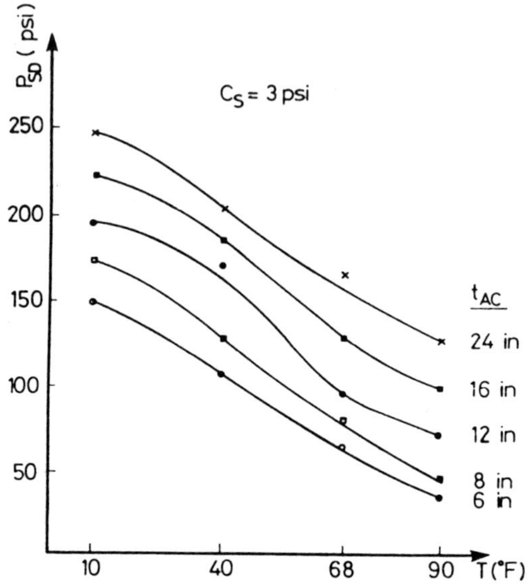

Figure 13. Variation of Shakedown Load for Soft Subgrade Conditions.

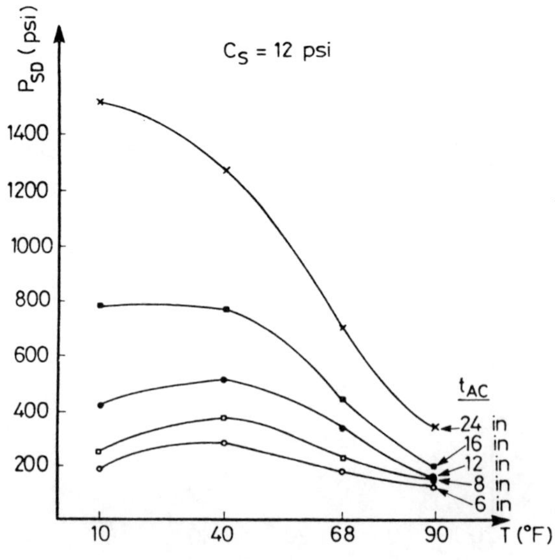

Figure 14. Variation of Shakedown Load for Stiff Subgrade Conditions.

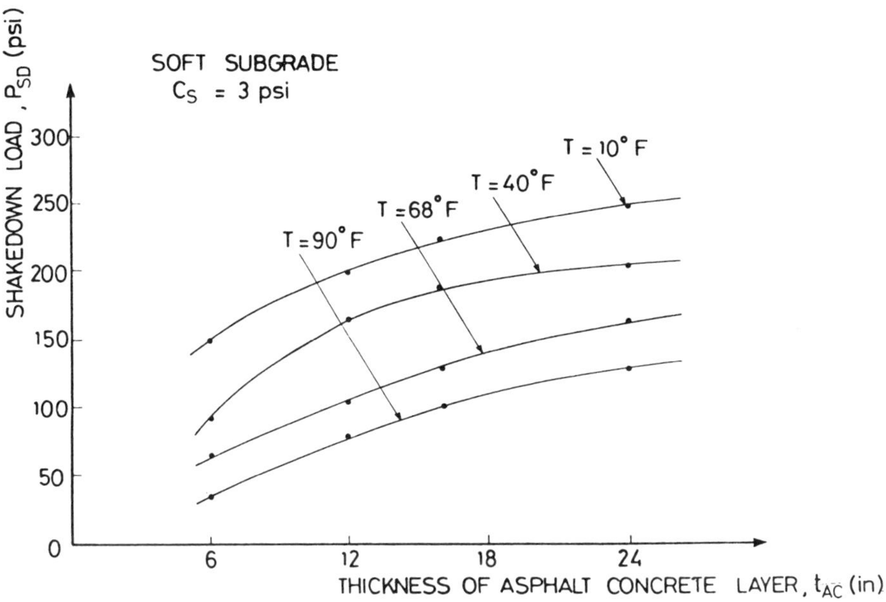

Figure 15. Influence of Asphalt Concrete Layer Thickness on Shakedown for Soft Subgrade.

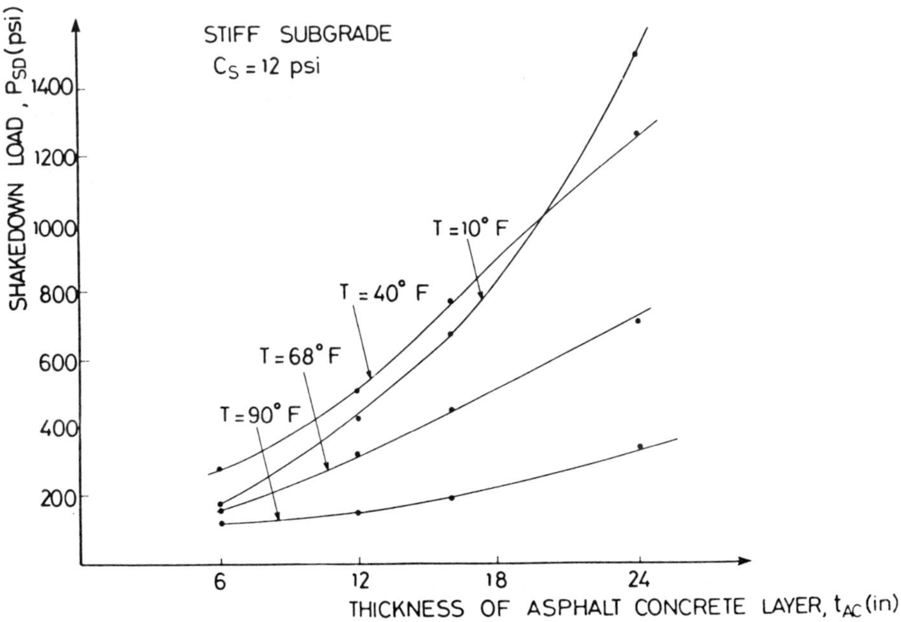

Figure 16. Influence of Asphalt Concrete Layer Thickness on Shakedown for Stiff Subgrade.

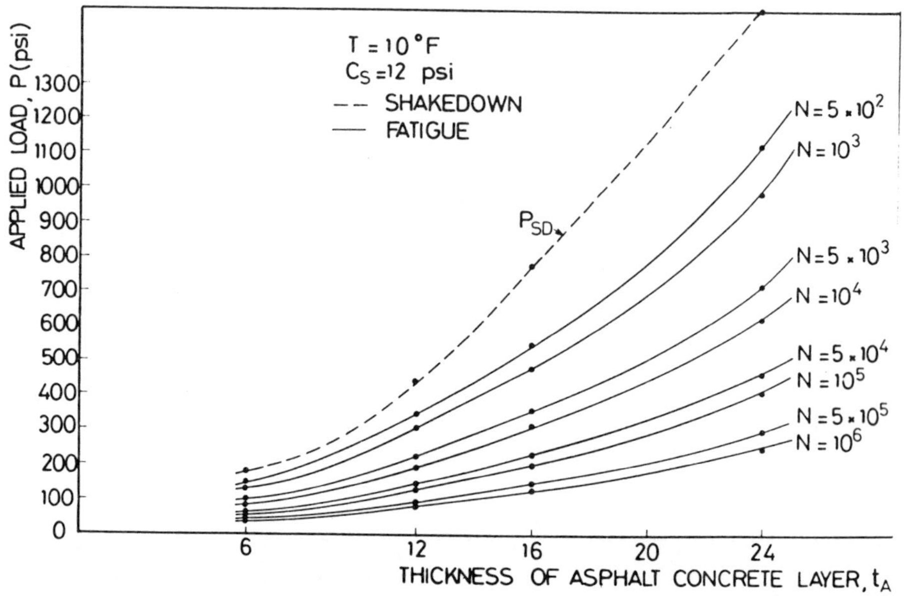

Figure 17. Fatigue and Shakedown Loadings for Stiff Subgrade and T = 10°F.

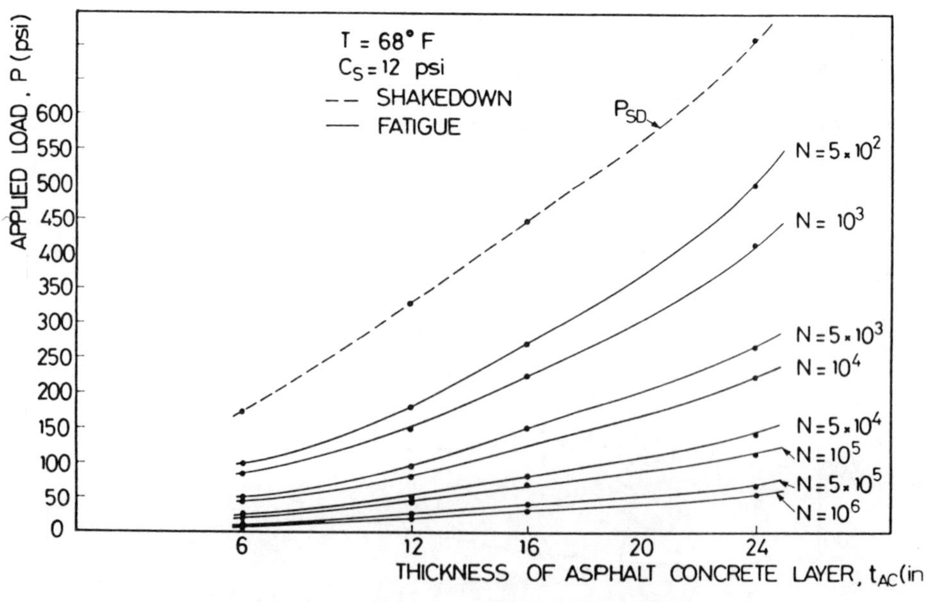

Figure 18. Fatigue and Shakedown Loadings for Stiff Subgrade and T = 68°F.

magnitude and repetitions of the applied load (Fig. 19 and 20). For a given applied load in this case, an increase in pavement thickness will increase the resistance to fatigue, and to potential incremetal collapse induced by the accumulation of plastic strains. Moreover, for a given temperature, pavement sections designed to carry a given number of load repetitions in fatigue may or may not shake down depending pavement thickness. For example, assuming a pavement temperature of 68°F and a design number of load repetitions equal to 10^4, a 6 in. thick asphalt concrete (design load equal to 35 psi) will stabilize and attain shakedown, whereas a 24 in. thick asphalt concrete layer (design load 195 psi) will not shake down and will exhibit incremental collapse (Fig. 20). Moreover, an increase in pavement temperature reduces both its fatigue and shakedown capacity.

Limiting Criteria

In many situations, limiting values in the critical response parameters have been proposed as means of predicting pavement performance. Specifically, in full-depth asphalt concrete pavements, the accumulation of plastic strains (i.e., rutting) has been linked to vertical stresses and /or vertical strains on the top of the subgrade. An attempt is made in this paper to determine limiting criteria associated with shakedown behavior. Analysis were conducted to determine the shakedown loads for asphalt concrete layer thicknesses of 6, 8, 12, 16, and 24 in., temperature 10°F, 40°F, 68°F, and 90°F, and for subgrade cohesion or undrained shear strength values of 3, 6, and 12 psi. Plane strain finite element analyses were then performed on the pavement section under the applied shakedown load using the nonlinear stress-dependent resilient properties of the subgrade illustrated in Fig. 3 for soft, medium, and stiff conditions, respectively. Limiting criteria in terms of subgrade vertical stress, σ_v, and subgrade vertical strain, ε_v, were determined. These critical stresses and strains occur at the top of the subgrade layer under the centerline of the applied surface load.

Results presented in Fig. 21 indicate that the ratio of subgrade normal stress, σ_v, to the subgrade cohesion or undrained shear strength, c_s, lies essentially between 3 and 4 for all the asphalt concrete layer thicknesses and stiffnesses considered, provided the asphalt concrete layer temperature is greater than 40°F. However, for cases where the temperature drops to 10°F and for asphalt concrete layer thickness less than 12 in., the ratio σ_v/c_s drops to a value between 2 and 3. This reflects the greater mobilization of tensile stresses in the surface layer that would result in lower values of vertical stresses acting on the top of the subgrade at shakedown. The average values of critical normal stress in this case are in the range of 10 to 40 psi.

Results of subgrade normal strains on top of the subgrade when the pavement is subjected to applied shakedown loading are presented in Fig. 22. Values indicate that the limiting vertical strain, ε_v is dependent on the stiffness of the subgrade. ε_v is given in this case as

$$1 \times 10^{-3} < \varepsilon_v < 2.5 \times 10^{-3} \quad \text{for soft subgrade} \qquad (15)$$

$$1.5 \times 10^{-3} < \varepsilon_v < 3.0 \times 10^{-3} \quad \text{for medium subgrade} \qquad (16)$$

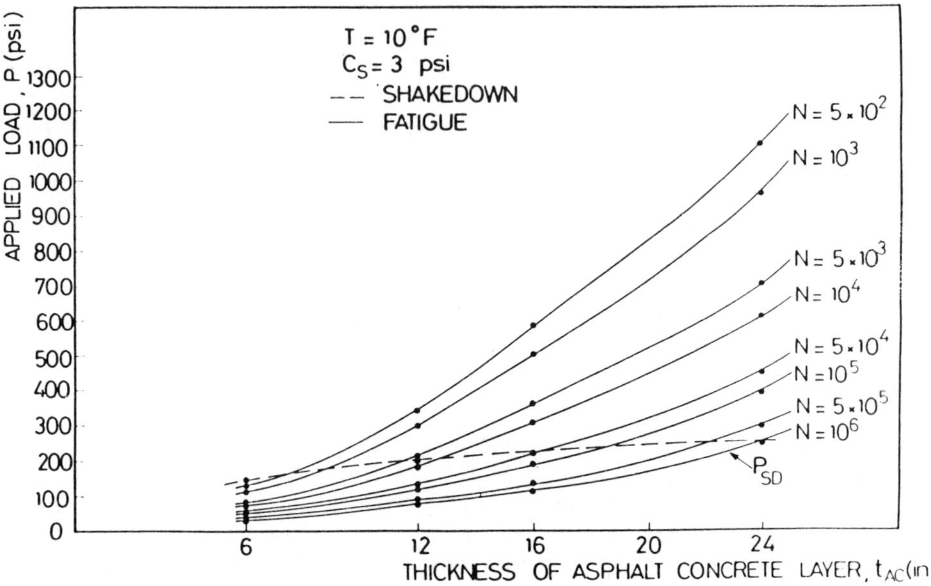

Figure 19. Fatigue and Shakedown Loadings for Soft Subgrade and T = 10°F.

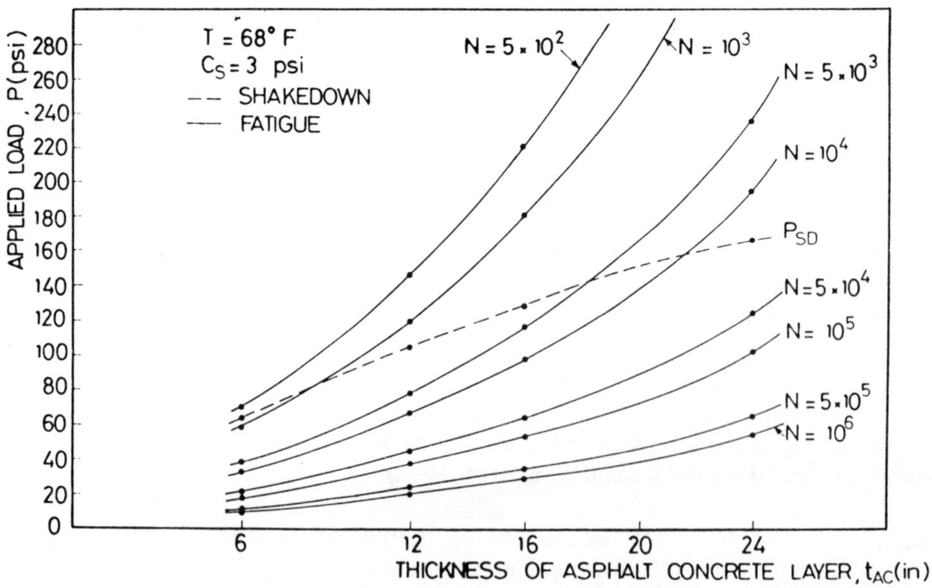

Figure 20. Fatigue and Shakedown Loadings for Soft Subgrade and T = 68°F.

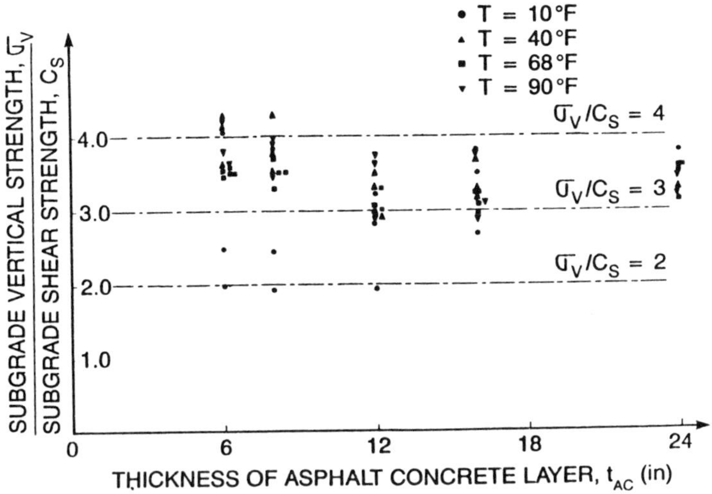

Figure 21. Limiting Subgrade Normal Stress Criteria.

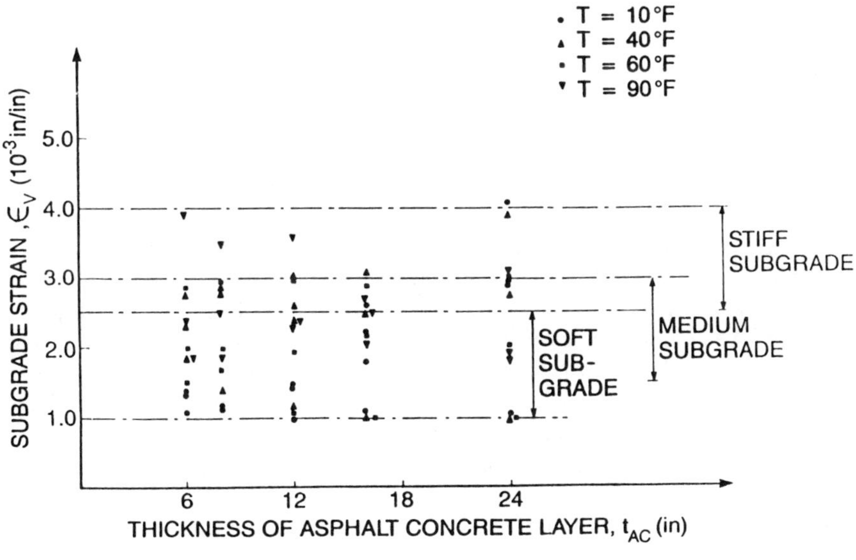

Figure 22. Limiting Subgrade Normal Strain Criteria.

$$2.5 \times 10^{-3} < \varepsilon_v < 4.0 \times 10^{-3} \quad \text{for stiff subgrade} \qquad (17)$$

The values for allowable subgrade stresses and strains in the pavement are generally less than those backcalculated using predicted shakedown loading conditions. The values of allowable stresses published in the literature range from 2 psi to 22 psi depending on subgrade strength, whereas the allowable vertical strains range from 0.10×10^{-3} to 1.5×10^{-3} (FIGUEROA [1979]).
The larger values of subgrade normal stress and strain obtained from shakedown analysis could indicate a more severe loading condition or could be a result of the conservative interpretation of pavement serviceability data when determining the existing criteria. Although shakedown loading could result in low pavement serviceability, it may nevertheless be used as an upper bound, which if exceeded, would lead to incremental collapse. Moreover, if the applied loads are kept below the shakedown limit, pavement maintenance will be more effective, because the rate of accumulation of plastic strains will eventually cease and the pavement system will exhibit a stable response.

SUMMARY

An improved numerical method for determining the shakedown behavior of pavement structures under long term repeated loading has been developed. The proposed method takes into consideration the variable elastic coefficients of both the coarse-grained and fine-grained soil layers in the pavement structure through the application of an extension to the general shakedown theorem.
The proposed method has been applied to study the shakedown behavior of typical three layer pavement s, and two layer systems consisting of full depth asphalt concrete surface overlying fine-grained subgrade. The influence of layer thickness, strength parameters of the coarse-grained base, temperature of the asphalt concrete surface, and subgrade stiffness and strength, on the shakedown capacity have been investigated. In addition, an attempt has been made to determine the effects of the coupled shakedown and fatigue behavior of these pavements. Results of analyses indicate that fatigue and shakedown criteria should be considered in the design and evaluation of pavement structures. A pavement designed to resist a number of load applications in fatigue would exhibit increased accumulation of plastic strains if the applied load exceeds the shakedown capacity, but the pavement would reach a stable response if the applied load were smaller than the shakedown capacity.

In addition to providing a basis for analyzing shakedown behavior for the conditions considered herein, improved versions for the proposed numerical model would provide a basis for analyzing pavements under more general and complex loading and environmental conditions. Moreover, experimental verification of numerical predictions is needed to calibrate the proposed models for practical applications.

REFERENCES

1936 MELAN, E., "Theorie Statisch unbestimmter Systeme aus ideal-plastischem Baustoff", Sitzungsber. Akad. der Wissenschaften in Wien, IIa, pp. 145 - 195.

1960 KOITER, W.T., "General Theorems for Elastic-Plastic Solids", in: SNEDDON, I.N. and HILL, R. (eds.), Progress in Solid Mechanics, North Holland, Amsterdam, pp. 165 - 221.

1961 HOOKE, R. and JEEVES, T.A., "Direct Search Solution of Numerical and Statistical Problems", J. Ass. Comp. Mach., 8, 212.

1969 KÖNIG, J.A., "Shakedown Theory of Plates", Archiwum Mechaniki Stosowanej, 21, 5, 623.

1972 BELYTSCHKO, T., "Plane Stress Shakedown Analysis by Finite Elements", Int. J. Mech. Sci. 14, 619.

1970 MONISMITH, C.L. et al., Asphalt Mixture Behavior In Repeated Flexure, Report No. TE-70-5, University of California, Berkeley, U.S.A.

1971 SALAM, Y.S. "Characterization of Deformation and Fracture of Asphalt Concrete", Ph.D. dissertation, University of California, Berkeley, U.S.A.

1974 MAIER, G. and CORRADI, L., "Upper Bounds on Dynamical Deformations of Elastoplastic Continua", Meccanica, IX, 30.

1977 KNUTSON, M.R. et al., Materials Foundation Study - Ballast and Foundation Materials Research Program, Tech. Rep. FRA-OR&D-77-02, Federal Railroad Administration, Washington, D.C., U.S.A.

1978 SHELL INTERNATIONAL COMPANY LIMITED, Shell Pavement Design Manual-Asphalt Pavements and Overlays for Road Traffic, London, U.K.

1979 FIGUEROA, J. "Resilient Based Flexible Design Procedures for Secondary Roads", Ph.D. dissertation, University of Illinois at Urbana-Champaign, U.S.A.

1979 THOMPSON, M.R. and ROBNETT, Q.L., "Resilient Properties of Subgrade Soils", J. Trans. Engg., ASCE, 105, TE1, 71.

1980 ABOUSTIT, B.L. and REDDY, D.V., "Finite Element Linear Programming Approach to Foundation Shakedown", in: Proc., Intern. Symp. on Soils Under Cyclic and Transient Loading, Swansea, U.K., pp. 727 - 738.

1980 ALWIS, W.M. and GRUNDY, P.,"Shakedown of Plates under Moving Loads", in: Proc., 7th Australian Conference on Mechanics of Structures and Materials, 1980, pp. 191 - 196.

1980 PANDE, G.N., ABDULLAH, W.S. and DAVIS, E.H., "Shakedown of Elasto-Plastic Continua with Special Reference to Soil-Rock Structures", in: Proc. Int. Symp. on Soils Under Cyclic and Transient Loading, Swansea, U.K., pp.39-74.

1980 RAAD, L. and FIGUEROA, J., "Load Response of Transportation Support systems". J. Trans. Engg., ASCE, 106, TE1, 111.

1984 SHARP, R.W. and BOOKER, J.R., "Shakedown of Pavements Under Moving Surface Loads", J. Trans. Eng., ASCE, 110, TE1, 1.

1985 SHARP, R.W., "Pavement Design Based on Shakedown Analysis", in: Transportation Research Record 1022, TRB, National Research Council, Washington, D.C., U.S.A., pp. 99 - 107.

1987 MAIER, G., "A Generalization to Non-Linear Hardening of the First Shakedown

Theorem for Discrete Elastic-Plastic Structural Models", Atti Acc. Lincei Ren. fis. 8, LXXXI, 161.

1988 RAAD, L., WEICHERT, D. and W. NAJM, W.,"Stability of Multilayer Systems Under Repeated Loads", in: Transportation Research Record 1207, TRB, National Research Council, Washington, D.C., U.S.A., pp. 181 - 186.

1988 WEICHERT, D. and GROSS-WEEGE, J., "The Numerical Assessment of Elastic-Plastic Sheets under Variable Mechanical and Thermal Loads Using A simplified Two-Yield Condition", In. J. Mech. Sc., 30, 10, 757.

1989a RAAD, L., WEICHERT, D. and HAIDAR, A., "Shakedown and Fatigue of Pavements with Granular Bases" in: Transportation Research Record 1227, National Research Council, Washington, D.C., U.S.A., pp. 159 - 172.

1989b RAAD, L., WEICHERT, D. and HAIDAR, A., "Analysis of Full-Depth Asphalt Concrete Pavements Using Shakedown Theory" in: Transportation Research Record 1227, National Research Council, Washington, D.C., U.S.A., pp. 53 - 65.

1990 STEIN, E., ZHANG, G. and KÖNIG, J.A., "Michromechanical Modelling and Computation of Shakedown with Nonlinear Kinematic Hardening Including Examples for 2-D Problems", in: AXELRAD, D.R. and MUSCHIK, W. (eds.), Recent Developments Micromechanics, Springer Verlag, Berlin.

1992 WEICHERT, D. and RAAD, L., "Extension of the Shakedown Theorem to a Certain Class of Materials with Variable Elastic Coefficients", Mech. Res. Com., 19, 6, 511.

Mechanics

SOLID MECHANICS AND ITS APPLICATIONS

Series Editor: G.M.L. Gladwell

Aims and Scope of the Series

The fundamental questions arising in mechanics are: *Why?, How?,* and *How much?* The aim of this series is to provide lucid accounts written by authoritative researchers giving vision and insight in answering these questions on the subject of mechanics as it relates to solids. The scope of the series covers the entire spectrum of solid mechanics. Thus it includes the foundation of mechanics; variational formulations; computational mechanics; statics, kinematics and dynamics of rigid and elastic bodies; vibrations of solids and structures; dynamical systems and chaos; the theories of elasticity, plasticity and viscoelasticity; composite materials; rods, beams, shells and membranes; structural control and stability; soils, rocks and geomechanics; fracture; tribology; experimental mechanics; biomechanics and machine design.

1. R.T. Haftka, Z. Gürdal and M.P. Kamat: *Elements of Structural Optimization.* 2nd rev.ed., 1990 ISBN 0-7923-0608-2
2. J.J. Kalker: *Three-Dimensional Elastic Bodies in Rolling Contact.* 1990 ISBN 0-7923-0712-7
3. P. Karasudhi: *Foundations of Solid Mechanics.* 1991 ISBN 0-7923-0772-0
4. *Not published*
5. *Not published.*
6. J.F. Doyle: *Static and Dynamic Analysis of Structures.* With an Emphasis on Mechanics and Computer Matrix Methods. 1991 ISBN 0-7923-1124-8; Pb 0-7923-1208-2
7. O.O. Ochoa and J.N. Reddy: *Finite Element Analysis of Composite Laminates.* ISBN 0-7923-1125-6
8. M.H. Aliabadi and D.P. Rooke: *Numerical Fracture Mechanics.* ISBN 0-7923-1175-2
9. J. Angeles and C.S. López-Cajún: *Optimization of Cam Mechanisms.* 1991 ISBN 0-7923-1355-0
10. D.E. Grierson, A. Franchi and P. Riva: *Progress in Structural Engineering.* 1991 ISBN 0-7923-1396-8
11. R.T. Haftka and Z. Gürdal: *Elements of Structural Optimization.* 3rd rev. and exp. ed. 1992 ISBN 0-7923-1504-9; Pb 0-7923-1505-7
12. J.R. Barber: *Elasticity.* 1992 ISBN 0-7923-1609-6; Pb 0-7923-1610-X
13. H.S. Tzou and G.L. Anderson (eds.): *Intelligent Structural Systems.* 1992 ISBN 0-7923-1920-6
14. E.E. Gdoutos: *Fracture Mechanics.* An Introduction. 1993 ISBN 0-7923-1932-X
15. J.P. Ward: *Solid Mechanics.* An Introduction. 1992 ISBN 0-7923-1949-4
16. M. Farshad: *Design and Analysis of Shell Structures.* 1992 ISBN 0-7923-1950-8
17. H.S. Tzou and T. Fukuda (eds.): *Precision Sensors, Actuators and Systems.* 1992 ISBN 0-7923-2015-8
18. J.R. Vinson: *The Behavior of Shells Composed of Isotropic and Composite Materials.* 1993 ISBN 0-7923-2113-8

Kluwer Academic Publishers – Dordrecht / Boston / London

Mechanics

SOLID MECHANICS AND ITS APPLICATIONS

Series Editor: G.M.L. Gladwell

19. H.S. Tzou: *Piezoelectric Shells.* Distributed Sensing and Control of Continua. 1993 ISBN 0-7923-2186-3
20. W. Schiehlen: *Advanced Multibody System Dynamics.* Simulation and Software Tools. 1993 ISBN 0-7923-2192-8
21. C.-W. Lee: *Vibration Analysis of Rotors.* 1993 ISBN 0-7923-2300-9
22. D.R. Smith: *An Introduction to Continuum Mechanics.* 1993 ISBN 0-7923-2454-4
23. G.M.L. Gladwell: *Inverse Problems in Scattering.* An Introduction. 1993 ISBN 0-7923-2478-1
24. G. Prathap: *The Finite Element Method in Structural Mechanics.* 1993 ISBN 0-7923-2492-7
25. J. Herskovits (ed.): *Structural Optimization '93.* 1995 (forthcoming) ISBN 0-7923-2510-9
26. M.A. González-Palacios and J. Angeles: *Cam Synthesis.* 1993 ISBN 0-7923-2536-2
27. W.S. Hall: *The Boundary Element Method.* 1993 ISBN 0-7923-2580-X
28. J. Angeles, G. Hommel and P. Kovács (eds.): *Computational Kinematics.* 1993 ISBN 0-7923-2585-0
29. A. Curnier: *Computational Methods in Solid Mechanics.* 1994 ISBN 0-7923-2761-6
30. D.A. Hills and D. Nowell: *Mechanics of Fretting Fatigue.* 1994 ISBN 0-7923-2866-3
31. B. Tabarrok and F.P.J. Rimrott: *Variational Methods and Complementary Formulations in Dynamics.* 1994 ISBN 0-7923-2923-6
32. E.H. Dowell, E.F. Crawley, H.C. Curtiss Jr., D.A. Peters, R. H. Scanlan and F. Sisto (eds.): *A Modern Course in Aeroelasticity.* Third Revised and Enlarged Edition. (forthcoming) ISBN 0-7923-2788-8; Pb: 0-7923-2789-6
33. A. Preumont: *Random Vibration and Spectral Analysis.* 1994 ISBN 0-7923-3036-6
34. J.N. Reddy (ed.): *Mechanics of Composite Materials.* Selected works of Nicholas J. Pagano. 1994 ISBN 0-7923-3041-2
35. A.P.S. Selvadurai (ed.): *Mechanics of Poroelastic Media.* 1995 (forthcoming) ISBN 0-7923-3329-2
36. Z. Mróz, D. Weichert, S. Dorosz (eds.): *Inelastic Behaviour of Structures under Variable Loads.* 1995 ISBN 0-7923-3397-7

Kluwer Academic Publishers – Dordrecht / Boston / London

Mechanics

From 1990, books on the subject of *mechanics* will be published under two series:

***FLUID* MECHANICS AND ITS APPLICATIONS**

Series Editor: R.J. Moreau

***SOLID* MECHANICS AND ITS APPLICATIONS**

Series Editor: G.M.L. Gladwell

Prior to 1990, the books listed below were published in the respective series indicated below.

MECHANICS: DYNAMICAL SYSTEMS

Editors: L. Meirovitch and G.Æ. Oravas

1. E.H. Dowell: *Aeroelasticity of Plates and Shells*. 1975 ISBN 90-286-0404-9
2. D.G.B. Edelen: *Lagrangian Mechanics of Nonconservative Nonholonomic Systems*. 1977 ISBN 90-286-0077-9
3. J.L. Junkins: *An Introduction to Optimal Estimation of Dynamical Systems*. 1978 ISBN 90-286-0067-1
4. E.H. Dowell (ed.), H.C. Curtiss Jr., R.H. Scanlan and F. Sisto: *A Modern Course in Aeroelasticity*. *Revised and enlarged edition see under Volume 11*
5. L. Meirovitch: *Computational Methods in Structural Dynamics*. 1980 ISBN 90-286-0580-0
6. B. Skalmierski and A. Tylikowski: *Stochastic Processes in Dynamics*. Revised and enlarged translation. 1982 ISBN 90-247-2686-7
7. P.C. Müller and W.O. Schiehlen: *Linear Vibrations*. A Theoretical Treatment of Multi-degree-of-freedom Vibrating Systems. 1985 ISBN 90-247-2983-1
8. Gh. Buzdugan, E. Mihăilescu and M. Radeş: *Vibration Measurement*. 1986 ISBN 90-247-3111-9
9. G.M.L. Gladwell: *Inverse Problems in Vibration*. 1987 ISBN 90-247-3408-8
10. G.I. Schuëller and M. Shinozuka: *Stochastic Methods in Structural Dynamics*. 1987 ISBN 90-247-3611-0
11. E.H. Dowell (ed.), H.C. Curtiss Jr., R.H. Scanlan and F. Sisto: *A Modern Course in Aeroelasticity*. Second revised and enlarged edition (of Volume 4). 1989 ISBN Hb 0-7923-0062-9; Pb 0-7923-0185-4
12. W. Szemplińska-Stupnicka: *The Behavior of Nonlinear Vibrating Systems*. Volume I: Fundamental Concepts and Methods: Applications to Single-Degree-of-Freedom Systems. 1990 ISBN 0-7923-0368-7
13. W. Szemplińska-Stupnicka: *The Behavior of Nonlinear Vibrating Systems*. Volume II: Advanced Concepts and Applications to Multi-Degree-of-Freedom Systems. 1990 ISBN 0-7923-0369-5
    Set ISBN (Vols. 12–13) 0-7923-0370-9

MECHANICS OF STRUCTURAL SYSTEMS

Editors: J.S. Przemieniecki and G.Æ. Oravas

1. L. Frýba: *Vibration of Solids and Structures under Moving Loads*. 1970 ISBN 90-01-32420-2
2. K. Marguerre and K. Wölfel: *Mechanics of Vibration*. 1979 ISBN 90-286-0086-8

Mechanics

3. E.B. Magrab: *Vibrations of Elastic Structural Members*. 1979 ISBN 90-286-0207-0
4. R.T. Haftka and M.P. Kamat: *Elements of Structural Optimization*. 1985
   Revised and enlarged edition see under Solid Mechanics and Its Applications, Volume 1
5. J.R. Vinson and R.L. Sierakowski: *The Behavior of Structures Composed of Composite Materials*. 1986 ISBN Hb 90-247-3125-9; Pb 90-247-3578-5
6. B.E. Gatewood: *Virtual Principles in Aircraft Structures*. Volume 1: Analysis. 1989
   ISBN 90-247-3754-0
7. B.E. Gatewood: *Virtual Principles in Aircraft Structures*. Volume 2: Design, Plates, Finite Elements. 1989 ISBN 90-247-3755-9
   Set (Gatewood 1 + 2) ISBN 90-247-3753-2

MECHANICS OF ELASTIC AND INELASTIC SOLIDS

Editors: S. Nemat-Nasser and G.Æ. Oravas

---

1. G.M.L. Gladwell: *Contact Problems in the Classical Theory of Elasticity*. 1980
   ISBN Hb 90-286-0440-5; Pb 90-286-0760-9
2. G. Wempner: *Mechanics of Solids with Applications to Thin Bodies*. 1981
   ISBN 90-286-0880-X
3. T. Mura: *Micromechanics of Defects in Solids*. 2nd revised edition, 1987
   ISBN 90-247-3343-X
4. R.G. Payton: *Elastic Wave Propagation in Transversely Isotropic Media*. 1983
   ISBN 90-247-2843-6
5. S. Nemat-Nasser, H. Abé and S. Hirakawa (eds.): *Hydraulic Fracturing and Geothermal Energy*. 1983 ISBN 90-247-2855-X
6. S. Nemat-Nasser, R.J. Asaro and G.A. Hegemier (eds.): *Theoretical Foundation for Large-scale Computations of Nonlinear Material Behavior*. 1984 ISBN 90-247-3092-9
7. N. Cristescu: *Rock Rheology*. 1988 ISBN 90-247-3660-9
8. G.I.N. Rozvany: *Structural Design via Optimality Criteria*. The Prager Approach to Structural Optimization. 1989 ISBN 90-247-3613-7

MECHANICS OF SURFACE STRUCTURES

Editors: W.A. Nash and G.Æ. Oravas

---

1. P. Seide: *Small Elastic Deformations of Thin Shells*. 1975 ISBN 90-286-0064-7
2. V. Panc: *Theories of Elastic Plates*. 1975 ISBN 90-286-0104-X
3. J.L. Nowinski: *Theory of Thermoelasticity with Applications*. 1978
   ISBN 90-286-0457-X
4. S. Łukasiewicz: *Local Loads in Plates and Shells*. 1979 ISBN 90-286-0047-7
5. C. Fiřt: *Statics, Formfinding and Dynamics of Air-supported Membrane Structures*. 1983 ISBN 90-247-2672-7
6. Y. Kai-yuan (ed.): *Progress in Applied Mechanics*. The Chien Wei-zang Anniversary Volume. 1987 ISBN 90-247-3249-2
7. R. Negruţiu: *Elastic Analysis of Slab Structures*. 1987 ISBN 90-247-3367-7
8. J.R. Vinson: *The Behavior of Thin Walled Structures*. Beams, Plates, and Shells. 1988
   ISBN Hb 90-247-3663-3; Pb 90-247-3664-1

Mechanics

MECHANICS OF FLUIDS AND TRANSPORT PROCESSES
Editors: R.J. Moreau and G.Æ. Oravas

1. J. Happel and H. Brenner: *Low Reynolds Number Hydrodynamics.* With Special Applications to Particular Media. 1983 ISBN Hb 90-01-37115-9; Pb 90-247-2877-0
2. S. Zahorski: *Mechanics of Viscoelastic Fluids.* 1982 ISBN 90-247-2687-5
3. J.A. Sparenberg: *Elements of Hydrodynamics Propulsion.* 1984 ISBN 90-247-2871-1
4. B.K. Shivamoggi: *Theoretical Fluid Dynamics.* 1984 ISBN 90-247-2999-8
5. R. Timman, A.J. Hermans and G.C. Hsiao: *Water Waves and Ship Hydrodynamics.* An Introduction. 1985 ISBN 90-247-3218-2
6. M. Lesieur: *Turbulence in Fluids.* Stochastic and Numerical Modelling. 1987 ISBN 90-247-3470-3
7. L.A. Lliboutry: *Very Slow Flows of Solids.* Basics of Modeling in Geodynamics and Glaciology. 1987 ISBN 90-247-3482-7
8. B.K. Shivamoggi: *Introduction to Nonlinear Fluid-Plasma Waves.* 1988 ISBN 90-247-3662-5
9. V. Bojarevičs, Ya. Freibergs, E.I. Shilova and E.V. Shcherbinin: *Electrically Induced Vortical Flows.* 1989 ISBN 90-247-3712-5
10. J. Lielpeteris and R. Moreau (eds.): *Liquid Metal Magnetohydrodynamics.* 1989 ISBN 0-7923-0344-X

MECHANICS OF ELASTIC STABILITY
Editors: H. Leipholz and G.Æ. Oravas

1. H. Leipholz: *Theory of Elasticity.* 1974 ISBN 90-286-0193-7
2. L. Librescu: *Elastostatics and Kinetics of Aniosotropic and Heterogeneous Shell-type Structures.* 1975 ISBN 90-286-0035-3
3. C.L. Dym: *Stability Theory and Its Applications to Structural Mechanics.* 1974 ISBN 90-286-0094-9
4. K. Huseyin: *Nonlinear Theory of Elastic Stability.* 1975 ISBN 90-286-0344-1
5. H. Leipholz: *Direct Variational Methods and Eigenvalue Problems in Engineering.* 1977 ISBN 90-286-0106-6
6. K. Huseyin: *Vibrations and Stability of Multiple Parameter Systems.* 1978 ISBN 90-286-0136-8
7. H. Leipholz: *Stability of Elastic Systems.* 1980 ISBN 90-286-0050-7
8. V.V. Bolotin: *Random Vibrations of Elastic Systems.* 1984 ISBN 90-247-2981-5
9. D. Bushnell: *Computerized Buckling Analysis of Shells.* 1985 ISBN 90-247-3099-6
10. L.M. Kachanov: *Introduction to Continuum Damage Mechanics.* 1986 ISBN 90-247-3319-7
11. H.H.E. Leipholz and M. Abdel-Rohman: *Control of Structures.* 1986 ISBN 90-247-3321-9
12. H.E. Lindberg and A.L. Florence: *Dynamic Pulse Buckling.* Theory and Experiment. 1987 ISBN 90-247-3566-1
13. A. Gajewski and M. Zyczkowski: *Optimal Structural Design under Stability Constraints.* 1988 ISBN 90-247-3612-9